AF324194

Series on Analysis, Applications and Computation – Vol. 3

Topics in Mathematical Analysis

Series on Analysis, Applications and Computation

Series on Analysis, Applications and Computation – Vol. 3

Topics in Mathematical Analysis

edited by

∘ Paolo Ciatti

∘ Eduardo Gonzalez

∘ Massimo Lanza de Cristoforis
Università di Padova, Italy

∘ Gian Paolo Leonardi
Università di Modena e Reggio Emilia, Italy

World Scientific

NEW JERSEY • LONDON • SINGAPORE • BEIJING • SHANGHAI • HONG KONG • TAIPEI • CHENNAI

Published by

World Scientific Publishing Co. Pte. Ltd.
5 Toh Tuck Link, Singapore 596224
USA office: 27 Warren Street, Suite 401-402, Hackensack, NJ 07601
UK office: 57 Shelton Street, Covent Garden, London WC2H 9HE

British Library Cataloguing-in-Publication Data
A catalogue record for this book is available from the British Library.

TOPICS IN MATHEMATICAL ANALYSIS
Series on Analysis, Applications and Computation — Vol. 3

ISBN-13 978-981-281-105-9
ISBN-10 981-281-105-2

Printed in Singapore by B & JO Enterprise

published with the contribution of

Università di Padova
Dipartimento di Metodi e Modelli Matematici
per le Scienze Applicate

Preface

The *Minicorsi of Mathematical Analysis* have been held at the University of Padova since 1998, and the subject of the Lectures ranges in various areas of Mathematical Analysis including Complex Variable, Differential Equations, Geometric Measure Theory, Harmonic Analysis, Potential Theory, Spectral Theory.

The purposes of the Minicorsi are:

- to provide an update on the most recent research themes in the field,
- to provide a presentation accessible also to beginners.

The Lecturers have been selected both on the basis of their outstanding scientific level, and on their clarity of exposition. Thus the Minicorsi and the present collection of Lectures are particularly indicated to young Researchers and to Graduate Students.

In this volume, the organizers have collected most of the lectures held in the years 2000–2003, and intend to provide the reader with material otherwise difficult to find and written in a way also accessible to nonexperts.

The organizers wish to express their sincere gratitude to the several participants who have contributed to the success of the Minicorsi.

The organizers are also indebted to the University of Padova, and in particular to the 'Dipartimento di Metodi e Modelli Matematici per le Scienze Applicate', and to the 'Dipartimento di Matematica Pura ed Applicata' of the University of Padova, both for the hospitality, and for the financial support. The organizers also acknowledge the financial support offered by the 'Gruppo Nazionale per l'Analisi Matematica, la Probabilità e le loro Applicazioni', and the European Commission IHP Network "Harmonic Analysis and Related Problems".

P. Ciatti, E. Gonzalez, M. Lanza de Cristoforis, G.P. Leonardi

Contents

PART 1
Complex variables and potential theory

Chapter 1

Integral representations in complex, hypercomplex and Clifford analysis*

Heinrich Begehr

Freie Universität Berlin, Mathematisches Institut
14195 Berlin, Germany
begehr@math.fu-berlin.de

Contents

1.1. Introduction

Integral representations are one of the main tools in analysis. They are useful to determine properties of the functions represented such as smoothness, differentiability, boundary behaviour etc. They serve to reduce boundary value problems etc. for differential equations to integral equations and thus lead to existence and uniqueness proofs. Well–known representation formulas are the Cauchy formula for analytic functions and the Green representation for harmonic functions. Both these formulas are consequences from

*From lecture notes of the minicorsi at Padova University in June 2000 and presented at AMADE 2001 in February 2001. Published in Integral Transformations and Special Functions 13 (2002), 223–241.
See also http://www.tandf.co.uk/journals/titles/10652469.asp.
The author is grateful to Taylor and Francis for permitting republication.

the Gauss divergence theorem where the area integral disappears because homogeneous equations (Cauchy–Riemann and Laplace, respectively) are considered. In the cases of inhomogeneous Cauchy–Riemann equations and the Poisson equation, the area integrals appearing lead to area integral operators of the Pompeiu type. They determine particular solutions to the inhomogeneous equation under consideration.

Now we make the following simple observation. Let ∂ be a linear differential operator and T be its related Pompeiu integral operator. Then ∂T is the identity mapping for a proper function space. For any power ∂^k, $k \in \mathbb{N}$, then the iteration T^k obviously is its right inverse, $\partial^k T^k$ is the identity again. More generally for two such differential operators ∂_1, ∂_2 with right inverses T_1, T_2 then the iteration $T_2 T_1$ is right inverse to $\partial_1 \partial_2$.

On this basis particular solutions for higher order differential operators can be constructed leading also to fundamental solutions. Moreover, these integral operators are useful for determining particular solutions to any higher order differential equation the leading term of which is related to them. In fact, one can solve boundary value problems to these higher order equations if, besides the particular solution for the leading term through the Pompeiu operator, the general solution to the related homogeneous leading term operator equation is taken into consideration.

This sketched procedure can be followed in complex, hypercomplex and Clifford analysis. But the resulting representation formulas of Cauchy-Pompeiu type do not automatically give solutions to related boundary value problems. This, however, is the case whenever these problems are solvable. This phenomenon is known already from the Cauchy formula. Not all functions on the boundary of a domain are boundary values of the analytic functions determined by their Cauchy integrals. In particular solvability conditions have to be observed in the theory of several complex variables where also compatibility conditions for the systems considered are important.

In these lectures the hierarchy of Pompeiu integral operators in the complex case will be presented and some higher order Cauchy–Pompeiu representation formulas given. As an application some orthogonal decompositions of the Hilbert space $L_2(G; \mathbb{C})$, $G \subset \mathbb{C}$ a regular domain, are given. For several complex variables only some results on bidomains are included. As the theory in hypercomplex analysis is analogue to the complex case only some references [2, 9] are given. The situation in Clifford analysis is shortly explained.

1.2. Complex case

Gauss divergence theorem. Let $D \subset \mathbb{R}^2$ be a regular domain, $f, g \in C^1(D; \mathbb{R}) \cap C(\overline{D}; \mathbb{R})$. Then

$$\int_D (f_x + g_y)\, dx\, dy = \int_{\partial D} \{ f\, dy - g\, dx \}\,.$$

Complex forms: $z = x + iy$, $\partial_z = \frac{1}{2}(\partial_x - i\partial_y)$, $\partial_{\bar{z}} = \frac{1}{2}(\partial_x + i\partial_y)$, $w = u + iv \in C^1(D; \mathbb{C}) \cap C(\overline{D}; \mathbb{C})$. Then

$$\int_D w_{\bar{z}} dx\, dy = \frac{1}{2i} \int_{\partial D} w\, dz\,, \qquad \int_D w_z dx\, dy = -\frac{1}{2i} \int_{\partial D} w\, d\bar{z}\,.$$

Cauchy–Pompeiu Representation. Let $w \in C^1(D; \mathbb{C}) \cap C(\overline{D}; \mathbb{C})$. Then with $\zeta = \xi + i\eta$

$$w(z) = \frac{1}{2\pi i} \int_{\partial D} w(\zeta) \frac{d\zeta}{\zeta - z} - \frac{1}{\pi} \int_D w_{\bar{\zeta}}(\zeta) \frac{d\xi\, d\eta}{\zeta - z}\,,$$

$$w(z) = -\frac{1}{2\pi i} \int_{\partial D} w(\zeta) \frac{d\bar{\zeta}}{\overline{\zeta - z}} - \frac{1}{\pi} \int_D w_\zeta(\zeta) \frac{d\xi\, d\eta}{\overline{\zeta - z}}\,.$$

Pompeiu Operator. Let $f \in L_1(D; \mathbb{C})$. Then

$$Tf(z) = -\frac{1}{\pi} \int_D f(\zeta) \frac{d\xi\, d\eta}{\zeta - z}\,, \qquad \overline{T}f(z) = -\frac{1}{\pi} \int_D f(\zeta) \frac{d\xi\, d\eta}{\overline{\zeta - z}}\,.$$

Properties of these operators are developed in [13], see also [1]. Important are

$$\partial_{\bar{z}} Tf = f\,, \qquad \partial_z Tf = \Pi f\,, \qquad f \in L_1(D; \mathbb{C})\,,$$

where

$$\Pi f(z) = -\frac{1}{\pi} \int_D f(\zeta) \frac{d\xi\, d\eta}{(\zeta - z)^2}$$

is a singular integral operator of Calderon–Zygmund type to be taken as a Cauchy principal integral. Here the derivatives are taken in the weak sense.

1.2.1. Complex first order systems

Theorem 1.1. *Any solution to $w_{\bar{z}} = f$ in D, $f \in L_1(D; \mathbb{C})$, is representable via $w = \varphi + Tf$ where φ is analytic in D.*

Proof.   (1) Obviously $\varphi + Tf$ with $\varphi_{\bar{z}} = 0$ is a solution.

(2) If w is a solution then $(w - Tf)_{\bar{z}} = 0$ i.e. is analytic. $\square$

Generalized Beltrami equation:

$$w_{\bar{z}} + \mu_1 w_z + \mu_2 \overline{w_z} + aw + b\overline{w} = f, \quad |\mu_1(z)| + |\mu_2(z)| \le q_0 < 1.$$

Find a particular solution in the form $w = T\rho$! Then ρ must satisfy the singular integral equation

$$\rho + \mu_1 \Pi \rho + \mu_2 \overline{\Pi \rho} + a\, T\rho + b\, \overline{T\rho} = f.$$

As $\mu_1 \Pi \rho + \mu_2 \overline{\Pi \rho}$ is contractive and $a\, T\rho + b\, \overline{T\rho}$ is compact this problem is solvable.

1.2.2. *Complex second order equations*

There are two principally different second order elliptic differential operators the main part of which is either the Laplace or the Bitsadze operator. As in the case of the generalized Beltrami equation the solutions to the inhomogeneous Laplace and Bitsadze equations can be used to solve the general equations of second order.

1.2.2.1. *Poisson equation $w_{z\bar{z}} = f$*

The Cauchy–Pompeiu formulas

$$w(z) = \frac{1}{2\pi i} \int_{\partial D} w(\zeta) \frac{d\zeta}{\zeta - \tau} - \frac{1}{\pi} \int_D w_{\bar{\zeta}}(\tilde{\zeta}) \frac{d\tilde{\xi}\, d\tilde{\eta}}{\tilde{\zeta} - z},$$

$$w_{\bar{\zeta}}(\tilde{\zeta}) = -\frac{1}{2\pi i} \int_{\partial D} w_{\bar{\zeta}}(\zeta) \frac{d\bar{\zeta}}{\zeta - \tilde{\zeta}} - \frac{1}{\pi} \int_D w_{\zeta\bar{\zeta}}(\zeta) \frac{d\xi\, d\eta}{\zeta - \tilde{\zeta}}$$

lead to

$$w(z) = \frac{1}{2\pi i} \int_{\partial D} \left\{ \frac{w(\zeta)}{\zeta - z}\, d\zeta - \psi(\zeta, z)\, w_{\bar{\zeta}}(\zeta)\, d\bar{\zeta} \right\} \tag{1.1}$$

$$-\frac{1}{\pi} \int_D \psi(\zeta, z)\, w_{\zeta\bar{\zeta}}(\zeta)\, d\xi\, d\eta$$

with

$$\psi(\zeta, z) = \frac{1}{\pi} \int_D \frac{1}{\tilde{\bar{\zeta}} - \zeta} \frac{d\tilde{\xi}\, d\tilde{\eta}}{\tilde{\zeta} - z}.$$

Applying the Cauchy–Pompeiu formula to $\log|\zeta - z|^2$ in the domain $D_\varepsilon = D\backslash\{z : |z - \zeta| \leq \varepsilon\}$ for sufficiently small positive ε gives

$$\log|\zeta - z|^2 = \frac{1}{2\pi i}\int_{\partial D_\varepsilon}\log|\zeta - \tilde{\zeta}|^2\frac{d\tilde{\zeta}}{\tilde{\zeta} - z} - \frac{1}{\pi}\int_{D_\varepsilon}\frac{1}{\overline{\tilde{\zeta} - \zeta}}\frac{d\tilde\xi\,d\tilde\eta}{\tilde{\zeta} - z}.$$

As for $\zeta \neq z$

$$\frac{1}{2\pi i}\int_{|\tilde\zeta - \zeta| = \varepsilon}\log|\zeta - \tilde{\zeta}|^2\frac{d\tilde{\zeta}}{\tilde{\zeta} - z} = \frac{\varepsilon\log\varepsilon}{\pi}\int_0^{2\pi}\frac{e^{it}}{\varepsilon e^{it} + \zeta - z}\,dt$$

and

$$\frac{1}{\pi}\int_{|\tilde\zeta - \zeta| < \varepsilon}\frac{1}{\overline{\tilde{\zeta} - \zeta}}\frac{d\tilde\xi\,d\tilde\eta}{\tilde{\zeta} - z} = \frac{1}{\pi}\int_0^{2\pi}\int_0^{\varepsilon}e^{it}\frac{dt\,dr}{re^{it} + \zeta - z}$$

tend to zero with ε then

$$\log|\zeta - z|^2 = \tilde{\psi}(\zeta, z) - \psi(\zeta, z) \tag{1.2}$$

with

$$\tilde{\psi}(\zeta, z) = \frac{1}{2\pi i}\int_{\partial D}\log|\tilde{\zeta} - \zeta|^2\frac{d\tilde{\zeta}}{\tilde{\zeta} - z}.$$

Because for $z, \zeta \in D$

$$\partial_\zeta\tilde{\psi}(\zeta, z) = -\frac{1}{2\pi i}\int_{\partial D}\frac{d\tilde{\zeta}}{(\overline{\tilde{\zeta} - \zeta})(\tilde{\zeta} - z)}$$

$$= -\frac{1}{2\pi i}\int_{\partial D}\left(\frac{1}{\overline{\tilde{\zeta} - \zeta}} - \frac{1}{\overline{\tilde{\zeta} - z}}\right)\frac{d\tilde{\zeta}}{\zeta - z} = 0$$

from the Gauss theorem

$$\frac{1}{2\pi i}\int_{\partial D}\tilde{\psi}(\zeta, z)\,w_{\overline{\zeta}}(\zeta)\,d\overline{\zeta} + \frac{1}{\pi}\int_D\tilde{\psi}(\zeta, z)\,w_{\zeta\overline{\zeta}}(\zeta)\,d\xi\,d\eta = 0.$$

Adding this to the right–hand side of (1.1) and observing (1.2) show

$$w(z) = \frac{1}{2\pi i}\int_{\partial D}\frac{w(\zeta)}{\zeta - z}\,d\zeta + \frac{1}{2\pi i}\int_{\partial D}\log|\zeta - z|^2 w_{\overline{\zeta}}(\zeta)\,d\overline{\zeta}$$
$$+ \frac{1}{\pi}\int_D\log|\zeta - z|^2 w_{\zeta\overline{\zeta}}(\zeta)\,d\xi\,d\eta. \tag{1.3}$$

As is well known $2/\pi\log|\zeta - z|$ is the fundamental solution to the Laplacian $\partial_z\partial_{\overline{z}}$. The representation (1.3) has the form

$$w = \varphi + \overline{\psi} + T_{1,1}f, \qquad f = w_{z\overline{z}},$$

with analytic functions φ and ψ.

1.2.2.2. *Bitsadze equation* $w_{\bar{z}\bar{z}} = f$

Similarly to the preceding subsection the Cauchy–Pompeiu formulas

$$w(z) = \frac{1}{2\pi i} \int_{\partial D} w(\zeta) \frac{d\zeta}{\zeta - z} - \frac{1}{\pi} \int_D w_{\bar{\zeta}}(\tilde{\zeta}) \frac{d\tilde{\xi}\, d\tilde{\eta}}{\tilde{\zeta} - z},$$

$$w_{\bar{\zeta}}(\tilde{\zeta}) = \frac{1}{2\pi i} \int_{\partial D} w_{\bar{\zeta}}(\zeta) \frac{d\zeta}{\zeta - \tilde{\zeta}} - \frac{1}{\pi} \int_D w_{\bar{\zeta}\bar{\zeta}}(\zeta) \frac{d\xi\, d\eta}{\zeta - \tilde{\zeta}}$$

imply

$$w(z) = \frac{1}{2\pi i} \int_{\partial D} \left\{ \frac{w(\zeta)}{\zeta - z} + \psi(\zeta, z)\, w_{\bar{\zeta}}(\zeta) \right\} d\zeta - \frac{1}{\pi} \int_D \psi(\zeta, z)\, w_{\bar{\zeta}\bar{\zeta}}(\zeta)\, d\xi\, d\eta \tag{1.4}$$

with

$$\psi(\zeta, z) = \frac{1}{\pi} \int_D \frac{1}{\tilde{\zeta} - \zeta} \frac{d\tilde{\xi}\, d\tilde{\eta}}{\tilde{\zeta} - z}.$$

Applying the Cauchy–Pompeiu formula to $\overline{\frac{\zeta - z}{\zeta - z}}$ in the domain $D_\varepsilon = D \backslash \{z : |z - \zeta| \leq \varepsilon\}$ for sufficiently small positive ε gives

$$\overline{\frac{\zeta - z}{\zeta - z}} = \frac{1}{2\pi i} \int_{\partial D_\varepsilon} \overline{\frac{\zeta - \tilde{\zeta}}{\zeta - \tilde{\zeta}}} \frac{d\tilde{\zeta}}{\tilde{\zeta} - z} + \frac{1}{\pi} \int_{D_\varepsilon} \frac{1}{\zeta - \tilde{\zeta}} \frac{d\tilde{\xi}\, d\tilde{\eta}}{\tilde{\zeta} - z}.$$

Observing that when ε tends to zero

$$\frac{1}{2\pi i} \int_{|\tilde{\zeta} - \zeta| = \varepsilon} \overline{\frac{\tilde{\zeta} - \zeta}{\tilde{\zeta} - \zeta}} \frac{d\tilde{\zeta}}{\tilde{\zeta} - z} = \frac{\varepsilon}{2\pi} \int_0^{2\pi} e^{-it} \frac{dt}{\varepsilon e^{it} + \zeta - z}$$

and

$$\frac{1}{\pi} \int_{|\tilde{\zeta} - \zeta| < \varepsilon} \frac{1}{\tilde{\zeta} - \zeta} \frac{d\tilde{\zeta}\, d\tilde{\eta}}{\tilde{\zeta} - z} = \frac{1}{\pi} \int_0^{2\pi} \int_0^\varepsilon e^{-it} \frac{dt\, dr}{r e^{it} + \zeta - z}$$

tend to zero then

$$\overline{\frac{\zeta - z}{\zeta - z}} = \tilde{\psi}(\zeta, z) - \psi(\zeta, z) \tag{1.5}$$

with

$$\tilde{\psi}(\zeta, z) = \frac{1}{2\pi i} \int_{\partial D} \overline{\frac{\tilde{\zeta} - \zeta}{\tilde{\zeta} - \zeta}} \frac{d\tilde{\zeta}}{\tilde{\zeta} - z}.$$

As for $z, \zeta \in D$

$$\partial_{\bar{\zeta}} \tilde{\psi}(\zeta, z) = -\frac{1}{2\pi i} \int_{\partial D} \frac{1}{\tilde{\zeta} - \zeta} \frac{d\tilde{\zeta}}{\tilde{\zeta} - z} = 0$$

from the Gauss theorem

$$\frac{1}{2\pi i}\int_{\partial D}\tilde{\psi}(\zeta,z)\,w_{\overline{\zeta}}(\zeta)\,d\zeta - \frac{1}{\pi}\int_{D}\tilde{\psi}(\zeta,z)\,w_{\overline{\zeta}\,\zeta}(\zeta)\,d\xi\,d\eta = 0\,.$$

Subtracting this from the right–hand side of (1.4) and observing (1.5) lead to

$$w(z) = \frac{1}{2\pi i}\int_{\partial D}\frac{w(\zeta)}{\zeta - z}\,d\zeta - \frac{1}{2\pi i}\int_{\partial D}\frac{\overline{\zeta - z}}{\zeta - z}\,w_{\overline{\zeta}}(\zeta)\,d\zeta \qquad (1.6)$$
$$+\frac{1}{\pi}\int_{D}\frac{\overline{\zeta - z}}{\zeta - z}\,w_{\overline{\zeta}\,\zeta}(\zeta)\,d\xi\,d\eta\,.$$

The kernel $(\overline{\zeta - z})/[(\zeta - z)\pi]$ is the fundamental kernel to the Bitsadze operator $\partial_{\overline{z}}^2$. Representation (1.6) is of the form

$$w = \varphi + \overline{z}\psi + T_{0,2}f\,, \qquad f = w_{\overline{z}\,\overline{z}}\,,$$

with analytic functions φ und ψ.

1.2.2.3. *General complex second order equations*

A general second order equation with leading term $\partial_{\overline{z}}^2$ has the form

$$w_{\overline{z}\,\overline{z}} + \mu_1 w_{z\overline{z}} + \mu_2 \overline{w_{z\overline{z}}} + a_1 w_{\overline{z}} + a_2 \overline{w_{\overline{z}}} + b_1 w_z + b_2 \overline{w_z} + c_1 w + c_2 \overline{w} = d$$

where $|\mu_1(z)|+|\mu_2(z)| \le q_0 < 1$. Setting $w_{\overline{z}\,\overline{z}} = f$ so that $w = \varphi + \overline{z}\psi + T_{0,2}f$ leads to a singular integral equation for f. With proper integral operators $T_{\mu,\nu}$, see the following section, it is, similarly to the generalized Beltrami equation,

$$f + \mu_1 T_{-1,1}f + \mu_2 T_{1,-1}\overline{f} + a_1 T_{0,1}f + a_2 T_{1,0}\overline{f} + b_1 T_{-1,2}f + b_2 T_{2,-1}\overline{f}$$
$$+c_1 T_{0,2}f + c_2 T_{2,0}\overline{f} = d - \mu_1 \psi' - \mu_2 \overline{\psi'} - a_1 \psi - a_2 \overline{\psi} - b_1 (\varphi' + \overline{z}\psi')$$
$$-b_2 (\overline{\varphi'} + z\overline{\psi'}) - c_1 (\varphi + \overline{z}\psi) - c_2 (\overline{\varphi} + z\overline{\psi})\,.$$

Here $T_{-1,1}$ and $T_{1,-1}$ are singular integral operators while the other ones are just weakly singular and give a compact operator.

1.2.3. *Complex higher order equations*

Continuing in the way indicated in the preceding subsections the prototyp $\partial_z^m \partial_{\overline{z}}^n w = f$ can be treated in regular domains $D \subset \mathbb{C}$.

Definition 1.1. Let for $m, n \in \mathbb{Z}$, $0 \leq m + n$, $1 \leq m^2 + n^2$

$$
K_{m,n}(z) = \begin{cases}
\dfrac{(-m)!(-1)^m}{(n-1)!\,\pi}\, z^{m-1}\bar{z}^{n-1}, & m \leq 0, \\[2ex]
\dfrac{(-n)!\,(-1)^n}{(m-1)!\,\pi}\, z^{m-1}\bar{z}^{n-1}, & n \leq 0, \\[2ex]
\dfrac{z^{m-1}\bar{z}^{n-1}}{(m-1)!\,(n-1)!\,\pi}\left[\log|z|^2 - \sum_{\mu=1}^{m-1}\frac{1}{\mu} - \sum_{\nu=1}^{n-1}\frac{1}{\nu}\right], & 1 \leq m, n.
\end{cases}
$$

These kernel functions determine fundamental solutions to $\partial_z^m \partial_{\bar{z}}^n$ for $0 \leq m, n$, $0 < m^2 + n^2$, see [8]. Their essential properties are

$$
K_{m,n} = \partial_z K_{m+1,n} = \partial_{\bar{z}} K_{m,n+1}
$$

and

$$
\int_{|z|<R} |K_{m,n}(z)|\, dx\, dy < +\infty \quad \text{for} \ \ 0 < m + n, \ 0 < R.
$$

Definition 1.2. For $D \subset \mathbb{C}$ a domain, $f \in L_1(D; \mathbb{C})$ and $m, n \in \mathbb{Z}$ with $0 \leq m + n$

$$
T_{m,n}f(z) = \int_D K_{m,n}(z - \zeta)\, f(\zeta)\, d\xi\, d\eta \quad \text{if} \ \ 1 \leq m^2 + n^2,
$$

$$
T_{0,0}f(z) = f(z).
$$

This is a hierarchy of integral operators with the Pompeiu operators as basic elements, namely

$$
T_{0,1} = T, \quad T_{1,0} = \overline{T}, \quad T_{-1,1} = \Pi, \quad T_{1,-1} = \overline{\Pi}.
$$

$T_{m,n}$ is a weakly singular integral operator for $0 < m + n$ but strongly singular of Calderon–Zygmund type to be understood as a Cauchy principal value integral operator if $m + n = 0$ but $0 < m^2 + n^2$. Moreover,

$$|T_{m,n} f(z)| \leq M \, \|f\|_p, \quad 1 < p, \quad 1 \leq m + n, \quad |z| \leq R,$$

$$|T_{m,n} f(z_1) - T_{m,n} f(z_2)| \leq M \, |z_1 - z_2|^\alpha, \quad |z_1|, |z_2| \leq R,$$

$$\alpha = \begin{cases} (p-2)/p, & m + n = 1, \quad 2 < p, \\ 1, & m + n = 2, \quad 2 < p; \quad 3 \leq m + n, \quad 1 \leq p, \end{cases}$$

$$\|T_{m,-m} f\|_{p,\mathbb{C}} \leq M(p) \, \|f\|_{p,D}, \quad m \neq 0,$$

$$\|T_{m,-m} f\|_{2,\mathbb{C}} \leq \|f\|_{2,\mathbb{C}}.$$

For details see [8].

There is a higher order Cauchy–Pompeiu formula related to the differential operator $\partial_z^m \partial_{\bar z}^n$. For simplicity only a particular case is presented here.

A higher–order Cauchy Pompeiu formula *Let $D \subset \mathbb{C}$ be a regular domain and $w \in C^m(D; \mathbb{C}) \cap C^{m-1}(\overline{D}; \mathbb{C})$, $1 \leq m$. Then*

$$w(z) = \sum_{\mu=0}^{m-1} \frac{1}{2\pi i} \int_{\partial D} \frac{1}{\mu!} \frac{(\overline{z - \zeta})^\mu}{\zeta - z} \partial_{\bar\zeta}^\mu w(\zeta) \, d\zeta$$

$$- \frac{1}{\pi} \int_D \frac{1}{(m-1)!} \frac{(\overline{z - \zeta})^{m-1}}{\zeta - z} \partial_{\bar\zeta}^m w(\zeta) \, d\xi \, d\eta.$$

Proof. (1) For $m = 1$ the formula coincides with one of the basic Cauchy–Pompeiu formulas.

(2) Assuming the formula holds for some m, $1 \leq m$, from the Gauss

theorem

$$\frac{1}{\pi}\int_D \frac{1}{m!}\frac{(\overline{z-\zeta})^m}{\zeta-z}\,\partial_{\overline{\zeta}}^{m+1}w(\zeta)\,d\xi\,d\eta$$

$$= \frac{1}{\pi}\int_D \left\{\partial_{\overline{\zeta}}\left[\frac{1}{m!}\frac{(\overline{z-\zeta})^m}{\zeta-z}\,\partial_{\overline{\zeta}}^m w(\zeta)\right]\right.$$

$$\left. +\frac{1}{(m-1)!}\frac{(\overline{z-\zeta})^{m-1}}{\zeta-z}\,\partial_{\overline{\zeta}}^m w(\zeta)\right\}\,d\xi\,d\eta$$

$$= \frac{1}{2\pi i}\int_{\partial D}\frac{1}{m!}\frac{(\overline{z-\zeta})^m}{\zeta-z}\,\partial_{\overline{\zeta}}^m w(\zeta)\,d\zeta$$

$$+\frac{1}{\pi}\int_D \frac{1}{(m-1)!}\frac{(\overline{z-\zeta})^{m-1}}{\zeta-z}\,\partial_{\overline{\zeta}}^m w(\zeta)\,d\xi\,d\eta\,.$$

This identity gives the formula for $m+1$ rather than m. $\square$

Corollary 1.1. *Any $w \in C^m(D;\mathbb{C})\cap C^{m-1}(\overline{D};\mathbb{C})$, $1\le m$, with $\partial_{\overline{z}}^m w = f$ is representable as*

$$w(z) = \sum_{\mu=0}^{m-1}\varphi_\mu(z)\,\overline{z}^\mu + T_{0,m}f(z)\,. \tag{1.7}$$

Here φ_μ, $0 \le \mu \le m-1$, is analytic. $\sum_{\mu=0}^{m-1}\varphi_\mu(z)\,\overline{z}^\mu$ is as a polyanalytic function of order m, the general solution to the homogeneous equation $\partial_{\overline{z}}^m\varphi = 0$.

A complex m–th order equation of the form

$$\partial_{\overline{z}}^m w + \sum_{\substack{\rho+\sigma=m\\ \sigma\ne 0}}\mu_{\rho\sigma}\partial_{\overline{z}}^\rho\partial_z^\sigma w = F(z,\partial_{\overline{z}}^\rho\partial_z^\sigma\,w\,(0\le \rho+\sigma < m))$$

with

$$\sum_{\substack{\rho+\sigma=m\\ \sigma\ne 0}}|\mu_{\rho\sigma}(z)| \le q_0 < 1$$

is transformed through (1.7) into a singular integral equation of the form

$$f + \sum_{\substack{\rho+\sigma=m\\ \sigma\ne 0}}\mu_{\rho\sigma}T_{-\sigma,m-\rho}f - F(z,T_{-\sigma,m-\rho}f\,(0\le \rho+\sigma < m))$$

$$= G(\varphi_\mu(0\le \mu \le m-1))\,.$$

Proper boundary conditions serve to determine the φ_μ through f and the boundary data.

1.2.4. *Orthogonal decomposition of $L_2(D;\mathbb{C})$*

The inner product for the Hilbert space of square integrable functions in D is defined as

$$(f,g) = \int_D \overline{f(z)}\, g(z)\, dx\, dy\,.$$

As before $D \subset \mathbb{C}$ is a regular domain.

Definition 1.3. The subset of polyholomorphic functions of order $k \geq 1$ in $L_2(D;\mathbb{C})$ is

$$\mathcal{O}_{k,2}(D;\mathbb{C}) = \{f : f \in L_2(D;\mathbb{C}),\ \partial_{\bar{z}}^k f = 0 \ \text{in}\ D\}\,.$$

Its orthogonal complement is denoted by

$$\mathcal{O}_{k,2}^{\perp}(D;\mathbb{C}) = \{g : g \in L_2(D;\mathbb{C}),\ (g,f) = 0 \ \text{for all}\ f \in \mathcal{O}_{k,2}(D;\mathbb{C})\}\,.$$

As usual

$$\overset{\circ}{W}{}_2^k(D;\mathbb{C}) = \{f : f \in W_2^k(D;\mathbb{C}),\ \partial_z^\nu f = 0 \ \text{on}\ \partial D,\ 0 \leq \nu \leq k - 1\}$$

denotes the subspace of functions with vanishing boundary data of the Sobolev space $W_2^k(D;\mathbb{C})$. To the latter belong all functions with weak derivatives up to the k–th order in $L_2(D;\mathbb{C})$.

Lemma 1.1. *For $q \in \mathcal{O}_{k,2}^{\perp}(D;\mathbb{C})$ the problem*

$$\partial_{\bar{z}}^k r = q \ \text{in}\ D\,, \qquad \partial_z^\nu r = 0 \ \text{on}\ \partial D \ \text{for}\ 0 \leq \nu \leq k - 1$$

is equivalent to the problem

$$\Delta^k r = 4^k \partial_{\bar{z}}^k q \ \text{in}\ D,\ \partial_z^\nu r = 0 \ \text{on}\ \partial D \ \text{for}\ 0 \leq \nu \leq k - 1\,.$$

Proof. (1) Applying $4^k \partial_z^k$ to $\partial_{\bar{z}}^k r = q$ shows $\Delta^k r = 4^k \partial_z^k q$.
(2) A solution r to the second problem satisfies

$$\partial_z^k(\partial_{\bar{z}}^k r - q) = 0\,.$$

Hence, $\partial_{\bar{z}}^k r - q \in \mathcal{O}_{k,2}(D;\mathbb{C})\,.$

Let now $r \in \overset{\circ}{W}{}_2^k(D;\mathbb{C})$ and $\varphi \in \mathcal{O}_{k,2}(D;\mathbb{C})$ then

$$(\partial_z^k r, \varphi) = \int_D \partial_{\bar z}^k \overline{r(z)}\, \varphi(z)\, dx\, dy$$

$$= \int_D \left\{ \partial_{\bar z}[\partial_{\bar z}^{k-1}\overline{r(z)}\, \varphi(z)] - \partial_{\bar z}^{k-1}\overline{r(z)}\, \partial_{\bar z}\varphi(z) \right\} dx\, dy$$

$$= \frac{1}{2i} \int_{\partial D} \overline{\partial_z^{k-1} r(z)}\, \varphi(z)\, dz - \int_D \partial_{\bar z}^{k-1}\overline{r(z)}\, \partial_{\bar z}\varphi(z)\, dx\, dy = \cdots$$

$$= (-1)^k \int_D \overline{r(z)}\, \partial_{\bar z}^k \varphi(z)\, dx\, dy = 0\,.$$

Thus, $\partial_z^k r \in \mathcal{O}_{k,2}^\perp(D;\mathbb{C})$. As also $q \in \mathcal{O}_{k,2}^\perp(D;\mathbb{C})$ then $\partial_z^k r - q \in \mathcal{O}_{k,2}^\perp(D;\mathbb{C})$. Therefore $\partial_z^k r - q = 0$. $\qquad\square$

Remark. While the problem

$$\partial_z^k r = q \ \text{ in } \ D\,, \quad \partial_z^\nu r = 0 \ \text{ on } \ \partial D \ \text{ for } 0 \le \nu \le k-1$$

is overdetermined, the problem

$$\Delta^k r = \widetilde{q} \ \text{ in } \ D\,, \quad \partial_z^\nu r = 0 \ \text{ on } \ \partial D \ \text{ for } 0 \le \nu \le k-1$$

is well–posed.

Theorem 1.2. *For regular domains* $D \subset \mathbb{C}$

$$\mathcal{O}_{k,2}^\perp(D;\mathbb{C}) = \partial_z^k \overset{\circ}{W}{}_2^k(D;\mathbb{C})\,.$$

Proof. (1) Consider for $q \in \mathcal{O}_{k,2}^\perp(D;\mathbb{C})$ the problem

$$\partial_z^k r = q \ \text{ in } \ D\,, \quad \partial_z^\nu r = 0 \ \text{ on } \ \partial D \ \text{ for } \ 0 \le \nu \le k-1\,.$$

This problem is solvable according to the preceding lemma. The solution is representable via the Cauchy–Pompeiu formula

$$r(z) = -\sum_{\mu=0}^{k-1} \frac{1}{2\pi i} \int_{\partial D} \frac{1}{\mu!} \frac{(z-\zeta)^\mu}{\zeta - z}\, \partial_\zeta^\mu r(\zeta)\, d\bar\zeta$$

$$-\frac{1}{\pi} \int_D \frac{1}{(k-1)!} \frac{(z-\zeta)^{k-1}}{\zeta - z}\, \partial_\zeta^k r(\zeta)\, d\xi\, d\eta$$

so that

$$r(z) = -\frac{1}{\pi} \int_D \frac{1}{(k-1)!} \frac{(z-\zeta)^{k-1}}{\overline{\zeta-z}} \, q(\zeta) \, d\xi \, d\eta \,.$$

The Green function

$$G_k(z,\zeta) = \frac{1}{(k-1)!^2} |\zeta-z|^{2(k-1)} [\log|\zeta-z|^2 - 2\sum_{\rho=1}^{k-1}\frac{1}{\rho}] + h_k(z,\zeta)$$

of the differential operator Δ^k where $h_k \in C^{2k}(D\times D; \mathbb{C}) \cap C^{2k-1}(\overline{D}\times\overline{D}; \mathbb{C})$ with $\Delta_z^k h_k(z,\zeta) = 0$ in $D\times D$, $\partial_z^\rho G_k(z,\zeta) = 0$, $\partial_{\overline{z}}^\rho G_k(z,\zeta) = 0$ on $\partial D \times D$ for $0 \le \rho \le k-1$ satisfies

$$\partial_{\overline{\zeta}}^k G_k(z,\zeta) = \frac{(\zeta-z)^{k-1}}{(k-1)!} \frac{1}{\overline{\zeta-z}} + \partial_{\overline{\zeta}}^k h_k(z,\zeta) \,.$$

From the Gauss theorem

$$\int_D \partial_{\overline{\zeta}}^k h_k(z,\zeta) \, \partial_{\overline{\zeta}}^k r(\zeta) \, d\xi \, d\eta$$

$$= \int_D \{\partial_{\overline{\zeta}}[\partial_{\overline{\zeta}}^k h_k(z,\zeta) \, \partial_{\overline{\zeta}}^{k-1} r(\zeta)] - \partial_{\overline{\zeta}} \partial_{\overline{\zeta}}^k h_k(z,\zeta) \, \partial_{\overline{\zeta}}^{k-1} r(\zeta)\} \, d\xi \, d\eta$$

$$= -\frac{1}{2i} \int_{\partial D} \partial_{\overline{\zeta}}^k h_k(z,\zeta) \, \partial_{\overline{\zeta}}^{k-1} r(\zeta) \, d\overline{\zeta} - \int_D \partial_{\overline{\zeta}} \partial_{\overline{\zeta}}^k h_k(z,\zeta) \, \partial_{\overline{\zeta}}^{k-1} r(\zeta) \, d\xi \, d\eta$$

$$= \cdots = (-1)^k \int_D \partial_{\overline{\zeta}}^k \partial_{\overline{\zeta}}^k h_k(z,\zeta) \, r(\zeta) \, d\xi \, d\eta = 0$$

follows. Adding this to the representation of r gives

$$r(z) = \frac{(-1)^k}{\pi} \int_D \partial_{\overline{\zeta}}^k G_k(z,\zeta) \, q(\zeta) \, d\xi \, d\eta$$

$$= -\frac{1}{\pi} \int_D \frac{1}{(k-1)!} \frac{(z-\zeta)^{k-1}}{\overline{\zeta-z}} \, q(\zeta) \, d\xi \, d\eta \,.$$

By differentiation

$$\partial_z^{k-\nu} r(z) = \frac{(-1)^k}{\pi} \int_D \partial_{\overline{\zeta}}^k \partial_z^{k-\nu} G_k(z,\zeta) \, q(\zeta) \, d\xi \, d\eta$$

$$= -\frac{1}{\pi} \int_D \frac{1}{(\nu-1)!} \frac{(z-\zeta)^{\nu-1}}{\overline{\zeta-z}} \, q(\zeta) \, d\xi \, d\eta$$

follows for $1 \le \nu \le k$. From here

$$\|r\|_{W_2^k} \le M \, \|q\|_{L_2}$$

can be shown. As $r \in \overset{\circ}{W}{}_2^k(D;\mathbb{C})$ satisfies $\partial_z^k r = q$ one has $q \in \partial_z^k \overset{\circ}{W}{}_2^k(D;\mathbb{C})$.

(2) Let $q \in \partial_z^k \overset{\circ}{W}{}_2^k(D;\mathbb{C})$. Then there exists an $r \in \overset{\circ}{W}{}_2^k(D;\mathbb{C})$ such that $\partial_z^k r = q$ in D, $\partial_z^\nu r = 0$ on ∂D for $0 \leq \nu \leq k-1$. Let now $\varphi \in \mathcal{O}_{k,2}(D;\mathbb{C})$. Then as before

$$(q,\varphi) = \int_D \partial_{\bar{z}}^k \overline{r(z)}\, \varphi(z)\, dx\, dy = (-1)^k \int_D \overline{r(z)}\, \partial_{\bar{z}}^k \varphi(z)\, dx\, dy = 0\,.$$

Thus $q \in \mathcal{O}_{k,2}^\perp(D;\mathbb{C})$. $\qquad\qquad\qquad\qquad\qquad\qquad\qquad\qquad\square$

Corollary 1.2. *For regular domains $D \subset \mathbb{C}$*

$$L_2(D;\mathbb{C}) = \mathcal{O}_{k,2}(D;\mathbb{C}) \oplus \partial_z^k \overset{\circ}{W}{}_2^k(D;\mathbb{C})\,.$$

By interchanging the roles of z and $\bar{z}$ a dual result is available. In the same way $L_2(D;\mathbb{C})$ can be orthogonally decomposed with respect to polyharmonic functions.

Definition 1.4. The subset of polyharmonic functions of order $k \geq 1$ in $L_2(D;\mathbb{C})$ is

$$\mathbb{H}_{k,2}(D;\mathbb{C}) = \{f : f \in L_2(D;\mathbb{C}), \partial_z^k \partial_{\bar{z}}^k f = 0 \ \text{ in } \ D\}\,.$$

Its orthogonal complement is denoted by

$$\mathbb{H}_{k,2}^\perp(D;\mathbb{C}) = \{g : g \in L_2(D;\mathbb{C}), (g,f) = 0 \ \text{ for all } \ f \in \mathbb{H}_{k,2}(D;\mathbb{C})\}\,.$$

Moreover,

$$\overset{\circ}{W}{}_{\Delta^k,2}^{2k}(D;\mathbb{C}) = \{f : f \in W_2^{2k}(D;\mathbb{C}), \partial_z^\nu \partial_{\bar{z}}^\nu f = 0,\ \partial_z^{\nu+1} \partial_{\bar{z}}^\nu f = 0,$$

$$\partial_z^\nu \partial_{\bar{z}}^{\nu+1} f = 0 \ \text{ on } \ \partial G \ \text{ for } \ 0 \leq \nu \leq k-1\}\,.$$

Theorem 1.3. *For regular domains $D \subset \mathbb{C}$*

$$\mathbb{H}_{k,2}^\perp(D;\mathbb{C}) = \partial_z^k \partial_{\bar{z}}^k \overset{\circ}{W}{}_{\Delta^k,2}^{2k}(D;\mathbb{C})$$

and

$$L_2(D;\mathbb{C}) = \mathbb{H}_{k,2}(D;\mathbb{C}) \oplus \partial_z^k \partial_{\bar{z}}^k \overset{\circ}{W}{}_{\Delta^k,2}^{2k}(D;\mathbb{C})\,.$$

For a proof see [7].

1.3. Several complex variables

Quite a natural extension of the representation formulas for one complex variable to higher dimensions is available for polydomains. In principle the unit ball can be treated too, see [4, 5], but because of the complicated structure of the Pompeiu operator, see [12], iterations cannot yet be given explictly. In order to present the concept of the procedure for polydomains just bidomains are studied in $\mathbb{C}^2$. The extension to higher dimensions is then obvious, see [6].

Theorem 1.4. *Let D_1 and D_2 be regular domains in $\mathbb{C}$ and $D = D_1 \times D_2 \subset \mathbb{C}^2$ be the bidomain composed by D_1 and D_2 and $\partial_0 D = \partial D_1 \times \partial D_2$ the distinguished boundary of D. Any $w \in C^1(D; \mathbb{C}) \cap C(D \cup \partial_0 D; \mathbb{C})$ can be represented as*

$$
\begin{aligned}
w(z_1, z_2) = {} & \frac{1}{(2\pi i)^2} \int_{\partial D_1} \int_{\partial D_2} w(\zeta_1, \zeta_2) \frac{d\zeta_1}{\zeta_1 - z_1} \frac{d\zeta_2}{\zeta_2 - z_2} \\
& - \frac{1}{\pi} \int_{D_1} w_{\overline{\zeta_1}}(\zeta_1, z_2) \frac{d\xi_1 d\eta_1}{\zeta_1 - z_1} - \frac{1}{\pi} \int_{D_2} w_{\overline{\zeta_2}}(z_1, \zeta_2) \frac{d\xi_2 d\eta_2}{\zeta_2 - z_2} \\
& - \frac{1}{\pi^2} \int_{D_1} \int_{D_2} w_{\overline{\zeta_1}\,\overline{\zeta_2}}(\zeta_1, \zeta_2) \frac{d\xi_1 d\eta_1}{\zeta_1 - z_1} \frac{d\xi_2 d\eta_2}{\zeta_2 - z_2} .
\end{aligned}
\tag{1.8}
$$

Proof.   Applying the Cauchy–Pompeiu formula for D_1 and then for D_2 gives

$$
\begin{aligned}
w(z_1, z_2) = {} & \frac{1}{2\pi i} \int_{\partial D_1} w(\zeta_1, z_2) \frac{d\zeta_1}{\zeta_1 - z_1} - \frac{1}{\pi} \int_{D_1} w_{\overline{\zeta_1}}(\zeta_1, z_2) \frac{d\xi_1 d\eta_1}{\zeta_1 - z_1} \\
= {} & \frac{1}{(2\pi i)^2} \int_{\partial D_1} \int_{\partial D_2} w(\zeta_1, \zeta_2) \frac{d\zeta_1}{\zeta_1 - z_1} \frac{d\zeta_2}{\zeta_2 - z_2} \\
& - \frac{1}{2\pi^2 i} \int_{\partial D_1} \int_{D_2} w_{\overline{\zeta_2}}(\zeta_1, \zeta_2) \frac{d\xi_2 d\eta_2}{\zeta_2 - z_2} \frac{d\zeta_1}{\zeta_1 - z_1} \\
& - \frac{1}{2\pi^2 i} \int_{D_1} \int_{\partial D_2} w_{\overline{\zeta_1}}(\zeta_1, \zeta_2) \frac{d\zeta_2}{\zeta_2 - z_2} \frac{d\xi_1 d\eta_1}{\zeta_1 - z_1} \\
& + \frac{1}{\pi^2} \int_{D_1} \int_{D_2} w_{\overline{\zeta_1}\,\overline{\zeta_2}}(\zeta_1, \zeta_2) \frac{d\xi_1 d\eta_1}{\zeta_1 - z_1} \frac{d\xi_2 d\eta_2}{\zeta_2 - z_2} .
\end{aligned}
$$

18   *H. Begehr*

With

$$\frac{1}{\pi}\int_{D_1} w_{\overline{\zeta_1}}(\zeta_1, z_2)\frac{d\xi_1 d\eta_1}{\zeta_1 - z_1} = \frac{1}{2\pi^2 i}\int_{D_1}\int_{\partial D_2} w_{\overline{\zeta_1}}(\zeta_1, \zeta_2)\frac{d\zeta_2}{\zeta_2 - z_2}\frac{d\xi_1 d\eta_1}{\zeta_1 - z_1}$$

$$-\frac{1}{\pi^2}\int_{D_1}\int_{D_2} w_{\overline{\zeta_1}\,\overline{\zeta_2}}(\zeta_1, \zeta_2)\frac{d\xi_1 d\eta_1}{\zeta_1 - z_2}\frac{d\xi_2 d\eta_2}{\zeta_2 - z_2},$$

$$\frac{1}{\pi}\int_{D_2} w_{\overline{\zeta_2}}(z_1, \zeta_2)\frac{d\xi_2 d\eta_2}{\zeta_2 - z_2} = \frac{1}{2\pi^2 i}\int_{\partial D_1}\int_{D_2} w_{\overline{\zeta_2}}(\zeta_1, \zeta_2)\frac{d\xi_2 d\eta_2}{\zeta_2 - z_2}\frac{d\zeta_1}{\zeta_1 - z_1}$$

$$-\frac{1}{\pi^2}\int_{D_1}\int_{D_2} w_{\overline{\zeta_1}\,\overline{\zeta_2}}(\zeta_1, \zeta_2)\frac{d\xi_1 d\eta_1}{\zeta_1 - z_2}\frac{d\xi_2 d\eta_2}{\zeta_2 - z_2}$$

the above formula follows. $\square$

Remarks. For polydomains $C^k(D; \mathbb{C})$ denotes the set of complex functions in D having continuous derivatives with respect to any single variable up to order k. E.g. for a bidomain D and $k = 1$ the functions w have continuous derivatives w_{z_1}, $w_{\overline{z_1}}$, w_{z_2}, $w_{\overline{z_2}}$, $w_{z_1 z_2}$, $w_{z_1 \overline{z_2}}$, $w_{\overline{z_1} z_2}$, $w_{\overline{z_1}\,\overline{z_2}}$. Other representation formulas are available by replacing ζ_k by $\overline{\zeta_k}$ for one index or both indices $k = 1, 2$. The respective representation in the case of $\mathbb{C}^n$ is

$$w(z) = \frac{1}{(2\pi i)^n}\int_{\partial_0 D} w(\zeta)\prod_{\nu=1}^{n}\frac{d\zeta_\nu}{\zeta_\nu - z_\nu}$$

$$-\sum_{\nu=1}^{n}\sum_{1\le\rho_1<\cdots<\rho_\nu\le n}\frac{1}{\pi^\nu}\int_{D_{\rho_1}}\cdots\int_{D_{\rho_\nu}} w_{\overline{\zeta_{\rho_1}}\cdots\overline{\zeta_{\rho_\nu}}}(\zeta)\prod_{\mu=1}^{\nu}\frac{d\xi_{\rho_\mu}d\eta_{\rho_\mu}}{\zeta_{\rho_\mu} - z_{\rho_\mu}}.$$

In this formula components of ζ not integrated upon have to be interpreted as z–components.

Iterating (1.8) leads to a second order representation.

Theorem 1.5. *Any $w \in C^2(D; \mathbb{C}) \cap C^1(D \cup \partial_0 D; \mathbb{C})$ can be represented as well as*

$$
w(z_1, z_2) = \frac{1}{(2\pi i)^2} \int_{\partial D_1} \int_{\partial D_2} \left\{ \frac{w(\zeta_1, \zeta_2)}{(\zeta_1 - z_1)(\zeta_2 - z_2)} - \frac{w_{\overline{\zeta_1}}(\zeta_1, \zeta_2)}{\zeta_2 - z_2} \frac{\overline{\zeta_1 - z_1}}{\zeta_1 - z_1} \right.
$$

$$
\left. - \frac{w_{\overline{\zeta_2}}(\zeta_1, \zeta_2)}{\zeta_1 - z_1} \frac{\overline{\zeta_2 - z_2}}{\zeta_2 - z_2} + w_{\overline{\zeta_1}\,\overline{\zeta_2}}(\zeta_1, \zeta_2) \frac{\overline{\zeta_1 - z_1}}{\zeta_1 - z_1} \frac{\overline{\zeta_2 - z_2}}{\zeta_2 - z_2} \right\} d\zeta_1 d\zeta_2
$$

$$
+ \frac{1}{\pi} \int_{D_1} w_{\overline{\zeta_1}\,\overline{\zeta_1}}(\zeta_1, z_2) \frac{\overline{\zeta_1 - z_1}}{\zeta_1 - z_1} \, d\xi_1 d\eta_1 \tag{1.9}
$$

$$
+ \frac{1}{\pi} \int_{D_2} w_{\overline{\zeta_2}\,\overline{\zeta_2}}(z_1, \zeta_2) \frac{\overline{\zeta_2 - z_2}}{\zeta_2 - z_2} \, d\xi_2 d\eta_2
$$

$$
- \frac{1}{\pi^2} \int_{D_1} \int_{D_2} w_{\overline{\zeta_1}\,\overline{\zeta_1}\,\overline{\zeta_2}\,\overline{\zeta_2}}(\zeta_1, \zeta_2) \frac{\overline{\zeta_1 - z_1}}{\zeta_1 - z_1} \frac{\overline{\zeta_2 - z_2}}{\zeta_2 - z_2} \, d\xi_1 d\eta_1 d\xi_2 d\eta_2
$$

as via

$$
w(z_1, z_2) = \frac{1}{(2\pi i)^2} \int_{\partial D_1} \int_{\partial D_2} \left\{ \frac{w(\zeta_1, \zeta_2)}{(\zeta_1 - z_1)(\zeta_2 - z_2)} - \frac{w_{\overline{\zeta_1}}(\zeta_1, \zeta_2)}{\zeta_2 - z_2} \frac{\overline{\zeta_1 - z_1}}{\zeta_1 - z_1} \right.
$$

$$
\left. - \frac{w_{\overline{\zeta_2}}(\zeta_1, \zeta_2)}{\zeta_1 - z_1} \frac{\overline{\zeta_2 - z_2}}{\zeta_2 - z_2} \right\} d\zeta_1 d\zeta_2 \tag{1.10}
$$

$$
+ \frac{1}{\pi} \int_{D_1} w_{\overline{\zeta_1}\,\overline{\zeta_1}}(\zeta_1, z_2) \frac{\overline{\zeta_1 - z_1}}{\zeta_1 - z_1} \, d\xi_1 d\eta_1
$$

$$
+ \frac{1}{\pi} \int_{D_2} w_{\overline{\zeta_2}\,\overline{\zeta_2}}(z_1, \zeta_2) \frac{\overline{\zeta_2 - z_2}}{\zeta_2 - z_2} \, d\xi_2 d\eta_2
$$

$$
+ \frac{1}{\pi^2} \int_{D_1} \int_{D_2} \left\{ \frac{w_{\overline{\zeta_1}\,\overline{\zeta_2}}(\zeta_1, \zeta_2)}{(\zeta_1 - z_1)(\zeta_2 - z_2)} + \frac{w_{\overline{\zeta_1}\,\overline{\zeta_1}\,\overline{\zeta_2}}(\zeta_1, \zeta_2)}{\zeta_2 - z_2} \frac{\overline{\zeta_1 - z_1}}{\zeta_1 - z_1} \right.
$$

$$
\left. + \frac{w_{\overline{\zeta_1}\,\overline{\zeta_2}\,\overline{\zeta_2}}(\zeta_1, \zeta_2)}{\zeta_1 - z_1} \frac{\overline{\zeta_1 - z_2}}{\zeta_2 - z_2} \right\} d\xi_1 d\eta_1 d\xi_2 d\eta_2 \, .
$$

Proof. Iterating instead of (1.8) the respective second order representations for one complex variable applied to z_1 and z_2 leads to (1.10). The equivalent formula (1.10) then follows as by the Gauss theorem

$$\frac{1}{(2\pi i)^2} \int_{\partial D_1} \int_{\partial D_2} w_{\overline{\zeta_1}\,\overline{\zeta_2}}(\zeta_1, \zeta_2) \, \frac{\overline{\zeta_1 - z_1}}{\zeta_1 - z_1} \frac{\overline{\zeta_2 - z_2}}{\zeta_2 - z_2} \, d\zeta_1 d\zeta_2$$

$$= \frac{1}{\pi^2} \int_{D_1} \int_{D_2} \partial_{\overline{\zeta_1}} \partial_{\overline{\zeta_2}} \left[w_{\overline{\zeta_1}\,\overline{\zeta_2}}(\zeta_1, \zeta_2) \, \frac{\overline{\zeta_1 - z_1}}{\zeta_1 - z_1} \frac{\overline{\zeta_2 - z_2}}{\zeta_2 - z_2} \right] d\xi_1 d\eta_1 d\xi_2 d\eta_2 \, .$$

$$\square$$

Remarks There are other second order representations by interchanging $\overline{\zeta_k}$–derivatives with ζ_k–derivatives for $k = 1$ and/or $k = 2$ and at the same time the $(\overline{\zeta_k - z_k})/(\zeta_k - z_k)$–kernels with the $\log|\zeta_k - z_k|^2$–kernels.

Although these representation formulas express the function through boundary values of proper lower–order derivatives this does not imply that the related boundary value problem is solvable and the solution is given by this formula. Only if the solvability is guaranteed then this representation formula may be used for representing the solution.

Of course, there are also mixed order representations. Again only one example is formulated.

Theorem 1.6. *Let w be defined and complex–valued in the regular bidomain D such that $w_{\overline{z_1}}$ and $w_{\overline{z_1}\,\overline{z_2}}$ are continuous and $w_{\overline{z_1}\,\overline{z_2}\,z_2} = w_{\overline{z_2}\,\overline{z_2}\,\overline{z_1}}$. Then*

$$w(z_1, z_2)$$

$$= \frac{1}{(2\pi i)^2} \int_{\partial D_1} \int_{\partial D_2} \left\{ \frac{w(\zeta_1, \zeta_2)}{(\zeta_1 - z_1)(\zeta_2 - z_2)} - \frac{w_{\overline{\zeta_2}}(\zeta_1, \zeta_2)}{\zeta_1 - z_1} \frac{\overline{\zeta_2 - z_2}}{\zeta_2 - z_2} \right\} d\zeta_1 d\zeta_2$$

$$- \frac{1}{\pi} \int_{D_1} w_{\overline{\zeta_1}}(\zeta_1, z_2) \, \frac{d\xi_1 d\eta_1}{\zeta_1 - z_1} + \frac{1}{\pi} \int_{D_2} w_{\overline{\zeta_2}\,\overline{\zeta_2}}(z_1, \zeta_2) \, \frac{\overline{\zeta_2 - z_2}}{\zeta_2 - z_2} \, d\xi_2 d\eta_2$$

$$+ \frac{1}{\pi^2} \int_{D_1} \int_{D_2} \frac{w_{\overline{\zeta_1}\,\overline{\zeta_2}\,\overline{\zeta_2}}(\zeta_1, \zeta_2)}{\zeta_1 - z_1} \frac{\overline{\zeta_2 - z_2}}{\zeta_2 - z_2} \, d\xi_1 d\eta_1 d\xi_2 d\eta_2 \, . \tag{1.11}$$

The proof again follows by proper iteration of the Cauchy–Pompeiu formulas. Other formulas of this kind with ∂_ζ–operators instead of $\partial_{\overline{\zeta}}$–operators are available. Generalization to more than two variables are obvious, see [4, 6].

1.4. Clifford analysis

Let $\{e_k : 1 \le k \le m\}$ be an orthonormal basis of $\mathbb{R}^m$ with $2 \le m$ such that $x \in \mathbb{R}^m$ is represented as $x = \sum_{\mu=1}^{m} x_k e_k$. Introducing a multiplication via

$$e_1 = 1, \quad e_j e_k + e_k e_j = -2\delta_{jk}, \quad 2 \le j, k \le m,$$

a Clifford algebra $\mathbb{C}_m$ is introduced as the set of elements $a = \sum_A a_A e_A$ where $a_A \in \mathbb{C}$

$$e_A = 1 \ \text{ if } \ A = \emptyset, \quad e_A = e_{\alpha_1} e_{\alpha_2} \cdots e_{\alpha_k} \ \text{ if } \ A = \{\alpha_1, \alpha_2, \cdots, \alpha_k\}$$

with $2 \le \alpha_1 < \alpha_2 < \cdots < \alpha_k \le m$ and the sum is taken over all subsets A of $\{2, 3, \ldots, m\}$.

If $m = 2$ the multiplication is commutative and $\mathbb{C}_2$ with $e_2 = i$ is just the field of complex numbers $\mathbb{C}$. Otherwise $\mathbb{C}_m$ is a noncommutative algebra over $\mathbb{C}$ of dimension 2^{m-1}. By

$$\overline{a} = \sum_A \overline{a_A}\, \overline{e_A} \ \text{ for } \ a = \sum_A a_A e_A$$

with

$$\overline{e_1} = e_1, \quad \overline{e_k} = -e_k, \quad 2 \le k \le m, \quad \overline{e_A e_B} = \overline{e_B}\, \overline{e_A}$$

a complex conjugation is introduced. Denoting

$$|a| = \left(\sum_A |a_A|^2\right)^{1/2} \ \text{ for } \ a = \sum_A a_A e_A$$

via $|a|_0 = 2^{m/2}|a|$ an algebra norm is defined. Identifying $x = \sum_{\mu=1}^{m} x_k e_k \in \mathbb{R}^m$ with $z = \sum_{\mu=1}^{m} x_k e_k \in \mathbb{C}_m$ the space $\mathbb{R}^m$ is embedded into $\mathbb{C}_m$. These elements satisfy $z\overline{z} = \overline{z}z = |z|^2$.

A natural basic first order differential operator for functions defined on subsets of $\mathbb{R}^m$ with values in $\mathbb{C}_m$ is the so–called Dirac operator $\partial = \sum_{\mu=1}^{m} e_k \partial_{x_k}$ and its complex conjugate $\overline{\partial} = \partial_{x_1} - \sum_{\mu=2}^{m} e_k \partial_{x_k}$. For $m = 2$ they essentially coincide with the Cauchy–Riemann operator and its complex conjugate $2\partial_{\overline{z}} = \partial_x + i\partial_y$, $2\partial_z = \partial_x - i\partial_y$. The importance of these operators is the connection with the Laplace operator $\partial\overline{\partial} = \overline{\partial}\partial = \Delta$. This factorization makes Clifford analysis important for mathematical physics.

Some basic differentiation rules are

$$\partial z = z\partial = 2 - m, \quad \partial \overline{z} = \overline{z}\partial = m, \quad \partial |z|^2 = |z|^2 \partial = 2z,$$

$$\partial |z|^\alpha = |z|^\alpha \partial = \alpha\, |z|^{\alpha-2} z, \quad \partial(\overline{z}/|z|^m) = (\overline{z}/|z|^m)\,\partial = 0.$$

The last rule identifies $\bar{z}/|z|^m$ as the fundamental solution to the Dirac equation. For more detailed information see [10, 11].

Basic for representation formulas for functions in Clifford algebra is a version of the Gauss theorem. Because of the anticommutativity of multiplication two functions are involved here.

Gauss Theorem. *Let* $D \subset \mathbb{R}^m$ *be a regular domain and* $f, g \in C^1(D; \mathbb{C}_m) \cap C(\overline{D}; \mathbb{C}_m)$. *Then*

$$\int_D [(f\partial)\,g + f(\partial g)]\,dv = \int_{\partial G} f\,d\vec{\sigma}\,g\,,$$

$$\int_D [(f\overline{\partial})\,g + f(\overline{\partial} g)]\,dv = \int_{\partial G} f\,d\overline{\vec{\sigma}}\,g\,.$$

Here dv *denotes the volume element of* D, $d\sigma$ *the area element of* ∂D, $n = (n_1, \cdots, n_m)$ *the outward normal vector on* ∂D, $\vec{n} = \sum_{\mu=1}^{m} n_\mu e_\mu$ *the corresponding element in* $\mathbb{C}_m$, *and* $d\vec{\sigma} = d\sigma\,\vec{n}$ *the directed area element on* ∂D, $d\overline{\vec{\sigma}} = d\sigma\,\overline{\vec{n}}$ *its complex conjugate.*

Proof. (1) Let $f, g \in C^1(D; \mathbb{C}) \cap C(\overline{D}; \mathbb{C})$. Then from the classical Gauss theorem

$$\int_D [f_{x_\mu} g + f g_{x_\mu}]\,dv = \int_D \partial_{x_\mu}(fg)\,dv = \int_{\partial D} fg\,n_\mu d\sigma$$

multiplication by e_μ and adding up gives

$$\int_D [(\partial f)\,g + f\,(\partial g)]\,dv = \int_{\partial D} fg\,d\vec{\sigma}\,.$$

Rearranging this formula and replacing f by f_A and g by g_B, $A, B \subset \{2, \ldots, m\}$ gives

$$\int_D [(f_A\partial)\,g_B + f_A(\partial g_B)]\,dv = \int_{\partial D} f_A d\vec{\sigma}\,g_B\,.$$

Multiplying with e_A from the left and e_B on the right and taking sums give

$$\int_D [(f\partial)\,g + f\,(\partial g)]\,dv = \int_{\partial D} f\,d\vec{\sigma}\,g\,.$$

The second formula follows analogously or by complex conjugation of the first one. $\square$

From this Gauss theorem Cauchy–Pompeiu representation formulas follow in the same way as in the complex case.

Cauchy–Pompeiu representation *Any $w \in C^1(D; \mathbb{C}_m) \cap C(\overline{D}; \mathbb{C}_m)$ can be represented as*

$$w(z) = \frac{1}{\omega_m} \int_{\partial D} \frac{\overline{\zeta - z}}{|\zeta - z|^m} \, d\vec{\sigma}(\zeta) \, w(\zeta) - \frac{1}{\omega_m} \int_D \frac{\overline{\zeta - z}}{|\zeta - z|^m} \, \partial w(\zeta) \, dv(\zeta),$$

$$w(z) = \frac{1}{\omega_m} \int_{\partial D} \frac{\zeta - z}{|\zeta - z|^m} \, d\overline{\vec{\sigma}(\zeta)} \, w(\zeta) - \frac{1}{\omega_m} \int_D \frac{\zeta - z}{|\zeta - z|^m} \, \overline{\partial} w(\zeta) \, dv(\zeta).$$

Here ω_m denotes the area of the unit sphere in $\mathbb{R}^m$. There are dual formulas where the function and its derivative, respectively and the kernel function are changing their positions with one another.

Proof.   Let $0 < \varepsilon$ be so small that $D_\varepsilon = \{\zeta : \zeta \in D, \varepsilon < |\zeta - z|\}$ is a regular domain. Then from the Gauss theorem

$$\int_{\partial D_\varepsilon} \frac{\overline{\zeta - z}}{|\zeta - z|^m} \, d\vec{\sigma}(\zeta) \, w(\zeta) = \int_{D_\varepsilon} \frac{\overline{\zeta - z}}{|\zeta - z|^m} \, \partial w(\zeta) \, dv(\zeta).$$

As

$$\frac{1}{\omega_m} \int_{|\zeta - z| = \varepsilon} \frac{\overline{\zeta - z}}{|\zeta - z|^m} \, d\vec{\sigma}(\zeta) \, w(\zeta)$$

$$= \frac{1}{\omega_m} \int_{|\omega| = 1} \varepsilon^{1-m} \overline{\omega} \, \varepsilon^{m-1} \omega \, d\sigma(\omega) \, w(z + \varepsilon \omega)$$

$$= \frac{1}{\omega_m} \int_{|\omega| = 1} w(z + \varepsilon \omega) \, d\sigma(\omega)$$

tends to $w(z)$ with ε tending to 0 and

$$\left| \int_{|\zeta - z| < \varepsilon} \frac{\overline{\zeta - z}}{|\zeta - z|^m} \, \partial w(\zeta) \, dv(\zeta) \right|_0$$

$$\leq 2^m \int_0^\varepsilon \int_{|\omega| = 1} t^{1-m} |\partial w(z + t\omega)| \, t^{m-1} dt \, d\sigma(\omega)$$

tends to 0 with ε, the first representation formula follows. The second can be deduced similarly. It also could be attained from the dual of the first formula where the places of the kernel and the function w and $w\partial$, respectively are interchanged with one another, by complex conjugation.

Iterating the first formula leads to higher order Cauchy–Pompeiu formulas as before. $\qquad\qquad\square$

Theorem 1.7. *Let* $w \in C^k(D; \mathbb{C}_m) \cap C^{k-1}(\overline{D}, \mathbb{C}_m)$ *for* $1 \leq k$. *Then*

$$w(z) = \sum_{\mu=0}^{k-1} \frac{1}{\omega_m} \int_{\partial D} \frac{(\overline{\zeta - z})(\overline{z - \zeta} + z - \zeta)^\mu}{2^\mu \mu!} \frac{1}{|\zeta - z|^m} \, d\vec{\sigma}(\zeta) \, \partial^\mu w(\zeta)$$

$$- \frac{1}{\omega_m} \int_D \frac{(\overline{\zeta - z})(\overline{z - \zeta} + z - \zeta)^{k-1}}{2^{k-1}(k-1)!} \frac{1}{|\zeta - z|^m} \, \partial^k w(\zeta) \, dv(\zeta) \, .$$

Also the two Cauchy–Pompeiu formulas can be iterated with one another leading to a formula related to the Laplacian $\Delta = \partial \overline{\partial}$.

Theorem 1.8. *Let* $w \in C^2(D; \mathbb{C}_m) \cap C^1(\overline{D}; \mathbb{C}_m)$. *Then*

$$w(z) = \frac{1}{\omega_m} \int_{\partial D} \frac{\overline{\zeta - z}}{|\zeta - z|^m} \, d\vec{\sigma}(\zeta) \, w(\zeta) - \frac{1}{\omega_m} \int_{\partial D} \frac{|\zeta - z|^{2-m}}{2-m} \, \overline{d\vec{\sigma}(\zeta)} \, \partial w(\zeta)$$

$$+ \frac{1}{\omega_m} \int_D \frac{|\zeta - z|^{2-m}}{2-m} \, \Delta w(\zeta) \, dv(\zeta) \, .$$

Proof. From the Cauchy–Pompeiu formulas

$$w(z) = \frac{1}{\omega_m} \int_{\partial D} \frac{\zeta - z}{|\zeta - z|^m} \, d\vec{\sigma}(\zeta) \, w(\zeta) - \frac{1}{\omega_m} \int_D \frac{\overline{\widetilde{\zeta} - z}}{|\widetilde{\zeta} - z|^m} \, \partial w(\widetilde{\zeta}) \, dv(\widetilde{\zeta}) \, ,$$

$$\partial w(\widetilde{\zeta}) = \frac{1}{\omega_m} \int_{\partial D} \frac{\zeta - \widetilde{\zeta}}{|\zeta - \widetilde{\zeta}|^m} \, d\vec{\sigma}(\zeta) \, \partial w(\zeta) - \frac{1}{\omega_m} \int_D \frac{\zeta - \widetilde{\zeta}}{|\zeta - \widetilde{\zeta}|^m} \, \Delta w(\zeta) \, dv(\zeta)$$

it follows

$$w(z) = \frac{1}{\omega_m} \int_{\partial D} \frac{\overline{\zeta - z}}{|\zeta - z|^m} \, d\vec{\sigma}(\zeta) \, w(\zeta) + \frac{1}{\omega_m} \int_{\partial D} \psi(\zeta, z) \, \overline{d\vec{\sigma}(\zeta)} \, \partial w(\zeta)$$

$$- \frac{1}{\omega_m} \int_D \psi(\zeta, z) \, \Delta w(\zeta) \, dv(\zeta)$$

where

$$\psi(\zeta, z) = \frac{1}{\omega_m} \int_D \frac{\overline{\widetilde{\zeta} - \zeta}}{|\widetilde{\zeta} - z|^m} \, \frac{\widetilde{\zeta} - \zeta}{|\widetilde{\zeta} - \zeta|^m} \, dv(\zeta) \, .$$

By an analogue argumentation as for (2)

$$\frac{|\zeta - z|^{2-m}}{2-m} = \frac{1}{\omega_m} \int_{\partial D} \frac{\overline{\widetilde{\zeta} - z}}{|\widetilde{\zeta} - z|^m} \, d\vec{\sigma}(\widetilde{\zeta}) \, \frac{|\widetilde{\zeta} - \zeta|^{2-m}}{2-m} \tag{1.12}$$

$$- \frac{1}{\omega_m} \int_D \frac{\overline{\widetilde{\zeta} - z}}{|\widetilde{\zeta} - z|^m} \, \frac{\widetilde{\zeta} - \zeta}{|\widetilde{\zeta} - \zeta|^m} \, dv(\zeta) = \widetilde{\psi}(\zeta, z) - \psi(\zeta, z) \, .$$

Applying the Gauss theorem for $\{\widetilde{\zeta} : \widetilde{\zeta} \in D, \, \varepsilon < |\widetilde{\zeta} - z|, \, \varepsilon < |\widetilde{\zeta} - \zeta|\}$ with proper positive ε it follows

$$\widetilde{\psi}(\zeta, z) \, \overline{\partial} = \frac{1}{\omega_m} \int_{\partial D} \frac{\overline{\widetilde{\zeta} - z}}{|\widetilde{\zeta} - z|^m} \, d\vec{\sigma}(\zeta) \, \frac{\zeta - \widetilde{\zeta}}{|\zeta - \widetilde{\zeta}|^m}$$

$$= \frac{\overline{\zeta - z}}{|\zeta - z|^m} - \frac{\overline{\zeta - z}}{|\zeta - z|^m}$$

$$+ \frac{1}{\omega_m} \int_D \left\{ \left(\frac{\overline{\widetilde{\zeta} - z}}{|\widetilde{\zeta} - z|^m} \, \partial_{\widetilde{\zeta}} \right) \frac{\overline{\zeta - \widetilde{\zeta}}}{|\zeta - \widetilde{\zeta}|^m} \right.$$

$$\left. - \frac{\overline{\widetilde{\zeta} - z}}{|\widetilde{\zeta} - z|^m} \left(\partial_\zeta \frac{\overline{\widetilde{\zeta} - \zeta}}{|\widetilde{\zeta} - \zeta|^m} \right) \right\} dv(\zeta) = 0$$

when ε tends to zero. Hence, again applying the Gauss formula

$$- \frac{1}{\omega_m} \int_{\partial D} \widetilde{\psi}(\zeta, z) \, d\overline{\vec{\sigma}(\zeta)} \, \partial w(\zeta) + \frac{1}{\omega_m} \int_D \widetilde{\psi}(\zeta, z) \, \Delta w(\zeta) \, dv(\zeta) = 0 \, .$$

Adding this to the representation of w and observing (1.12) proves the representation claimed. Iterating the representation in Theorem 1.8 with itself gives the next formula. $\qquad \square$

Theorem 1.9. *Let* $w \in C^{2k}(D; \mathbb{C}_m) \cap C^{2k-1}(\overline{D}; \mathbb{C}_m)$ *for* $1 \le k$ *if* m *is odd* *and for $1 \leq 2k < m$ if m is even. Then*

$$w(z) = \sum_{\mu=1}^{k} \left\{ \frac{1}{\omega_m} \int_{\partial D} \frac{(\overline{\zeta - z})\,|\zeta - z|^{2(\mu-1)-m}}{2^{\mu-1}(\mu - 1)!\,\prod_{\nu=1}^{\mu-1}(2\nu - m)}\, d\vec{\sigma}(\zeta)\,\Delta^{\mu-1}w(\zeta) \right.$$

$$\left. - \frac{1}{\omega_m} \int_{\partial D} \frac{|\zeta - z|^{2\mu-m}}{2^{\mu-1}(\mu - 1)!\,\prod_{\nu=1}^{\mu}(2\nu - m)}\, d\overline{\vec{\sigma}}(\zeta)\,\partial\Delta^{\mu-1}w(\zeta) \right\}$$

$$+ \frac{1}{\omega_m} \int_D \frac{|\zeta - z|^{2k-m}}{2^{k-1}(k - 1)!\,\prod_{\nu=1}^{k}(2\nu - m)}\, \Delta^k w(\zeta)\,dv(\zeta).$$

For the proofs of these results see [3]. Further representation formulas for $m \leq 2k$ if m is even are given in [19]. Those related to operators of the form $\partial^\ell \overline{\partial}^k$ are the subject of a forthcoming thesis of Heinz Otto.

References

[1] H. Begehr, *Complex analytic methods for partial differential equations.* An introductory text. World Scientific, Singapore, 1994.

[2] H. Begehr, *Second order Cauchy–Pompeiu representations.* Complex methods for partial differential equations. Eds. H. Begehr, A. Celebi, W. Tutschke. Kluwer, Dordrecht, 1999, 165–188.

[3] H. Begehr, *Iterated integral operators in Clifford analysis.* ZAA **18** (1999), 361–377.

[4] H. Begehr, *Representations in polydomains.* Acta Math. Vietnamica 27 (2002), 271–282.

[5] H. Begehr, *Integral representations for differentiable functions.* Problemi attuali dell'analisi e della fisica matematica dedicato alla memoria di Gaetano Fichera, Taormina, 1998. Ed. P.E. Ricci. Aracne, Roma, 2000, 111–130.

[6] H. Begehr, D.Q. Dai and X. Li, *Integral representation formulas in polydomains.* Complex Var. Theory Appl. 47 (2002), 463–484.

[7] H. Begehr and Ju. Dubinskii, *Orthogonal decompositions of Sobolev spaces in Clifford analysis.* Ann. Mat. Pura Appl. 181 (2002), 55–71.

[8] H. Begehr and G.N. Hile, *A hierarchy of integral operators.* Rocky Mountain J. Math. **27** (1997), 669–706.

[9] H. Begehr and G.C. Wen, *Some second order systems in the complex plane.* Revue Roumaine Math. Pure Appl. **44** (1999), 521–554.

[10] F. Bracks, R. Delanghe and F. Sommen, *Clifford analysis.* Pitman, London, 1982.

[11] E. Obolashvili, *Partial differential equations in Clifford analysis.* Addison Wesley Longman, Harlow, 1998.

[12] W. Rudin, *Function theory in the unit ball of* $\mathbb{C}^n$. Springer–Verlag, New York etc., 1980.

[13] I.N. Vekua, *Generalized analytic functions.* Pergamon Press, Oxford, 1962.

Further Bibliography

[14] H. Begehr, *Cauchy–Pompeiu representations.* Problems in Differential Equations, Analysis and Algebra, Actobe, 15.–19.9.1999. Eds. K.K. Kenchebaev et al. Actobe Univ., Actobe, 2000, 218–221

[15] H. Begehr, *Hypercomplex Bitsadze systems.* Boundary value problems, integral equations and related problems. J.K. Lu, G.C. Wen (eds.), World Scientific, Singapore, 2000, 33–40.

[16] H. Begehr, *Pompeiu operators in complex, hypercomplex and Clifford analysis.* Revue Roumaine Math. Pure Appl. 46 (2001), 1–11.

[17] H. Begehr, *Integral decomposition of differentiable functions.* Proc. 2nd ISAAC Congress Fukuoka 1999. Vol 2. Eds. H. Begehr, R.P. Gilbert, J. Kajiwara. Kluwer, Dordrecht, 2000, 1301–1312.

[18] H. Begehr, *Orthogonal decompositions of the function space $L_2(\overline{D}; \mathbb{C})$.* J. Reine Angew. Math. 549 (2002), 191–219.

[19] H. Begehr, *Representation formulas in Clifford Analysis.* Acoustics, Mechanics and the Related Topics of Mathematical Analysis. Ed. A. Wirgin. World Sci., New Jersey, 2002, 8-13.

[20] H. Begehr, D.Q. Dai, A. Dzhuraev, *Integral representations for inhomogeneous overdetermined second order systems of several complex equations.* Functional analytic and complex methods, their interactions, and applications to partial differential equations. Eds. H. Florian et al. World Sci. Singapore, 2001, 378–393.

[21] H. Begehr, J.-Y. Du, Z.-X. Zhang, *On Cauchy-Pompeiu formula for functions with values in a universal Clifford algebra.* Acta Math. Scientia (China) 23B (2003), 95-103

[22] H. Begehr, F. Gackstatter, A. Krausz, *Integral representations in octonionic analysis.* Proc. 10th Intern. Conf. Complex Analysis. Eds. J. Kajiwara, K.W. Kim, K.H. Shon. Busan, Korea, 2002, 1-7.

Chapter 2

Nonlinear potential theory in metric spaces

Olli Martio

Department of Mathematics and Statistics
University of Helsinki, Finland
olli.martio@helsinki.fi

Contents

2.1. Introduction

The classical Potential Theory is the theory of harmonic functions. For
modern abstract linear potential theories, developed by M. Brelot, H. Bauer
etc, no differential equations are needed. There is a corresponding nonlinear
abstract potential theory, see [HKM], but to model interesting new nonlinear theories the derivative of a function and quasilinear elliptic partial dif-

29

ferential equations or variational integrals are employed. In a metric space (X, d) with a measure μ we introduce two counterparts for the derivative of a function and then investigate a potential theory based on minimizers or quasiminimizers of variational integrals.

The main emphasize in these notes is on the definitions of a Sobolev space on X and on the concepts which can be used to build a Potential Theory on X.

2.2. Hajłasz space $M^{1,p}(X)$

2.2.1. *Definition*

Let (X, d) be a metric space with a measure μ (no extra assumptions). Let $u : X \to [-\infty, \infty]$. We let $D(u)$ denote the set of all μ–measurable functions $g \geq 0$ such that

$$|u(x) - u(y)| \leq d(x, y)(g(x) + g(y)) \tag{2.1}$$

holds $\mu-$ a.e. $x, y \in X$. This means the following: There is a set $C \subset X$ such that $\mu(C) = 0$ and (2.1) holds for every $x, y \in X \setminus C$ (observe that $\infty - \infty$ and 0∞ are not defined, however for $x = y$ we always interpret (2.1) as a triviality).

For $1 \leq p < \infty$ the Hajłasz–space $M^{1,p}(X)$ is defined as

$$M^{1,p}(X) = \{u \in L^p(X) : \exists g \in D(u) \cap L^p(X)\}.$$

If $u \in M^{1,p}(X)$, then $|u| < \infty$ $\mu-$ a.e. and hence we can redefine u and $g \in D(u) \cap L^p(X)$ so that (2.1) holds for every $x, y \in X$.

The space $M^{1,p}(X)$ was introduced by P. Hajłaz [H].

For $u \in M^{1,p}(X)$ we set

$$\|u\|_{M^{1,p}(X)} = \|u\|_{1,p} = \|u\|_{L^p(X)} + \inf_{g \in D(u)} \|g\|_{L^p(X)}.$$

Theorem 2.1. *The space $M^{1,p}(X)$ is a Banach space with the norm $\| \ \|_{1,p}$, $1 \leq p < \infty$.*

Proof. Clearly $M^{1,p}(X)$ is a linear space and $\|u\|_{1,p}$ is a norm in $M^{1,p}(X)$.

To show that $M^{1,p}(X)$ is a Banach space, let (u_i) be a Cauchy sequence in $M^{1,p}(X)$. Then u_i is a Cauchy sequence in $L^p(X)$ and hence

$$u_i \to u \in L^p(X) \text{ in } L^p(X)$$

because $L^p(X)$ is a Banach space.

It remains to show that $u \in M^{1,p}(X)$ and that

$$u_i \to u \quad \text{in} \quad M^{1,p}(X).$$

Passing to a subsequence we may assume

$$\|u_{i+1} - u_i\|_{1,p} < 2^{-i}$$

and that $u_i \to u$ $\mu-$ a.e. Thus there is $g_i \in L^p(X) \cap D(u_{i+1} - u_i)$ with

$$|(u_{i+1} - u_i)(x) - (u_{i+1} - u_i)(y)| \leq d(x,y)(g_i(x) + g_i(y))$$

$\mu-$ a.e. and

$$\|g_i\|_{L^p} < 2^{-i}.$$

For each $k \geq 0$

$$|(u_{i+k} - u_i)(x) - (u_{i+k} - u_i)(y)|$$
$$\leq d(x,y)(\sum_{j=i}^{\infty}(g_j(x) + g_j(y))), \tag{2.2}$$

and letting $k \to \infty$ we obtain

$$|(u - u_i)(x) - (u - u_i)(y)| \leq d(x,y)(g(x) + g(y))$$

where

$$g = \sum_{j=i}^{\infty} g_j \; ;$$

note that

$$\|g\|_{L^p(X)} < 2^{-i+1}$$

and hence $g \in L^p(X)$. From this it follows that

$$|u(x) - u(y)| \leq |u_i(x) - u_i(y)| + d(x,y)(g(x) + g(y))$$

which clearly implies that $u \in M^{1,p}(X)$. Moreover, the above inequalities also yield that $u_i \to u$ in $M^{1,p}(X)$.

Since this reasoning can be applied to any subsequence of (u_i), it follows that $u_i \to u$ in $M^{1,p}(X)$; note that the limit function u is independent of the subsequence. $\qquad\square$

Example. Let

$$X = [-1, 0] \cup \bigcup_{i=1}^{\infty} [a_i, b_i] \subset \mathbb{R}$$

where $b_i = 1/2^i$, $i = 1, 2, \ldots$, and the intervals $[a_i, b_i]$ are disjoint and so short that for a fixed p, $1 \leq p < \infty$,

$$\int_{\cup [a_i, b_i]} \frac{1}{x^p} dx < \infty.$$

For example

$$\sum_{i=1}^{\infty} \frac{b_i - a_i}{a^{ip}} < \infty$$

suffices.

Next let

$$u(x) = \begin{cases} 0, & x \in [-1, 0] \\ 1, & x \in \cup [a_i, b_i] \end{cases}$$

and let X be equipped with the Lebesgue measure and $d(x, y) = |x - y|$. Now $u \in M^{1,p}(X)$. Indeed, let

$$g(x) = \begin{cases} 0, & x \in [-1, 0] \\ \frac{1}{x}, & x \in \cup [a_i, b_i]. \end{cases}$$

Then $g \in L^p(X)$ and since $u \in L^p(X)$, it suffices to show that $g \in D(u)$.

The inequality

$$|u(x) - u(y)| \leq |x - y|(g(x) + g(y)) \tag{2.3}$$

is trivial for $x, y \in [-1, 0]$ and $x, y \in \cup [a_i, b_i]$. Let $y \in [-1, 0]$ and $x \in \cup [a_i, b_i]$. Then

$$|u(x) - u(y)| = 1 \leq |x - y|/|x| = |x - y|g(x)$$
$$\leq |x - y|(g(x) + g(y))$$

as required and hence (2.3) holds, by symmetry, for all $x, y \in X$.

Observe that X is compact and u is not continuous at 0. Similar examples can be constructed on $[-1, 1]$ if the Lebesgue measure is modified and then restricted to X.

2.2.2. *Poincaré inequality*

Theorem 2.2. *Let (X, d) be a metric space with $0 < \mu(X) < \infty$ and $u \in M^{1,p}(X)$, $1 \le p < \infty$. Then*

$$\int_X |u - u_X|^p d\mu \le 2^p (\operatorname{diam} X)^p \int_X g^p d\mu \qquad (2.4)$$

where $g \in D(u)$ and

$$u_X = \fint_X u \, d\mu = \frac{1}{\mu(X)} \int_X u \, d\mu.$$

Remarks.

(a) Note that

$$\int_X |u| d\mu < \infty$$

by the Hölder inequality since $\mu(X) < \infty$.

(b) The inequality (2.4) is called a Poincaré inequality, sometimes a (p, p)–Poincaré inequality, see subsection 3.5.

Proof of Theorem 2.2. Since

$$|u(x) - u_X| = |\fint_X (u(x) - u(y)) d\mu(y)|$$

$$\le \operatorname{diam}(X) \fint_X (g(x) + g(y)) d\mu(y) = \operatorname{diam}(X)(g(x) + g_X),$$

we obtain by integrating over x

$$\int_X |u(x) - u_X|^p d\mu \le \operatorname{diam}(X)^p \int_X (g(x) + g_Y)^p d\mu$$

$$= \operatorname{diam}(X)^p 2^{p-1} \left[\int_X g^p d\mu + \int_X g_X^p d\mu \right]$$

where we have used the inequality $(a+b)^p \le 2^{p-1}(a^p + b^p)$ valid for $a, b \ge 0$ and $p \ge 1$. Now

$$\int_X g_X^p d\mu = \mu(X)(\fint_X g \, d\mu)^p \le \mu(X) \fint_X g^p d\mu = \int_X g^p d\mu$$

          O. Martio

by the Hölder inequality and we obtain

$$\int_X |u(x) - u_X|^p d\mu \leq \text{diam } (X)^p 2^{p-1} \cdot 2 \int_X g^p d\mu$$

as required.

2.2.3. *Approximation by lipschitz functions*

A function $u : X \to \mathbb{R}$ is L–lipschitz, $0 \leq L < \infty$, in a metric space (X, d) if

$$|u(x) - u(y)| \leq Ld(x, y)$$

for all $x, y \in X$.

We study first an extension problem for L–lipschitz maps.

Lemma 2.1. *Suppose that $\mathcal{M}$ is a family of L–lipschitz functions $f : X \to \mathbb{R}$. Then*

$$g(x) = \inf\{f(x) : f \in \mathcal{M}\}$$

is L–lipschitz in X provided that $g(x) \in \mathbb{R}$ for some $x \in X$.

Proof. If $g(x) \in \mathbb{R}$, then $\mathcal{M} \neq \emptyset$ and $g(y) \in \mathbb{R}$ for all $y \in X$. Indeed, let $y \in X$ and suppose that $g(y) = -\infty$. Then for all $M > 0$ there is $f \in \mathcal{M}$ s.t. $f(y) < -M$. Now

$$f(x) - f(y) \leq |f(x) - f(y)| \leq Ld(x, y)$$

and hence

$$-M > f(y) \geq f(x) - Ld(x, y) \geq g(x) - Ld(x, y) > -\infty.$$

Letting $M \to \infty$ we obtain a contradiction. Thus $g(y) \in \mathbb{R}$ for all $y \in X$.

Hence $g : X \to \mathbb{R}$ and it remains to show that g is L–lipschitz. Fix $x, y \in X$ and let $\varepsilon > 0$. We may suppose that $g(x) \leq g(y)$. Next choose $f \in \mathcal{M}$ such that $g(x) > f(x) - \varepsilon$. Now

$$|g(x) - g(y)| = g(y) - g(x) \leq f(y) - f(x) + \varepsilon \leq |f(y) - f(x)| + \varepsilon$$
$$\leq Ld(y, x) + \varepsilon.$$

Letting $\varepsilon \to 0$ we obtain $|g(x) - g(y)| \leq Ld(y, x)$ as required. $\square$

Theorem 2.3. *(McShane) Let $A \subset X$ and $u : A \to \mathbb{R}$ L–lipschitz. Then there is an L–lipschitz function $u^* : X \to \mathbb{R}$ such that $u^*|A = u$.*

Proof. For given $a \in A$ the function

$$f_a(x) = u(a) + Ld(x,a)$$

is an L–lipschitz function on X (triangle inequality). Define

$$u^*(x) = \inf\{f_a(x) : a \in A\}.$$

Then $u^*(x) = u(x)$ for all $x \in A$ and by Lemma 2.1, u^* is L–lipschitz. $\square$

Next we show that a function $u \in M^{1,p}(X)$ can be approximated by lipschitz functions in the $\| \; \|_{1,p}$–norm.

Theorem 2.4. *Let $u \in M^{1,p}(X)$, $1 \le p < \infty$, and $\varepsilon > 0$. Then there is a lipschitz function $\varphi : X \to \mathbb{R}$ such that*
(i) $\mu(\{x \in X : u(x) \ne \varphi(x)\}) < \varepsilon$,
(ii) $\|u - \varphi\|_{1,p} < \varepsilon$.

Proof. Let $\varepsilon > 0$ and choose $g \in D(u)$ s.t.

$$\|u\|_{1,p} \ge \|u\|_p + \|g\|_p - \varepsilon/2.$$

Write $E_\lambda = \{x \in X : |u(x)| \le \lambda, \; g(x) \le \lambda\}$. Then

$$\lambda^p \mu(X \setminus E_\lambda) \to 0 \quad \text{as} \quad \lambda \to \infty \tag{2.5}$$

because

$$\infty > \int_X |u|^p d\mu = \int_{|u(x)| \le \lambda} |u|^p d\mu + \int_{|u(x)| > \lambda} |u|^p d\mu$$

$$\ge \int_{|u(x)| \le \lambda} |u|^p d\mu + \lambda^p \mu(|u(x)| > \lambda)$$

and since

$$\int_{|u(x)| \le \lambda} |u|^p d\mu \to \int_X |u|^p d\mu, \quad \lambda \to \infty,$$

we have

$$\lambda^p \mu(|u(x)| > \lambda) \to 0 \quad \text{as} \quad \lambda \to \infty.$$

Similarly $\lambda^p \mu(g(x) > \lambda) \to 0$ as $\lambda \to 0$ and now

$$\mu(X \setminus E_\lambda) \le \mu(|u(x)| > \lambda) + \mu(g(x) > \lambda)$$

gives (2.5). The function u is 2λ–lipschitz on E_λ since for $x, y \in E_\lambda$

$$|u(x) - u(y)| \le d(x,y)(g(x) + g(y)) \le 2\lambda d(x,y);$$

we have assumed that the inequality

$$|u(x) - u(y)| \leq d(x,y)(g(x) + g(y))$$

holds everywhere in X.

By Theorem 2.3 $u|E_\lambda$ can be extended to a 2λ–bilipschitz function $v_\lambda :$ $X \to \mathbb{R}$. Now v_λ need not belong to $L^p(X)$ and we set

$$u_\lambda = \operatorname{sgn} v_\lambda \ \min \ (|v_\lambda|, \lambda).$$

Then u_λ is again a 2λ–lipschitz function on X and $u_\lambda = u$ on E_λ. Moreover,

$$\|u_\lambda - u\|_p \to 0 \ \text{ as } \ \lambda \to \infty,$$

since

$$\|u_\lambda - u\|_p^p \leq \int_{X \backslash E_\lambda} |u_\lambda - u|^p d\mu \leq 2^{p-1} \Big[\int_{X \backslash E_\lambda} |u_\lambda|^p d\mu$$

$$+ \int_{X \backslash E_\lambda} |u|^p d\mu \Big] \leq 2^{p-1} \Big[\lambda^p \mu(X \backslash E_\lambda) + \int_{X \backslash E_\lambda} |u|^p d\mu \Big] \to 0 \text{ as } \lambda \to \infty.$$

Finally the function

$$g_\lambda(x) = 0, \ \ x \in E_\lambda, \ \ g_\lambda(x) = g(x) + 3\lambda, \ \ x \in X \backslash E_\lambda,$$

belongs to $D(u - u_\lambda)$. To see this note that the inequality

$$|(u - u_\lambda)(x) - (u - u_\lambda)(y)| \leq d(x,y)(g_\lambda(x) + g_\lambda(y))$$

is trivial for $x, y \in E_\lambda$. If $x, y \in X \backslash E_\lambda$, then

$$|(u - u_\lambda)(x) - (u - u_\lambda)(y)|$$
$$\leq |u(x) - u(y)| + |u_\lambda(x) - u_\lambda(y)|$$
$$\leq d(x,y)(g(x)\chi_{X \backslash E_\lambda}(x) + g(y)\chi_{X \backslash E_\lambda}(y) + 2\lambda\chi_{X \backslash E_\lambda}(x))$$
$$\leq d(x,y)(g_\lambda(x) + g_\lambda(y)).$$

If $x \in E_\lambda$ and $y \in X \backslash E_\lambda$, then

$$|(u - u_\lambda)(x) - (u - u_\lambda)(y)| \leq |u(x) - u(y)| + |u_\lambda(x) - u_\lambda(y)|$$
$$\leq d(x,y)(g(x) + g(y)) + d(x,y)2\lambda$$
$$\leq d(x,y)(\lambda + g(y) + 2\lambda) \leq d(x,y)(g_\lambda(x) + g_\lambda(y)),$$

On the other hand,

$$\int_X g_\lambda^p d\mu = \int_{X \setminus E_\lambda} (g(x) + 3\lambda)^p d\mu$$

$$\leq 2^{p-1} \left(\int_{X \setminus E_\lambda} g^p d\mu + \underbrace{3^p \lambda^p \mu(X \setminus E_\lambda)}_{\to 0 \ \text{ as } \ \lambda \to \infty} \right)$$

and the first term goes to 0 as well because $\mu(X \setminus E_\lambda) \to 0$. The proof is complete. $\qquad\square$

2.2.4. Properties of functions u in $M^{1,p}(X)$

Theorem 2.5. *Suppose that $u \in M^{1,p}(X)$, $1 < p < \infty$. Then there is a function $g \in D(u)$ with the least L^p–norm. Moreover, g is unique.*

Proof. Standard (based on the Mazur lemma and strict convexity of the L^p–norm for $p > 1$). $\qquad\square$

There are results, called Sobolev imbedding theorems, which tell that the functions $u \in M^{1,p}(X)$ have additional regularity properties; for the proofs see [H].

The measure μ is said to be *s–regular*, $s > 0$, if there is $c > 0$ such that

$$cr^s < \mu(B(x,r)) < \infty$$

for all $x \in X$ and all $0 < r \leq \operatorname{diam}(X)$.

Let $1 < p < s$. The exponent

$$p^* = \frac{sp}{s - p}$$

is called the *Sobolev conjugate of p*. Note that $p^* \to \infty$ as $p \nearrow s$.

Theorem 2.6. *Suppose that μ is s–regular, $\mu(X) < \infty$. Let $u \in M^{1,p}(X)$, $p > 1$.*

(a) If $1 < p < s$, then $u \in L^{p^*}(X)$ and

$$\|u\|_{L^{p^*}} \leq C[\operatorname{diam}(X)^{-1}\|u\|_{L^p} + \inf_{g \in D(u)} \|g\|_{L^p}],$$

$$\|u - u_X\|_{L^{p^*}} \leq C \inf_{g \in D(u)} \|g\|_{L^p}.$$

(b) If $p > s$, then

$$\|u - u_X\|_{L^\infty} \leq C\mu(X)^{\frac{1}{s}-\frac{1}{p}} \inf_{g\in D(u)} \|g\|_{L^p}$$

$$|u(x) - u(y)| \leq C\mu(X)^{\frac{1}{s}-\frac{1}{p}} \inf_{g\in D(u)} \|g\|_{L^p}.$$

Remarks. (a) There is a version of Theorem 2.6 for $p = s$.

(b) With some extra assumptions on X in the case (b) the function u is Hölder continuous.

2.2.5. *The case $X = \mathbb{R}^n$ and $\mu = m$*

Let Ω be an open set in $\mathbb{R}^n$. There are many different ways to characterize the classical first order Sobolev space $W^{1,p}(\Omega)$, $1 \leq p \leq \infty$.

(A) $u \in W^{1,p}(\Omega)$ iff (definition) $u \in L^p(\Omega)$ (Lebesgue measure) and there are functions $g_1, \ldots, g_n \in L^p(\Omega)$ (the distributional derivatives of u) such that

$$\int_\Omega g_i \varphi \, dm = - \int_\Omega u \partial_i \varphi \, dm$$

for every $\varphi \in C_0^1(\Omega)$ (integration by parts).

(B) $u \in W^{1,p}(\Omega)$ iff (equivalent definition) $u \in L^p(\Omega)$ and there is a sequence of functions $u_j \in C^1(\Omega)$ such that

$$u_j \to u \ \text{ in } \ L^p(\Omega),$$

$(\partial_i u_j)$ is a Cauchy sequence in $L^p(\Omega)$ for each $i = 1, \ldots, n$.

The space $W^{1,p}(\Omega)$ is equipped with the norm

$$\|u\|_{W^{1,p}(\Omega)} = \|u\|_{L^p} + \||\nabla u|\|_{L^p},$$

$\nabla u = (g_1, \ldots, g_n)$.

If $\Omega = \mathbb{R}^n$, then the Hajłasz space $M^{1,p}(\mathbb{R}^n)$, $p > 1$, $\mu = m$, is exactly the space $W^{1,p}(\mathbb{R}^n)$; this is not true for an arbitrary open set $\Omega \subset \mathbb{R}^n$. For the proof of Theorem 2.7 see [H].

Theorem 2.7. *Let $1 < p < \infty$.*

(a) If $u \in W^{1,p}(\mathbb{R}^n)$, then

$$|u(x) - u(y)| \leq C|x - y|(M(|\nabla u|)(x) + M(|\nabla u|)(y))$$

for a.e. $x, y \in \mathbb{R}^n$. Here

$$M(|\nabla u|)(x) = \sup_{r>0} \fint_{B(x,n)} |\nabla u| \, dm$$

is the Hardy–Littlewood maximal function of $|\nabla u|$ (note that for $p > 1$, $M(v) \in L^p(\mathbb{R}^n)$ if $v \in L^p(\mathbb{R}^n)$ and $\|M(v)\|_{L^p} \leq C\|v\|_{L^p}$; $C = C(n, p)$).

(b) If $u \in M^{1,p}(\mathbb{R}^n)$, then $u \in W^{1,p}(\mathbb{R}^n)$ and u has $g \in D(u)$ with

$$\|g\|_{L^p} \leq C(n, p)\||\nabla u|\|_{L^p}.$$

Remark. In fact $W^{1,p}(\mathbb{R}^n) = M^{1,p}(\mathbb{R}^n)$ for all $1 \leq p < \infty$ but the "pointwise" inequality in (a) does not give the information needed.

If $u \in C^1(\Omega)$, then $\nabla u = 0$ in each open set $V \subset \Omega$ where $u = \text{const}$. More generally, if $u \in W^{1,p}(\Omega)$, then $|\nabla u| = 0$ a.e. in the set where $u = \text{const}$. This does not hold in the space $M^{1,p}(X)$. Because of this I have not yet seen any serious development for the Potential Theory in the space $M^{1,p}(X)$. However, for many fractal sets X of $\mathbb{R}^n$ the space $M^{1,p}(X)$ with a natural Hausdorff measure is a natural space to study minimizers of variational integrals and it would be interesting to know if a potential theory similar to the theory based on the Newtonian space can be constructed on $M^{1,p}(X)$.

2.3. Newtonian space $N^{1,p}(X)$

This is the second choice to form a function space in X similar to the classical first order Sobolev space $W^{1,p}$. In some sense it is closer to the classical Sobolev space $W^{1,p}$ than $M^{1,p}$.

2.3.1. *Line integrals*

Let (X, d) be a metric space. A path $\gamma : [a, b] \to X$ is a continuous map. The length of the path γ is

$$l(\gamma) = \sup \sum_{i=1}^{n} d(\gamma(t_i), \gamma(t_{i+1}))$$

where the supremum is taken over all sequences

$$a = t_1 \leq t_2 \leq \ldots \leq t_{n+1} = b.$$

A curve is rectifiable if $l(\gamma) < \infty$. In the sequel we assume that all paths are non–degenerate, i.e. $\gamma([a, b])$ is not a point, unless otherwise stated.

Two important concepts are associated with a rectifiable path γ: the length function $S_\gamma : [a, b] \to \mathbb{R}$ and parametrization by arc length. The function S_γ is defined as

$$S_\gamma(t) = l(\gamma|[a, t]), \quad a \le t \le b,$$

and the path $\tilde{\gamma} : [0, l(\gamma)] \to X$ is the unique 1–lipschitz continuous map such that

$$\gamma = \tilde{\gamma} \circ S_\gamma.$$

In particular, $l(\tilde{\gamma}|[0, t]) = t$, $0 \le t \le l(\gamma)$, and $\tilde{\gamma}$ is obtained from γ by an increasing change of parameter. The path $\tilde{\gamma}$ is called the parametrization of γ by arc length.

If $\gamma : [a, b] \to X$ is a path, then the set

$$|\gamma| = \{\gamma(t) : t \in [a, b]\}$$

is called a (closed) curve. We shall not distinguish paths and curves. In general, this is dangerous.

If γ is a rectifiable curve in X, then the line integral over γ of a Borel function $\rho : X \to [0, \infty]$ is

$$\int_\gamma \rho \, ds = \int_\gamma \rho |dx| = \int_0^{l(\gamma)} \rho(\tilde{\gamma}(t)) dt.$$

The geometric meaning of the line integral is that it represents the total area of the fence whose height is $\rho(\gamma(t))$ at the point t. Note that if γ travels back and forth on $|\gamma|$, then the area is computed accordingly.

Remark. If $X = \mathbb{R}^n$ and if $\gamma : [a, b] \to \mathbb{R}^n$ is absolutely continuous (each coordinate function γ_i of γ is absolutely continuous), then

$$\int_\gamma \rho \, ds = \int_a^b \rho(\gamma(t)) |\gamma'(t)| dt,$$

$\gamma'(t) = (\gamma_1'(t), \ldots, \gamma_n'(t)).$

2.3.2. *p-modulus*

We assume in the rest of this chapter that μ is a regular (each μ–measurable set is contained in a Borel set of equal measure) Borel measure in X. For a given curve family Γ in X and $p \geq 1$ we define the p–modulus of Γ by

$$M_p(\Gamma) = \inf \int_X \rho^p d\mu$$

where the infimum is taken over all Borel functions $\rho : X \to [0, \infty]$ such that

$$\int_\gamma \rho ds \geq 1 \tag{2.6}$$

for all (rectifiable) curves $\gamma \in \Gamma$. Functions ρ that satisfy (2.6) are called admissible functions for the family Γ.

By definition, the modulus of all curves in X that are not rectifiable is zero. If Γ contains a constant curve, then there are no admissible functions and hence the modulus is $+\infty$. The following properties are easily verified:

$$M_p(\emptyset) = 0, \tag{2.7}$$

$$M_p(\Gamma_1) \leq M_p(\Gamma_2), \quad \text{if } \Gamma_1 \subset \Gamma_2, \tag{2.8}$$

$$M_p\left(\bigcup_{i=1}^{\infty} \Gamma_i\right) < \sum_{i=1}^{\infty} M_p(\Gamma_i), \tag{2.9}$$

$$M_p(\Gamma) \leq M_p(\Gamma_0) \tag{2.10}$$

if each curve $\gamma \in \Gamma$ has a subcurve $\gamma_0 \in \Gamma_0$.

These properties show that M_p is an outer measure on the set of all rectifiable curves in X.

Observe that ρ needs to be a Borel function since otherwise the line integral (2.6) may be undefined. However, the situation can be corrected in many cases because of the following lemma (the proof is based on the Borel regularity of μ):

Lemma 2.2. *Let $\rho : X \to [0, \infty]$ be a μ–measurable function. Then there is a Borel function $\rho^* : X \to [0, \infty]$ such that $\rho^*(x) \geq \rho(x)$ for all $x \in X$ and $\rho^* = \rho \ \mu$-a.e.*

In general, it is difficult to compute $M_p(\Gamma)$ for a given curve family Γ. Upper bounds for $M_p(\Gamma)$ are (usually) easy to obtain.

Lemma 2.3. *Suppose that the curves γ of a family Γ lie in a Borel set $A \subset X$ and that $l(\gamma) \geq r > 0$ for each $\gamma \in \Gamma$. Then $M_p(\Gamma) \leq \mu(A)/r^p$.*

Proof.   Put $\rho(x) = 1/r$ for $x \in A$ and $\rho(x) = 0$ for $x \in X \setminus A$. Then ρ is admissible for Γ and the inequality follows. $\square$

Example. Let Γ be the family of curves in a Borel cylinder G of $\mathbb{R}^n$ which join the bases of the cylinder,

$$G = \{x \in \mathbb{R}^n : (x_1, \ldots, x_{n-1}) \in E, \ 0 \leq x_n \leq h\}$$

where $E \subset \mathbb{R}^{n-1}$ is a Borel set. In $\mathbb{R}^n$ we use the Lebesgue measure m. Now

$$M_p(\Gamma) = \frac{m_{n-1}(E)}{h^{p-1}} = \frac{m(G)}{h^p}.$$

To see this note that $M_p(\Gamma) \leq m(G)/h^p$ follows from Lemma 2.3. For the opposite inequality let ρ be an arbitrary admissible function for Γ. For each $y \in E$ let $\gamma_y : [0, h] \to \mathbb{R}^n$ be the vertical segment $\gamma_y(t) = y + te_n$. Then $\gamma_y \in \Gamma$ and assuming $p > 1$ we obtain from the Hölder inequality

$$1 \leq \left(\int_{\gamma_y} \rho ds \right)^p \leq h^{p-1} \int_0^h \rho(y + te_n)^p dt.$$

Integration over $y \in E$ yields by the Fubini theorem

$$m_{n-1}(E) \leq h^{p-1} \int_E dm_{n-1} \int_0^h \rho(y + te_n)^p dt$$

$$= h^{p-1} \int_G \rho^p dm \leq h^{p-1} \int_{\mathbb{R}^n} \rho^p dm.$$

Since this holds for every admissible ρ, $M_p(\Gamma) \geq m_{n-1}(E)/h^{p-1}$ as required. The case $p = 1$ is similar.

Lemma 2.4. *Let $\gamma : [a, b] \to \mathbb{R}$, $\gamma(t) = t$, and $\Gamma = \{\gamma\}$. Then $M_p(\Gamma) = 1/(b-a)^{p-1}$ (in $\mathbb{R}$ the Lebesgue measure is used),*

Proof.   This easily follows from the previous method. $\square$

A comprehensive treatment of the p–modulus in euclidean spaces can be found in [V].

Note that if a family Γ of paths lies in a set E with $\mu(E) = 0$, then $M_p(\Gamma) = 0$. Thus $M_p(\{\gamma\}) > 0$ holds in a few cases only, the situation in Lemma 2.4 being a typical example. A more general example than $\mathbb{R}$ is a graph.

2.3.3. *Upper gradient and* ACC$_p$ *functions*

A C^1–function u in a domain Ω of $\mathbb{R}^n$ satisfies

$$u(x) - u(y) = \int_\gamma \nabla u \cdot d\bar{s} = \int_0^{l(\gamma)} \nabla u(\tilde\gamma(s)) \cdot \tilde\gamma'(s) ds$$

where γ is any rectifiable path in Ω with endpoints x and y. This leads to

$$|u(x) - u(y)| \leq \int_\gamma |\nabla u| dx.$$

We will now see that this inequality is almost as useful as the previous equality. We extend the latter inequality to a metric space (X, d).

A Borel function $g = g_u : X \to [0, \infty]$ is said to be an upper gradient of $u : X \to \mathbb{R}$ if

$$|u(x) - u(y)| \leq \int_\gamma g ds \tag{2.11}$$

for each rectifiable path γ joining x and y in X.

Every function has an upper gradient, namely $g \equiv \infty$, and upper gradients are never unique. Each L–lipschitz function has an upper gradient $g \equiv L$ but this is rarely an optimal choice. The constant function has an upper gradient $= 0$.

A function $u : X \to [-\infty, \infty]$ is said to be ACC$_p$, $p \geq 1$, if $u \circ \tilde\gamma$ is absolutely continuous on $[0, l(\gamma)]$ for p–almost every rectifiable curve γ in X (this means that the property holds for all (rectifiable) paths except possibly for a family Γ_0 with $M_p(\Gamma_0) = 0$).

Definition. Let $u : X \to [-\infty, \infty]$ and let $g : X \to [0, \infty]$ be a Borel function. If there exists a family Γ of paths such that $M_p(\Gamma) = 0$ and the inequality (2.11) is true for all (rectifiable) paths $\gamma \notin \Gamma$, then g is said to be a p–weak upper gradient of u.

The exponent $p \geq 1$ is usually fixed and we simply call g a weak upper gradient of u.

2.3.4. *Newtonian space $N^{1,p}(X)$*

Let $\tilde{N}^{1,p}(X) = \tilde{N}^{1,p}(X, d, \mu)$ is the set of all functions $u : X \to \mathbb{R}$ which belong to $L^p(X)$, $p \geq 1$, and have a p–weak upper gradient $g \in L^p(X)$. The space $\tilde{N}^{1,p} = \tilde{N}^{1,p}(X)$ is called a prenewtonian space. It is a vector space, since if $\alpha, \beta \in \mathbb{R}$ and $u_1, u_2 \in \tilde{N}^{1,p}$ with weak upper gradients g_1 and g_2, then $|\alpha|g_1 + |\beta|g_2$ is an upper gradient for $\alpha u_1 + \alpha u_2$. For $u \in \tilde{N}^{1,p}$ we let

$$\|u\|_{N^{1,p}} = \|u\|_{L^p(X)} + \inf_g \|g\|_{L^p(X)}$$

where the infimum is taken over all weak upper gradients of u. It is easy to see that $\|u\|_{N^{1,p}}$ satisfies the triangle inequality.

If $u, v \in \tilde{N}^{1,p}$ let

$$u \approx v \Leftrightarrow \|u - v\|_{N^{1,p}} = 0.$$

Then $\approx$ is an equivalence relation in $\tilde{N}^{1,p}$, partitioning $\tilde{N}^{1,p}$ into equivalence classes. The collection of equivalence classes, under the $\|u\|_{N^{1,p}}$–norm, is a normed vector space

$$N^{1,p}(X) = \tilde{N}^{1,p}(X)/\approx$$

called the Newtonian space corresponding to the exponent p, $1 \leq p < \infty$.

Example. If X contains no rectifiable paths, then $N^{1,p}(X) = L^p(X)$.

Lemma 2.5. *If $u \in \tilde{N}^{1,p}$, then u is* ACC_p.

Proof. Since $u \in \tilde{N}^{1,p}$, u has a weak upper gradient $g \in L^p(X)$. Let Γ be the collection of all paths for which (3.6) does not hold; then $M_p(\Gamma) = 0$. Let Γ_1 be the collection of paths that have some subpath belonging to Γ. Then

$$M_p(\Gamma_1) \leq M_p(\Gamma) = 0.$$

Let Γ_2 be the family of paths γ such that $\int_\gamma g ds = \infty$. As $g \in L^p$, $M_p(\Gamma_2) = 0$ and hence

$$M_p(\Gamma_1 \cup \Gamma_2) = 0.$$

If $\gamma \notin \Gamma_1 \cup \Gamma_2$, then γ has no subpath in Γ and hence for all $x, y \in |\gamma|$

$$|u(x) - u(y)| \leq \int_{\gamma_{xy}} g ds < \infty.$$

Hence if (a_i, b_i), $i = 1, 2, \ldots, n$, are disjoint intervals in $[0, l(\gamma)]$, then

$$\sum |u(\tilde{\gamma}(b_i)) - u(\tilde{\gamma}(a_i))| \le \int_{\bigcup_i (a_i, b_i)} g(\tilde{\gamma}(s))ds$$

and this shows that $u \circ \tilde{\gamma}$ is absolutely continuous on $[0, l(\gamma)]$ as required. $\square$

Note that the above lemma remains valid if the function u is required only to have a weak upper gradient in L^p.

Remark. The above proof reveals the following: If u is a function with a weak upper gradient $g \in L^p$, then there is a family Γ of paths such that $M_p(\Gamma) = 0$ and for all rectifiable paths $\gamma \notin \Gamma$, u satisfies

$$|u(x) - u(y)| \le \int_{\gamma'} gds$$

for each subpath γ' of γ.

Remarks. (a) Lemma 2.5 has a converse: If $u \in L^p$ and there is a Borel function $g \in L^p$ such that for p–almost every path γ in X the function $s \xmapsto{f} u(\tilde{\gamma}(s))$ is absolutely continuous on $[0, l(\gamma)]$ and $|f'(s)| \le g(\tilde{\gamma}(s))$ a.e. in $[0, l(\gamma)]$, then $u \in \tilde{N}^{1,p}$.

(b) Observe also that if u has a weak upper gradient $g \in L^p$, then for p–almost every path γ in X

$$|f'(s)| \le g(\tilde{\gamma}(s)) \quad \text{a.e. in} \quad [0, l(\gamma)].$$

These observations are useful in proving the following lattice property of $\tilde{N}^{1,p}$ (and $N^{1,p}$):

Lemma 2.6. *If* $u, v \in \tilde{N}^{1,p}$, $\lambda \ge 0$, *then the functions*
   (a) $\min(u, \lambda)$
   (b) $|u|$
   (c) $\min(u, v)$
   (d) $\max(u, v)$
belong to $\tilde{N}^{1,p}$.

Remark. As a byproduct of the proofs of the above statements one obtains the following important fact: if g_u and g_v are weak upper gradients of u

and v, then $\max(u, v)$ has a weak upper gradient g such that

$$
g(x) = \begin{cases} g_u(x), & \text{if } u(x) \geq v(x) \\ g_v(x), & \text{if } v(x) > u(x) \end{cases}
$$

μ $-$ a.e. in X. Note that the function on the right hand side need not be a Borel function, however, the situation can be corrected, see Lemma 2.2.

Theorem 2.8. *$N^{1,p}$ is a Banach space.*

For the proof see [Sh] (the proof is not trivial). The proof uses the concept of p–capacity of a set $E \subset X$. This is defined as

$$
\mathrm{Cap}_p E = \inf_u \|u\|_{N^{1,p}}
$$

where the infimum is taken over all $u \in \tilde{N}^{1,p}$ such that $u \geq 1$ on E. Then $\mathrm{Cap}_p E \in [0, \infty]$ and $\mathrm{Cap}_p E = \infty$ if there is no $u \in \tilde{N}^{1,p}$ with $u \geq 1$ on E. The proof for Theorem 2.8 reveals the following

Lemma 2.7. *If $E \subset X$, $\mathrm{Cap}_p E = 0$ and $u \in N^{1,p}(X \setminus E)$, then u has a uniquely determined extension u^* to E such that $u^* \in N^{1,p}(X)$.*

As a corollary from the proof of Theorem 2.8 one obtains:

Corollary 2.1. *If (u_i) is a Cauchy sequence in $N^{1,p}$, then it has a subsequence which converges pointwise outside a set of p–capacity zero. Moreover, this subsequence can be chosen so that it converges uniformly off a set of arbitrary small p–capacity.*

Remarks. (a) A function u is called p–quasicontinuous if for each $\varepsilon > 0$ there is a set $E \subset X$ such that $u | X \setminus E$ is continuous and $\mathrm{Cap}_p(E) < \varepsilon$. Corollary 2.1 makes it possible to study p–quasicontinuity properties of functions in $N^{1,p}(X)$.

(b) Lemma 2.7 can also be used to show that if $u, v \in \tilde{N}^{1,p}(X)$ and $u = v$ μ $-$ a.e., then $u \approx v$.

The following property is important in applications. The proof is based on the strict convexity of the L^p–norm, $p > 1$.

Lemma 2.8. *If $u \in \tilde{N}^{1,p}$, $p > 1$, then u has a minimal weak upper gradient $g \in L^p$ in the sense that if $\tilde{g}$ is another weak upper gradient for u, then $g \leq \tilde{g}$ μ $-$ a.e..*

2.3.5. *Doubling measure and Poincaré inequality*

A measure μ in X is said to be doubling if

$$\mu(2B) \leq C\mu(B)$$

for all balls B in X; this usually includes the assumption $0 < \mu(B) < \infty$. If μ is a doubling measure, then

$$\mu(B(z,r))/\mu(B(z,R)) \geq C'(r/R)^Q$$

where C' and Q depend on C only. The number Q can be viewed as a kind of dimension of X.

If a metric space does not contain rectifiable paths, then $N^{1,p}(X) = L^p(X)$. The existence of rectifiable paths is guaranteed by the Poincaré inequality.

The space X is said to support a weak $(1,q)$–Poincaré inequality if there are constants $c < \infty$ and $\tau \geq 1$ such that

$$\fint_{B(z,r)} |u - u_{B(x,r)}| d\mu \leq cr \left(\fint_{B(z,\tau r)} g^q d\mu \right)^{1/q} \tag{2.12}$$

wherever $u \in L^1(B(a, \tau r))$ and g is a q–weak upper gradient of u. Observe that (2.12) is assumed to hold for all balls $B(z,r)$ in X. The word weak refers to the possibility that $\tau > 1$; for $\tau = 1$ the inequality is called a $(1,q)$–Poincaré inequality.

If μ is doubling and if X satisfies some natural topological assumptions, see Section 5, then (2.12) implies a weak (t,q)–inequality for some $t > q$

$$\left(\fint_{B(z,r)} |u - u_{B(z,r)}|^t d\mu \right)^{1/t} \leq C'c \left(\fint_{B(z,\tau' r)} g^q d\mu \right)^{1/q}.$$

Thus, if μ is doubling and X supports a weak $(1,p)$–inequality, then under natural topological assumptions X supports a weak (p,p)–Poincaré inequality

$$\left(\fint_{B(z,r)} |u - u_{B(z,r)}|^p d\mu \right)^{1/p} \leq cr \left(\fint_{B(z,\tau r)} g^p d\mu \right)^{1/p} ; \tag{2.13}$$

this follows from the Hölder–inequality. Also it follows from the Hölder inequality that if X supports a weak $(1,q)$–Poincaré inequality for some $1 < q < p$, then it supports a weak $(1,p)$–Poincaré inequality. Hence a reasonable assumption is a weak $(1,q)$–Poincaré inequality $1 < q < p$. For a treatment of the Poincaré inequality see [HK].

Remark. If μ is a doubling measure and X supports a weak $(1,p)$–Poincaré inequality, $1 < p < \infty$, then $N^{1,p}(X) = M^{1,p}(X)$ and the norms are equivalent. Lipschitz functions are dense in $N^{1,p}(X)$ in this case.

If $X = \Omega$ an open set in $\mathbb{R}^n$, $d(x,y) = |x - y|$ and $\mu = m$, then $N^{1,p}(X)$ is the classical Sobolev space and the norms are equivalent (note that m need not be doubling in Ω). For the result see [Sh].

2.4. Zero boundary values and local Newtonian spaces

In order to be able to compare the boundary values of functions in $N^{1,p}$ we need the concept of Sobolev spaces with zero boundary values in a metric measure space.

Remark. If $\Omega \subset \mathbb{R}^n$ is open, then the classical Sobolev space $W_0^{1,p}(\Omega)$ with zero boundary values is the closure of the space $C_0^1(\Omega)$ in the norm $\|u\|_{L^p} + \||\nabla u|\|_{L^p}$.

Let E be an arbitrary subset of X. The set $\tilde{N}_0^{1,p}(E)$ is the set of all functions $u : E \to [-\infty.\infty]$ for which there is a function $\tilde{u} \in \tilde{N}^{1,p}(X)$ s.t. $\tilde{u} = u \; \mu - $ a.e. in E and

$$C_p([x \in X \setminus E : \tilde{u}(x) \neq 0\}) = 0.$$

Then $N_0^{1,p}(E) = \tilde{N}_0^{1,p}(E)/ \sim$ where $u \sim v$ means that $u = v \; \mu - $ a.e. in E. The norm in $N_0^{1,p}(E)$ is defined as

$$\|u\|_{N_0^{1,p}(E)} = \|\tilde{u}\|_{\tilde{N}^{1,p}(X)}.$$

The space $N_0^{1,p}(E)$ is called the *space with zero boundary values.*

We are mainly interested in local properties of minimizers of variational integrals. Thus we need the notation of a local Newton space.

Let Ω be an open set in X. We say that a subset A of Ω is compactly contained in Ω if $\overline{A}$ is a compact subset of Ω, abbreviated $A \subset\subset \Omega$ (in many cases it suffices that A is bounded and $d(A, X \setminus \Omega) > 0$).

We say that u belongs to the local Newtonian space $N_{\text{loc}}^{1,p}(\Omega)$ if $u \in N^{1,p}(A)$ for every measurable set $A \subset\subset \Omega$.

Note that for $1 < p < \infty$ each function $u \in N_{\text{loc}}^{1,p}(\Omega)$ has a minimal p–weak upper gradient $g = g_u$ in Ω in the following sense: If $\Omega' \subset\subset \Omega$ is an open set and g is the minimal upper gradient of u in Ω', then $g = g_u$ $\mu - $ a.e. in Ω'.

Remark. Suppose that X is proper (closed and bounded subsets of X are compact), μ is doubling and X supports a $(1,p)$–Poincaré inequality.

If $E \subset X$ is given, then each function $u \in N_0^{1,p}(E)$ can be approximated in $N_0^{1,p}(E)$ by lipschitz functions which vanish in $X \setminus E$. Moreover, the functions can be chosen to have compact support in E.

The following inequality is sometimes called a Sobolev–Poincaré inequality. The above discussion shows that it holds with the exponent p in both sides. This is the most useful inequality for functions u in $N_0^{1,p}(\Omega)$ when $\mu(X \setminus \Omega) > 0$ and Ω is bounded. Then u is $= 0$ in the large subset of $X \setminus \Omega$.

Theorem 2.9. *Let X be a doubling metric space supporting a weak (t,q)–Poincaré inequality for some $1 < q < p$ and $t > 1$. Suppose that $u \in N^{1,p}(X)$ and let $A = \{y \in B(x,R) : |u(y)| > 0\}$. If $\mu(A) \le \gamma\mu(B(x,R))$ for some γ with $0 < \gamma < 1$, then there is a constant $c > 0$ so that*

$$\left(\fint_{B(x,R)} |u|^t d\mu \right)^{1/t} \le cR\left(\fint_{B(x,\tau'R)} g_u^q d\mu \right)^{1/q}.$$

The constant c is independent of u.

Proof. By the Minkowski inequality and by the weak (t,q)–Poincaré inequality we have

$$\left(\fint_{B(x,R)} |u|^t d\mu \right) \le \left(\fint_{B(x,R)} |u - u_{B(x,R)}|^t d\mu \right)^{1/t} + |u_{B(x,R)}|$$

$$\le cR\left(\fint_{B(x,\tau'R)} g_u^q d\mu \right)^{1/q} + |u_{B(x,R)}|.$$

The Hölder inequality implies that

$$|u_{B(x,R)}| \le \left(\frac{\mu(A)}{\mu(B(x,R))} \right)^{1-1/t} \left(\fint_{B(x,R)} |u|^t d\mu \right)^{1/t}$$

$$\le \gamma^{1-1/t} \left(\fint_{B(x,R)} |u|^t d\mu \right)^{1/t}.$$

Hence we obtain

$$(1 - \gamma^{1-1/t})\left(\fint_{B(x,R)} |u|^t d\mu \right)^{1/t} \le cR\left(\fint_{B(x.\tau'R)} g_u^q d\mu \right)^{1/q},$$

from which the claim follows since $0 < \gamma < 1$. $\qquad\square$

From Theorem 2.9 we obtain

Corollary 2.2. *Suppose that $\Omega \subset X$ is bounded and $\mu(X \setminus \Omega) > 0$. Then*

$$\int_{\Omega} |u|^p d\mu \leq c \int_{\Omega} g_u^p d\mu$$

for every function $u \in N_0^{1,p}(\Omega)$. The constant c is independent of u.

2.5. Minimizers and quasiminimizers in Newtonian spaces

2.5.1. *The obstacle problem*

The obstacle method is the most important method in the nonlinear potential theory.

Let $1 < p < \infty$ and let $\Omega \subset X$ be an open set. Suppose that $\theta \in N^{1,p}(\Omega)$ and $\psi : \Omega \to [-\infty, \infty]$. Write

$$\mathcal{K}_{\psi,\theta} = \mathcal{K}_{\psi,\theta}(\Omega) = \{v \in N^{1,p}(\Omega) : v - \theta \in N_0^{1,p}(\Omega), \ \ v \geq \psi \ \mu - \text{a.e. in } \Omega\}$$

A function $u \in \mathcal{K}_{\psi,\theta}$ is a *solution to the $\mathcal{K}_{\psi,\theta}-$obstacle problem* (with the obstacle ψ and boundary values θ) if

$$\int_{\Omega} g_u^p d\mu \leq \int_{\Omega} g_v^p d\mu, \ \ \forall v \in \mathcal{K}_{\psi,\theta}.$$

Here g_u and g_v are the minimal upper gradients of u and v in Ω. If $\psi \equiv -\infty$, then the obstacle has no effect, and a solution of the $\mathcal{K}_{-\infty,\theta}-$obstacle problem is said to be the *minimizer* (with boundary values θ) in Ω.

The $\mathcal{K}_{\psi,\theta}(\Omega)$–obstacle problem can be studied on an arbitrary μ–measurable set $\Omega \subset X$.

Let $K \geq 1$ and $\theta \in N^{1,p}(\Omega)$. A function $u \in N^{1,p}(\Omega)$ is called a K–quasiminimizer with the boundary values θ if

    (a) $u - \theta \in N_0^{1,p}(A)$
    (b) $\int_A g_u^p d\mu \leq K \int_A g_v^p d\mu$

holds for all open (measurable) $A \subset \Omega$ and all $v \in N^{1,p}(\Omega)$ such that $u - v \in N_0^{1,p}(A)$. Then u is a minimizer in Ω iff u is a 1–quasiminimizer in Ω.

From now on we make the following standard assumptions and keep the exponent p, $1 < p < \infty$, fixed.

(i) μ is a doubling regular Borel measure such that $\mu(\Omega) > 0$ if $\emptyset \neq \Omega$ is open and $\mu(\Omega) < \infty$ if Ω is bounded,

(ii) closed and bounded sets are compact,

(iii) X supports a weak $(1,q)$–Poincaré inequality for some $q \in (1,p)$ (too strong in most cases but needed to pass from the weak $(1,q)$–Poincaré inequality to the weak (p,p)–Poincaré inequality).

Theorem 2.10. *Let $\Omega \subset X$ be a bounded open set with $\mu(X \setminus \Omega) > 0$. If $\mathcal{K}_{\psi,\theta}(\Omega) \neq \emptyset$, then there is a unique solution to the $\mathcal{K}_{\psi,\theta}(\Omega)$–obstacle problem.*

Proof. Set

$$I = \inf_{v} \int_{\Omega} g_v^p d\mu, \quad v \in \mathcal{K}_{\psi,\theta}.$$

Now $0 \leq I < \infty$ and let $u_i \in \mathcal{K}_{\psi,\theta}$, $i = 1, 2, \ldots$, be a minimizing sequence. This implies that the sequence g_{u_i} is bounded in $L^p(\Omega)$. Since Ω is bounded and $\mu(X \setminus \Omega) > 0$, we have

$$\int_{\Omega} |u_i - \theta|^p d\mu \leq C \int_{\Omega} g_{u_i - \theta}^p d\mu$$

where we have used Corollary 2.2. This yields

$$\int_{\Omega} |u_i - \theta|^p d\mu < C \int_{\Omega} g_{u_i}^p + C \int_{\Omega} g_\theta^p d\mu$$

and hence the sequence $(u_i - \theta)$ is uniformly bounded in $N_0^{1,p}(\Omega)$. Then there is a subsequence (u_{i_j}) and $u \in N^{1,p}(\Omega)$ such that $u - \theta \in N_0^{1,p}(\Omega)$, $u_{i_j} \to u$ weakly in $L^p(\Omega)$ and

$$\int_{\Omega} g_u^p d\mu \leq \liminf_{j \to \infty} \int_{\Omega} g_{u_{i_j}}^p d\mu \leq I.$$

(this requires some work). Hence u has the minimizing property.

Next we show that $u \in \mathcal{K}_{\psi,\theta}$. It suffices to show that $u \geq \psi \; \mu - $a.e. The Mazur lemma implies that a sequence (v_j) of convex combinations of u_{i_j} converges to u in $L^p(\Omega)$, and hence passing to a subsequence we may assume that $v_j \to u \; \mu - $a.e. Since $u_{i_j} \geq \psi$, $v_j \geq \psi \; \mu - $a.e. as well and consequently $u \geq \psi \; \mu - $a.e.

To prove the uniqueness is more difficult than in the classical case since derivation is not a linear operation: If $u_1, u_2 \in \mathcal{K}_{\psi,\theta}(\Omega)$ are solutions such that $g_{u_1} \neq g_{u_2}$ on a set of positive measure, then $v = (u_1 + u_2)/2 \in \mathcal{K}_{\psi,\theta}$ and by the strict convexity

$$\int_\Omega g_v^p d\mu < \frac{1}{2} \int_\Omega g_{u_1}^p d\mu + \frac{1}{2} \int_\Omega g_{u_2}^p d\mu = I,$$

a contradiction. Hence $g_{u_1} = g_{u_2}$ and it remains to show that $u_1 = u_2$.

To this end suppose that $u_1 \neq u_2$. We may now assume that the set $V = \{x \in \Omega : u_2(x) > u_1(x)\}$ has positive measure. Set $g = g_{u_1} = g_{u_2}$ in Ω and write $W = \{x \in V : g > 0\}$. If $\mu(W) > 0$, then for some constant c the set $W_1 = \{x \in W : u_1(x) < c < u_2(x)\}$ satisfies $\mu(W_1) > 0$. Define v in Ω as follows

$$v(x) = \begin{cases} u_1(x), & u_1(x) \leq c, \\ c, & u_1(x) < c < u_2(x), \\ u_2(x), & u_2(x) \geq c. \end{cases}$$

Now $v \in \mathcal{K}_{\psi,\theta}(\Omega)$ and the function $g_v = 0$ in W_1 and $g_v = g$ in $\Omega \setminus W_1$ belongs to $D(v)$. Thus

$$\int_\Omega g^p d\mu \leq \int_\Omega g_v^p d\mu < \int_\Omega g^p d\mu$$

is a contradiction since $\mu(W_1) > 0$ and $g > 0$ in W_1.

Hence $\mu(W) = 0$ and this implies that either $u_1(x) = u_2(x)$ or $g(x) = 0$ for μ−a.e. $x \in \Omega$. In the set $u_1 = u_2$ we can take $g_{u_2-u_1} = 0$ and since $g = 0$ in the set $u_2 \neq u_1$, we also conclude that $g_{u_2-u_1} = 0$ in this set because $g_{u_2-u_1} \leq g_{u_2} + g_{u_1}$ there. Hence $g_{u_2-u_1} = 0$ and we obtain, as in the beginning of the proof, using the Sobolev–Poincaré inequality (Corollary 19) that $u_1 = u_2$ in Ω; note that $u_1 - u_2 \in N_0^{1,p}(\Omega)$. This completes the proof. $\qquad\qquad\square$

We are mostly interested in local properties of solutions of the obstacle problem and hence solutions should be defined without boundary values in Ω.

A function $u \in N_{\text{loc}}^{1,p}(\Omega)$ is a *minimizer* in Ω, if u satisfies (b) for $K = 1$ in every open (measurable) $A \subset\subset \Omega$. Similarly, u is a *K–quasiminimizer* if (b) holds true for given K.

A function $u \in N_{\text{loc}}^{1,p}(\Omega)$ is a *superminimizer* in Ω if u is a solution to the $\mathcal{K}_{u,u}(\Omega')$–obstacle problem for each open $\Omega' \subset\subset \Omega$.

A solution of the $\mathcal{K}_{\psi,\theta}(\Omega)$–obstacle problem is always a superminimizer in Ω. If u and $-u$ are superminimizers, then u is a minimizer (exercise).

A connection to the classical superharmonic case is as follows: Let $\Omega \subset \mathbb{R}^n$ open and $u : \Omega \to \mathbb{R} \cup \{\infty\}$ superharmonic in the classical sense, i.e. $u \not\equiv \infty$ in any component of Ω and

(A) u is lower semicontinuous

(B) $u(x_0) \geq \fint_{B(x_0,r)} u\,dm$ for each ball $B(x_0, r) \subset\subset \Omega$.

If $u : \Omega \to \mathbb{R}$ is superharmonic and locally bounded, then u is a superminimizer for the integral

$$\int |\nabla u|^2 dm.$$

See [HKM] for details,

Lemma 2.9. *If u_1 and u_2 are superminimizers in Ω, then $\min(u_1, u_2)$ is a superminimizer in Ω.*

Proof. Somewhat tricky but simple, see [KM2]. $\square$

Remark. There exists an interesting class of K–superquasiminimizers: A function $u \in N_{\mathrm{loc}}^{1,p}(\Omega)$ is called a K–superquasiminimizer in Ω if (b) holds for all v such that $v - u \in N_0^{1,p}(A)$, $A \subset\subset \Omega$ open and $v \geq u$ μ – a.e. Lemma 21 holds for K_i–superquasiminimizers u_i, $i = 1, 2$, in the form that $\min(u_1, u_2)$ is a $\min(K_1 K_2, K_1 + K_2)$–superquasiminimizer.

2.5.2. *Regularity theory for minimizers and superminimizers*

This is a technical part. Here the celebrated De Giorgi method plays an essential role.

A function $u \in N_{\mathrm{loc}}^{1,p}(\Omega)$ belongs to the De Giorgi class $DG_p(\Omega, k_0)$, $k_0 \in \mathbb{R}$, if there is a constant $c < \infty$ such that for all $k \geq k_0$, $z \in \Omega$ and $0 < \rho < R$ for which $B(z.R) \subset\subset \Omega$ the function u satisfies the Caccioppoli type estimate

$$\int_{A_z(k,\rho)} g_u^p d\mu \leq c(R - \rho)^{-p} \int_{A_z(k,R)} (u - k)^p d\mu \qquad (*)$$

where

$$A_z(k, r) = \{x \in B(z, r) : u(x) > k\}.$$

If (*) holds for all $k \in \mathbb{R}$, then we simply write $u \in DG_p(\Omega)$.

The following lemma is not difficult to prove [KM2], [KS]:

Lemma 2.10. *(a) If u is a superminimizer in Ω then $-u \in DG_p(\Omega)$ (corollary: a minimizer belongs to $DG_p(\Omega)$).*

(b) If $k_0 = \text{ess sup}_\Omega \psi < \infty$, then the solution u to the $\mathcal{K}_{\psi,\theta}(\Omega)$–obstacle problem belongs to $DG_p(\Omega, k_0)$.

The De Giorgi method was developed in order to prove the following result, see [G].

Lemma 2.11. *If $u \in DG_p(\Omega)$, then u is locally Hölder continuous in Ω. More precisely,*

$$\text{osc}(u, B(z, \rho)) \leq c(\frac{\rho}{R})^\alpha \text{osc}(u, B(z, R))$$

whenever $0 < \rho \leq R$ and $B(z, 2R) \subset\subset \Omega$; c and $\alpha > 0$ depend on data but not on u.

Remark. The proof in [G] is in $\mathbb{R}^n$ but with minor modifications it can be extended to the Newtonian case.

The De Giorgi method also provides the following results:

(I) If u is a superminimizer in Ω, and $u_- = -\min(u, 0)$, then

$$\text{ess}\inf_{B(z,R)} u \geq -c\Big(\fint_{B(z,2R)} u_-^p \, d\mu \Big)^{1/p}$$

whenever $B(z, 3R) \subset\subset \Omega$. In particular, u is locally bounded below.

(II) If u is a superminimizer, $u \geq 0$, then

$$\Big(\fint_{B(z,R)} u^\sigma d\mu \Big)^{1/\sigma} \leq c \, \text{ess}\inf_{B(z,3R)} u$$

whenever $B(z, 5R) \subset\subset \Omega$; $\sigma > 0$ depends on data. This is sometimes called a weak Harnack inequality.

(III) If u is a solution of the $\mathcal{K}_{\psi,\theta}(\Omega)$–obstacle problem, $k_0 = \text{ess sup}_\Omega \psi$ and $k \geq k_0$. then

$$\text{ess}\sup_{B(z,R/2)} u \leq k + c\Big(\fint_{B(z,R)} (u - k)_+^p \, d\mu \Big)^{1/p}.$$

These results imply that a superminimizer is locally bounded below (it need not be locally bounded above unless $p > Q$). If the obstacle is locally bounded from above, then (III) implies that a solution of the $\mathcal{K}_{\psi,\theta}$–obstacle problem is also locally bounded above.

The property (I) can be used to prove

Theorem 2.11. *Let u be a superminimizer in Ω. Then the function*

$$u^*(x) = \operatorname{ess} \lim_{y \to x} \inf u(y) = \lim_{r \to 0} \operatorname{ess} \inf_{B(x,r)} u$$

is lower semicontinuous (this is automatically true) in Ω and $u^* \approx u$ in $\tilde{N}^{1,p}_{\mathrm{loc}}(\Omega)$ (more difficult to prove).

This result shows that the lower semicontinuity assumption for a superharmonic functions is a natural assumption.

The De Giorgi method can also be used to prove that minimizers satisfy the Harnack inequality. This is an idea of Di Benedetto–Trudinger [DT] and the proof in this setup is rather complicated, see [KS]; the proof employs the Krylov–Safonov covering argument in metric spaces with a doubling measure.

Theorem 2.12. *Suppose that $u \geq 0$ is a minimizer. Then there is a constant $c < \infty$ such that*

$$\sup_{B(x,R)} u \leq c \inf_{B(x,R)} u \tag{2.14}$$

for every ball $B(x,R)$ with $B(x,5R) \subset \Omega$ (the constant c is independent of $B(x,R)$ and u).

Remark. Theorems 2.12 and 2.11 remain true for K–quasiminimizers and K–superquasiminimizers, respectively, see [KM2] and [KS]. A consequence of this is, for example, that the integral condition in the definition for a K–quasiminimizer, $K > 1$, does not have a local character only because it is easy to give examples of local K–quasiminimizers for any $K > 1$ that do not satisfy the "global" Harnack inequality (2.14).

Since every solution to the $\mathcal{K}_{\psi,\theta}(\Omega)$–obstacle problem is a superminimizer, the above theory (with some extra work, see [KM2]) leads to

Theorem 2.13. *Suppose that $\psi : \Omega \to (-\infty, \infty)$ is continuous. Then the solution u of the $\mathcal{K}_{\psi,\theta}(\Omega)$–obstacle problem is continuous.*

2.5.3. *Comments on superharmonic functions in the metric setup*

In addition to the supermeanvalue property of classical superharmonic functions there are several other equivalent definitions for superharmonicity: Let $\Omega \subset \mathbb{R}^n$ be domain and $u : \Omega \to (-\infty, \infty]$ a lower semicontinuous function, $u \not\equiv \infty$. Then u is superharmonic (in the classical sense) iff u satisfies one of the conditions:

(a) For each open set $D \subset\subset \Omega$ u satisfies the comparison principle: If $h \in C(\overline{D})$ and h is harmonic in D with $u \geq h$ on ∂D, then $u \geq h$ in D.

(b) u is the limit of an increasing sequence of superminimizers (for the integral

$$\int |\nabla u|^2 dm).$$

Both these approaches can be used in the nonlinear potential theory see [HKM] but metric spaces offer topological and analytic difficulties. A substitute for (a) in the metric setup is the following: Let $\Omega \subset X$ be open. A function $u : \Omega \to (-\infty, +\infty]$ is called $(p-)$ superharmonic in Ω if

($*$) u is lower semicontinuous and not identically $+\infty$ in any component of Ω,

($**$) for every open $\Omega' \subset\subset \Omega$ the following comparison principle holds: if $v \in C(\overline{\Omega}') \cap N^{1,p}(\Omega')$ and $v \leq u$ on $\partial\Omega'$, then the minimizer $h = h(v)$ with boundary values v in $\partial\Omega'$ satisfies $h \leq v$ in Ω'.

The condition ($**$) is stronger than (a). In [KM2] it is then proved that ($*$) and ($**$) imply (b) in the metric setup.

Many potential theoretic problems remain open in metric spaces. For example: Characterize the set $E \subset X$ where a superharmonic function $u : X \to (-\infty, +\infty]$ takes the value $+\infty$.

2.6. Graphs

Potential Theory on graphs has been intensively studied recently. We indicate a connection to the theory developed in the previous chapters.

A graph G consists of a set V (the vertex set) and a neighborhood relation $x \sim y$; we say that x and y are neighbors. The metric on G is obtained as follows: If x and y are vertices, then $d(x, z)$ denotes the smallest number of vertices $x = x_0, x_1, \ldots, x_{n-1}, x_n = y$ needed so that $x_i \sim x_{i+1}$. If U is a subset of G. then the measure $\#U$ of U is simply the number of vertices in U. This measure is doubling if

$$\#B(x, 2r) \leq c\#B(x, r)$$

with the constant $c < \infty$ independent of $x \in V$ and $r > 0$. The $(1, p)$–Poincaré inequality takes the following form

$$\frac{1}{\#B} \sum_{x \in B} |u(x) - u_B| \leq c \operatorname{diam} (B)(\frac{1}{\#B} \sum_{x \in B, y \sim x} |u(x) - u(y)|^p)^{1/p}$$

whenever $u : V \to R$.

Let $U \subset V$. We denote by ∂U the (outer) boundary of U, i.e. the set of all vertices $x \in V \setminus U$ which have at least one neighbor in U.

For a function $u : U \cup \partial U \to \mathbb{R}$, $U \subset V$, the p–th power of the "gradient" at $x \in V$ and the p–Dirichlet sum (integral) over a set $U \subset V$ are defined as

$$|Du(x)|^p = \sum_{y \sim x} |u(y) - u(x)|^p,$$

$$I_p(u, U) = \sum_{x \in U} |Du(x)|^p.$$

As in the previous sections a function u is a p–harmonic in U (U finite) if u is a minimizer of $I_p(u, U)$ among all functions in $U \cup \partial U$ with the same values in ∂U. In other words, if

$$I_p(u, U) \leq I_p(v, U)$$

whenever $v : U \cup \partial U \to R$ with $v = u$ in ∂U.

In this case there is an Euler equation for p–harmonic functions, see [HS]. A function u is p–harmonic in U iff

$$\Delta_p u(x) = \sum_{y \sim x} |u(y) - u(x)|^{p-2}(u(y) - u(x)) = 0 \qquad (2.15)$$

for every $x \in U \cup \partial U$. Note that for $p = 2$ this leads to

$$\Delta_2 u(x) = \sum_{y \sim x} (u(y) - u(x)) = 0 \ \ \forall x \in U \cup \partial U$$

which is the mean value property at x.

The "weak" formulation for (2.15) is

$$\sum_{x \in U, y \sim x} |u(y) - u(x)|^{p-2}(u(y) - u(x))(w(y) - w(x)) = 0 \qquad (2.16)$$

for every function $w : V \to \mathbb{R}$ with finite support in $U \cup \partial U$ (finite support means: zero except at a finite set).

Note that (2.16) corresponds to the Euler equation of the variational integral

$$\int_U |\nabla u|^p dm \to \min$$

in the weak formulation:

$$\int_U |\nabla u|^{p-2}\nabla u \cdot \nabla w \, dm = 0$$

for all $w \in W_0^{1,p}(U)$ $(= N_0^{1,p}(U))$.

In [HS] it was shown that if the graph is doubling and supports a $(1,p)$–Poincaré inequality, then p–harmonic functions satisfy the Harnack inequality and, in particular, such graphs do not carry positive non–constant p–harmonic functions. The proofs require rather extensive computation. It is possible, however, to take a continuous approach.

Associate to G "a connected graph" G_0 as follows: If $x \sim y$, then connect x and y by a closed interval I of unit length. Associate a measure μ on G_0 as

$$\mu(U) = \sum_I \text{length } (I \cap U), \quad U \subset G_0$$

where the sum is extended over all intervals I in G_0.

Now one can study absolutely continuous functions u on the intervals I and use $|\nabla u| = |u'|$ as an upper gradient, but $u' = \nabla u$ can also be used as an "ordinary" derivative. Observe that there is no need to consider the derivation, or the upper gradient, at the vertices since they represent a set of linear measure zero on each curve in G_0. The following lemma is not difficult, see [KS].

Lemma 2.12. *Suppose that G has a doubling measure and supports the $(1,p)$–Poincaré inequality. Then the measure μ on G_0 is doubling and G_0 supports an ordinary Poincaré inequality*

$$\fint_B |u - u_B| d\mu \leq cr \left(\int_B |\nabla u|^p d\mu \right)^{1/p}, \quad B = B(x,r).$$

The next step is to prove that if u is p–harmonic in the sense of (2.16) (or (2.15)), then the linear extension u_0 of u to G_0 satisfies the weak Euler equation

$$\int\limits_{U_0} |\nabla u_0|^{p-2}\nabla u_0 \cdot \nabla w\, d\mu = 0 \tag{2.17}$$

for every lipschitz function w with finite support in U. The linear extension u_0 of u is defined as follows: $u_0(x) = u(x)$ for each vertex $x \in V$ and on the unit interval I connecting x to y $(x \sim y)$ we have

$$u_0(t) = (u(y) - u(x))t + u(x)$$

(this is a piecewise linear extension of u).

Theorem 2.14. *Suppose that U is connected (every vertex in U can be connected to every other vertex in U by a finite sequence of vertex "neighborhoods"). Let u be p–harmonic in U. Then the "linear" extension u_0 of u is p–harmonic in the sense of (2.17) in U_0 (U_0 is obtained from U as G_0 is obtained from G).*

Proof. Let u (or u_0) satisfy (2.16). Let w be a lipschitz function with finite support in U. Associate with w the function w_1 on G_0 so that for each vertex x we have $w_1(x) = w(x)$ and w_1 is "linear" on each edge I. Then $w_2 = w - w_1$ is lipschitz and vanishes on the vertex set V. By (2.16)

$$0 = \sum_{x \in U, x \sim y} |u(y) - u(x)|^{p-2}(u(y) - u(x))(w(y) - w(x))$$

$$- \int\limits_{U_0} |\nabla u_0|^{p-2}\nabla u_0 \cdot \nabla w_1\, d\mu. \tag{2.18}$$

Since the linear extension u_0 of u is p–harmonic on each edge I of U_0 and w_2 vanishes at the end points of I, we have

$$\int\limits_{I} |\nabla u_0|^{p-2}\nabla u_0 \cdot \nabla w_2\, d\mu = 0$$

(this is simply $\int\limits_{I} |u_0'(t)|^{p-2}u_0'(t)w_2'(t)dt$.) Thus

$$\sum_{I} \int\limits_{I} |\nabla \mu_0|^{p-2}\nabla u_0 \cdot \nabla w_2\, d\mu = 0$$

where the sum is taken over all edges I with one endpoint in U. But this, together with (2.18), means that

$$\int\limits_{U_0} |\nabla u_0|^{p-2}\nabla u_0 \cdot \nabla w\, d\mu = 0$$

as required. $\qquad\square$

Remark. In this discrete setup the distances between vertices are $= 1$. Arbitrary distances can also be considered. I have not seen any "deep" use of the modulus of a family of curves in graphs although it is clear that this concept has an influence in the potential theory on graphs. Note that this method has been applied to Riemann surfaces.

References

[AT] Ambrosio, L. and Tilli, P., *Selected topics on analysis on metric spaces,* Oxford Univ. Press, 2004.

[C] Cheeger, J., *Differentiability of Lipschitz functions on metric spaces,* GAFA, Geom. funct. anal. 9 (1999), 428–517.

[DT] Di Benedetto, E. and Trudinger, N.S., *Harnack inequalities for quasi-minima of variational integrals,* Ann. de l'Inst. H. Poincaré, Analyse non linéare 1 (1984), 295–308.

[G] Giaquinta, M., *Introduction to Regularity Theory for Nonlinear Elliptic Systems,* Birkhauser, 1993.

[H] Hajłasz, P., *Sobolev spaces on an arbitrary metric space,* Potential Anal. 5 (1996), 403–415.

[HK] Hajłasz, P. and Koskela, P., *Sobolev Met Poincaré,* Memoirs of AMS, vol. 145, 2000.

[He] Heinonen, J., *Lectures of Analysis on Metric Spaces,* Springer 2000 (Universitext).

[HKM] Heinonen, J., Kilpeläinen, T. and Martio, O., *Nonlinear Potential Theory of Degenerate Elliptic Equations,* Oxford Univ. Press, 1993.

[HS] Holopainen, I. and Soardi, P., *A strong Liouville theorem for p–harmonic functions in graphs,* Ann. Acad. Sci. Fenn. Math. 22 (1997), 205–226.

[KKM] Kilpeläinen. T., Kinnunen, J. and Martio, O., *Sobolev spaces with zero boundary values on metric spaces,* Potential Anal. 12 (2000), 233–247.

[KM1] Kinnunen, J. and Martio, O., *Sobolev capacity on metric spaces,* Ann. Acad. Sci. Fenn. Math. 21, (1996), 367–382.

[KM2] Kinnunen, J. and Martio, O., *Nonlinear potential theory on metric spaces,* Illinois J. Math. 46 (2002), 857–863.

[KS] Kinnunen, J. and Shanmugalingam, N., *Regularity of quasiminimizers on metric spaces,* Manuscripta Math 100 (2001), 401–423.

[Sh] Shanmugalingam. N., Newtonian spaces: *An extension of Sobolev spaces to metric measure spaces,* Rev. Math. Iberoamericana 16 (2000), 243–279.

[V] Väisälä, J., *Lectures on n–dimensional quasiconformal mappings,* Lecture Notes in Mathematics 229 (1971), Springer–Verlag.

PART 2

Differential equations and nonlinear analysis

Chapter 3

An introduction to mean curvature flow

Giovanni Bellettini

Dipartimento di Matematica
Università di Roma "Tor Vergata"
via della Ricerca Scientifica 00133 Roma, Italy
belletti@mat.uniroma2.it

Contents

3.1. Introduction

This paper is a slightly extended version of the five hours course that I delivered in the summer of 2003 for the *Minicorsi di Analisi Matematica* at the University of Padova. Those lectures were conceived as a brief and elementary introduction to classical solutions to mean curvature flow of boundaries. The present paper, which is far from being a complete survey on the subject, illustrates some of the many ideas that have been developed in this field of research in the last few years. Beside the pioneering book of Brakke [27], other references on this topic are for instance [65], [78], [5], [47],

[16], [85], [34], where the reader can find more informations on geometric evolution problems, weak solutions and regularity.

The content of the paper is the following. In Section 3.2 we give the main notation used in the sequel. In Section 3.3 we recall some of the properties of the signed distance function d from the boundary ∂E of a smooth open set E, and its relations with the second fundamental form of ∂E. In particular, in Theorem 3.3 we recall the expansion of the Hessian of d in a tubular neighbourhood of ∂E. Remark 3.4 is concerned with the square distance function from ∂E, and is related with the discussion in Section 3.11 concerning mean curvature flow in arbitrary codimension. More details on the contents of Section 3.3 can be found in [5, Section 4]. In Section 3.4 we define smooth mean curvature flow of boundaries. After some preliminaries, in Definition 3.4 we define mean curvature flow of boundaries using the signed distance function. Mean curvature flow using level sets and mean curvature flow of graphs are considered in Examples 3.2 and 3.3, respectively. The short time existence and uniqueness theorem is briefly described in Subsection 3.5, where we follow the approach of [41]. In Section 3.6 we illustrate some special solutions to the flow, such as self-similar solutions. The comparison principle between smooth compact mean curvature flows is discussed in some detail in Section 3.7. As an application of the comparison principle, following [38] and [50] we show how to derive estimates from above and below on the lifespan of a classical solution (Theorem 3.7). In Section 3.8 we illustrate Huisken monotonicity formula [56], see Theorem 3.8. In Section 3.9 we derive the gradient estimate of Ecker-Huisken for mean curvature evolution of graphs [35], see Theorem 3.9. Section 3.10 is devoted to the description of an example of Grayson [53] of singularity of the mean curvature flow of a surface. Following [32], in Section 3.11 we briefly recall the definition of mean curvature flow in arbitrary codimension, using the square distance function. A more detailed discussion on the contents of Section 3.11 can be found in [8], [5]. We conclude the paper with Section 3.12, where we give some references on weak solutions for geometric evolution problems and on regularity.

3.2. List of notations

If $x \in \mathbb{R}^n$ and $\rho > 0$, we let $B_\rho(x) := \{y \in \mathbb{R}^n : |y - x| < \rho\}$.
$\langle \cdot, \cdot \rangle$ is the scalar product in $\mathbb{R}^n$. If $a, b \in \mathbb{R}^n$, $a = (a_1, \ldots, a_n)$, $b = (b_1, \ldots, b_n)$, $a \otimes b$ is the matrix whose (i, j)-entry is $a_i b_j$. $\mathrm{Id} = (\delta_{ij})$ is the identity matrix in $\mathbb{R}^n$.

If $F \subseteq \mathbb{R}^n$ and $x \in \mathbb{R}^n$, we let $\mathrm{dist}(x, F) := \inf\{|y - x| : y \in F\}$. $|F|$ is the Lebesgue measure of F.

$\mathcal{H}^\alpha$ is the α-dimensional Hausdorff measure in $\mathbb{R}^n$, $\alpha \in [0, n]$.

∇ (resp. ∇^2, Δ) is the gradient (resp. the Hessian, the Laplacian) in $\mathbb{R}^n$. If $i \in \{1, \ldots, n\}$ we indicate by e_i the canonical basis of $\mathbb{R}^n$ and by ∇_i the partial derivative with respect to x_i. Hence $\nabla^2 = (\nabla_{ij})$.

In the sequel the symbol E denotes an open subset of $\mathbb{R}^n$ with boundary ∂E of class C^∞,

$$d^E(x) := \mathrm{dist}(x, E) - \mathrm{dist}(x, \mathbb{R}^n \setminus E), \qquad x \in \mathbb{R}^n \tag{3.1}$$

is the signed distance function from ∂E negative inside E. Whenever no confusion is possible, the function d^E will be denoted by d. Moreover, when integrating a function f on ∂E we omit the symbol $d\mathcal{H}^{n-1}$, thus writing $\int_{\partial E} f$ in place $\int_{\partial E} f \, d\mathcal{H}^{n-1}$.

ν^E is the unit normal vector to $\partial E = \{d = 0\}$ pointing outward of E.

If f (resp. X) is a function (resp. a vector field) which is smooth in a neighbourhood of ∂E and $x \in \partial E$, $\nabla_\tau f(x)$ (resp. $\mathrm{div}_\tau X(x)$) is the tangential gradient of f (resp. the tangential divergence of X) at x, i.e.,

$$\nabla_\tau f(x) := \left(\mathrm{Id} - \nu^E(x) \otimes \nu^E(x)\right) \nabla f(x) \tag{3.2}$$

$$\left(\ \mathrm{resp.} \ \ \mathrm{div}_\tau X(x) := \mathrm{tr}\left(\left(\mathrm{Id} - \nu^E(x) \otimes \nu^E(x)\right) \nabla X(x)\right) \ \right). \tag{3.3}$$

If the function f (resp. the vector field X) is defined only on ∂E, then $\nabla_\tau f$ (resp. $\mathrm{div}_\tau X$) is defined by replacing ∇f (resp. ∇X) in (3.2) (resp. in (3.3)) by $\nabla \overline{f}$ (resp. $\nabla \overline{X}$), where $\overline{f}$ (resp. $\overline{X}$) is an arbitrary smooth extension of f (resp. of X) on a neighbourhood of ∂E.

$\Delta_\tau f$ is the Laplace-Beltrami operator on ∂E, i.e., $\Delta_\tau f := \mathrm{div}_\tau(\nabla_\tau f)$.

We will adopt the convention of implicit summation over repeated indices.

If $x \in \partial E$, we denote by $T_x(\partial E)$ (resp. $N_x(\partial E)$) the tangent (resp. the normal) space to ∂E at x.

We denote by $\kappa_1^E, \ldots, \kappa_{n-1}^E$ the principal curvatures of ∂E (positive for convex sets), and by $\mathbf{H}^E$ the mean curvature vector of ∂E, i.e.,

$$\mathbf{H}^E = -\sum_{i=1}^{n-1} \kappa_i^E \, \nu^E = -\mathrm{div}_\tau \nu^E \, \nu^E \qquad \text{on } \partial E.$$

With our conventions, $\mathbf{H}^E$ points inside E if ∂E is a sphere. We set

$$H^E := \sum_{i=1}^{n-1} \kappa_i^E. \tag{3.4}$$

Remark 3.1. If $E = \{u < 0\}$ and $\partial E = \{u = 0\}$, where $u : \mathbb{R}^n \to \mathbb{R}$ is a function which is smooth in a neighbourhood U of $\{u = 0\}$, and $u^2 + |\nabla u|^2 > 0$ in U, then

$$\mathbf{H}^E = -\mathrm{div}\left(\frac{\nabla u}{|\nabla u|}\right)\frac{\nabla u}{|\nabla u|} \tag{3.5}$$

$$= -\frac{1}{|\nabla u|}\mathrm{tr}\left(\left(\mathrm{Id} - \frac{\nabla u \otimes \nabla u}{|\nabla u|^2}\right)\nabla^2 u\right)\frac{\nabla u}{|\nabla u|} \qquad \text{on } \{u = 0\}.$$

Notice that the application $\mathrm{Id} - \frac{\nabla u \otimes \nabla u}{|\nabla u|^2}$ is, on $\{u = 0\}$, the orthogonal projection on the tangent space to $\{u = 0\}$. Notice also that multiplying u by a nonzero scalar factor does not change the set $\{u = 0\}$, and this is reflected by the zero homogeneity of the right hand side of (3.5) with respect to u. Using the rotation and translation invariance of the right hand side of (3.5), if we assume that $x = 0 \in \{u = 0\}$ and $\nabla u(0) = e_n$, then (3.5) becomes $\mathbf{H}^E(0) = -e_n \sum_{i=1}^{n-1} \nabla_{ii} u(0)$. If in addition $|\nabla u|^2 = 1$ in a neighbourhood of 0, then $\mathbf{H}^E(0) = -e_n \Delta u(0)$.

We recall the integration by parts formula on ∂E:

$$\int_{\partial E} \mathrm{div}_\tau X = -\int_{\partial E} \langle X, \mathbf{H}^E \rangle \qquad X \in C_c^1(\mathbb{R}^n; \mathbb{R}^n), \tag{3.6}$$

see for instance [72], [6]. In particular

$$\int_{\partial E} \phi \Delta_\tau \psi = \int_{\partial E} \psi \Delta_\tau \phi \qquad \phi, \psi \in C_c^2(\mathbb{R}^n). \tag{3.7}$$

One of the motivations for studying mean curvature flow is given by the following classical result, which is the computation of the first variation of area. As a consequence, mean curvature flow can be interpreted as the gradient flow of the area functional: the sense in which one has to take the gradient flow is made rigorous in the paper [1].

Theorem 3.1. *Let $\psi \in C^\infty(\mathbb{R}^{n+1}; \mathbb{R}^n)$ and set $\psi_\lambda(x) := \psi(\lambda, x)$ for $\lambda \in \mathbb{R}$ and $x \in \mathbb{R}^n$. Assume*

(i) $\psi_0 = \mathrm{Id}$ *on* $\mathbb{R}^n$;
(ii) $\psi_\lambda = \mathrm{Id}$ *out of a compact set of* $\mathbb{R}^n$, *for* $|\lambda|$ *small enough.*

Define $E_\lambda := \psi_\lambda(E)$. Then ∂E_λ is smooth for $|\lambda|$ small enough, and

$$\frac{d}{d\lambda}\mathcal{H}^{n-1}(\partial E_\lambda)_{|\lambda=0} = -\int_{\partial E} \langle X, \mathbf{H}^E \rangle, \tag{3.8}$$

where $X(\cdot) := \frac{\partial \psi_\lambda}{\partial \lambda}(\cdot)_{|\lambda=0}$.

3.3. The distance function

We briefly list the main properties of the distance function from a smooth compact boundary needed in the sequel. Recall that ∂E is smooth and that the function d^E in (3.1) is also denoted by d.

Theorem 3.2. *Assume that ∂E is compact. Then there exists $\rho > 0$ such that, setting $U := \{y \in \mathbb{R}^n : |d(y)| < \rho\}$, the following properties hold:*

(i) *$d \in C^\infty(U)$ and satisfies the eikonal equation*

$$|\nabla d|^2 = 1 \qquad \text{in } U. \tag{3.9}$$

(ii) *If $y \in U$, the point $\mathrm{pr}(y) := y - d(y)\nabla d(y)$ belongs to ∂E and is the unique solution of*

$$\min \{|x - y| : x \in \partial E\},$$

i.e. $\mathrm{pr}(y)$ is the orthogonal projection of y on ∂E. Moreover

$$\nabla d(y) = \nabla d(\mathrm{pr}(y)). \tag{3.10}$$

Proof. See [5], [51, Appendix B]. $\square$

Remark 3.2. Observe that

(i) from (3.9) it follows $\nabla d = \nu^E$ on ∂E and, by differentiation,

$$\nabla d \in \ker(\nabla^2 d) \qquad \text{on } U. \tag{3.11}$$

Hence ∇d is a unit zero eigenvector of $\nabla^2 d$, and therefore, given $x \in \partial E$, we can choose an orthonormal basis $\{v_1, \ldots, v_n\}$ of $\mathbb{R}^n$ which diagonalizes $\nabla^2 d(x)$ and such that $v_n = \nabla d(x)$.

(ii) From (3.11) it follows that

$$\mathrm{div}_\tau(\nabla d) = \mathrm{div}(\nabla d) = \Delta d \qquad \text{on } \partial E, \tag{3.12}$$

hence

$$\Delta d = H^E \quad \text{and} \quad -\Delta d \nabla d = \mathbf{H}^E \qquad \text{on } \partial E. \tag{3.13}$$

(iii) Differentiating twice identity (3.9) we get

$$\nabla_{ijk} d \nabla_k d = -\nabla_{jk} d \nabla_{ik} d \qquad \text{on } U, \tag{3.14}$$

for any $i, j, k \in \{1, \ldots, n\}$.

The following result describes the expansion of the eigenvalues of $\nabla^2 d$ on the whole of U.

Theorem 3.3. *Let E, U and d be as in Theorem 3.2. Let $y \in U$ and let $x := \mathrm{pr}(y)$ be the orthogonal projection of y on ∂E. Fix an orthonormal basis $\{v_1, \ldots, v_n\}$ of $\mathbb{R}^n$ in which $\nabla^2 d(x)$ is diagonal, such that $v_n = \nabla d(x)$ and*

$$\nabla^2 d(x)v_i = \kappa_i^E(x)v_i, \qquad i = 1, \ldots, n, \tag{3.15}$$

where $\kappa_n^E(x) := 0$. Then $v_n \in \mathrm{Ker}(\nabla^2 d(y))$, the basis $\{v_1, \ldots, v_n\}$ diagonalizes $\nabla^2 d(y)$, and if we denote by $\mu_i(y)$ the eigenvalue corresponding to v_i for $i = 1, \ldots, n$, then

$$\mu_i(y) = \frac{\kappa_i^E(\mathrm{pr}(y))}{1 + d(y)\kappa_i^E(\mathrm{pr}(y))}. \tag{3.16}$$

Proof. We follow [5, Theorem 3]. Define

$$B(s) := \nabla^2 d\left(x + s\nabla d(x)\right)$$

for $|s|$ small enough in such a way that $x + s\nabla d(x) \in U$. Observe that

$$B(0) = \sum_{l=1}^{n} \kappa_l^E(x)v_l \otimes v_l. \tag{3.17}$$

Fix $i, j \in \{1, \ldots, n\}$. Consider the (i,j)-th entry $B_{ij}(s)$ of $B(s)$. Then, using (3.14) we get

$$B_{ij}'(s) = \nabla_{ijk}d(x + s\nabla d(x))\nabla_k d(x) = -(B^2(s))_{ij}, \tag{3.18}$$

hence $B'(s) = -B^2(s)$. The solution of this system of ODEs with initial condition (3.17) is $B(s) = \sum_{l=1}^{n} \frac{\kappa_l^E(x)}{1+s\kappa_i^E(x)} v_l \otimes v_l$. $\square$

Remark 3.3. In the statement of Theorem 3.2 the number $\rho > 0$ is small enough in such a way that, in particular, $1 + d(y)\kappa_i^E(\mathrm{pr}(y)) > 0$ for any $y \in U$.

Remark 3.4. Let E and U be as in Theorem 3.3, and set $\eta := d^2/2$. Then $\eta \in C^\infty(U)$, $\eta = 0$ on ∂E, and

$$\nabla\eta = d\nabla d = 0 \qquad \text{on } \partial E. \tag{3.19}$$

Moreover, if $i, j \in \{1, \ldots, n\}$ we have $\nabla_{ij}\eta = \nabla_i d\nabla_j d + d\nabla_{ij}d$ on U, so that

$$\nabla^2\eta = \nabla d \otimes \nabla d \qquad \text{on } \partial E. \tag{3.20}$$

Hence $\nabla^2\eta(x)$ is the orthogonal projection on $N_x(\partial E)$ at $x \in \partial E$. Finally, if $i,j,k \in \{1,\ldots,n\}$ we have

$$\nabla_{ijk}\eta = \nabla_i d\nabla_{jk}d + \nabla_j d\nabla_{ik}d + \nabla_k d\nabla_{ij}d + d\nabla_{ijk}d \qquad (3.21)$$

on U. Therefore, recalling (3.11), we find

$$\Delta\nabla\eta = \Delta d\nabla d = -\mathbf{H}^E \qquad \text{on } \partial E. \qquad (3.22)$$

Remark 3.5. Observe that

(i) from (3.16) we obtain

$$\kappa_i^E(\mathrm{pr}(y)) = \frac{\mu_i(y)}{1 - d(y)\mu_i(y)} \qquad i = 1,\ldots,n. \qquad (3.23)$$

(ii) For any $i = 1,\ldots,n$ we have $\frac{\mu_i(y)}{1-d(y)\mu_i(y)} \leq \mu_i(y)$ in $U \cap \overline{E}$, hence

$$\sum_{i=1}^n \frac{\mu_i(y)}{1 - d(y)\mu_i(y)} \leq \sum_{i=1}^n \mu_i(y) \qquad \text{in } U \cap \{d \leq 0\}.$$

Similarly $\frac{\mu_i(y)}{1-d(y)\mu_i(y)} \geq \mu_i(y)$ in $U \cap (\mathbb{R}^n \setminus E)$, hence

$$\sum_{i=1}^n \frac{\mu_i(y)}{1 - d(y)\mu_i(y)} \geq \sum_{i=1}^n \mu_i(y) \qquad \text{in } U \cap \{d \geq 0\}.$$

3.4. Smooth mean curvature flows

Let us recall the definition of smooth flow and of normal velocity, see for instance [5, Section 5]. We begin by looking at the flow as a smooth family of smooth immersions of a given boundary.

Definition 3.1. Assume that ∂E is connected and compact. Let $T > 0$ and let $\Gamma(t)$ be a subset of $\mathbb{R}^n$ for any $t \in [0, T]$. We say that $(\Gamma(t))_{t\in[0,T]}$ is a smooth flow on $[0, T]$ starting from $\partial E = \partial E(0)$ if there exists a map $\phi : \partial E \times [0, T] \to \mathbb{R}^n$ of class C^∞ such that

(i) $\phi(y, 0) = y$ for any $y \in \partial E$;
(ii) $\phi(\cdot, t)$ is a bijection between ∂E and $\Gamma(t)$ for any $t \in [0, T]$;
(iii) the $(n-1)$-dimensional tangential Jacobian $J_y\phi(y,t)$ [5] is not zero for any $y \in \partial E$ and any $t \in [0, T]$.

Observe that $\Gamma(t)$ is a smooth, compact, connected hypersurface without boundary.

Definition 3.2. Let $y \in \partial E$, $t \in [0,T]$ and set $x := \phi(y,t) \in \Gamma(t)$. The normal velocity $\mathbf{V}(y,t)$ of $\Gamma(t)$ at x is defined as the orthogonal projection of $\frac{\partial \phi}{\partial t}(y,t)$ on the normal space $N_x(\Gamma(t))$ to $\Gamma(t)$ at x.

The normal velocity $\mathbf{V}(y,t)$ depends only on the set $\Gamma(t)$ and not on the way $\Gamma(t)$ is parameterized, since reparameterizations add only tangential components to the velocity.

Definition 3.3. If $E(t)$ is a family of subsets of $\mathbb{R}^n$ parameterized by $t \in [0,T]$ we let

$$d^{E(t)}(x) := \operatorname{dist}(x, E(t)) - \operatorname{dist}(x, \mathbb{R}^n \setminus E(t)), \qquad t \in [0,T], \ x \in \mathbb{R}^n \quad (3.24)$$

the signed distance function negative inside $E(t)$. Whenever no confusion is possible, the function $d^{E(t)}(x)$ will be denoted by $d(x,t)$.

The symbols ∇d and Δd stand for the gradient and the Laplacian of d with respect to x. By ∇_τ (resp. div_τ, Δ_τ) on $\partial E(t)$ we mean the tangential gradient (resp. the tangential divergence, the Laplace-Beltrami operator) on $\partial E(t)$.

The following result shows the connection between the normal velocity and the gradient of the function $d(x,t)$.

Theorem 3.4. *Let $(\partial E(t))_{[0,T]}$ be a smooth flow on $[0,T]$ starting from $\partial E = \partial E(0)$. Then there exists an open set A such that $A \supset \partial E(t)$ for any $t \in [0,T]$ and $d \in C^\infty(A \times [0,T])$, and*

$$\frac{\partial d}{\partial t}(x,t)\, \nabla d(x,t) = -\mathbf{V}(y,t), \qquad x := \phi(y,t) \in \partial E(t) \quad (3.25)$$

for any $y \in \partial E$ and any $t \in [0,T]$.

SKETCH OF PROOF. We prove only (3.25). We know that $d(\phi(y,t),t) = 0$ for any $y \in \partial E$ and any $t \in [0,T]$. Hence, differentiating with respect to t and setting $x := \phi(y,t)$, we get

$$\frac{\partial d}{\partial t}(x,t) + \left\langle \nabla d(x,t), \frac{\partial \phi}{\partial t}(y,t) \right\rangle = 0. \quad (3.26)$$

Then, using the definition of normal velocity and (3.26) we get

$$\mathbf{V}(y,t) = \left\langle \nabla d(x,t), \frac{\partial \phi}{\partial t}(y,t) \right\rangle \nabla d(x,t) = -\frac{\partial d}{\partial t}(x,t) \nabla d(x,t),$$

i.e. (3.25). $\square$

We are now in a position to define classical mean curvature flow of boundaries using the signed distance function d defined in (3.24).

Definition 3.4. We say that $(E(t))_{t\in[0,T]}$ is a smooth mean curvature flow on $[0,T]$ starting from $E = E(0)$ if

   (i) there exists an open set A such that $A \supset \partial E(t)$ for any $t \in [0,T]$ and $d \in C^\infty(A \times [0,T])$;

 (ii) we have

$$\frac{\partial d}{\partial t}(x,t)\nabla d(x,t) = \Delta d(x,t)\nabla d(x,t), \qquad t \in [0,T],\ x \in \partial E(t).$$

$$(3.27)$$

Sometimes we will also say that $(\partial E(t))_{t\in[0,T]}$ is a smooth mean curvature flow on $[0,T]$ starting from $\partial E = \partial E(0)$. Note that (3.27) means that $\mathbf{V}(y,t) = \mathbf{H}^{E(t)}(x)$, for $x := \phi(y,t) \in \partial E(t)$.

Remark 3.6. If ∂E is compact, then the open set A in Definition 3.4 can be taken as a suitable tubular neighbourhood of ∂E.

Remark 3.7. The system in (3.27) is equivalent to

$$\frac{\partial d}{\partial t}(x,t) = \Delta d(x,t), \qquad t \in [0,T],\ x \in \partial E(t) \tag{3.28}$$

which, in turn, is equivalent to the system

$$\begin{cases} \dfrac{\partial d}{\partial t} = \Delta d, \\[2mm] d(\cdot,t) = 0, \end{cases} \qquad t \in [0,T]. \tag{3.29}$$

Remark 3.8. If we define $\eta(x,t) := \frac{1}{2}d(x,t)^2$, then recalling (3.22), we have that (3.27) can be written also as

$$\frac{\partial \nabla \eta}{\partial t}(x,t) = \Delta \nabla \eta(x,t), \qquad t \in [0,T],\ x \in \partial E(t). \tag{3.30}$$

Example 3.1. Let $R_0 > 0$; the mean curvature flow starting from the sphere $B_{R_0}(0)$ is the sphere $B_{R(t)}(0)$, where

$$R(t) = \sqrt{R_0^2 - 2(n-1)t}, \qquad t \in [0,t^\dagger[, \qquad t^\dagger := \frac{R_0^2}{2(n-1)}.$$

Indeed $d^{B_{R(t)}(x_0)}(x) = d(x,t) = |x| - R(t)$, hence $\frac{\partial d}{\partial t}(x,t) = -\dot{R}(t)$, and

$$\nabla d(x,t) = \frac{x}{|x|}, \quad \nabla^2 d(x,t) = \frac{1}{|x|}\left(\mathrm{Id} - \frac{x}{|x|} \otimes \frac{x}{|x|}\right), \quad \Delta d(x,t) = \frac{n-1}{|x|}.$$

Hence (3.28) becomes

$$\dot{R}(t) = -\frac{n-1}{R(t)}.$$

Coupled with $R(0) = R_0$, the solution is $R(t) = \sqrt{R_0^2 - 2(n-1)t}$.

Example 3.2. Assume that $E(t) = \{x \in \mathbb{R}^n : u(x,t) < 0\}$ and $\partial E(t) = \{x \in \mathbb{R}^n : u(x,t) = 0\}$, where $u : \mathbb{R}^n \times [0,T] \to \mathbb{R}$ is a function which is smooth in $A \times [0,T]$, where A is an open set containing $\{u(\cdot,t) = 0\}$ for any $t \in [0,T]$, and such that $u^2 + |\nabla u|^2 > 0$ in $A \times [0,T]$. Then, letting $u_t := \frac{\partial u}{\partial t}$, (3.27) can be written as

$$\frac{u_t}{|\nabla u|}\frac{\nabla u}{|\nabla u|} = \mathrm{div}\left(\frac{\nabla u}{|\nabla u|}\right)\frac{\nabla u}{|\nabla u|} \qquad \text{on } \{u(\cdot,t) = 0\}, \qquad (3.31)$$

which is usually rewritten in the equivalent scalar form as

$$u_t = |\nabla u|\mathrm{div}\left(\frac{\nabla u}{|\nabla u|}\right) = \Delta u - \frac{\nabla_i u \nabla_j u \nabla_{ij} u}{|\nabla u|^2} \qquad \text{on } \{u(\cdot,t) = 0\}. \quad (3.32)$$

Indeed, since $u(\phi(y,t),t) = 0$, differentiating with respect to t yields

$$\frac{\partial u}{\partial t}(x,t) + \langle \nabla u(x,t), \frac{\partial \phi}{\partial t}(y,t) \rangle = 0, \qquad x := \phi(y,t).$$

Hence the normal velocity $\mathbf{V}(y,t)$ of $x \in \{u(\cdot,t) = 0\}$ is given by $-\frac{u_t(x,t)}{|\nabla u(x,t)|}\frac{\nabla u(x,t)}{|\nabla u(x,t)|}$, so that (3.31) follows from (3.5).

Remark 3.9. If $|\nabla u|^2 = 1$ in a neighbourhood of $\{(x,t) : u(x,t) = 0\}$ then problem (3.32) reduces to

$$\begin{cases} \dfrac{\partial u}{\partial t} = \Delta u, \\ u(\cdot,t) = 0, \end{cases} \qquad (3.33)$$

i.e., (3.29).

The partial differential equation

$$u_t = |\nabla u|\mathrm{div}\left(\frac{\nabla u}{|\nabla u|}\right) \qquad (3.34)$$

is fully nonlinear parabolic, and is degenerate where $|\nabla u| = 0$. It has been extensively studied, in particular in the framewok of viscosity solutions, see for instance [40], [42], [43], [30], [49]. If u is a solution to (3.34) which is smooth in a space time region around one of its level sets $\{u(\cdot, t) = \lambda\}$, then this level set flows smoothly by mean curvature.

Example 3.3. Assume that $E(t) = \{(x', x_n) \in \mathbb{R}^{n-1} \times \mathbb{R} : x_n > f(x', t)\}$ for a smooth function $f : \mathbb{R}^{n-1} \times [0, T] \to \mathbb{R}$. Then $\partial E(t) = \{(x', x_n) : u(x', x_n, t) := f(x', t) - x_n = 0\}$, hence by (3.5) it follows

$$\mathbf{H}^E = -\operatorname{div}\left(\frac{\nabla f}{\sqrt{1 + |\nabla f|^2}}\right)\frac{(\nabla f, -1)}{\sqrt{1 + |\nabla f|^2}} \tag{3.35}$$

Since the flow can be parameterized as $(x', t) \to (x', f(x', t))$, we have $\frac{\partial \phi}{\partial t} = (0, \frac{\partial f}{\partial t})$, which is in the "vertical" direction e_n and not in normal direction. The normal velocity is then

$$\mathbf{V} = \langle \frac{\partial \phi}{\partial t}, \nu^{E(t)} \rangle \nu^{E(t)} = \frac{f_t}{1 + |\nabla f|^2}(-\nabla f, 1). \tag{3.36}$$

The mean curvature flow of the graph of f therefore reads as

$$f_t = \sqrt{1 + |\nabla f|^2} \operatorname{div}\left(\frac{\nabla f}{\sqrt{1 + |\nabla f|^2}}\right) \tag{3.37}$$

on $\mathbb{R}^{n-1} \times [0, T]$.

Equation (3.37) is a quasilinear parabolic equation, and has been studied in [35], [36].

Remark 3.10. As a consequence of (3.8), (3.25) and (3.27), we have

$$\frac{d}{dt}\mathcal{H}^{n-1}(\partial E(t)) = -\int_{\partial E(t)} (\Delta d)^2, \tag{3.38}$$

which shows how the area of $\partial E(t)$ is decreasing along the flow.

Example 3.4. Assume that $\partial E \subset \mathbb{R}^3$ is the rotation around the x-axis of the graph of a smooth function $f = f(x) : \mathbb{R} \to]0, +\infty[$, precisely $E = \{(x, y, z) \in \mathbb{R}^3 : y^2 + z^2 < f^2(x)\}$. Define $u(x, y, z) := y^2 + z^2 - f^2(x)$, for $(x, y, z) \in \mathbb{R}^3$. Then $\{u = 0\} = \partial E$ and $\{u < 0\} = E$, and $\nabla u = 2(-f(x)f'(x), y, z)$. Hence a direct computation gives

$$H^E = \operatorname{div}\left(\frac{\nabla u}{|\nabla u|}\right) = -\frac{f''}{(1 + f'^2)^{3/2}} + \frac{1}{f(1 + f'^2)^{1/2}} \quad \text{on } \partial E. \tag{3.39}$$

3.5. Short time existence

Short time existence and uniqueness of the solution to (3.28), starting from a compact smooth initial datum $\partial E(0)$, is a consequence of a general theorem proved in [46]. A proof using the signed distance function was originally given in [41] (see also [66]). In order to illustrate briefly the approach of [41] we need the following lemma.

Lemma 3.1. *Let d, A and T be as in Definition 3.4. Let $z \in A$ and $t \in [0,T]$. Let $x \in \partial E(t)$ be the projection point on $\partial E(t)$ of $z \in A$. Then*

$$\frac{\partial d}{\partial t}(x,t) = \frac{\partial d}{\partial t}(w,t), \tag{3.40}$$

for any w belonging to the closed segment connecting x with z.

Proof. We follow [5]. Let $s \in \mathbb{R}$ be such that $|s|$ is small enough, in such a way that $y^s(x,t) := x + s\nabla d(x,t) \in A \times [0,T]$. Using (3.9) we have

$$|\nabla d(y^s(x,t),t)|^2 = 1. \tag{3.41}$$

Differentiating (3.41) with respect to t yields

$$0 = \langle \nabla \frac{\partial d}{\partial t}(y^s(x,t),t), \nabla d(y^s(x,t),t) \rangle \tag{3.42}$$
$$+ s\langle \nabla^2 d(y^s(x,t),t)\nabla \frac{\partial d}{\partial t}(y^s(x,t),t), \nabla d(y^s(x,t),t) \rangle.$$

Using (3.11) we have $\langle \nabla^2 d\nabla \frac{\partial d}{\partial t}, \nabla d \rangle = \langle \nabla \frac{\partial d}{\partial t}, \nabla^2 d\nabla d \rangle = 0$ at the point $(y^s(x,t),t)$. Therefore, from (3.42) we deduce

$$0 = \langle \nabla \frac{\partial d}{\partial t}(y^s(x,t),t), \nabla d(y^s(x,t),t) \rangle. \tag{3.43}$$

Define now

$$b(s) := \frac{\partial d}{\partial t}(y^s(x,t),t).$$

By (3.10) and (3.43) we deduce

$$b'(s) = \langle \nabla \frac{\partial d}{\partial t}(y^s(x,t),t), \nabla d(x,t) \rangle$$
$$= \langle \nabla \frac{\partial d}{\partial t}(y^s(x,t),t), \nabla d(y^s(x,t),t) \rangle = 0.$$

Hence b is constant, and this concludes the proof. $\square$

In view of Lemma 3.1 and (3.23), we have that (3.28) is equivalent to a single equation on the *fixed* set $A \times [0, T]$, which reads as

$$\frac{\partial d}{\partial t} = \sum_{i=1}^{n} \frac{\mu_i}{1 - d\mu_i} \qquad \text{in } A \times [0, T]. \tag{3.44}$$

Given a smooth set $E = E(0)$ with compact boundary, the idea of [41] is therefore to prove local existence and uniqueness of the solution v (in appropriate functional spaces) to the problem

$$\begin{cases} \dfrac{\partial v}{\partial t} := F(\nabla^2 v, v) & \text{in } A \times \,]0, T[, \\ v(x, 0) = d^E(x) & x \in A, \\ |\nabla v|^2 = 1 & \text{on } \partial A \times [0, T], \end{cases} \tag{3.45}$$

where $F(R, z) := \text{tr}(R(\text{Id} - zR)^{-1})$ for a symmetric matrix R and a vector z, and to show that v satisfies the eikonal equation $|\nabla v|^2 = 1$ in the whole of $A \times [0, T]$. Once this is proved, the authors recover the (unique local) classical mean curvature flow starting from E as $\partial E(t) := \{x \in A : v(x, t) = 0\}$ for any $t \in [0, T]$.

Remark 3.11. Thanks to (3.44) and (ii) of Remark 3.5, we have

$$\begin{cases} \dfrac{\partial d}{\partial t} - \Delta d \leq 0 & \text{in } A \cap \{d \leq 0\}, \\[4mm] \dfrac{\partial d}{\partial t} - \Delta d \geq 0 & \text{in } A \cap \{d \geq 0\}. \end{cases} \tag{3.46}$$

Inequalities (3.46) are useful when studying mean curvature flow with the level set method (using viscosity solutions) and with the reaction-diffusion equations, see for instance [39], [76] and references therein.

Let us show how to find the evolution law of the normal vector field [55] using the signed distance function.

Lemma 3.2. *Assume that $\partial E = \partial E(0)$ is compact. Let $(E(t))_{t \in [0,T]}$ be the smooth mean curvature flow on $[0, T]$ starting from E. Then*

$$\frac{\partial}{\partial t} \nabla d = \nabla_\tau \Delta d \qquad \text{on } \partial E(t). \tag{3.47}$$

Proof. Let A be the open set in Definition 3.4. Define $G(x, t) := (\text{Id} - d(x, t)\nabla^2 d(x, t))^{-1}$ for $(x, t) \in \mathbb{R}^n \times [0, T]$. Recalling (3.44), we have

$$\frac{\partial d}{\partial t} = \text{tr}(G \, \nabla^2 d) \qquad \text{in } A \times [0, T]. \tag{3.48}$$

Let $G_{ij}(x,t)$ be the (i,j)-th entry of $G(x,t)$ and let $k \in \{1,\dots,n\}$. Observe that

$$G_{ij} = \delta_{ij} + d\nabla_{ij}d + o(d), \qquad \nabla_k G_{ij} = \nabla_k d\nabla_{ij}d + O(d), \qquad (3.49)$$

in $A \times [0,T]$. Differentiating (3.48) with respect to x_k and using (3.49) yields, in $A \times [0,T]$,

$$\begin{aligned}
\nabla_k \frac{\partial d}{\partial t} &= \frac{\partial}{\partial t}\nabla_k d = \nabla_k(G_{ij}\nabla_{ji}d) = \nabla_k G_{ij}\nabla_{ji}d + G_{ij}\nabla_{ijk}d \\
&= \nabla_{ijk}d\,\delta_{ij} + \nabla_k d\nabla_{ij}d\nabla_{ji}d + O(d) = \nabla_{iik}d + \nabla_k d|\nabla^2 d|^2 + O(d) \\
&= \nabla_k\Delta d + \nabla_k d\,|\nabla^2 d|^2 + O(d),
\end{aligned}$$

where $|\nabla^2 d|^2 := \mathrm{tr}(\nabla^2 d\nabla^2 d)$ is the square of the length of the second fundamental form. Observe that (3.14) (applied with $i = j$) implies $-|\nabla^2 d|^2 = \langle \nabla d, \nabla\Delta d\rangle$. Hence

$$\nabla\frac{\partial d}{\partial t} = \nabla\Delta d - \langle\nabla\Delta d, \nabla d\rangle\nabla d + O(d)$$

in $A \times [0,T]$, and (3.47) follows. $\square$

Remark 3.12. In general, the evolution equations satisfied by geometric quantities, such as the mean curvature or the square of the second fundamental form, is crucial to derive informations on the flow. See for instance the papers [46], [52] for the evolution of curves, the paper [55] for evolution of boundaries of convex sets.

Curvature flow of curves, called also curves-shortening, was studied in [45], [46] under convexity assumptions, and in [52] in the general case. See also the papers [11], [12], [13] and references therein. In [52] Grayson was able to prove a global result, i.e., if $\partial E(0)$ is compact then the curvature flow $(\partial E(t))_{t\in[0,t\dagger[}$ starting from $\partial E(0)$ is smooth on a maximal time interval $[0,t^\dagger[$, the curves $\partial E(t)$ eventually become convex and contract, as $t \uparrow t^\dagger$, to a point. The time $t^\dagger$ is called the extinction time. Another global result in higher dimension was proved in [55]: if $E(0) \subset \mathbb{R}^n$ is bounded and convex, then the mean curvature flow $(\partial E(t))_{t\in[0,t\dagger[}$ starting from $\partial E(0)$ is smooth on a maximal time interval $[0,t^\dagger[$, the sets $E(t)$ are strictly convex for $t \in]0,t^\dagger[$, and converges to a point as $t \uparrow t^\dagger$. Moreover, when appropriately rescaled, the hypersurfaces $\partial E(t)$ converge to a sphere as $t \uparrow t^\dagger$. Other global results can be proved when ∂E is the graph of a function f satisfying suitable properties [35] (see also [17]): the mean curvature flow of graphs will be considered in Section 3.9.

3.6. Special solutions to mean curvature flow

Beside the shrinking sphere (Example 3.1) there are other interesting examples of classical mean curvature flow, see also [34].

Example 3.5. Let $h \in \{1, \ldots, n-1\}$ and $R_0 > 0$; the mean curvature evolution of the cylinder $\{(z,w) \in \mathbb{R}^{n-h} \times \mathbb{R}^h : |z| < R_0\}$ is the cylinder $\{(z,w) \in \mathbb{R}^{n-h} \times \mathbb{R}^h : |z| < R(t)\}$, where $R(t) = \sqrt{R_0^2 - 2(n-h-1)t}$, $t \in [0, R_0^2/(2(n-h-1)))[$.

Example 3.6. Assume that $H^E \equiv 0$. Then ∂E is a minimal surface, hence is a stationary solution to (3.28).

Let us define self-similar contracting solutions to mean curvature flow.

Definition 3.5. We say that E gives raise to a self-similar contracting solution to mean curvature flow if there exists $T > 0$ such that

$$E(t) = \alpha(t)E, \qquad t \in [0, T[, \qquad \alpha(t) := \sqrt{1 - \frac{t}{T}}, \qquad (3.50)$$

where $(E(t))_{t \in [0,T[}$ is the mean curvature flow starting from $E = E(0)$.

Sets E giving raise to self-similar contracting solutions have a boundary that must satisfy a suitable elliptic partial differential equation. More precisely, the following result holds.

Lemma 3.3. *The set E gives raise to a self-similar contracting solution to mean curvature flow if and only if*

$$\Delta d(x) = \frac{1}{2T}\langle \nabla d(x), x \rangle, \qquad x \in \partial E. \qquad (3.51)$$

Proof. Assume that E gives raise to a self-similar contracting solution to mean curvature flow. Let $x \in \mathbb{R}^n$. We have

$$\text{dist}\,(x, \alpha(t)E) = \inf_{y \in \alpha(t)E} |y - x| = \alpha(t) \inf_{y/\alpha(t) \in E} |y/\alpha(t) - x/\alpha(t)|$$

$$= \alpha(t)\text{dist}(x/\alpha(t), E).$$

Hence, if $d^{E(t)}(x)$ (resp. d) is the function defined in (3.24) (resp. in (3.1)), we have $d^{E(t)}(x) = \alpha(t)d(x/\alpha(t))$, so that $\nabla d^{E(t)}(x) = \nabla d(x/\alpha(t))$,

$$\Delta d^{E(t)}(x) = \frac{1}{\alpha(t)}\Delta d(x/\alpha(t)), \qquad (3.52)$$

$$\frac{\partial}{\partial t}d^{E(t)}(x) = \alpha'(t)d(x/\alpha(t)) - \frac{\alpha'(t)}{\alpha(t)}\langle \nabla d(x/\alpha(t)), x \rangle. \qquad (3.53)$$

Since $\partial E(t) = \{x : d^{E(t)}(x) = 0\} = \{x : d(x/\alpha(t)) = 0\}$, from (3.53) we deduce

$$\frac{\partial}{\partial t} d^{E(t)}(x) = -\alpha'(t)\langle \nabla d(x/\alpha(t)), x/\alpha(t)\rangle, \qquad x \in \partial E(t). \qquad (3.54)$$

Using (3.52), (3.54) and $\partial E(t) = \alpha(t)\partial E$, the equation (3.28) expressing mean curvature flow of $\partial E(t)$ becomes an equation for the function d on ∂E which reads as

$$-\alpha'(t)\langle \nabla d(x/\alpha(t)), x/\alpha(t)\rangle = \frac{1}{\alpha(t)}\Delta d(x/\alpha(t)), \qquad x/\alpha(t) \in \partial E,$$

i.e.,

$$\Delta d(z) = -\alpha'(t)\alpha(t)\langle \nabla d(z), z\rangle, \qquad z \in \partial E. \qquad (3.55)$$

Since $\alpha'(t)\alpha(t) = -\frac{1}{2T}$, equation (3.51) follows.

Conversely, if there exists $T > 0$ such that equation (3.51) holds, the family $(E(t))_{t\in[0,T[}$ defined in (3.50) is the mean curvature flow in $[0, T[$ starting from E. $\square$

An interesting problem is to classify all solutions ∂E to (3.51), see [57] for results in this direction.

Remark 3.13. Observe that (3.51) is the Euler-Lagrange equation of the functional

$$E \to \int_{\partial E} e^{-\frac{|x|^2}{4T}}.$$

Indeed, using the notation of Theorem 3.1, recalling (3.6) and setting $a(x) := e^{-\frac{|x|^2}{4T}}$, we have (see also formula (3.83) below)

$$\frac{d}{d\lambda}\int_{\partial E_\lambda} e^{-\frac{|x|^2}{4T}}\Big|_{\lambda=0} = \int_{\partial E}\Big(\langle X, \nabla d\rangle\Delta d\, a + \langle X, \nabla a - \nabla_\tau a\rangle\Big). \qquad (3.56)$$

Since

$$\nabla a - \nabla_\tau a = \langle \nabla a, \nabla d\rangle\nabla d = -\frac{1}{2T}e^{-\frac{|x|^2}{4T}}\langle x, \nabla d\rangle\nabla d \qquad \text{on } \partial E,$$

from (3.56) we deduce

$$\frac{d}{d\lambda}\int_{\partial E_\lambda} e^{-\frac{|x|^2}{4T}}\Big|_{\lambda=0} = \int_{\partial E}\langle X, \nabla d\rangle\left(\Delta d - \frac{1}{2T}\langle x, \nabla d\rangle\right)e^{-\frac{|x|^2}{4T}}.$$

Example 3.7. We look for a special solution to (3.37) of the form $f(x,t) = g(x) + t$, for some real valued function g to be determined. The function g must satisfy

$$\sqrt{1 + |\nabla g|^2} \, \operatorname{div} \left(\frac{\nabla g}{\sqrt{1 + |\nabla g|^2}} \right) = 1. \tag{3.57}$$

Assume $n = 2$. Then (3.57) reads as

$$\frac{g_{xx}}{1 + g_x^2} = (\operatorname{arctg}(g_x))_x = 1. \tag{3.58}$$

A solution to (3.58) is given by $g(x) = -\log(\cos x)$ for $x \in \left] -\frac{\pi}{2}, \frac{\pi}{2} \right[$. The corresponding solution $f(x,t) = -\log(\cos x) + t$, for $x \in \left] -\frac{\pi}{2}, \frac{\pi}{2} \right[$ and $t \in [0, +\infty[$ (called grim reaper) is said to be a translating solution to curvature flow.

3.7. The comparison principle between smooth compact flows

Let us recall the following form of the maximum principle.

Lemma 3.4. *Let A and B be two open sets with smooth boundary, and assume that there exist $x \in \mathbb{R}^n$ and $\rho > 0$ with the following properties:*

$$x \in \partial A \cap \partial B, \qquad A \cap B_\rho(x) \subseteq B \cap B_\rho(x).$$

Then $H^A(x) \geq H^B(x)$.

Proof. Since the mean curvature is rotationally invariant, we can assume that $\nu^A(x) = \nu^B(x) = -e_n$, x is the origin of the coordinates, $\partial A \cap B_\rho(x) = \operatorname{graph}(f)$, $\partial B \cap B_\rho(x) = \operatorname{graph}(g)$, where f and g are two smooth functions defined on an open set of $\mathbb{R}^{n-1} = \operatorname{span}\{e_1, \ldots, e_{n-1}\}$ such that $f \geq g$ locally around 0. Then $f - g$ has a local minimum at 0, so that $\nabla f(0) = \nabla g(0) = 0$ and $\Delta f(0) \geq \Delta g(0)$, see Figure 3.1. Then

$$H^A(x) = \Delta f(0) \geq \Delta g(0) = H^B(x). \square$$

The maximum principle is at the basis of the comparison principle for smooth compact mean curvature flows, which reads as follows.

Theorem 3.5. *Let $E = E(0)$ and $F = F(0)$ be two open sets with smooth compact boundary, and assume that*

$$E(0) \subseteq F(0).$$

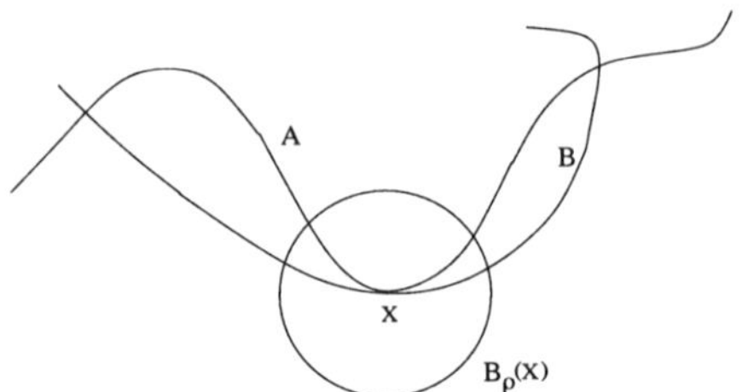

Figure 3.1. Locally around the point $x \in \partial A \cap \partial B$, the set A is contained in the set B. The tangent space to ∂A at x equals the tangent space to ∂B at x, and is horizontal. The mean curvature of ∂A at x is greater than or equal to the mean curvature of ∂B at x.

Let $(E(t))_{t \in [0,T]}$ and $(F(t))_{t \in [0,T]}$ be the two smooth mean curvature flows in a common time interval $[0, T]$ starting from $E(0)$ and $F(0)$ respectively. Then the function

$$\delta(t) := \operatorname{dist}(E(t), \mathbb{R}^n \setminus F(t)), \qquad t \in [0, T], \tag{3.59}$$

is nondecreasing. In particular

$$E(t) \subseteq F(t), \qquad t \in [0, T]. \tag{3.60}$$

SKETCH OF PROOF. Let $M := \partial E(0) \times \partial F(0)$. We divide the proof into three steps.

Step 1. Let $u : M \times [0, T] \to \mathbb{R}$ be a smooth function. Define, for $t \in [0, T]$,

$$m(t) := \min_{x \in M} u(x, t), \quad C(t) := \{x \in M : u(x, t) = m(t)\}. \tag{3.61}$$

Then for any $t \in [0, T[$ there exists $\lim_{\tau \to 0+} \frac{1}{\tau}(m(t + \tau) - m(t))$ and

$$\lim_{\tau \to 0+} \frac{1}{\tau}(m(t + \tau) - m(t)) = \min \left\{ \frac{\partial u}{\partial t}(x, t) : x \in C(t) \right\}.$$

For any $t \in [0, T[$, $x \in C(t)$, $\tau > 0$ small enough, we have

$$m(t + \tau) \leq u(x, t + \tau) = u(x, t) + \tau \frac{\partial u}{\partial t}(x, t) + o(\tau)$$

$$= m(t) + \tau \frac{\partial u}{\partial t}(x, t) + o(\tau).$$

Therefore $\limsup_{\tau \to 0+} \frac{1}{\tau}(m(t + \tau) - m(t)) \leq \frac{\partial u}{\partial t}(x, t)$, which implies

$$\limsup_{\tau \to 0+} \frac{1}{\tau}(m(t + \tau) - m(t)) \leq \min_{x \in C(t)} \frac{\partial u}{\partial t}(x, t). \tag{3.62}$$

Fix $\epsilon > 0$ and define $C_\epsilon(t) := \{x \in M : u(x,t) < m(t) + \epsilon\}$. Then $C_\epsilon(t)$ is a neighbourhood of $C(t)$. Let $L := \sup_{t \in]0,T[} \|\frac{\partial u}{\partial t}\|_{L^\infty(M)}$. For any $t \in [0,T[$, $x \in C_\epsilon(t)$ and $\tau > 0$ small enough, we have

$$u(x,t+\tau) = u(x,t) + \tau \frac{\partial u}{\partial t}(x,t) + o(\tau)$$

$$\geq m(t) + \tau \inf_{y \in C_\epsilon(t)} \frac{\partial u}{\partial t}(y,t) + o(\tau), \qquad (3.63)$$

while if $x \in M \setminus C_\epsilon(t)$ we have

$$u(x,t+\tau) \geq m(t) + \epsilon + \tau \frac{\partial u}{\partial t}(x,t) + o(\tau)$$

$$\geq m(t) + \epsilon - \tau L + o(\tau). \qquad (3.64)$$

For $\tau < \frac{\epsilon}{3L}$ the right hand side of (3.63) is smaller than the right hand side of (3.64). We deduce

$$u(x,t+\tau) \geq m(t) + \tau \inf_{y \in C_\epsilon(t)} \frac{\partial u}{\partial t}(y,t) + o(\tau)$$

for any $x \in M$ and $\tau < \frac{\epsilon}{3L}$. It follows

$$m(t+\tau) \geq m(t) + \tau \inf_{y \in C_\epsilon(t)} \frac{\partial u}{\partial t}(y,t) + o(\tau), \qquad \tau < \frac{\epsilon}{3L}.$$

Hence

$$\liminf_{\tau \to 0^+} \frac{1}{\tau} (m(t+\tau) - m(t)) \geq \inf_{y \in C_\epsilon(t)} \frac{\partial u}{\partial t}(y,t),$$

which implies

$$\liminf_{\tau \to 0^+} \frac{1}{\tau} (m(t+\tau) - m(t)) \geq \sup_\epsilon \inf_{y \in C_\epsilon(t)} \frac{\partial u}{\partial t}(y,t) = \min_{x \in C(t)} \frac{\partial u}{\partial t}(x,t),$$

where the last equality follows from the continuity of the function $\frac{\partial u}{\partial t}(\cdot,t)$. The proof of *step 1* is concluded.

Since $E(t)$ (resp. $F(t)$) is a smooth flow on $[0,T]$, there exists a map $\mathcal{E} : \partial E \times [0,T] \to \mathbb{R}^n$ (resp. $\mathcal{F} : \partial F \times [0,T] \to \mathbb{R}^n$) having the properties (i)-(iii) of the map ϕ in Definition 3.1. Observe that

$$\delta(t) = \min \{u(s,\sigma,t) : (s,\sigma) \in M\}, \qquad t \in [0,T], \qquad (3.65)$$

where $u : M \to [0,+\infty[$ is defined as

$$u(s,\sigma,t) := |\mathcal{E}(s,t) - \mathcal{F}(\sigma,t)|. \qquad (3.66)$$

Let $0 \leq t_1 \leq t_2 < T$. We have to prove that $\delta(t_1) \leq \delta(t_2)$. Assume for simplicity $t_1 = 0$. Assume also

$$\delta(0) > 0. \tag{3.67}$$

Thanks to the smoothness of the flows, there exists $a \in \,]0, T[$ such that $\delta(t) > 0$ in $[0, a]$. In particular, the function u defined in (3.66) is smooth on $M \times [0, a]$.

Step 2. We have

$$\lim_{\tau \to 0^+} \frac{1}{\tau}(\delta(t + \tau) - \delta(t)) \geq 0 \qquad t \in [0, a[. \tag{3.68}$$

From *step 1*, applied with $m := \delta$ and $T := a$, we get

$$\lim_{\tau \to 0^+} \frac{\delta(t + \tau) - \delta(t)}{\tau} \tag{3.69}$$

$$= \min_{(s, \sigma) \in M} \left\{ \frac{\partial u}{\partial t}(s, \sigma, t) : u(s, \sigma, t) = \delta(t) \right\}, \qquad t \in [0, a[.$$

At fixed $t \in [0, a[$, let $(\overline{s}, \overline{\sigma}) \in M$ be two minimizing parameters for the right hand side of (3.69). Set $z := \mathcal{E}(\overline{s}, t) \in \partial E(t)$ and $w := \mathcal{F}(\overline{\sigma}, t) \in \partial F(t)$, see Figure 3.2.

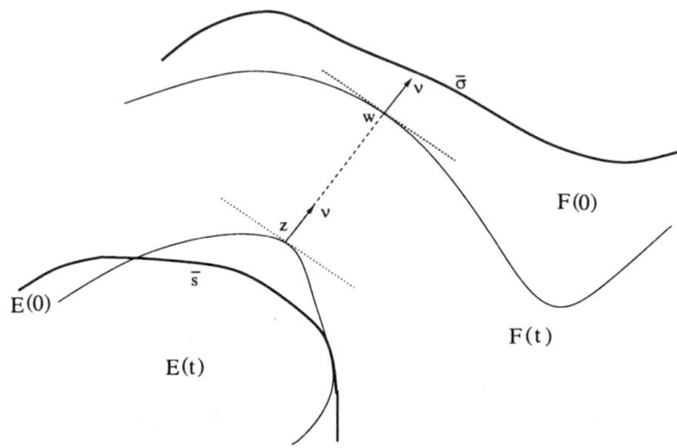

Figure 3.2.   The bold curves denote $\partial E(0)$ and $\partial F(0)$. The outward unit normals to $\partial E(t)$ at z and to $\partial F(t)$ at w are parallel (and denoted by ν), since $|z - w| = \delta(t)$.

Then

$$\frac{\partial u}{\partial t}(\overline{s}, \overline{\sigma}, t) = \langle \frac{w - z}{|w - z|}, \frac{\partial \mathcal{F}}{\partial t}(\overline{\sigma}, t) - \frac{\partial \mathcal{E}}{\partial t}(\overline{s}, t) \rangle. \tag{3.70}$$

As $\delta(t) = |w - z| > 0$, we have

$$\frac{w - z}{|w - z|} = \nu^{E(t)}(z) = \nu^{F(t)}(w) =: \nu. \tag{3.71}$$

Observe also that

$$\langle \nu, \frac{\partial \mathcal{F}(\overline{\sigma}, t)}{\partial t} \rangle \nu = \mathbf{V}^{\mathcal{F}}, \qquad \langle \nu, \frac{\partial \mathcal{E}(\overline{s}, t)}{\partial t} \rangle \nu = \mathbf{V}^{\mathcal{E}}, \tag{3.72}$$

where $\mathbf{V}^{\mathcal{F}}$ (resp. $\mathbf{V}^{\mathcal{E}}$) is the normal velocity of $\partial F(t)$ at w (resp. of $\partial E(t)$ at z), and

$$\langle \nu, \frac{\partial \mathcal{F}(\overline{\sigma}, t)}{\partial t} \rangle = -H^{F(t)}(w), \qquad \langle \nu, \frac{\partial \mathcal{E}(\overline{s}, t)}{\partial t} \rangle = -H^{E(t)}(z). \tag{3.73}$$

From (3.70), (3.71) and (3.73) we get

$$\lim_{\tau \to 0^+} \frac{1}{\tau}(\delta(t + \tau) - \delta(t)) = -H^{F(t)}(w) + H^{E(t)}(z). \tag{3.74}$$

Observe now that, being $\delta(t) = |w - z|$, if we consider the translated set $E_{\mathrm{tr}}(t) := E(t) + \delta(t)\nu$, then $E_{\mathrm{tr}}(t) \subseteq F(t)$, and $w \in \partial(E_{\mathrm{tr}}(t)) \cap \partial F(t)$. Hence, by Lemma 3.4, $H^{E_{\mathrm{tr}}(t)}(w) \geq H^{F(t)}(w)$. Since $H^{E_{\mathrm{tr}}(t)}(w) = H^{E(t)}(z)$, from (3.74) we then get (3.68). The proof of *step 2* is concluded.

Step 3. Assume (3.67). Then the function δ is nondecreasing in $[0, a]$. Assume by contradiction that there exist $0 \leq t_1 < t_2 \leq a$ such that $\delta(t_2) < \delta(t_1)$. Assume for simplicity that $t_1 = 0$. Let $P(s)$ be a linear decreasing function such that $P(0) = \delta(0)$ and $P(t_2) > \delta(t_2)$. Let

$$t^* := \sup\{t \in [0, a] : \delta(t) \leq P(t)\}.$$

Then $P(t^*) = \delta(t^*)$, $t^* < a$, and

$$\lim_{\tau \to 0^+} \frac{1}{\tau}(\delta(t^* + \tau) - \delta(t^*)) < P'(t^*) < 0,$$

a contradiction with *step 2*.

Once we assume (3.67), we get that δ is nondecreasing in $[0, a]$, hence the argument can be repeated also for times larger than a. Removing assumption (3.67) requires an approximation argument which is omitted. $\square$

The following theorem is proved in [41].

Theorem 3.6. *Let $(E(t))_{t \in [0,T]}$ and $(F(t))_{t \in [0,T]}$ be two smooth mean curvature flows in $[0, T]$ starting from $E(0)$ and $F(0)$ respectively. Assume $\partial E(0)$ and $\partial F(0)$ connected, and*

$$E(0) \subseteq F(0), \qquad E(0) \neq F(0).$$

Then

$$E(t) \subset F(t) \quad \text{and} \quad \partial E(t) \cap \partial F(t) = \emptyset \qquad t \in \,]0, T]. \tag{3.75}$$

3.7.1. *Estimates of the extinction time*

We begin with the following result, which is a consequence of the comparison principle.

Lemma 3.5.

Let ∂E be compact. Assume that the mean curvature flow starting from E is smooth up to the extinction time $t^\dagger$. Then

$$t^\dagger \leq \frac{(\mathrm{diam}(E))^2}{2(n-1)}. \tag{3.76}$$

Proof. Denote by $(E(t))_{t \in [0, t^\dagger[}$ the smooth mean curvature flow starting from $E = E(0)$. Let $R_0 := \mathrm{diam}(E)$. We have $E \subseteq B_R(p)$ for some point $p \in \mathbb{R}^n$. Therefore, by the comparison principle between smooth flows (Theorem 3.5) and Example 3.1 we have

$$E(t) \subseteq B_{R(t)}(p), \qquad R(t) = \sqrt{R_0^2 - 2(n-1)t}, \quad t \in \left[0, \frac{R_0^2}{2(n-1)}\right[.$$

Hence $t^\dagger \leq \frac{R_0^2}{2(n-1)}$. $\square$

The following theorem gives an estimate from above and from below of the extinction time for a smooth compact mean curvature flow. We refer to the papers [38], [50], [42] for all details.

Theorem 3.7. *Let ∂E be compact. Assume that the mean curvature flow starting from E is smooth up to the extinction time $t^\dagger$. Then*

$$\frac{2|E|^2}{\left(\mathcal{H}^{n-1}(\partial E)\right)^2} \leq t^\dagger \leq C \left(\mathcal{H}^{n-1}(\partial E)\right)^{2/(n-1)}, \tag{3.77}$$

where $C > 0$ is a constant depending only on the dimension.

Proof. Denote by $(E(t))_{t \in [0, t^\dagger[}$ the smooth mean curvature flow starting from $E = E(0)$. To prove the estimate on the right hand side of (3.77) we follow [38]. We recall the following inequality (see for instance [68, 1.4.1]): if $\phi \in C_c^\infty(\mathbb{R}^n)$, then

$$\left(\int_{\partial E} |\phi|^{\frac{n-1}{n-2}}\right)^{\frac{n-2}{n-1}} \leq c \int_{\partial E} \left[|\nabla_\tau \phi| + |H^E||\phi|\right], \tag{3.78}$$

where $c = \frac{(n-1)4^{(n-1)}}{(n-2)\omega_{n-1}^{1/(n-1)}}$ and $\omega_{n-1} := |\{x \in \mathbb{R}^{n-1} : |x| < 1\}|$. Applying (3.78) with $\phi \equiv 1$ in a neighbourhood of ∂E yields

$$\left(\mathcal{H}^{n-1}(\partial E)\right)^{\frac{n-2}{n-1}} \leq c \int_{\partial E} |H^E||\phi|$$

$$\leq c \left(\int_{\partial E} |H^E|^2\right)^{1/2} \left(\mathcal{H}^{n-1}(\partial E)\right)^{1/2}. \qquad (3.79)$$

Since $\frac{n-2}{n-1} - \frac{1}{2} = \frac{n-3}{2(n-1)}$, from (3.79) we get

$$\left(\mathcal{H}^{n-1}(\partial E)\right)^{\frac{n-3}{n-1}} \leq c \int_{\partial E} |H^E|^2. \qquad (3.80)$$

Recalling (3.38) and applying (3.80) with $E(t)$ in place of E we deduce

$$\frac{d}{dt}\mathcal{H}^{n-1}(\partial E(t)) = -\int_{\partial E(t)} |H^{E(t)}|^2 \leq -C \left(\mathcal{H}^{n-1}(\partial E(t))\right)^{(n-3)/(n-1)},$$

$$(3.81)$$

and integrating we have

$$\left(\mathcal{H}^{n-1}(\partial E(t))\right)^{2/(n-1)} - \left(\mathcal{H}^{n-1}(\partial E(0))\right)^{2/(n-1)} \leq -Ct,$$

where C denotes (possibly different) constants depending only on the dimension. Letting $t \uparrow t^\dagger$ we deduce $t^\dagger \leq C \left(\mathcal{H}^{n-1}(\partial E)\right)^{2/(n-1)}$.

Let us prove the left inequality in (3.77). We follow [50]. Set

$$v(t) := |E(t)|, \qquad a(t) := \mathcal{H}^{n-1}(\partial E(t)).$$

Recalling also (3.38) we find

$$-v'(t) = \int_{\partial E(t)} H^{E(t)} \leq \left(\int_{\partial E(t)} |H^{E(t)}|^2\right)^{1/2} \left(\mathcal{H}^{n-1}(\partial E(t))\right)^{1/2}$$

$$= (-a'(t)a(t))^{1/2} = \left(-\frac{(a^2(t))'}{2}\right)^{1/2},$$

where $'$ denotes the derivative with respect to t. Using Jensen's inequality, it follows

$$v(0) - v(t^\dagger) \leq \int_0^{t^\dagger} \left(-\frac{(a^2(t))'}{2}\right)^{1/2} dt \leq t^{\dagger 1/2} \left(\int_0^{t^\dagger} -\frac{(a^2(t))'}{2} dt\right)^{1/2}$$

$$= t^{\dagger 1/2} \left(\frac{a^2(0)}{2} - \frac{a^2(t^\dagger)}{2}\right)^{1/2}.$$

Recalling $a(t^\dagger) = v(t^\dagger) = 0$, we get $|E| \leq \left(t^\dagger \dfrac{(\mathcal{H}^{n-1}(\partial E))^2}{2} \right)^{1/2}$. $\square$

Remark 3.14. To show estimates on the lifespan of weak solutions requires refined tools, such as the clearing out lemma, see [27], [42].

3.8. Huisken monotonicity formula

Concerning the arguments of this section, we refer to [56], [57], [34] for a much wider discussion. Let us begin with the following result.

Lemma 3.6. *Assume that ∂E is compact. Let $(E(t))_{t \in [0,T]}$ be the smooth mean curvature flow on $[0,T]$ starting from $E = E(0)$. Let $\psi \in C^\infty(\mathbb{R}^n \times [0,T])$. Then*

$$\frac{d}{dt} \int_{\partial E(t)} \psi(x,t) = \int_{\partial E(t)} \left(-\psi(\Delta d)^2 - \langle \nabla\psi, \nabla d\rangle \Delta d + \frac{\partial\psi}{\partial t} \right). \qquad (3.82)$$

Proof. We recall that

$$\frac{d}{dt} \int_{\partial E(t)} \psi(x,t) = \int_{\partial E(t)} \left(\psi \operatorname{div}_\tau X + \langle \nabla\psi, X\rangle + \frac{\partial\psi}{\partial t} \right), \qquad (3.83)$$

where $X := -\Delta d \nabla d$ is the velocity field of $\partial E(t)$. Observe that on $\partial E(t)$ we have

$$-\operatorname{div}_\tau X = \langle \nabla_\tau \Delta d, \nabla d\rangle + \Delta d \operatorname{div}_\tau \nabla d = \Delta d \operatorname{div}_\tau \nabla d = (\Delta d)^2,$$

where the last equality follows from (3.12). Then (3.82) follows. $\square$

Remark 3.15. By the integration by parts formula (3.6) we have

$$\int_{\partial E(t)} \operatorname{div}_\tau \nabla\psi = \int_{\partial E(t)} \langle \nabla\psi, \nabla d\rangle \Delta d.$$

Hence (3.82) can be rewritten, for a test function $\psi > 0$, as

$$\frac{d}{dt} \int_{\partial E(t)} \psi(x,t) = -\int_{\partial E(t)} \psi \left(\Delta d + \frac{1}{\psi}\langle \nabla\psi, \nabla d\rangle \right)^2 \qquad (3.84)$$

$$+ \int_{\partial E(t)} \left(\frac{1}{\psi}\langle \nabla\psi, \nabla d\rangle^2 + \frac{\partial\psi}{\partial t} + \operatorname{div}_\tau \nabla\psi \right).$$

Observe that, in the particular case $\psi \equiv 1$, (3.84) coincides with (3.38).

Remark 3.16. Formula (3.83) is at the basis of Brakke's definition of weak motion by mean curvature, see [27].

The following result is Huisken's monotonicity formula [56]: it describes how the area of $\partial E(t)$ changes when weighted with a backward heat kernel in the ambient space. It is a parabolic version of the monotonicity formula for minimal surfaces, and is an important tool for the analysis of singularities in mean curvature flow (see for instance [56], [57]). We will use a similar monotonicity formula in Section 3.9.

Theorem 3.8. *Let $x_0 \in \mathbb{R}^n$, $t_0 \in [0, T]$ and set*

$$\rho(x,t) = \rho_{(x_0,t_0)}(x,t) := \frac{e^{-\frac{|x-x_0|^2}{4(t_0-t)}}}{(4\pi(t_0-t))^{\frac{n-1}{2}}}, \qquad x \in \mathbb{R}^n, \ t < t_0. \qquad (3.85)$$

We have

$$\frac{d}{dt} \int_{\partial E(t)} \rho = - \int_{\partial E(t)} \rho \left(\Delta d + \frac{1}{\rho} \langle \nabla\rho, \nabla d \rangle \right)^2 \leq 0. \qquad (3.86)$$

Proof. Since ρ is positive, we can use formula (3.84) with $\psi = \rho$. Then the equality in (3.86) will follow if we prove that

$$\frac{1}{\rho} \langle \nabla\rho, \nabla d \rangle^2 + \frac{\partial\rho}{\partial t} + \operatorname{div}_\tau \nabla\rho = 0. \qquad (3.87)$$

Assume for simplicity that $x_0 = 0$ and $t_0 = 0$. Then

$$\rho(x,t) = \frac{1}{(-4\pi t)^{\frac{n-1}{2}}} e^{\frac{|x|^2}{4t}}, \qquad x \in \mathbb{R}^n, \ t < 0.$$

Therefore $\nabla\rho = \dfrac{\rho}{2t} x$,

$$\frac{\partial\rho}{\partial t} = \left(\frac{-|x|^2}{4t^2} - \frac{n-1}{2t} \right) \rho, \qquad (3.88)$$

$$\nabla^2\rho = \left(\frac{1}{4t^2} x \otimes x + \frac{1}{2t} \operatorname{Id} \right) \rho,$$

$$\operatorname{div}_\tau \nabla\rho = \operatorname{tr}((\operatorname{Id} - \nabla d \otimes \nabla d)\nabla^2\rho) \qquad (3.89)$$

$$= \left(\frac{|x|^2}{4t^2} - \frac{\langle x, \nabla d \rangle^2}{4t^2} + \frac{n-1}{2t} \right) \rho.$$

Since $\frac{1}{\rho} \langle \nabla\rho, \nabla d \rangle^2 = \frac{1}{4t^2} \langle x, \nabla d \rangle^2 \rho$, from (3.88) and (3.89) we deduce (3.87). $\square$

3.9. The gradient estimate of Ecker-Huisken

Let $f_0 : \mathbb{R}^{n-1} \to \mathbb{R}$ be a smooth Lipschitz function. It is possible to prove that, under suitable growth assumptions at infinity on the graph of f, there exist $T > 0$ and a unique smooth function $f : \mathbb{R}^{n-1} \times \;]0, T[$, satisfying (3.37) in $\mathbb{R}^{n-1} \times \;]0, T[$, such that $\lim_{t \to 0+} f(x,t) = f_0(x)$ for any $x \in \mathbb{R}^{n-1}$, and

$$\sup_{x \in \mathbb{R}^{n-1}} |\nabla f(x,t)| < +\infty, \qquad t \in \;]0, T[. \tag{3.90}$$

One of the results proved by Ecker-Huisken in [35] is the following.

Theorem 3.9. *We have*

$$\sup_{x \in \mathbb{R}^{n-1}} |\nabla f(x,t)| \le \sup_{x \in \mathbb{R}^{n-1}} |\nabla f_0(x)|, \qquad t \in \;]0, T[. \tag{3.91}$$

Once (3.91) is established, the authors are able to prove the following global result for mean curvature evolution of Lipschitz graphs (see the original paper [35] for more details and estimates on derivatives of the normal vector to the graph of $f(\cdot, t)$).

Theorem 3.10. *Let $f_0 : \mathbb{R}^{n-1} \to \mathbb{R}$ be a Lipschitz function. Then there exists a unique smooth function $f : \mathbb{R}^{n-1} \times \;]0, +\infty[\to \mathbb{R}$ satisfying (3.37) in $\mathbb{R}^{n-1} \times \;]0, +\infty[$, such that $\lim_{t \to 0+} f(x', t) = f_0(x')$ for any $x' \in \mathbb{R}^{n-1}$, and*

$$\sup_{x' \in \mathbb{R}^{n-1}} |\nabla f(x',t)| \le \sup_{x' \in \mathbb{R}^{n-1}} |\nabla f_0(x')|, \qquad t \in \;]0, +\infty[. \tag{3.92}$$

SKETCH OF THE PROOF OF THEOREM 3.9. Set $\nu = (\nu_1, \ldots, \nu_n) = \frac{1}{\sqrt{1+|\nabla f|^2}}(-\nabla f, 1)$, where ν is evaluated at the points of $\mathrm{graph}(f(\cdot, t))$. We are interested in the last component

$$\nu_n(x', f(x',t)) = \frac{1}{\sqrt{1 + |\nabla f(x',t)|^2}}, \qquad x' \in \mathbb{R}^{n-1}, \; t \in [0, T].$$

Indeed, estimating the gradient of f is equivalent to estimating

$$w := \frac{1}{\nu_n} = \sqrt{1 + |\nabla f|^2}.$$

For notational simplicity set

$$\delta := \nabla_\tau = \nabla - \langle \nabla, \nu \rangle \nu$$

the tangential differential operator on the graph of $f(\cdot, t)$, see [69], [68], [51], [70]. Observe that

$$\delta_h w = -w^2 \delta_h \nu_n. \tag{3.93}$$

We divide the proof into four steps.

Step 1. We have

$$\frac{\partial w}{\partial t} = -w^2 \frac{\partial \nu_n}{\partial t} = -w^2 \delta_n \Delta d. \tag{3.94}$$

This follows as in the proof of (3.47) (applied with $E(t)$ replaced by the subgraph of $f(\cdot, t)$).

Step 2. We have

$$-w^2 \Delta_\tau \nu_n = \Delta_\tau w - 2w^{-1}|\delta w|^2. \tag{3.95}$$

From (3.93) it follows $|\delta w|^2 = w^4 \delta_h \nu_n \delta_h \nu_n$. Hence

$$\Delta_\tau w = \delta_h \delta_h w = \delta_h(-w^2 \delta_h \nu_n) = -w^2 \Delta_\tau \nu_n - 2w \delta_h w \delta_h \nu_n$$
$$= -w^2 \Delta_\tau \nu_n + 2w^3 \delta_h \nu_n \delta_h \nu_n = -w^2 \Delta_\tau \nu_n + 2w^{-1}|\delta w|^2,$$

and *step 2* follows.

Step 3. We have

$$w^2 \delta_n \Delta d = w^2 \Delta_\tau \nu_n + w|A|^2, \tag{3.96}$$

where $|A|^2 := \delta_h \nu_j \delta_h \nu_j$ is the square of the length of the second fundamental form.

Recall (see for instance [68]) that, given $h, k \in \{1, \ldots, n\}$, the following commutation rule holds:

$$\delta_h \delta_k - \delta_k \delta_h = (\nu_h \delta_k \nu_j - \nu_k \delta_h \nu_j)\delta_j. \tag{3.97}$$

Recalling (3.12) and applying (3.97) with $k = n$ we get

$$\delta_n \Delta d = \delta_n \delta_h \nu_h = \delta_h \delta_n \nu_n - \nu_h \delta_n \nu_j \delta_j \nu_h + \nu_n \delta_h \nu_j \delta_j \nu_h$$
$$= \delta_h \delta_n \nu_n + \nu_n \delta_j \nu_h \delta_j \nu_h = \delta_h \delta_n \nu_n + \nu_n |A|^2$$
$$= \delta_h \delta_h \nu_n + \nu_n |A|^2 = \Delta_\tau \nu_n + \frac{1}{w}|A|^2, \tag{3.98}$$

because $\nu_h \delta_j \nu_h = 0$ and $(\delta_i \nu_j)$ is symmetric. Then (3.96) follows by multiplying (3.98) by $|w|^2$.

Before passing to the next step we observe that from *steps 2,3* we deduce

$$-w^2 \delta_n \Delta d = \Delta_\tau w - 2w^{-1}|\delta w|^2 - w|A|^2.$$

Hence, from (3.94) we get

$$\frac{\partial w}{\partial t} - \Delta_\tau w = -2w^{-1}|\delta w|^2 - w|A|^2 \leq 0 \qquad \text{on graph}(f(\cdot, t)). \tag{3.99}$$

The following step is a weighted monotonicity formula (compare (3.86)) on unbounded boundaries. Recall that $\rho = \rho_{(x_0,t_0)}$ is defined in (3.85).

Step 4. Let $\phi \in C^\infty(\mathbb{R}^n \times [0,T])$ and $t < t_0$. Assume that ϕ satisfies proper growth conditions at infinity that ensure that all integrals that follow are finite. Then [35]

$$\frac{d}{dt} \int_{\text{graph}(f(\cdot,t))} \phi\rho = - \int_{\text{graph}(f(\cdot,t))} \rho\phi \left(\Delta d + \frac{1}{\rho}\langle \nabla\rho, \nabla d\rangle \right)^2 \qquad (3.100)$$
$$+ \int_{\text{graph}(f(\cdot,t))} \rho \left(\frac{\partial\phi}{\partial t} - \langle \Delta d \nabla d, \nabla\phi\rangle - \Delta_\tau\phi \right).$$

In particular, if for such a ϕ we have also

$$\phi \geq 0, \qquad \frac{\partial\phi}{\partial t} - \langle \Delta d\nabla d, \nabla\phi\rangle - \Delta_\tau\phi \leq 0 \qquad \text{on graph}(f(\cdot,t)), \quad (3.101)$$

then

$$\frac{d}{dt} \int_{\text{graph}(f(\cdot,t))} \phi\rho \leq 0. \qquad (3.102)$$

Assume for simplicity that $(x_0, t_0) = (0,0)$. Recalling (3.83) we have

$$\frac{d}{dt} \int_{\text{graph}(f(\cdot,t))} \phi\rho = \int_{\text{graph}(f(\cdot,t))} \left[\rho \left(\frac{\partial\phi}{\partial t} - \langle \Delta d\nabla d, \nabla\phi\rangle \right) \right.$$
$$\left. + \phi \left(\frac{\partial\rho}{\partial t} - \langle \Delta d\nabla d, \nabla\rho\rangle \right) - (\Delta d)^2 \phi\rho \right].$$

Now, adding and subtracting $\int_{\partial E(t)} \rho\Delta_\tau\phi$ and using (3.7), we can write

$$\int_{\text{graph}(f(\cdot,t))} \rho \left(\frac{\partial\phi}{\partial t} - \langle \Delta d\nabla d, \nabla\phi\rangle \right)$$
$$= \int_{\text{graph}(f(\cdot,t))} \left[\rho \left(\frac{\partial\phi}{\partial t} - \langle \Delta d\nabla d, \nabla\phi\rangle - \Delta_\tau\phi \right) + \phi\Delta_\tau\rho \right].$$

Hence

$$\frac{d}{dt} \int_{\partial E(t)} \rho\phi = \int_{\text{graph}(f(\cdot,t))} \rho \left(\frac{\partial\phi}{\partial t} - \langle \Delta d\nabla d, \nabla\phi\rangle - \Delta_\tau\phi \right)$$
$$+ \int_{\text{graph}(f(\cdot,t))} \phi \left(\frac{\partial\rho}{\partial t} - \langle \Delta d\nabla d, \nabla\rho\rangle + \Delta_\tau\rho - (\Delta\rho)^2\rho \right).$$

Therefore, to prove (3.100) it remains to show

$$\frac{\partial\rho}{\partial t} - \langle \Delta d\nabla d, \nabla\rho\rangle + \Delta_\tau\rho - (\Delta\rho)^2\rho = -\rho \left(\Delta d + \frac{1}{\rho}\langle \nabla\rho, \nabla d\rangle \right)^2. \qquad (3.103)$$

Let us compute $\frac{\partial \rho}{\partial t} - \langle \Delta d \nabla d, \nabla \rho \rangle + \Delta_\tau \rho$. Using (3.87) and the equality $\operatorname{div}_\tau \nabla \rho = \Delta_\tau \rho + \langle \nabla \rho, \nabla d \rangle \Delta d$, we have

$$\frac{\partial \rho}{\partial t} - \langle \Delta d \nabla d, \nabla \rho \rangle + \Delta_\tau \rho = -2 \langle \nabla d, \nabla \rho \rangle \Delta d - \frac{1}{\rho} \langle \nabla \rho, \nabla d \rangle^2,$$

and (3.103) follows. The proof of *step 4* is concluded.

Let us now conclude the proof. Define $L := \sup_{x' \in \mathbb{R}^{n-1}} w(x', 0) < +\infty$. Let $p > 2$ and define

$$\psi := g(w), \qquad g(s) := [\max(s, L) - L]^p .$$

Observe that g is convex and of class $C^2(\mathbb{R})$. Using (3.99) it is possible to prove that

$$\frac{\partial \psi}{\partial t} - \langle \Delta d \nabla d, \nabla \psi \rangle - \Delta_\tau \psi \leq 0 \qquad \text{on graph}(f(\cdot, t)). \tag{3.104}$$

Applying (3.102) with $\phi := \psi \geq 0$ we deduce

$$\frac{d}{dt} \int_{\text{graph}(f(\cdot,t))} \psi \rho \leq 0 \qquad t \in \,]0, T[. \tag{3.105}$$

Since $\int_{\partial E(0)} \psi \rho = 0$ and $\int_{\partial E(t)} \psi \rho \geq 0$, from (3.105) it necessarily follows $\int_{\partial E(t)} \psi \rho = 0$ in $]0, T[$. Hence $\psi \equiv 0$ in $]0, T[$, which is equivalent to (3.91). $\square$

3.10. Formation of singularities: the example of Grayson

We have seen that the sphere of radius R_0 shrinks to a point in the finite time $R_0^2/(2(n-1))$. This behaviour can be interpreted as a singularity of the flow; in this section we show a more interesting example of singularity, which is due to Grayson in [53].

Let r and l be two positive numbers such that

$$2l - \pi r > 0. \tag{3.106}$$

Let $g : \mathbb{R} \to \,]0, +\infty[$ be a function whose graph has the shape depicted in Figure 3.3. Let us rotate the graph of g around the x-axis, and denote by $D \subset \mathbb{R}^3$ the solid set inside the rotated graph, see Figure 3.4.

It is possible to choose the function g smooth enough and such that the mean curvature of ∂D is nonnegative everywhere. Choose now a real number R with

$$R^2 > \frac{8lr^2}{2l - \pi r}, \tag{3.107}$$

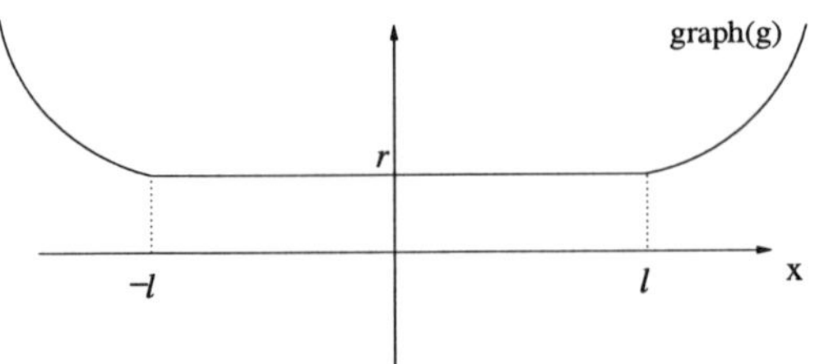

Figure 3.3. Qualitative shape of $g : \mathbb{R} \to \mathbb{R}$.

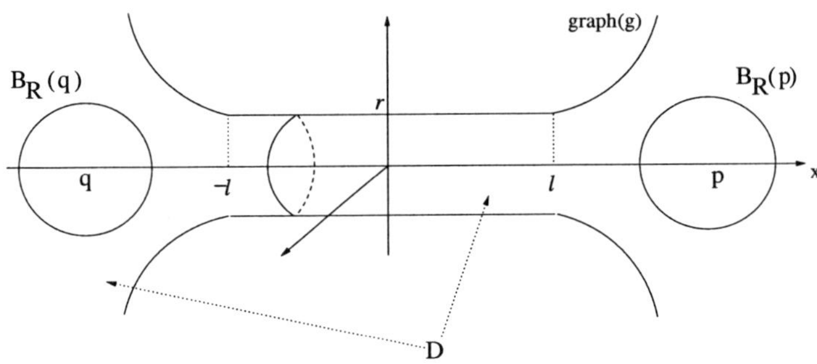

Figure 3.4. The set D, which is the interior of the rotation of the graph of g, contains the union of the two balls. The function g is chosen in such a way that the mean curvature of ∂D is nonnegative.

and two points p, q inside D far enough from the origin so that $\overline{B}_R(q) \cup \overline{B}_R(p) \subset D$ as in Figure 3.4. The result of Grayson is the following.

Theorem 3.11. *Let l, r, R satisfy (3.106) and (3.107). Assume that the mean curvature of ∂D is nonnegative everywhere. Let $E \subset \mathbb{R}^3$ be a bounded connected open set with smooth boundary, and assume*

$$B_R(q) \cup B_R(p) \subset E \subset D. \tag{3.108}$$

Then the mean curvature flow starting from ∂E develops a singularity before its extinction time.

Proof. Take a (comparison) surface Σ generated by the rotation of a smooth periodic graph $y = y(x) : \mathbb{R} \to]0, +\infty[$ around the x-axis such that

$$\Sigma \subset D, \qquad \text{int}(\Sigma) \supset E, \tag{3.109}$$

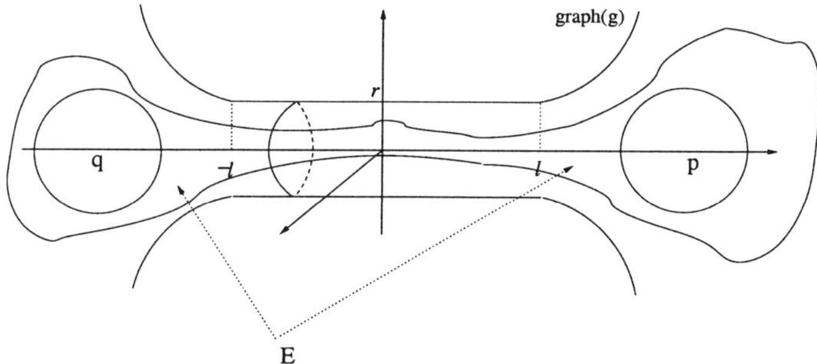

Figure 3.5. The initial set E contains the union of the two balls and must be connected and contained in D.

see Figure 3.6. Here int(Σ) denotes the rotation around the x-axis of the open subgraph of y.

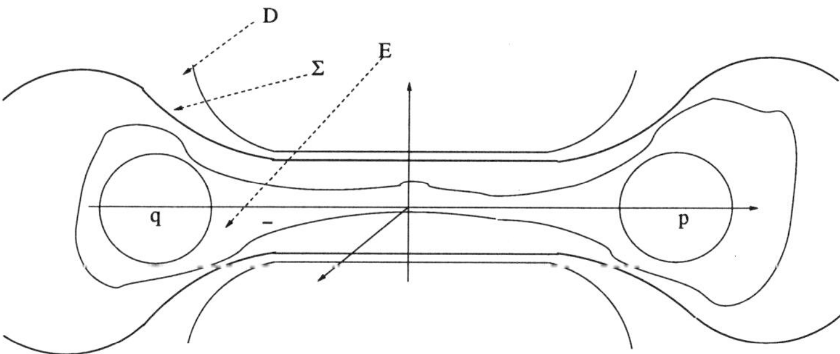

Figure 3.6. The surface of revolution Σ is in between ∂E and ∂D. It is not compact, but is periodic.

Since Σ is a surface of revolution, its mean curvature H^Σ is

$$H^\Sigma = \frac{y_{xx}}{(1+y_x^2)^{3/2}} - \frac{1}{y(1+y_x^2)^{1/2}},$$

recall Example 3.4. It follows that the mean curvature flow $\Sigma(t)$ of Σ is the rotationally symmetric graph, obtained by rotating around the x-axis the graph of the smooth periodic function $y(\cdot,t)$, with $y : \mathbb{R} \times [0,T] \to [0,+\infty[$ solving

$$\frac{\partial y}{\partial t} = \frac{y_{xx}}{1+y_x^2} - \frac{1}{y}, \tag{3.110}$$

(recall Example 3.2), for some $T > 0$. It is possible to prove that $y(x, t)$ remains smooth till $y > 0$, i.e. the only way for $\Sigma(t)$ to singularize is when it intersects the x-axis. By the comparison principle (observe that $\Sigma(t)$ is unbounded, but periodic)

$$B_{R(t)}(q) \cup B_{R(t)}(p) \subset E(t) \subset \text{int}(\Sigma(t)) \subset \partial D, \qquad (3.111)$$

where $R(t) = \sqrt{R^2 - 4t}$ and $\text{int}(\Sigma(t))$ denotes the rotation around the x-axis of the open subgraph of $y(\cdot, t)$. The last inclusion in (3.111) follows because the mean curvature of ∂D is nonnegative everywhere. From (3.111) it follows that the extinction time $t^\dagger(E)$ of E satisfies

$$t^\dagger(E) \geq R^2/4. \qquad (3.112)$$

Define now

$$A(t) := \int_{-l}^{l} y(x, t) \, dx.$$

By (3.110) we have

$$A'(t) = \int_{-1}^{1} \frac{\partial y}{\partial t}(x, t) \, dx = \int_{-l}^{l} \left(\frac{y_{xx}}{1 + y_x^2} - \frac{1}{y} \right) \, dx$$

$$= \int_{\text{graph}(y(\cdot, t))} \kappa - \int_{-l}^{l} \frac{1}{y} \, dx,$$

where κ indicates the curvature of the plane curve $\text{graph}(y(\cdot, t))$. Hence

$$A'(t) \leq \pi - \int_{-l}^{l} \frac{1}{y} \, dx. \qquad (3.113)$$

From (3.111) we have $y(x, t) \leq r$ for any $x \in [-l, l]$ so that $-1/y(x, t) \leq -1/r$, and therefore from (3.113) we deduce $A'(t) \leq \pi - 2l/r$. Since $A(0) \leq 2lr$, we find

$$A(t) \leq (\pi - 2l/r)t + A(0) \leq (\pi - 2l/r)t + 2rl. \qquad (3.114)$$

The right hand side at (3.114) vanishes for $t = \frac{2lr^2}{2l - \pi r}$ which, in view of (3.107), is strictly less than $R^2/4$. Hence, at time $t^* = R^2/4$ (which is smaller than the extinction time of E thanks to (3.112)) we have $A(t^*) = 0$, so that $\Sigma(t)$ must shrink around the x-axis for some time t less than or equal to t^*. The assertion of the theorem then follows by the comparison principle. $\square$

3.11. Manifolds with arbitrary codimension

Following the ideas in [31], [32], in this section we briefly show how to define mean curvature flow in arbitrary codimension using the square distance function (see also [8]). In the sequel Γ denotes a k-dimensional embedded compact connected manifold of class C^∞ without boundary in $\mathbb{R}^n$, $k \in \{1, \ldots, n-1\}$, and

$$\eta^\Gamma(x) := \frac{1}{2}\mathrm{dist}(x, \Gamma)^2, \qquad x \in \mathbb{R}^n.$$

If $x \in \Gamma$ we let $N_x\Gamma$ (resp. $T_x\Gamma$) be the normal (resp. tangent) space to Γ at x. Concerning the following result, we refer to [5, Section 4] for the proof and the expansion of $\nabla^2\eta^\Gamma$ out of Γ. We refer to [7], [67], [37] for more details on the relations between the derivatives of $\nabla^2\eta^\Gamma$ and the second fundamental form of Γ.

Theorem 3.12. *There exists $\rho > 0$ such that, setting $U := \{x \in \mathbb{R}^n : \mathrm{dist}(x, \Gamma) < \rho\}$, the following properties hold:*

(i) $\eta^\Gamma \in C^\infty(U)$ and $|\nabla\eta^\Gamma|^2 = 2\eta^\Gamma$ in U;

(ii) if $y \in U$ the point $\mathrm{pr}(y) := y - \nabla\eta(y)$ belongs to Γ and is the unique solution of $\min\{|y - x| : x \in \Gamma\}$. Moreover, for any $x \in \Gamma$, the matrix $\nabla^2\eta^\Gamma(x)$ is the orthogonal projection on $N_x(\Gamma)$;

(iii) $-\Delta\nabla\eta^\Gamma$ is, on Γ, the mean curvature of Γ.

In (iii), $\Delta\nabla\eta^\Gamma$ is the row vector whose i-th component is the Laplacian of $\frac{\partial}{\partial x_i}\eta^\Gamma$.

Remark 3.17. If we define $\zeta(y) := \frac{1}{2}|y|^2 - \eta^\Gamma(y)$, then $\nabla\zeta(y)$ gives the projection $\mathrm{pr}(y)$ of y on Γ and $\nabla^2\eta^\Gamma(x)$ is the orthogonal projection on $T_x(\Gamma)$. Observe also that, when $k = n - 1$, (ii) (resp. (iii)) of Theorem 3.12 is expressed by (ii) of Theorem 3.2 and (3.20) (resp. (3.22)).

Also in arbitrary codimension, one of the motivations of studying mean curvature flow is given by the first variation of area.

Theorem 3.13. *Let ψ and X be as in Theorem 3.1. Let $\Gamma_\lambda := \psi_\lambda(\Gamma)$. Then Γ_λ is smooth for small $|\lambda|$, and*

$$\frac{d}{d\lambda}\mathcal{H}^k(\Gamma_\lambda)_{|\lambda=0} = \int_{\partial\Gamma} \langle X, \Delta\nabla\eta^\Gamma\rangle \, d\mathcal{H}^k.$$

If $\Gamma(t)$ is a family of subsets of $\mathbb{R}^n$ parameterized by $t \in [0,T]$ we let

$$\eta^{\Gamma(t)}(x) := \frac{1}{2}\text{dist}(x, \Gamma(t))^2, \qquad x \in \mathbb{R}^n, \ t \in [0,T]. \qquad (3.115)$$

Since no confusion is possible, the function $\eta^{\Gamma(t)}(x)$ will be denoted by $\eta(x,t)$; $\nabla\eta$ (resp. $\nabla^2\eta$) stands for the gradient (resp. the Hessian) of η with respect to x.

The definition of smooth flow in arbitrary codimension is the one in Definition 3.1 replacing ∂E with $\Gamma = \Gamma(0)$, and where in (iii) we take k in place of $n-1$. The definition of velocity is given in Definition 3.2. In order to define mean curvature flow using the function η we need some preliminaries.

Theorem 3.14. *Let $(\Gamma(t))_{t\in[0,T]}$ be a smooth flow on $[0,T]$ starting from $\Gamma = \Gamma(0)$. Then there exists an open set A such that $A \supset \Gamma(t)$ for any $t \in [0,T]$ and $\eta \in C^\infty(A \times [0,T])$, and*

$$\frac{\partial}{\partial t}\nabla\eta(x,t) = -\mathbf{V}(y,t), \qquad x := \phi(y,t) \in \partial\Gamma(t). \qquad (3.116)$$

SKETCH OF PROOF. We prove only (3.116). Let $k \in \{1,\dots,n\}$. From the equality $\nabla_k\eta(\phi(y,t),t) = 0$ for any $y \in \Gamma$ and $t \in [0,T]$ we get, for $x := \phi(y,t)$,

$$\nabla_k\frac{\partial\eta}{\partial t}(x,t) + \langle\nabla\nabla_k\eta(x,t), \frac{\partial\phi}{\partial t}(y,t)\rangle = 0.$$

Recalling that $\nabla^2\eta(x,t)$ is the orthogonal projection on $N_x(\Gamma(t))$, formula (3.116) follows. $\square$

Definition 3.6. We say that $(\Gamma(t))_{t\in[0,T]}$ is a smooth k-dimensional mean curvature flow in $[0,T]$ starting from $\Gamma = \Gamma(0)$ if the following conditions hold:

(i) there exists an open set $A \subset \mathbb{R}^n$ containing $\Gamma(t)$ for any $t \in [0,T]$ such that $\eta \in C^\infty(A \times [0,T])$ and rank $(\nabla^2\eta(x,t)) = n - k$ for any $t \in [0,T]$ and any $x \in \Gamma(t)$;

(ii) we have

$$\frac{\partial}{\partial t}\nabla\eta(x,t) = \Delta\nabla\eta(x,t), \qquad t \in [0,T], \ x \in \Gamma(t). \qquad (3.117)$$

Condition (i) implies that $\Gamma(t)$ is a smooth embedded compact connected k-dimensional manifold without boundary, smoothly evolving in time [32], [8], [5]. Condition (ii) implies that the normal velocity at each point x of $\Gamma(t)$ equals the mean curvature of $\Gamma(t)$ at x.

Remark 3.18. The proof of a short time existence and uniqueness theorem for mean curvature flow in arbitrary codimension follows from [46].

Mean curvature flow in arbitrary codimension has been the subject of several papers, starting from the work of Brakke [27], see for instance [2], [4], [65], [20], [81], [80], [73]. Interesting connections between the gradient flow of the Ginzburg-Landau functionals and motion by mean curvature in higher codimension have been proved in [9], [26].

3.12. Final comments

The discussion in Section 3.10 is the motivation for studying global solutions to mean curvature flow, defined beyond singularities. There are several notions of weak solutions present in the literature, which may differ after the onset of singularities. In addition, the mutual relations between these solutions are not yet completely classified. We refer to [27], [40], [30], [74], [64], [62], [63], [65], [31], [32], [1] and to the bibliographies of [5] and [19] for a (possibly incomplete) list of papers in this direction.

Related to the previous comment, one natural issue is studying the qualitative behaviour of the evolving manifold at the singularity time and, in general, the regularity of weak solutions. In this respect we refer to [27], [18], [3], [35], [56], [71], [57], [75], [23], [15], [24], [33], [44], [59], [60], [82], [14], [83], [84], [34] and references therein (again the present list of references is far from being complete).

We conclude the paper by recalling that mean curvature flow is an example of geometric evolution of manifolds. Other interesting examples are for instance the Ricci flow [54], the evolution of harmonic maps [77], the anisotropic mean curvature flow [25], the crystalline mean curvature flow [79], [29], [48], [22], the curvature flow of networks [28], the inverse mean curvature flow [58], the Gauss mean curvature flow [10], and the flow of the Willmore functional [61].

References

[1] F.J. Almgren, J.E. Taylor and L. Wang, *Curvature-driven flows: a variational approach*, SIAM J. Control Optim., **31** (1993), 387–437.

[2] S. Altschuler, *Singularities of the curve shortening flow for space curves*, J. Differential Geom., **34** (1991), 499–514.

[3] S. Altschuler, S. Angenent and Y. Giga, *Mean curvature flow through singularities for surfaces of rotation*, J. Geom. Anal., **5** (1995), 293–358.

[4] S. Altschuler and M. A. Grayson, *Shortening space curves and flow through singularities*, J. Differential Geom., **35** (1992), 283–398.

[5] L. Ambrosio, *Lecture notes on geometric evolution problems, distance function and viscosity solutions*, In: Calculus of Variations and Partial Differential Equations. Topics on Geometrical Evolution Problems and Degree Theory, 5–94, Springer-Verlag, 1999.

[6] L. Ambrosio, N. Fusco and D. Pallara, Functions of Bounded Variation and Free Discontinuity Problems, Clarendon Press (Oxford), 2000.

[7] L. Ambrosio and C. Mantegazza, *Curvature and distance function from a manifold*, J. Geom. Anal., **8** (1998), 719–744.

[8] L. Ambrosio and H.-M. Soner, *A level set approach to the evolution of surfaces of any codimension*, J. Differential Geom., **43** (1996), 693–737.

[9] L. Ambrosio and H.-M. Soner, *A measure theoretic approach to higher codimension mean curvature flow*, Ann. Scuola Norm. Sup. Pisa Cl. Sci. (4), **25** (1997), 27–49.

[10] B. Andrews, *Gauss curvature flow: the fate of the rolling stones*, Invent. Math., **138** (1999), 151–161.

[11] S. Angenent, *Parabolic equations for curves and surfaces Part I. Curves with p-integrable curvature*, Ann. of Math., **132** (1990), 451–483.

[12] S. Angenent, *On the formation of singularities in the curve shortening flow*, J. Differential Geom., **33** (1991) 601–633.

[13] S. Angenent, *Parabolic equations for curves and surfaces Part II. Intersections, blow-up and generalized solutions*, Ann. of Math., **133** (1991), 171–215.

[14] S. Angenent *Some recent results on mean curvature flow*, RAM Res. Appl. Math., **30** (1994), 1–18.

[15] S. Angenent, D.L. Chopp and T. Ilmanen, *A computed example of nonuniqueness of mean curvature flow in R^3*, Comm. Partial Differential Equations, **20** (1995), 1937–1958.

[16] M. Bardi, M.C. Crandall, L.C. Evans, H.-M. Soner and P.E. Souganidis, Viscosity Solutions and Applications (Montecatini Terme 1995), edited by I. Capuzzo Dolcetta and P.L. Lions. Lectures Notes in Math., 1660, Springer-Verlag, Berlin 1997.

[17] G. Barles, S. Biton and O. Ley, *A geometrical approach to the study of unbounded solutions of quasilinear parabolic equations*, Arch. Rational Mech. Anal., **162** (2002), 287–325.

[18] G. Barles, H.-M. Soner and P.E. Souganidis, *Front propagation and phase field theory*, SIAM J. Control Optim., **31** (1993), 439–469.

[19] G. Bellettini and M. Novaga, *Some aspects of De Giorgi's barriers for geometric evolutions*, In Calculus of Variations and Partial Differential Equations. Topics on Geometrical Evolution Problems and Degree Theory, Springer-Verlag, 1999, 115–151.

[20] G. Bellettini and M. Novaga, *A result on motion by mean curvature in arbitrary codimension*, Diff. Geom. Appl., **11** (1999), 205–220.

[21] G. Bellettini and M. Novaga and M. Paolini, *An example of three-dimensional fattening for linked space curves evolving by curvature*, Comm.

Partial Differential Equations, **23** (1998), 1475–1492.

[22] G. Bellettini, M. Novaga and M. Paolini *Facet-breaking for three-dimensional crystals evolving by mean curvature,* Interfaces Free Bound. **1** (1999), 39–55.

[23] G. Bellettini and M. Paolini, *Two examples of fattening for the curvature flow with a driving force,* Atti Accad. Naz. Lincei Cl. Sci. Fis. Mat. Natur. Rend. Lincei (9) Mat., **5** (1994) 229–236.

[24] G. Bellettini and M. Paolini, *Some results on minimal barriers in the sense of De Giorgi applied to driven motion by mean curvature,* Rend. Atti Acc. Naz. Sci. XL Mem. Mat., **XIX**, (1995), 43–67. Errata Corrige vol. XXVI (2002) pag. 161–165.

[25] G. Bellettini and M. Paolini, *Anisotropic motion by mean curvature in the context of Finsler geometry,* Hokkaido Math. J., **25**, (1996), 537–566.

[26] F. Bethuel and G. Orlandi and D. Smetz, *Convergence of the parabolic Ginzburg-Landau equation to motion by mean curvature,* Ann. Math. (2) 163 (2006), 37–163.

[27] K.A. Brakke, The Motion of a Surface by its Mean Curvature, Princeton Univ. Press, 1978, Princeton.

[28] L. Bronsard and F. Reitich, *On three-phase boundary motion and the singular limit of a vector-valued Ginzburg-Landau equation,* Arch. Rational Mech. Anal. **124** (1993), 355–379.

[29] J.W. Cahn, C.A. Handwerker and J.E. Taylor, *Geometric models of crystal growth,* Acta Metall. Mater. **40** (1992), 1443–1474.

[30] Y.-G. Chen, Y. Giga and S. Goto, *Uniqueness and existence of viscosity solutions of generalized mean curvature flow equations,* J. Differential Geom., **33** (1991), 749–786.

[31] E. De Giorgi, *Congetture riguardanti barriere, superfici minime, movimenti secondo la curvatura media,* Manuscript, Lecce November 4, (1993).

[32] E. De Giorgi, *Barriers, boundaries, motion of manifolds,* Conference held at Department of Mathematics of Pavia, March 18, (1994).

[33] E. De Giorgi, *Congetture riguardanti alcuni problemi di evoluzione - A paper in honour of John Nash,* Duke Math. J., **81-2** (1996), 255–268.

[34] K. Ecker, Regularity Theory for Mean Curvature Flow, Birkhäuser, Boston (2004).

[35] K. Ecker and G. Huisken, *Mean curvature flow of entire graphs,* Ann. of Math.(2), **130** (1989), 453–471.

[36] K. Ecker and G. Huisken, *Interior estimates for hypersurfaces moving by mean curvature,* Invent. Math., **103** (1991), 547–569.

[37] M. Eminenti and C. Mantegazza, *Some properties of the distance function and a conjecture of De Giorgi,* J. Geom. Anal. 14 (2004), 267–279.

[38] L.C. Evans, *Mean curvature motion,* in Viscosity solutions and applications (Montecatini Terme 1995), edited by I. Capuzzo Dolcetta and P.L. Lions. Lectures Notes in Math., 1660, Springer-Verlag, Berlin (1997), 115–133.

[39] L.C. Evans, H.-M. Soner and P.E. Souganidis, *Phase transitions and generalized motion by mean curvature,* Comm. Pure Appl. Math., **45** (1992), 1097–1123.

[40] L.C. Evans and J. Spruck, *Motion of level sets by mean curvature I,* J.

Differential Geom., **33** (1991), 635–681.

[41] L.C. Evans and J. Spruck, *Motion of level sets by mean curvature II*, Trans. Amer. Math. Soc., **330** (1992), 321–332.

[42] L.C. Evans and J. Spruck, *Motion of level sets by mean curvature III*, J. Geom. Anal. **2** (1992), 121–160.

[43] L.C. Evans and J. Spruck, *Motion of level sets by mean curvature IV*, J. Geom. Anal. **5** (1995), 77–114.

[44] F. Fierro and M. Paolini, *Numerical evidence of fattening for the mean curvature flow* Math. Models Methods Appl. Sci., **6** (1996), 793–813.

[45] M.E. Gage, *Curve-shortening makes convex curves circular*, Invent. Math., **76** (1984), 357–364.

[46] M.E. Gage and R.S. Hamilton, *The heat equation shrinking convex plane curves*, J. Differential Geom., **23** (1986), 69–96.

[47] Y. Giga, Surface Evolution Equations - a Level Set Method, Monographs in Mathematics 99, Basel, Birkhäuser (2006).

[48] M.-H. Giga and Y. Giga, *Evolving graphs by singular weighted curvature*, Arch. Rational Mech. Anal., **141** (1998), 117–198.

[49] Y. Giga, S. Goto, H. Ishii and M.-H. Sato, *Comparison principle and convexity preserving properties for singular degenerate parabolic equations on unbounded domains*, Indiana Univ. Math. J., **40** (1991), 443-470.

[50] Y. Giga and K. Yama-uchi, *On a lower bound for the extinction time of surfaces moved by mean curvature*, Calc. Var. Partial Differential Equations, **1** (1993), 417–428.

[51] E. Giusti, Minimal Surfaces and Functions of Bounded Variation, Birkhäuser, Boston (1984).

[52] M.A. Grayson, *The heat equation shrinks embedded plane curves to round points*, J. Differential Geom., **26** (1987), 285–314.

[53] M.A. Grayson, *A short note on the evolution of a surface by its mean curvature*, Duke Math. J., **58** (1989), 555–558.

[54] R.S. Hamilton, *Three manifolds with positive Ricci curvature*, J. Differential Geom., **17** (1982), 255–306.

[55] G. Huisken, *Flow by mean curvature of convex surfaces into spheres*, J. Differential Geom., **20** (1984), 237–266.

[56] G. Huisken, *Asymptotic behaviour for singularities of the mean curvature flow*, J. Differential Geom., **31** (1990),

[57] G. Huisken, *Local and global behaviour of hypersurfaces moving by mean curvature*, Proc. of Symp. Pure Math., **54** (1993), 175–191.

[58] G. Huisken and T. Ilmanen, *The inverse mean curvature flow and the Riemannian Penrose inequality*, J. Differential Geom. **59** (2001), 353–437.

[59] G. Huisken and C. Sinestrari, *Mean curvature flow singularities for mean convex surfaces*, Calc. Var. Partial Differential Equations, **8** (1999), 1–14.

[60] G. Huisken and C. Sinestrari, *Convexity estimates for mean curvature flow and singularities of mean convex surfaces*, Acta Math., **183** (1999), 45–70.

[61] E. Kuwert and R. Schätzle, *Gradient flow for the Willmore functional*, Comm. Anal. Geom., **10** (2002), 307–339.

[62] T. Ilmanen, *Generalized flow of sets by mean curvature on a manifold*, Indi-

ana Univ. Math. J., **41** (1992), 671–705.

[63] T. Ilmanen, *The level set flow on a manifold*, Differential Geometry: partial differential equations on manifolds (Los Angeles, CA 1990), 193–204, Proc. Sympos. Pure Math., **54**, Part 1, Amer. Math. Soc., Providence, RI, 1993.

[64] T. Ilmanen, *Convergence of the Allen-Cahn equation to Brakke's motion by mean curvature*, J. Differential Geom., **38** (1993) 417–461.

[65] T. Ilmanen, Elliptic Regularization and Partial Regularity for Motion by Mean Curvature, Mem. Amer. Math. Soc., **108** (1994).

[66] A. Lunardi, Analytic Semigroups and Optimal Regularity in Parabolic Problems, Birkhäuser, 1995, Boston.

[67] C. Mantegazza, *Smooth geometric evolutions of hypersurfaces*, Geom. Funct. Anal., **12** (2002), 138–182.

[68] U. Massari and M. Miranda, Minimal Surfaces in Codimension One, Notas de Matemática, North-Holland, Amsterdam 1984.

[69] M. Miranda, *Una maggiorazione integrale per le curvature delle ipersuperficie minimali*, Rend. Sem. Mat. Univ. Padova, **38** (1967), 91–102.

[70] M. Miranda, *Movimento di superfici: approccio intrinseco*, Rend. Sem. Mat. Fis. Milano, **LXII** (1992), 127–136.

[71] M. Paolini and C. Verdi, *Asymptotic and numerical analyses of the mean curvature flow with a space-dependent relaxation parameter*, Asymptotic Anal., **5** (1992), 553–574.

[72] L. Simon, Lectures on Geometric Measure Theory, Proc. of the Centre of Math. Anal., Australian National Univ., vol. 3, 1983.

[73] D. Slepcev, *On level set approach to motion of manifolds of arbitrary codimension*, Interfaces Free Bound. 5 (2003), 417-458.

[74] H.-M. Soner, *Motion of a set by the curvature of its boundary*, J. Differential Equations, **101** (1993), 313–372.

[75] H.-M. Soner and P.E. Souganidis, *Singularities and uniqueness of cylindrically symmetric surfaces moving by mean curvature*, Comm. Partial Differential Equations, **18** (1993), 859–894.

[76] P.E. Souganidis, *Front propagation: theory and application*, in Viscosity solutions and applications (Montecatini Terme 1995), edited by I. Capuzzo Dolcetta and P.L. Lions. Lectures Notes in Math., 1660, Springer-Verlag, Berlin 1997, 186–242.

[77] M. Struwe, *On the evolution of harmonic maps in higher dimensions*, J. Differential Geom. **28** (1988), 485–502.

[78] M. Struwe, *Geometric evolution problems*, In Nonlinear Partial Differential Equations in Differential Geometry (Park City, UT, 1992), IAS/Park City Math. Ser. 2, Amer. Math. Soc., Providence, RI, (1996), 257–338.

[79] J.E. Taylor, *II-Mean curvature and weighted mean curvature*, Acta Metall. Mater., **40** (1992), 1475–1485.

[80] M.-T. Wang, *Gauss maps of the mean curvature flow*, Math. Res. Lett., **10** (2003), 287–299.

[81] M.-T. Wang, *Long-time existence and convergence of graphic mean curvature flow in arbitrary codimension*, Invent. Math., **148** (2002), 525–543.

[82] B. White, *The topology of hypersurfaces moving by mean curvature*, Comm.

Anal. Geom. **3** (1995), 317–333.

[83] B. White, *The size of the singular set in mean curvature flow of mean-convex sets*, J. Amer. Math. Soc. **13** (2000), 665–695.

[84] B. White, *The nature of singularities in mean curvature flow*, J. Amer. Math. Soc. **16** (2003), 123–138.

[85] X.P. Zhu, Lectures on Mean Curvature Flows, AMS/IP Studies in Advanced Mathematics, 32, International Press, 2002.

Chapter 4

Introduction to bifurcation theory

Pavel Drábek*

Centre of Applied Mathematics
University of West Bohemia
Univerzitní 22, 306 14 Plzeň
Czech Republic
pdrabek@kma.zcu.cz

Contents

4.1. Preface

These notes present three basic methods of the modern *Bifurcation Theory*. The topic was lectured by the author at Dipartimenti di Matematica

---

*This research has been supported by the Grant Agency of the Czech Republic, no. 201/03/0671.

103

Pura ed Applicata e Metodi e Modelli Matematici per le Scienze Applicate, Università di Padova, Italy, during June 11–15, 2001, in the framework of

"Minicorsi di Analisi Matematica".

The material of these notes is also a part of the syllabus of the course

"Modern Mathematical Methods"

at the Faculty of Applied Sciences, University of West Bohemia in Pilsen, Czech Republic.

This text was published as a textbook for the students of the University of West Bohemia in Pilsen, in 2002. The author is grateful to this institution for permitting republication.

The reader should keep in mind that this is just an introduction to the subject, only basic results are presented while many other methods are ignored here. For instance, the important and very interesting notion of the Hopf Bifurcation is not mentioned in these notes at all! However, if the reader wants to understand the present "state of the art" of the Bifurcation Theory, then s/he should master the methods mentioned in these notes.

I would like to thank Professor Massimo Lanza de Cristoforis for his kind invitation to Padova and to my young colleagues RNDr. Jiří Benedikt, Mgr. Marie Benediktová, Ing. Petr Girg, Ph.D., Ing. Gabriela Holubová, Ph.D., Ing. Aleš Matas, Ph.D., Ing. Petr Nečesal, Ph.D. and Lenka Stuchlová who all helped me to give the text this present form.

Padova – Pilsen, June 2001 – December 2003

4.2. Introduction, basic notation

The *Bifurcation Theory* is nowadays one of the most developing parts of the modern nonlinear analysis. Even if some related results in the finite dimension go back to the 19th century, modern methods dealing with the infinite dimension were developed in the second half of the 20th century. The Bifurcation Theory combines deep results of the linear functional analysis (especially the *Spectral Theory of linear compact operators* are often used here) and sophisticated methods of the nonlinear functional analysis (such as the *Implicit Function Theorem*, the *Degree Theory*, *Variational Principles*, etc.). There are numerous results in the literature which deal with (or refer to) the Bifurcation Theory. One of the main impulses to develop this theory were open problems in various fields of differential equations. That is why, besides of some motivation in the finite dimension, we apply abstract results to boundary value problems. We choose ordinary differential equations only in order to avoid necessary technical assumptions which would have to be done if we dealt with partial differential equations. However, it is clear that abstract results apply to more dimensional boundary value problems as well.

These notes actually follow selected parts of the text [8] by Stará and John which is available only in Czech. We want to concentrate here exclusively on three basic methods used in the Bifurcation Theory. On the other hand, to make the exposition brief and clear we cannot define every notion used here, and we have to assume that the reader is acquainted with some knowledge from both the linear and the nonlinear functional analysis. For the reader's convenience, we list the most important notions used in these notes and present the most frequent notation as well. It is assumed that the *reader will be able to master the notions presented below from other sources.* In order to keep the exposition as smooth as possible, we decided to postpone some important facts and proofs to appendices at the end of these notes.

Basic notions (*the order is determined by the exposition in Sections* 4.3–4.7)

- normed linear space, Banach space, Hilbert space, metric space
- linear operator
- eigenvalue, eigenvector
- Fréchet differential of the nonlinear operator, partial Fréchet derivative
- isomorphism
- spectrum of the linear operator (point spectrum)
- compact linear (nonlinear) operator
- regular (symmetric) matrix
- positive definite (semidefinite) matrix
- Fredholm operator (mapping), kernel and range (image) of the linear operator, index of the Fredholm operator
- co-dimension of the linear subspace
- linear projection
- expressions which are of order $o(\varepsilon)$ when $\varepsilon \to 0$
- Fredholm Alternative (for linear operators)

- Brouwer degree of the mapping
- Brouwer Fixed Point Theorem
- Leray–Schauder degree of the mapping
- Leray–Schauder index of the point (with respect to a given mapping), Leray–Schauder Index Formula
- algebraic and geometric multiplicity of the eigenvalue of the linear operator
- strong and weak convergence in the Banach (Hilbert) space
- Lebesgue space (Sobolev space) of functions on a given interval
- spaces of continuous and smooth functions
- continuity and compactness of the embedding
- derivatives in the distributional sense
- potential operator
- Riesz Representation Theorem
- Lagrange Multiplier Method
- saddle point
- initial value problem for the ordinary differential equation with values in the Hilbert space, existence and uniqueness of the solution
- Courant–Weinstein Variational Principle

Notation (*the order is determined by the exposition in Sections* 4.3–4.7)

- $\mathbb{R}^n$ $(\mathbb{R})$ n-dimensional Euclidean space $(n = 1)$
- $\boldsymbol{x}$ finite dimensional vector
- $\boldsymbol{A}$ finite dimensional matrix
- o zero element of the Banach space X
- $\boldsymbol{o}$ zero element (origin) of $\mathbb{R}^n$
- $X \times \mathbb{R}$ Cartesian product of the spaces X and $\mathbb{R}$
- $\dfrac{\partial F}{\partial x}, \dfrac{\partial^2 F}{\partial x \partial \lambda}, \ldots$ partial derivatives (in the finite dimension)
- F'_x (F'_1), F'_{xy} $(F'_{1,2})$ partial Fréchet derivatives (in the infinite dimension)
- $\sigma(T)$ spectrum of the linear operator T
- $\mathbb{N}$ set of all natural numbers $(1, 2, 3, \ldots)$
- C^p class of functions (operators, functionals), the derivatives of which are continuous up to the order p
- $\mathcal{S}$ solution set of the equation $F(x, \lambda) = o$
- $\det \boldsymbol{A}$ determinant of the matrix $\boldsymbol{A}$
- $F'(o)$ $(F''(o))$ the first (second) differential of F at o

- $Ker\, f'(o)$ kernel of the linear operator $f'(o)$
- $Im\, f'(o)$ image (range) of the linear operator $f'(o)$
- $\dim X_1$ dimension of the linear space X_1
- $\operatorname{codim} Y_2$ co-dimension of the linear subspace Y_2
- $Qf'(o)|_{X_2}$ restriction of the operator $Qf'(o)$ onto the subspace X_2
- $y^*(y_0)$ value of the continuous linear functional y^* at the element y_0
- $Lin\{x_1,\ldots,x_n\}$ linear hull of the elements $x_1,\ldots,x_n$
- $\tilde{x}'(s)|_{s=0}$ derivative of $\tilde{x}$ with respect to s at the point $s=0$
- $\deg\,[f;\mathcal{D},p]$ Leray–Schauder (L.–S.) degree of the mapping f with respect to the set $\mathcal{D}$ and the point p
- $i(x_0)$ Leray–Schauder index of the point x_0
- $B(x_0;\varepsilon)$ ball centred at x_0, with the radius ε
- $\overline{\mathcal{D}}$ closure of the set $\mathcal{D}$
- $\partial\mathcal{D}$ boundary of the set $\mathcal{D}$
- $P_\sigma(T)$ point spectrum of the operator T
- $x_n \to x$ sequence $\{x_n\}$ converges strongly to the element x
- $x_n \rightharpoonup x$ sequence $\{x_n\}$ converges weakly to the element x
- $L^1(0,\pi)$, $W_0^{1,2}(0,\pi)$ usual Lebesgue, Sobolev space of functions on $(0,\pi)$
- $C[0,\pi]$, $C^2[0,\pi]$ usual spaces of continuous functions and functions which have continuous second derivatives in $[0,\pi]$
- $X \subset\subset Y$ continuous embedding of X into Y
- $X \subset\subset\subset Y$ compact embedding of X into Y
- $\langle\cdot,\cdot\rangle$ scalar product in the Hilbert space
- $(\cdot,\cdot)$ scalar product in $\mathbb{R}^n$
- $S(o;r)\ (=\partial B(o;r))$ sphere centred at o, with the radius r

4.3. Motivation, examples

Many mathematical models are expressed in the form of the operator equation

$$F(x,\lambda) = o. \tag{4.1}$$

Here x is unknown (function, point in $\mathbb{R}^n,\ldots$) and λ is a parameter (real, complex) which characterizes the problem. The operator F is, in general, a nonlinear mapping.

We shall restrict our attention to problems where x is an element of some Banach space X,[a] and λ is a real parameter. The operator F will be a mapping from $X \times \mathbb{R}$ into another Banach space Y:

$$F \colon X \times \mathbb{R} \to Y.$$

By a *solution set* $\mathcal{S}$ of (4.1) we mean the set of all couples $(x, \lambda) \in X \times \mathbb{R}$ which verify (4.1). The structure of the solution set of (4.1) may be very complicated in general. Actually, for every closed set $\mathcal{S} \subset X \times \mathbb{R}$ there exists a continuous operator F such that $\mathcal{S}$ is the solution set of (4.1). For this reason it is impossible to say much about $\mathcal{S}$ without any additional specifications of F.

Set

$$\mathcal{S}_\lambda := \{x \in X : F(x, \lambda) = o\}.$$

Our aim is to study qualitative properties of the set $\mathcal{S}_\lambda$ depending on the parameter λ. We concentrate on the values of the parameter λ in the neighbourhoods of which $\mathcal{S}_\lambda$ essentially changes its character. This is possible to illustrate in a special case when F is a linear operator.

Example 4.1. Let $X = Y = \mathbb{R}^n$, $\boldsymbol{A}$ be a real matrix $n \times n$,

$$\boldsymbol{F}(\boldsymbol{x}, \lambda) = \lambda \boldsymbol{x} - \boldsymbol{A}\boldsymbol{x}, \qquad \boldsymbol{x} \in \mathbb{R}^n, \ \lambda \in \mathbb{R}.$$

It is easy to describe the set $\mathcal{S}_\lambda$ in this case:

- let λ be not an eigenvalue of $\boldsymbol{A}$, then $\boldsymbol{x} = \boldsymbol{o}$ is the only solution of (4.1), and so $\mathcal{S}_\lambda = \{\boldsymbol{o}\}$;
- let λ be an eigenvalue of $\boldsymbol{A}$ and $\boldsymbol{u}$ be a corresponding (normalized) eigenvector, then for any $t \in \mathbb{R}$, $\boldsymbol{x} = t\boldsymbol{u}$ is a solution of (4.1), and so

$$\mathcal{S}_\lambda = \{t\boldsymbol{u} : t \in \mathbb{R}, \ \boldsymbol{u} \text{ is an eigenvector associated with } \lambda\}.$$

The solution set $\mathcal{S}$ of (4.1) contains then the line

$$\{(\boldsymbol{o}, \lambda) : \lambda \in \mathbb{R}\}$$

of trivial solutions from which the "branches"

$$\{(t\boldsymbol{u}, \lambda_0) : t \in \mathbb{R} \setminus \{0\}, \ \lambda_0 \text{ is an eigenvalue}$$
$$\text{and } \boldsymbol{u} \text{ a corresponding eigenvector of } \boldsymbol{A}\}$$

of nontrivial solutions "bifurcate" at the eigenvalues λ_0 (see Figure 4.1).

It is then natural to say that the eigenvalues of $\boldsymbol{A}$ are "the points of bifurcation" of $\boldsymbol{F}(\boldsymbol{x}, \lambda) = \boldsymbol{o}$.

---

[a]In what follows all the linear spaces are assumed to be real.

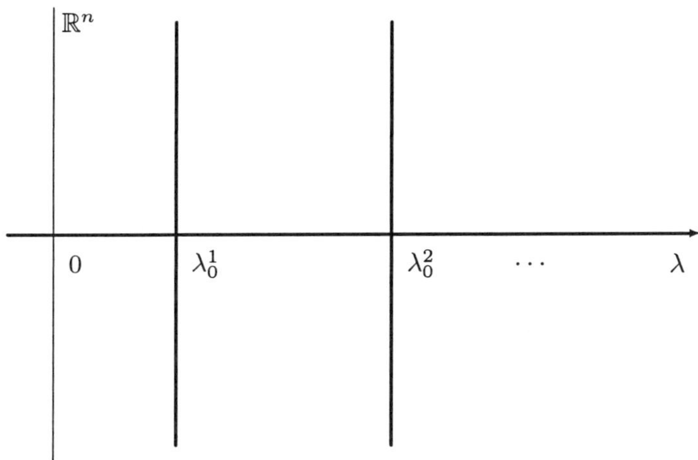

Figure 4.1.

Definition 4.1. Let $F(o, \lambda) = o$ for any $\lambda \in \mathbb{R}$. The point $\lambda_0 \in \mathbb{R}$ is then called a *point of bifurcation*, or a *bifurcation point* (of the solution set $\mathcal{S}$ subject to the line $\{(o, \lambda) : \lambda \in \mathbb{R}\}$, or shortly of (4.1)) if for any neighbourhood $\mathcal{U}$ of the point (o, λ_0) (in $X \times \mathbb{R}$) there exists at least one $(x, \lambda) \in \mathcal{U}$, $x \neq o$ such that $F(x, \lambda) = o$ (i.e. $(x, \lambda) \in \mathcal{S}$).

Example 4.2. Let $X = Y = \mathbb{R}$ and

$$F(x, \lambda) = x(\lambda - x^2). \tag{4.2}$$

Then the solution set $\mathcal{S}$ is sketched in Figure 4.2, and $\lambda_0 = 0$ is a point of bifurcation of $F(x, \lambda) = 0$.

Let us look at the point of bifurcation $\lambda_0 = 0$ from the following point of view:

"In any neighbourhood of $(0, \lambda_0)$ the solution set $\mathcal{S}$
cannot be expressed as a graph of a function of the variable λ."

If we confront this fact with the Implicit Function Theorem then we immediately get that the only "suspicious" points (to be bifurcation points) are those values of λ where

$$\frac{\partial F}{\partial x}(0, \lambda) = 0.$$

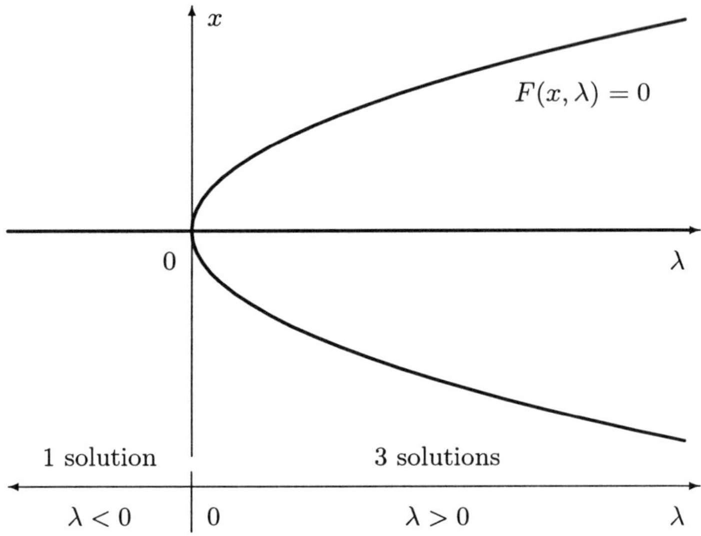

Figure 4.2.

In our case of F given by (4.2), we have

$$\frac{\partial F}{\partial x}(x, \lambda) = \lambda - 3x^2, \qquad \text{i.e.} \qquad \frac{\partial F}{\partial x}(0, \lambda) = \lambda.$$

Hence the only "suspicious" point is $\lambda = 0$ in this case.

In the general case $F\colon X \times \mathbb{R} \to Y$ where X and Y are Banach spaces, the condition $\frac{\partial F}{\partial x} \neq 0$ is substituted by the condition

$$F_x'(o, \lambda_0) \text{ is an isomorphism of } X \text{ onto } Y. \tag{4.3}$$

Thus we have the following *general necessary condition.*

Lemma 4.1. *Let F be continuously differentiable in some neighbourhood of (o, λ_0), $F(o, \lambda) = o$ for any $\lambda \in \mathbb{R}$. Let λ_0 be a point of bifurcation of $F(x, \lambda) = o$. Then $F_x'(o, \lambda_0)$ is not an isomorphism of X onto Y.*

Let us consider the special case $X = Y$ and

$$F(x, \lambda) = \lambda x - f(x).$$

Then

$$F_x'(o, \lambda_0) = \lambda_0 I - f'(o)$$

and the condition (4.3) does not hold if λ_0 belongs to the *spectrum* of the operator $f'(o)$. If, moreover, $f'(o)$ is *compact*, then the points of bifurcation *may be* only *eigenvalues* of the linear operator $f'(o)$ and $\lambda = 0$. On the other hand, if f is a compact linear operator, then every real point of its spectrum is a bifurcation point as well.

Theorem 4.1. *Let X be a real Banach space, $T\colon X \to X$ a compact linear operator,*

$$F(x,\lambda) = \lambda x - Tx, \qquad \lambda \in \mathbb{R}, \ x \in X.$$

Then λ_0 is a point of bifurcation of $F(x,\lambda) = o$ if and only if

$$\lambda_0 \in \sigma(T) \cap \mathbb{R}.$$

Proof.   It follows from Lemma 4.1 that if λ_0 is a point of bifurcation, then necessarily $\lambda_0 \in \sigma(T) \cap \mathbb{R}$.

Let us suppose now that $\lambda_0 \in \sigma(T) \cap \mathbb{R}$, and prove that λ_0 is a point of bifurcation of $F(x,\lambda) = o$. The fact $\lambda_0 \in \sigma(T) \cap \mathbb{R}$ implies that either λ_0 is an eigenvalue of T, or $\lambda_0 = 0$ and there is a sequence $\{\lambda_n\}$ of eigenvalues of T approaching 0. Assume first that λ_0 is an eigenvalue, and denote by u a corresponding eigenvector. For arbitrarily small neighbourhood $\mathcal{U}$ of the point (o,λ_0) there exists $t \neq 0$ (sufficiently small) such that

$$(tu,\lambda_0) \in \mathcal{U} \qquad \text{and} \qquad F(tu,\lambda_0) = t(\lambda_0 u - Tu) = o.$$

Hence λ_0 is a point of bifurcation.

Let now $0 = \lambda_0 = \lim\limits_{n\to\infty} \lambda_n$, λ_n be eigenvalues of T and u_n be corresponding eigenvectors. Let us fix some neighbourhood $\mathcal{U}$ of (o,λ_0). Then there exists $n_0 \in \mathbb{N}$ such that for all $n \geq n_0$ we have $(o,\lambda_n) \in \mathcal{U}$. Since $\mathcal{U}$ is an open set, we obtain $t_n \in \mathbb{R}$, $t_n \neq 0$, such that

$$(t_n u_n,\lambda_n) \in \mathcal{U} \qquad \text{and} \qquad F(t_n u_n,\lambda_n) = o.$$

Hence $\lambda_0 = 0$ is a bifurcation point. $\qquad\qquad\qquad\qquad\square$

Let us give examples which illustrate that the fact that T was a linear operator was essential in Theorem 4.1 (i.e. that $\sigma(T)$ cannot be replaced by $\sigma(T'(o))$ for T nonlinear).

Example 4.3. Let $f\colon \mathbb{R}^2 \to \mathbb{R}^2$ be defined by

$$f(x_1,x_2) = (x_1 + x_2^3, x_2 - x_1^3)$$

and

$$F(x,\lambda) = (\lambda I - f)(x), \qquad x = (x_1,x_2) \in \mathbb{R}^2.$$

Then

$$f'(z)x = (x_1 + 3z_2^2 x_2, x_2 - 3z_1^2 x_1), \qquad \text{i.e.} \qquad f'(o)x = (x_1, x_2).$$

Hence $f'(o) = I$, and $\lambda_0 = 1$ is its eigenvalue of multiplicity 2.

On the other hand, any solution (x, λ) of the equation $F(x, \lambda) = o$ satisfies

$$(\lambda - 1)x_1 - x_2^3 = 0,$$
$$(\lambda - 1)x_2 + x_1^3 = 0.$$

Multiplying the first equation by x_2, the second by x_1, and subtracting we get

$$x_1^4 + x_2^4 = 0, \qquad \text{i.e.} \qquad x_1 = x_2 = 0.$$

Hence there is no bifurcation point of $F(x, \lambda) = o$. This example shows that even if $\lambda = 1$ *is an eigenvalue of $f'(o)$ (of even multiplicity), it is not a point of bifurcation.*

Example 4.4. Let $f \colon \mathbb{R}^2 \to \mathbb{R}^2$ be defined by

$$f(x_1, x_2) = (2x_1 + 4x_1^3, 2x_2 + 4x_2^3),$$

and set

$$F(x, \lambda) = (\lambda I - f)(x), \qquad x = (x_1, x_2) \in \mathbb{R}^2.$$

Similarly as above we show that $\lambda_0 = 2$ is an eigenvalue of $f'(o)$ and its multiplicity is again 2.

Let us investigate now the solution set of $F(x, \lambda) = o$. The point (x, λ) solves $F(x, \lambda) = o$ if and only if

$$x_i(\lambda - 2 - 4x_i^2) = 0, \qquad i = 1, 2,$$

i.e.

$$x_i = 0 \qquad \text{or} \qquad x_i = \pm\frac{1}{2}\sqrt{\lambda - 2} \quad \text{for } \lambda \geq 2, \quad i = 1, 2.$$

Hence $\lambda_0 = 2$ is a point of bifurcation of $F(x, \lambda) = o$ (see Figure 4.3). Let us point out that $f = \Phi'$ where

$$\Phi(x_1, x_2) = x_1^2 + x_2^2 + x_1^4 + x_2^4.$$

This was not the case in the previous example.

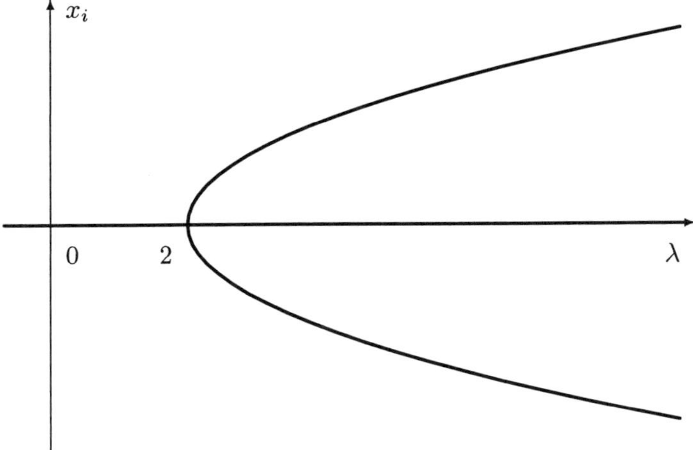

Figure 4.3.

The examples above illustrate the following *general scheme* for

$$F(x, \lambda) = \lambda x - f(x)$$

where f is compact and differentiable at o:

- f is *linear* — points of bifurcation coincide with $\sigma(f'(o)) \cap \mathbb{R}$ (Theorem 4.1),
- f is *nonlinear* — every eigenvalue of $f'(o)$ with odd algebraic multiplicity is a point of bifurcation,
  — an eigenvalue of $f'(o)$ with even multiplicity need not be a point of bifurcation (Example 4.3),
- f is *nonlinear and potential* (i.e., there is $\Phi \colon X \to \mathbb{R}$ such that $\Phi' = f$) — every *nonzero* eigenvalue of $f'(o)$ is a point of bifurcation (Example 4.4).

In the following sections we shall prove the latter two assertions in spaces of the infinite dimension.

4.4. Crandall–Rabinowitz Bifurcation Theorem

In this section we concentrate on a bifurcation result based on the Implicit Function Theorem. Let

$$F\colon \mathbb{R}^m \to \mathbb{R}, \quad m \in \mathbb{N},$$

be a function which is defined in some neighbourhood of the origin, and is of the class C^p $(p > 2)$. Assume that $F(o) = 0$ and o is a critical point of F, i.e. $F'(o) = o$. Then we have (see Appendix H)

$$F(x) = \frac{1}{2}(F''(o)x, x) + o(\|x\|^2), \qquad \|x\| \to 0.$$

If $F''(o)$ is a regular $m \times m$ matrix, then there is a neighbourhood of the origin in which the only critical point of F is o.

The following Morse Lemma says that in such a neighbourhood we can find a (nonlinear) transformation of variables $x \mapsto \xi$ for which

$$F(x) = \frac{1}{2}(F''(o)\xi, \xi),$$

i.e., the graph of F is a "quadratic manifold" in the new variables.

Lemma 4.2 (A. P. Morse). *Let* $F\colon \mathbb{R}^m \to \mathbb{R}$ *(*$m \in \mathbb{N}$*) be a function defined in some neighbourhood of the origin,* $F \in C^p$, $p > 2$. *Let*

$$F(o) = 0, \qquad F'(o) = o, \qquad F''(o) \text{ be a regular } m \times m \text{ matrix.}$$

Then there exists $\xi\colon \mathbb{R}^m \to \mathbb{R}^m$ *defined in some (smaller) neighbourhood of the origin,* $\xi \in C^{p-2}$, *and such that*

$$\xi(o) = o, \qquad \xi'(o) = I,$$
$$F(x) = \frac{1}{2}\big(F''(o)\xi(x), \xi(x)\big).$$

The proof of this assertion is based on the Implicit Function Theorem (see Appendix A), and we shall omit it here (see Appendix B). We concentrate now on the description of the solution set of the equation

$$G(x) = 0, \qquad x = (x_1, x_2) \tag{4.4}$$

where $G(x) = (Ax, x)$ and A is a regular symmetric matrix 2×2. Then (4.4) reads as

$$a_{11}x_1^2 + 2a_{12}x_1x_2 + a_{22}x_2^2 = 0. \tag{4.5}$$

If $a_{11} \neq 0$, then we can factorize the left hand side of (4.5) as follows

$$\left(a_{11}x_1 + \left(a_{12} - \sqrt{D}\right)x_2\right)\left(a_{11}x_1 + \left(a_{12} + \sqrt{D}\right)x_2\right) = 0$$

where

$$D = a_{12}^2 - a_{11}a_{22}.$$

If $\boldsymbol{A}$ is positive (negative) definite ($D < 0$), the solution set $\mathcal{S}$ of (4.4) consists of one point:

$$\mathcal{S} = \{(0,0)\}.$$

For $D = 0$ ($\boldsymbol{A}$ is semidefinite) we have that

$$\mathcal{S} = \{(x_1, x_2) \in \mathbb{R}^2 : a_{11}x_1 + a_{12}x_2 = 0\}.$$

The most interesting structure of $\mathcal{S}$ we have if $\boldsymbol{A}$ is an indefinite matrix ($D > 0$). In this case,

$$\mathcal{S} \text{ is a union of two lines which intersect at the origin.}$$

These facts combined with the Morse Lemma yield the following assertion.

Theorem 4.2. *Let $\mathcal{U}$ be some neighbourhood of $(0,0) \in \mathbb{R}^2$, $F \colon \mathbb{R}^2 \to \mathbb{R}$ is of the class $C^p(\mathcal{U})$, $p > 2$. Assume that*

$$F(\boldsymbol{o}) = 0, \qquad F'(\boldsymbol{o}) = \boldsymbol{o}, \tag{4.6}$$

$$\text{the symmetric matrix } F''(\boldsymbol{o}) \text{ is regular and indefinite.} \tag{4.7}$$

Then the solution set $\mathcal{S}$ of $F(\boldsymbol{x}) = 0$ in some (smaller) neighbourhood of $(0,0)$ is a union of two C^{p-2} curves Γ_1, Γ_2 which intersect transversally[b] just at the origin.

Proof. It follows from Lemma 4.2 that there exists $\boldsymbol{\xi} = (\xi_1, \xi_2) \colon \mathbb{R}^2 \to \mathbb{R}^2$ of the class C^{p-2} such that

$$\boldsymbol{\xi}(\boldsymbol{o}) = \boldsymbol{o}, \qquad \boldsymbol{\xi}'(\boldsymbol{o}) = \boldsymbol{I}, \tag{4.8}$$

$$F(\boldsymbol{x}) = \left(\frac{1}{2}F''(\boldsymbol{o})\boldsymbol{\xi}(\boldsymbol{x}), \boldsymbol{\xi}(\boldsymbol{x})\right)$$

for $\boldsymbol{x}$ from some neighbourhood of $(0,0)$. Denote

$$\boldsymbol{A} = \frac{1}{2}F''(\boldsymbol{o}).$$

---

[b]i.e., there are tangent lines of Γ_1, Γ_2 at $(0,0)$ which intersect with a nonzero angle.

If $a_{11} \neq 0$, the equation $(\boldsymbol{A}\boldsymbol{\xi}, \boldsymbol{\xi}) = 0$ is equivalent to

$$\left(a_{11}\xi_1 + \left(a_{12} - \sqrt{D}\right)\xi_2\right)\left(a_{11}\xi_1 + \left(a_{12} + \sqrt{D}\right)\xi_2\right) = 0. \qquad (4.9)$$

Indefiniteness of $\boldsymbol{A}$ yields

$$0 > \det \boldsymbol{A} = -D.$$

The solutions of (4.9) form then two lines

$$p_1 = \left\{(\xi_1, \xi_2) : a_{11}\xi_1 + \left(a_{12} - \sqrt{D}\right)\xi_2 = 0\right\},$$

$$p_2 = \left\{(\xi_1, \xi_2) : a_{11}\xi_1 + \left(a_{12} + \sqrt{D}\right)\xi_2 = 0\right\}$$

which intersect at $(0,0)$ with a nonzero angle $\alpha \neq 0$.

It follows now from (4.8) that the solution set of $F(\boldsymbol{x}) = 0$ (in some neighbourhood of the origin) is formed by two curves

$$\Gamma_1 = \left\{(x_1, x_2) : a_{11}\xi_1(\boldsymbol{x}) + \left(a_{12} - \sqrt{D}\right)\xi_2(\boldsymbol{x}) = 0\right\},$$

$$\Gamma_2 = \left\{(x_1, x_2) : a_{11}\xi_1(\boldsymbol{x}) + \left(a_{12} + \sqrt{D}\right)\xi_2(\boldsymbol{x}) = 0\right\}.$$

These curves are of the class C^{p-2}, and their tangents in $(0,0)$ are p_1 and p_2 (see Figure 4.4).

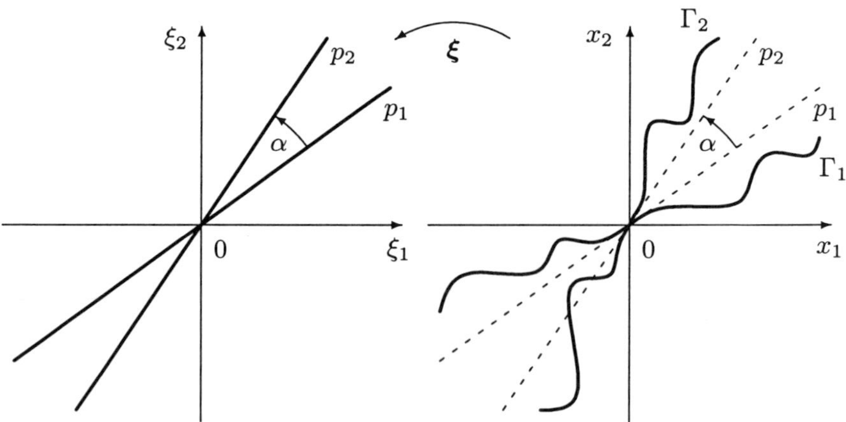

Figure 4.4.

The proof for $a_{11} = 0$ is similar (we would use $(2a_{12}\xi_1 + a_{22}\xi_2)\xi_2 = 0$ instead of (4.9)). $\qquad\square$

Corollary 4.1. *Let $f = f(x, \lambda)\colon \mathbb{R} \times \mathbb{R} \to \mathbb{R}$, $f \in C^p(\mathcal{U})$, $p > 2$, $\mathcal{U}$ is some neighbourhood of $(0,0) \in \mathbb{R}^2$. Let*

$$f(0,0) = \frac{\partial f}{\partial x}(0,0) = \frac{\partial f}{\partial \lambda}(0,0) = 0, \tag{4.10}$$

$$\frac{\partial^2 f}{\partial \lambda^2}(0,0) = 0, \qquad \frac{\partial^2 f}{\partial x \partial \lambda}(0,0) \neq 0. \tag{4.11}$$

Then the solution set of $f(x, \lambda) = 0$ in some (smaller) neighbourhood of $(0,0)$ consists of two curves

$$\Gamma_1 = \{(x, \lambda) : x = \hat{x}(\lambda),\ \lambda \in (-\varepsilon, \varepsilon)\}, \qquad \hat{x}(0) = \hat{x}'(0) = 0, \tag{4.12}$$

$$\Gamma_2 = \{(x, \lambda) : \lambda = \hat{\lambda}(x),\ x \in (-\varepsilon, \varepsilon)\}, \qquad \hat{\lambda}(0) = 0 \tag{4.13}$$

where $\Gamma_i \in C^{p-2}$, $i = 1, 2$ (see Figure 4.5).

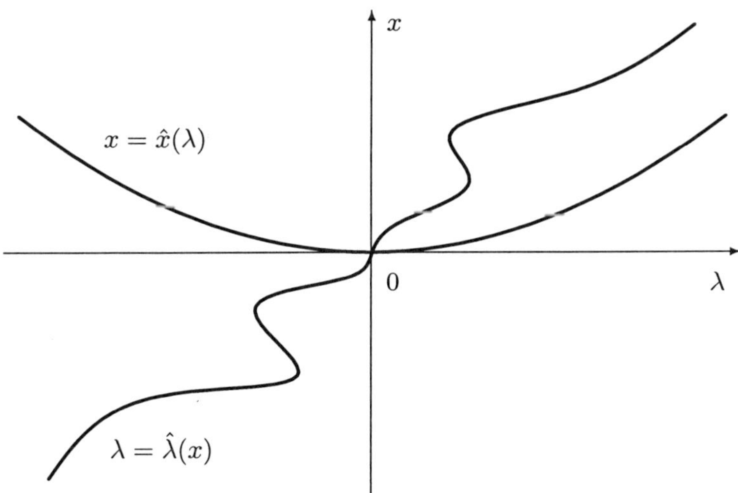

Figure 4.5.

Proof. The function f satisfies the assumptions of Theorem 4.2. Indeed, (4.6) follows from (4.10), and due to (4.11) we have that the determinant

118   *P. Drábek*

of the matrix

$$f''(0,0) = \begin{pmatrix} \dfrac{\partial^2 f}{\partial x^2}(0,0) & \dfrac{\partial^2 f}{\partial x \partial \lambda}(0,0) \\[2ex] \dfrac{\partial^2 f}{\partial \lambda \partial x}(0,0) & \dfrac{\partial^2 f}{\partial \lambda^2}(0,0) \end{pmatrix}$$

is

$$\det f''(0,0) = -\left[\frac{\partial^2 f}{\partial x \partial \lambda}(0,0) \right]^2 < 0.$$

Assume first $\frac{\partial^2 f}{\partial x^2}(0,0) \neq 0$. By the same reason as in the proof of Theorem 4.2 the set of all zero points of f in some neighbourhood of $(0,0)$ is the union of the curves

$$\Gamma_1 = \{(x,\lambda) : \xi_1(x,\lambda) = 0\}$$

and

$$\Gamma_2 = \left\{ (x,\lambda) : \frac{\partial^2 f}{\partial x^2}(0,0)\xi_1(x,\lambda) + 2\frac{\partial^2 f}{\partial x \partial \lambda}(0,0)\xi_2(x,\lambda) = 0 \right\}.$$

It follows from the Implicit Function Theorem that we can parametrize them as in (4.12) and (4.13).

If $\frac{\partial^2 f}{\partial x^2}(0,0) = 0$, then the equation

$$\left(\frac{1}{2}f''(0,0)\boldsymbol{\xi}(x,\lambda), \boldsymbol{\xi}(x,\lambda) \right) = 0$$

takes the form

$$\frac{\partial^2 f}{\partial x \partial \lambda}(0,0)\xi_1(x,\lambda)\xi_2(x,\lambda) = 0,$$

and the set of all zero points is formed by

$$\Gamma_1 = \{(x,\lambda) : \xi_1(x,\lambda) = 0\} \quad \text{and} \quad \Gamma_2 = \{(x,\lambda) : \xi_2(x,\lambda) = 0\}.$$

Applying the Implicit Function Theorem, we arrive again at (4.12), (4.13). $\square$

The following assertion is a simple *one-dimensional* version of the *Crandall–Rabinowitz Bifurcation Theorem.*

Corollary 4.2. *Let $f = f(x, \lambda)\colon \mathbb{R} \times \mathbb{R} \to \mathbb{R}$ be of the class C^p, $p > 2$, in some neighbourhood $\mathcal{U}$ of the point $(0, \lambda_0)$. Let*

$$f(0, \lambda) = 0 \qquad \forall \lambda \in \mathbb{R}, \tag{4.14}$$

$$\frac{\partial f}{\partial x}(0, \lambda_0) = 0, \tag{4.15}$$

$$\frac{\partial^2 f}{\partial x \partial \lambda}(0, \lambda_0) \neq 0. \tag{4.16}$$

Then λ_0 is a point of bifurcation of $f(x, \lambda) = 0$. In some (smaller) neighbourhood of $(0, \lambda_0)$ nonzero solutions form the set $\Gamma \setminus \{(0, \lambda_0)\}$ where

$$\Gamma = \{(x, \lambda) : \lambda = \hat{\lambda}(x),\ x \in (-\varepsilon, \varepsilon)\},$$

$\Gamma \in C^{p-2}$, and $\hat{\lambda}(0) = \lambda_0$ (see Figure 4.6).

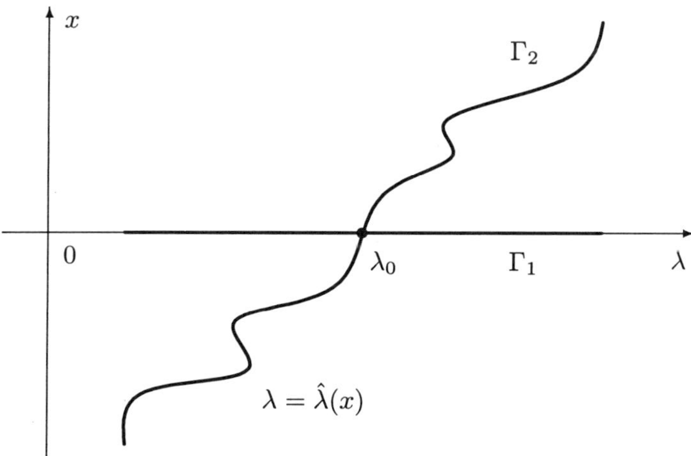

Figure 4.6.

Proof. We can assume, without loss of generality, that $\lambda_0 = 0$. Thanks to (4.14) we have

$$\frac{\partial f}{\partial \lambda}(0, 0) = \frac{\partial^2 f}{\partial \lambda^2}(0, 0) = 0.$$

Hence (4.10), (4.11) are fulfilled. In this case, however,

$$\Gamma_1 = \{(x, \lambda) : x = 0\}.$$

The nonzero points of the other curve $\Gamma_2 = \Gamma$ are the only nontrivial solutions of $f(x, \lambda) = 0$ in some neighbourhood of $(0, 0)$. $\square$

Remark 4.1. The set $\Gamma \setminus \{(0, \lambda_0)\}$ will be called a *branch of nontrivial solutions* of $f(x, \lambda) = 0$.

Remark 4.2. In particular, the condition (4.16) guarantees that Γ_1 and Γ_2 intersect transversally. It follows from here that if Γ_1 is formed by trivial solutions, the set $\Gamma_2 \setminus \{(0, \lambda_0)\}$ is a branch of nontrivial solutions of $f(x, \lambda) = 0$ in some neighbourhood of $(0, \lambda_0)$.

Exercise 4.3. In the following cases decide if $\lambda_0 = 0$ is a point of bifurcation of $f(x, \lambda) = 0$:

(1) $f(x, \lambda) = x^3 - \lambda x,$
(2) $f(x, \lambda) = x^3 - \lambda x - \sin \lambda x,$
(3) $f(x, \lambda) = x^3 - \lambda x + \sin \lambda x,$
(4) $f(x, \lambda) = x^3 - \lambda^2 x.$

In the following exposition we shall present the so called *Lyapunov--Schmidt Reduction* which is one of the main tools used to study nonlinear equations and boundary value problems.

Let $f\colon X \to Y$, X and Y be Banach spaces, and $f(o) = o$. Let us look for all solutions of the equation

$$f(x) = o \tag{4.17}$$

in some neighbourhood $\mathcal{U}$ of $x = o$. Assume that $f \in C^p(\mathcal{U})$, $p \in \mathbb{N}$. Let us denote

$$X_1 = Ker\, f'(o), \qquad Y_1 = Im\, f'(o),$$

and assume that $f'(o)$ is a *Fredholm operator*, i.e.,

(1) $\dim X_1 = d < \infty,$
(2) Y_1 is a closed subspace of Y,
(3) $\operatorname{codim} Y_1 = m < \infty.$

It follows from 1–3 that there are closed subspaces $X_2 \subset X$, $Y_2 \subset Y$, $\dim Y_2 = m < \infty$, such that $X = X_1 \oplus X_2$, $Y = Y_1 \oplus Y_2$. Let

$$P\colon X \to X_1, \qquad Q\colon Y \to Y_1$$

be the projections of X and Y onto X_1 and Y_1, respectively (see Figure 4.7).

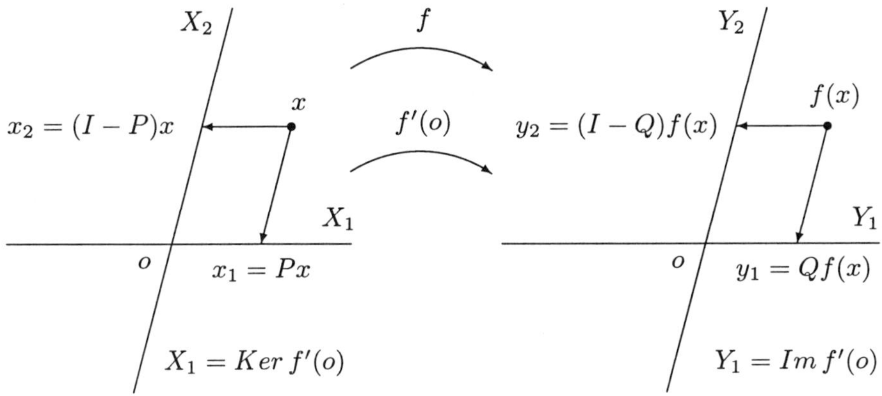

Figure 4.7.

Set $x_1 = Px$, $x_2 = (I - P)x$, i.e. $x = x_1 + x_2$. The equation (4.17) is equivalent to the system of equations

$$Qf(x_1 + x_2) = o, \tag{4.18}$$

$$(I - Q)f(x_1 + x_2) = o. \tag{4.19}$$

Let us denote $G(x_1, x_2) = Qf(x_1 + x_2)$. Then $G\colon X_1 \times X_2 \to Y_1$ and $G \in C^p(\mathcal{U})$ where $\mathcal{U}$ is some neighbourhood of the zero element in $X_1 \times X_2$, and $G(o, o) = o$. Moreover, $G_2'(o, o) = Qf'(o)|_{X_2}$ is an isomorphism of X_2 and Y_1.

Now it follows from the Implicit Function Theorem that there are neighbourhoods $\mathcal{W}_1$ of o in X_1, $\mathcal{W}_2$ of o in X_2 and a unique mapping $\hat{x}_2\colon \mathcal{W}_1 \to \mathcal{W}_2$ such that $\hat{x}_2(o) = o$ and

$$G(x_1, \hat{x}_2(x_1)) = o \qquad \forall x_1 \in \mathcal{W}_1.$$

We thus reduced the equation (4.17) to

$$(I - Q)f(x_1 + \hat{x}_2(x_1)) = o \tag{4.20}$$

on $\mathcal{W}_1 \subset X_1$.

We can formulate now the following assertion.

                                        P. Drábek

Theorem 4.4. *Solutions x of (4.17) in some neighbourhood of o in X and solutions x_1 of (4.20) in the corresponding neighbourhood of o in X_1 are in one to one correspondence:*

$$x = x_1 + \hat{x}_2(x_1).$$

Remark 4.3. The equation (4.20) will be called a *guiding equation* for (4.17). Under the assumptions 1–3 this is a finite dimensional equation because

$$H(x_1) := (I - Q)f(x_1 + \hat{x}_2(x_1))$$

maps X_1 ($\dim X_1 = d < \infty$) into Y_2 ($\dim Y_2 = m < \infty$). From this point of view, (4.20) is simpler than (4.17). At the same time, the reader should bear in mind that the solution set of the equation

$$\boldsymbol{H(x_1) = o} \qquad \text{for} \quad \boldsymbol{H} \colon \mathbb{R}^d \to \mathbb{R}^m$$

might be rather complicated even for $m = 1$!

The following assertion provides more information about the mapping $x_2 = \hat{x}_2(x_1)$.

Lemma 4.3. *Let $x_2 = \hat{x}_2(x_1) \colon \mathcal{W}_1 \to \mathcal{W}_2$ be as above. Then*

$$\hat{x}_2'(o) = o.$$

Proof. Due to the Implicit Function Theorem

$$\hat{x}_2'(o) = -[G_2'(o,o)]^{-1} \circ G_1'(o,o) \qquad \text{where} \quad G_1'(o,o) = Qf'(o)|_{X_1}.$$

Since $X_1 = Ker\, f'(o)$, the mapping $f'(o)|_{X_1}$ is the zero mapping. The mapping $G_2'(o,o)$ is an isomorphism, and so the assertion is proved. $\qquad \square$

Remark 4.4. The Lyapunov–Schmidt Reduction (Method) can be generalized also for the equation

$$f(x, \lambda) = o, \qquad x \in X,\ \lambda \in \mathbb{R} \tag{4.21}$$

where $f \colon X \times \mathbb{R} \to Y$, $f(o, \lambda) = o$ for all $\lambda \in \mathbb{R}$. We investigate (4.21) in some neighbourhood of $(x, \lambda) = (o, 0)$. The only difference consists in the fact that instead of X_1 we shall consider $X_1 \times \mathbb{R}$. The Implicit Function Theorem applied on

$$Qf(x, \lambda) = o$$

then yields a C^p-function $x_2 = \hat{x}_2(x_1, \lambda)$ such that

$$Qf(x_1 + \hat{x}_2(x_1, \lambda), \lambda) = o \qquad \forall (x_1, \lambda) \in \mathcal{W}_1 \times \Lambda$$

where $\mathcal{W}_1$ is a neighbourhood of o in $Ker\, f_1'(o,0)$, and Λ is a neighbourhood of 0 in $\mathbb{R}$.

In what follows we shall assume that X and Y are Banach spaces,

$$f\colon X \times \mathbb{R} \to Y$$

is a map of the class $C^p(\mathcal{U})$, $p > 2$, where $\mathcal{U}$ is some neighbourhood of $(o,0)$ in $X \times \mathbb{R}$, $f(o,0) = o$. We assume that $f_1'(o,0) \in \mathscr{L}(X,Y)$ is a Fredholm operator. Denote

$$X_1 = Ker\, f_1'(o,0), \qquad Y_1 = Im\, f_1'(o,0).$$

Let $X_2 \subset X$, $Y_2 \subset Y$ be complements of X_1, Y_1 such that

$$X = X_1 \oplus X_2, \quad Y = Y_1 \oplus Y_2.$$

Assume that

(1) $\dim X_1 = 1$, $\quad X_1 = Lin\{x_0\}$,
(2) $\mathrm{codim}\, Y_1 = 1$.

Since $\dim Y_2 = 1$ (by 2), there exists $y_0 \notin Y_1$ such that $Y_2 = Lin\{y_0\}$ (see Figure 4.8), and there exists a continuous linear functional $y^* \in Y^*$ such that[c]

$$y^*(y_0) = 1, \qquad y^*(y) = 0 \quad \forall y \in Y_1.$$

From the identity

$$y = Qy + (I - Q)y$$

then follows

$$(I - Q)y = y^*(y)y_0, \qquad Y_1 = Ker\, y^*. \tag{4.22}$$

Theorem 4.5 (Crandall–Rabinowitz). *In addition to* $\dim X_1 = 1$, $\mathrm{codim}\, Y_1 = 1$, *assume that*

$$f(o,\lambda) = o \qquad \forall \lambda \in \mathbb{R}, \tag{4.23}$$

$$f_{1,2}''(o,0)x_0 \notin Y_1. \tag{4.24}$$

---

[c]The functional y^* is easily constructed as a linear function on Y_2 extended to Y with the kernel equal to Y_1.

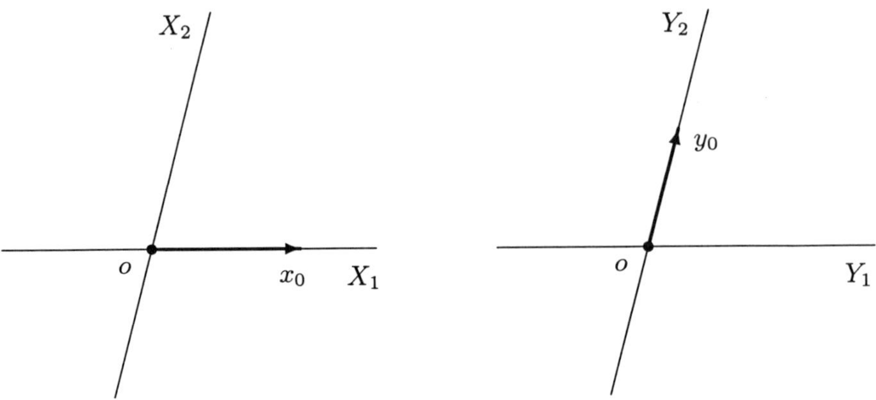

Figure 4.8.

Then $\lambda_0 = 0$ is a point of bifurcation of $f(x, \lambda) = o$. The nontrivial solutions emanating from the point (o, λ_0) form (in some neighbourhood of this point) a C^{p-2}-branch

$$\{(x, \lambda) : x = sx_0 + \tilde{x}_2(s), \ \lambda = \hat{\lambda}(s), \ s \in (-\varepsilon, \varepsilon)\} \setminus \{(o, \lambda_0)\} \qquad (4.25)$$

where

$$\tilde{x}_2(0) = o, \qquad \tilde{x}_2'(0) = o, \qquad \hat{\lambda}(0) = \lambda_0 = 0. \qquad (4.26)$$

Proof.   Let us derive the guiding equation for $f(x, \lambda) = o$. We identify X_1 with $\mathbb{R}$ using the following isomorphism:

$$(sx_0 =) \ x_1 \mapsto s.$$

It follows from Remark 4.4 that there exists a C^p-function $x_2 = \hat{x}_2(s, \lambda)$ which maps some neighbourhood of the zero in $\mathbb{R} \times \mathbb{R}$ into some neighbourhood of the zero element in X_2, such that

$$\hat{x}_2(0, \lambda) = o \qquad \forall \lambda \in \mathbb{R}, \qquad (4.27)$$

$$Qf(sx_0 + \hat{x}_2(s, \lambda), \lambda) = o \qquad (4.28)$$

for all (s, λ) from the neighbourhood of zero in $\mathbb{R} \times \mathbb{R}$. Lemma 4.3 then yields

$$\frac{\partial \hat{x}_2}{\partial s}(0, 0) = o. \qquad (4.29)$$

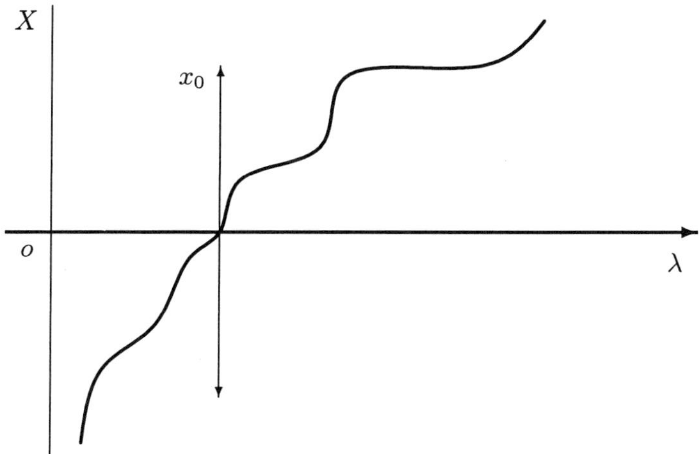

Figure 4.9.

The guiding equation for $f(x, \lambda) = o$ can be then written (see (4.22)) as

$$F(s, \lambda) = y^* \big(f(sx_0 + \hat{x}_2(s, \lambda), \lambda) \big) = 0. \tag{4.30}$$

Now we apply Corollary 4.2 to $F(s, \lambda)$ with $\lambda_0 = 0$. It follows from (4.23) and (4.27) that

$$F(0, \lambda) = 0 \qquad \forall \lambda \in \mathbb{R}.$$

Hence (4.14) holds true. Further, we have

$$\frac{\partial F}{\partial s}(s, \lambda) = y^* \left(f_1'(sx_0 + \hat{x}_2(s, \lambda), \lambda) \left(x_0 + \frac{\partial \hat{x}_2}{\partial s}(s, \lambda) \right) \right), \tag{4.31}$$

and so

$$\frac{\partial F}{\partial s}(0, 0) = y^*(f_1'(o, 0)x_0) = 0$$

(cf. (4.29)). Hence also (4.15) is verified. It follows from (4.31) that

$$\frac{\partial^2 F}{\partial s \partial \lambda}(0, 0) = \lim_{\lambda \to 0} \frac{\frac{\partial F}{\partial s}(0, \lambda)}{\lambda} = y^* \left(\lim_{\lambda \to 0} \frac{f_1'(o, \lambda)(x_0 + \frac{\partial \hat{x}_2}{\partial s}(0, \lambda))}{\lambda} \right) =$$

$$= y^*(f_{1,2}''(o, 0)x_0) + y^* \left(f_1'(o, 0)\frac{\partial^2 \hat{x}_2}{\partial s \partial \lambda}(0, 0) \right). \tag{4.32}$$

We have $f_1'(o,0)h \in Y_1$ for all $h \in X$, i.e.

$$y^*(f_1'(o,0)h) = 0$$

for all $h \in X$. Hence the second term in (4.32) is equal to 0. The first term in (4.32) is different from zero. Otherwise we should have $f_{1,2}''(o,0)x_0 \in Y_1$ which contradicts (4.24). Hence

$$\frac{\partial^2 F}{\partial s \partial \lambda}(0,0) \neq 0$$

which verifies (4.16).

It follows from Corollary 4.2 that nontrivial solutions of $F(s,\lambda) = 0$ form (in some neighbourhood of zero) a C^{p-2}-branch

$$\{(s,\lambda) : \lambda = \hat{\lambda}(s), \ s \in (-\varepsilon, \varepsilon)\} \setminus \{(0,0)\} \tag{4.33}$$

where $\hat{\lambda}(0) = 0$.

Now we apply an analogue of Theorem 4.4: the set of all nontrivial solutions of $f(x,\lambda) = o$ in some neighbourhood of $(o,0) \in X \times \mathbb{R}$ forms the set

$$\{(x,\lambda) : x = sx_0 + \hat{x}_2(s,\hat{\lambda}(s)), \ \lambda = \hat{\lambda}(s), \ s \in (-\varepsilon,\varepsilon)\} \setminus \{(o,0)\}$$

where $\hat{\lambda}(0) = 0$. Set $\tilde{x}_2(s) = \hat{x}_2(s,\hat{\lambda}(s))$. Then obviously $\tilde{x}_2(0) = o$ and

$$\tilde{x}_2'(s)|_{s=0} = \frac{\partial \hat{x}_2}{\partial s}(0,0) + \frac{\partial \hat{x}_2}{\partial \lambda}(0,0)\hat{\lambda}'(0) = o$$

according to (4.27) and (4.29). This proves (4.26). $\qquad\qquad \square$

Remark 4.5.

(1) It follows from the previous theorem that the essential information about the bifurcation branch is contained in the space $X_1 \times \mathbb{R}$. The part from X_2 is just a perturbation $\tilde{x}_2(s)$ which is of order $o(s)$ as $s \to 0$. From this point of view, the parametrization contains all the essential information about the branch.

(2) Assume that the function $\lambda = \hat{\lambda}(s)$ from (4.33) has the form

$$\hat{\lambda}(s) = \hat{\lambda}'(0)s + \frac{1}{2}\hat{\lambda}''(0)s^2 + \cdots,$$

and assume that

$$\hat{\lambda}^{(p)}(0) = 0 \qquad \text{for} \quad p = 1, 2, \ldots, q-1,$$
$$\hat{\lambda}^{(q)}(0) \neq 0.$$

If q is odd, we speak about a *transcritical* bifurcation, if q is even and $\hat{\lambda}^{(q)}(0) > 0$, we speak about a *supercritical* bifurcation, and finally, if q is even and $\hat{\lambda}^{(q)}(0) < 0$, the bifurcation is called *subcritical* (see Figure 4.10).

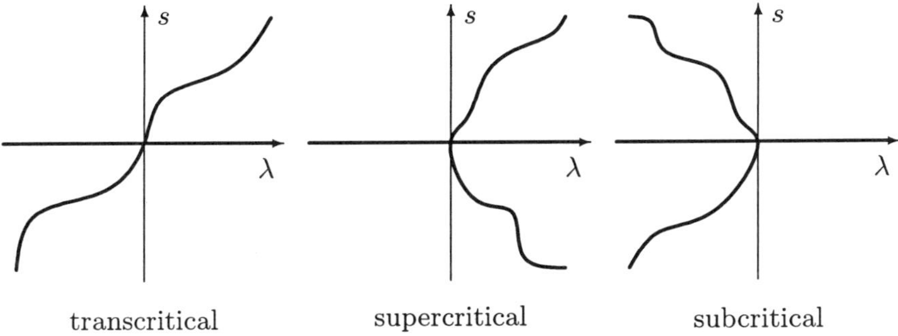

Figure 4.10.

Example 4.5. (Application of the Crandall–Rabinowitz Bifurcation Theorem) We shall study bifurcation points of the periodic problem

$$\begin{cases} u''(x) + \lambda u(x) + g(x, u(x), u'(x), \lambda) = 0 & \text{in} \quad [0, 2\pi], \\ u(0) = u(2\pi), \quad u'(0) = u'(2\pi). \end{cases} \tag{4.34}$$

In this example we shall concentrate on the point $\lambda = 0$ which is an eigenvalue of the associated eigenvalue problem

$$\begin{cases} u''(x) + \lambda u(x) = 0 & \text{in} \quad [0, 2\pi], \\ u(0) = u(2\pi), \quad u'(0) = u'(2\pi) \end{cases} \tag{4.35}$$

of the multiplicity 1.

We shall denote

$$X = C^2 P := \{u \colon \mathbb{R} \to \mathbb{R} : u \text{ is } 2\pi\text{-periodic}, \ u \in C^2(\mathbb{R})\},$$

$$Y = CP := \{u \colon \mathbb{R} \to \mathbb{R} : u \text{ is } 2\pi\text{-periodic}, \ u \in C(\mathbb{R})\}.$$

Then Y and X equipped with the norms

$$\|u\|_Y = \max\{|u(x)| : x \in [0, 2\pi]\}, \qquad \|u\|_X = \|u\|_Y + \|u'\|_Y + \|u''\|_Y,$$

respectively, are Banach spaces.

Let us define $F \colon X \times \mathbb{R} \to Y$ by

$$F(u, \lambda)(x) = u''(x) + \lambda u(x) + g(x, u(x), u'(x), \lambda)$$

where the function

$$g = g(x, \tau, t, \lambda)$$

satisfies the following hypotheses:

(1) g is 2π-periodic in x and continuous with respect to all four variables (as a function from $\mathbb{R}^4$ into $\mathbb{R}$);
(2) the derivatives of g with respect to τ, t, λ up to the order p $(p > 2)$ are continuous functions from $\mathbb{R}^4$ into $\mathbb{R}$;
(3) $g(x, 0, 0, \lambda) = 0$ for all $x, \lambda \in \mathbb{R}$;
(4) $g'_\tau(x, 0, 0, \lambda) = g'_t(x, 0, 0, \lambda) = 0$ for all $x, \lambda \in \mathbb{R}$.

It follows from 3 that $F(o, \lambda) = o$ for all $\lambda \in \mathbb{R}$. Moreover, thanks to 4 we have

$$F'_u(o, \lambda)w = w'' + \lambda w,$$

and so we conclude

$$\dim Ker\, F'_u(o, 0) = 1.$$

It follows from 2 that $F \in C^p$. The Fredholm Alternative (for linear operators) yields

$$Y_1 = Im\, F'_u(o, 0) = \left\{ \tilde{w} \in Y : \int_0^{2\pi} \tilde{w}(x)\, \mathrm{d}x = 0 \right\}.$$

Hence $Im\, F'_u(o, 0)$ is a closed subspace of Y of the co-dimension 1. Set

$$x_0 = 1, \qquad X_1 = Lin\{1\}, \qquad X_2 = \left\{ \tilde{u} \in X : \int_0^{2\pi} \tilde{u}(x)\, \mathrm{d}x = 0 \right\}.$$

Since

$$F''_{u\lambda}(o, 0)1 = 1 \qquad \text{and} \qquad 1 \notin Im\, F'_u(o, 0),$$

the transversality condition (4.24) is verified, too.

It follows from the Crandall–Rabinowitz Bifurcation Theorem (Theorem 4.5) that $\lambda = 0$ is a point of bifurcation of (4.34). In particular, the point $(o, 0) \in X \times \mathbb{R}$ belongs to the branch of trivial solutions (o, λ), but also to the branch

$$\Gamma = \{(s + \tilde{u}(s), \hat{\lambda}(s)) : s \in (-\varepsilon, \varepsilon)\}, \qquad \tilde{u}(0) = o,\ \tilde{u}'_s(0) = o,\ \hat{\lambda}(0) = 0.$$

Hence for any $s \in (-\varepsilon, \varepsilon)$, $s \neq 0$, the nontrivial solution $s + \tilde{u}(s)$ is the sum of a constant function (with respect to x) and the perturbed function $\tilde{u}(s)$ (which depends on x) such that $\tilde{u}(s)$ belongs to X_2.

Remark 4.6. Theorem 4.5 can be generalized for the situation when

$$0 < \dim X_1 = d < \infty.$$

The proof then requires a generalization of the Morse Lemma (Lemma 4.2) and some other properties of Fredholm mappings (see, e.g., [9]).

4.5. Local bifurcation theorems

The basic assertion of this section is the following theorem. For the notions "degree of the mapping" and "index" see Appendix C.

Theorem 4.6. *Let X be a Banach space, Ω an open set in $X \times \mathbb{R}$, and define a mapping $f \colon \Omega \to X$ by*

$$f(x, \lambda) = x - h(x, \lambda), \quad (x, \lambda) \in \Omega. \tag{4.36}$$

Assume that

$$h \text{ is a compact (nonlinear) mapping from } \Omega \text{ into } X, \tag{4.37}$$

$$h(o, \lambda) = o \quad \forall (o, \lambda) \in \Omega,$$

$$(o, \lambda_0) \in \Omega. \tag{4.38}$$

Let for all $\lambda \in (\lambda_0 - \varepsilon_0, \lambda_0 + \varepsilon_0)$, with $\varepsilon_0 > 0$ sufficiently small, the Leray–Schauder index of o with respect to $f(\cdot, \lambda)$ and the point o be defined (and we denote it by $i_\lambda(o)$). Assume

$$i_{\lambda_0 + \varepsilon}(o) \neq i_{\lambda_0 - \varepsilon}(o) \tag{4.39}$$

for all $\varepsilon \in (0, \varepsilon_0)$. Then λ_0 is a point of bifurcation of

$$f(x, \lambda) = o.$$

Proof. We proceed via contradiction. Assume that λ_0 is not a bifurcation point. Then there exists a neighbourhood $\mathcal{U}$ of (o, λ_0) such that the only solutions of $f(x, \lambda) = o$ belonging to $\mathcal{U}$ are of the form (o, λ). Take $\varepsilon > 0$ so small that (4.39) is satisfied, and, moreover,

$$\overline{B(o; \varepsilon)} \times [\lambda_0 - \varepsilon, \lambda_0 + \varepsilon] \subset \mathcal{U}$$

(see Figure 4.11).

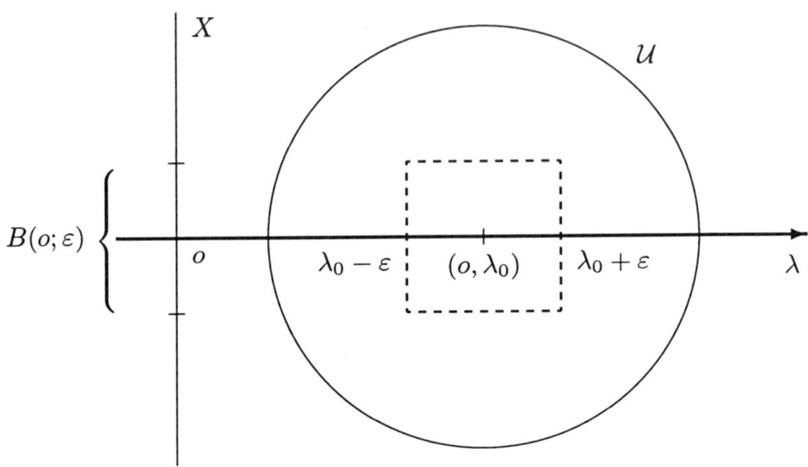

Figure 4.11.

Then for any $\lambda \in [\lambda_0 - \varepsilon, \lambda_0 + \varepsilon]$ and $x \in \partial B(o; \varepsilon)$ we have that

$$f(x, \lambda) \neq o.$$

It follows from the homotopy invariance property of the Leray–Schauder degree (Appendix C) that (thanks to (4.37))

$$i_{\lambda_0 + \varepsilon}(o) = \deg \left[f(\cdot, \lambda_0 + \varepsilon); B(o; \varepsilon), o \right] =$$

$$= \deg \left[f(\cdot, \lambda_0 - \varepsilon); B(o; \varepsilon), o \right] = i_{\lambda_0 - \varepsilon}(o)$$

which contradicts (4.39). $\qquad \square$

The proof of the above theorem is very simple. However, the basic question connected with its application consists in the assumption (4.39) which may be very difficult to verify. Below we present a lemma which states how to verify (4.39) for a *special form of* $f = f(x, \lambda)$.

Lemma 4.4. *Let X be a Banach space, Ω an open set in $X \times \mathbb{R}$. Assume*

$$T \text{ is a linear compact operator from } X \text{ into } X, \qquad (4.40)$$

$$G \text{ is a compact (nonlinear) operator from } \Omega \text{ into } X, \qquad (4.41)$$

$$\lim_{\|x\| \to 0} \frac{G(x, \lambda)}{\|x\|} = o \text{ holds uniformly with respect to } \lambda \in \mathcal{J} \qquad (4.42)$$
$$\text{for any bounded set } \mathcal{J} \subset \mathbb{R},$$

$$\lambda_0 \neq 0, \qquad (o, \lambda_0) \in \Omega. \qquad (4.43)$$

Let λ_0 be an eigenvalue of T of odd (algebraic) multiplicity. Set

$$f(x, \lambda) = \lambda x - Tx + G(x, \lambda) = \lambda(x - h(x, \lambda))$$

with

$$h(x, \lambda) = \frac{1}{\lambda}(Tx - G(x, \lambda)).$$

Then for any $\varepsilon > 0$ small enough the index of o with respect to $\frac{1}{\lambda_0 \pm \varepsilon} f(\cdot, \lambda_0 \pm \varepsilon)$ and the point o (we denote it by $i_{\lambda_0 \pm \varepsilon}(o)$) is well defined, and, moreover,

$$i_{\lambda_0+\varepsilon}(o) \neq i_{\lambda_0-\varepsilon}(o). \tag{4.44}$$

Note that $\lambda_0 \neq 0$ guarantees that $\varepsilon > 0$ can be chosen so small that $0 \notin [\lambda_0 - \varepsilon, \lambda_0 + \varepsilon]$. The index of o with respect to $\frac{1}{\lambda_0 \pm \varepsilon} f(\cdot, \lambda_0 \pm \varepsilon)$ and the point o, and the solution set of $f(x, \lambda \pm \varepsilon) = o$ then coincide with the index of o with respect to $I - h(\cdot, \lambda_0 \pm \varepsilon)$ and the point o, and with the solution set of $x - h(x, \lambda \pm \varepsilon) = o$.

Proof. [of Lemma **4.4**] We can assume, without loss of generality, that $\lambda_0 > 0$, and choose $\varepsilon > 0$ so small that $\lambda_0 - \varepsilon > 0$ and

$$P_\sigma(T) \cap [\lambda_0 - \varepsilon, \lambda_0 + \varepsilon] = \{\lambda_0\}$$

(here $P_\sigma(T)$ is the point of spectrum of T). Now we apply the Leray–Schauder Index Formula (Theorem 4.16). By (4.40) and (4.42) we have

$$f_x'(o, \lambda)y = \lim_{t \to 0} \frac{\lambda ty - T(ty) + G(ty, \lambda)}{t} = \lambda y - Ty.$$

According to the choice of ε, the mapping $\frac{1}{\lambda} f(\cdot, \lambda)$ satisfies the assumptions of Theorem 4.16 for any λ such that

$$0 < |\lambda - \lambda_0| \leq \varepsilon.$$

Then

$$i_{\lambda_0+\varepsilon}(o) = \deg\left[\frac{1}{\lambda_0 + \varepsilon} f(\cdot, \lambda_0 + \varepsilon); B(o; \varepsilon), o\right] = (-1)^{\beta_+}$$

where β_+ is the sum of algebraic multiplicities of all real eigenvalues μ of $\frac{1}{\lambda_0+\varepsilon} T$ which are greater than 1. However, μ is an eigenvalue of $\frac{1}{\lambda_0+\varepsilon} T$ if and only if $\lambda = \mu(\lambda_0 + \varepsilon)$ is an eigenvalue of T. Hence $\mu > 1$ if and only if λ (as an eigenvalue of T) is greater than λ_0. So

$$\beta_+ = \sum_{\substack{\lambda \in P_\sigma(T) \cap \mathbb{R} \\ \lambda > \lambda_0}} n_\lambda \qquad \text{where} \quad n_\lambda = \dim \bigcup_{k=1}^{\infty} Ker\,(\lambda I - T)^k$$

(algebraic multiplicity of λ). In the same way we prove

$$i_{\lambda_0-\varepsilon}(o) = (-1)^{\beta_-} \qquad \text{where} \quad \beta_- = \sum_{\substack{\lambda \in P_\sigma(T) \cap \mathbb{R} \\ \lambda \geq \lambda_0}} n_\lambda = \beta_+ + n_{\lambda_0}.$$

Since n_{λ_0} is an odd number, we have

$$i_{\lambda_0-\varepsilon}(o) = (-1)^{\beta_-} \neq (-1)^{\beta_+} = i_{\lambda_0+\varepsilon}(o). \qquad \square$$

The following example shows that in the previous lemma it is necessary to consider the algebraic multiplicity of λ_0, and so, in particular, the assumption that $\dim Ker\,(\lambda I - T)$ is odd is not enough.

Example 4.6. Let $X = \mathbb{R}^2$, $\boldsymbol{T} = \begin{pmatrix} 1 & 1 \\ 0 & 1 \end{pmatrix}$, $\boldsymbol{G}\colon \mathbb{R}^2 \to \mathbb{R}^2$, $\boldsymbol{G} = (g_1, g_2)$ where $\boldsymbol{x} = (x_1, x_2)$,

$$g_1(x_1, x_2) = -x_2^3, \qquad g_2(x_1, x_2) = x_1^3.$$

The mapping $\boldsymbol{f} = (f_1(\boldsymbol{x}, \lambda), f_2(\boldsymbol{x}, \lambda))$ is defined as

$$f_1(\boldsymbol{x}, \lambda) = (\lambda - 1)x_1 - x_2 - x_2^3, \qquad f_2(\boldsymbol{x}, \lambda) = (\lambda - 1)x_2 + x_1^3.$$

The matrix $\boldsymbol{T}$ has just one eigenvalue $\lambda = 1$,

$$\dim Ker\,(\boldsymbol{I} - \boldsymbol{T}) = 1, \qquad \dim Ker\,(\boldsymbol{I} - \boldsymbol{T})^2 = 2.$$

The equation $\boldsymbol{f}(\boldsymbol{x}, \lambda) = \boldsymbol{o}$ has for any $\lambda \in \mathbb{R}$ only the trivial solution $\boldsymbol{x} = \boldsymbol{o}$.

We have also the following *necessary condition* for $\lambda_0 \neq 0$ to be a point of bifurcation.

Lemma 4.5. *Let $\lambda_0 \neq 0$ be a point of bifurcation of $f(x, \lambda) = o$, and assume (4.40)–(4.43). Then λ_0 is an eigenvalue of T.*

Proof. If λ_0 is a point of bifurcation of $f(x, \lambda) = o$, then there exists a sequence $\{(x_n, \lambda_n)\} \subset \Omega$ such that $o \neq x_n \to o$, $\lambda_n \to \lambda_0$, $f(x_n, \lambda_n) = o$. Set

$$v_n = \frac{x_n}{\|x_n\|}.$$

Then $\{v_n\}$ is a bounded sequence, and the compactness of T implies that we can select a convergent subsequence from $\{Tv_n\}$ (denoted again by $\{Tv_n\}$). Dividing $f(x_n, \lambda_n) = o$ by $\|x_n\|$ we obtain

$$\lambda_n v_n = T v_n - \frac{G(x_n, \lambda_n)}{\|x_n\|}. \tag{4.45}$$

Due to (4.42) and the convergence of $\{Tv_n\}$ we have from (4.45) that $\{\lambda_n v_n\}$ converges to an element $\lambda_0 v$, $v \in X$ (we have $\lambda_0 \neq 0$). Taking the limits in (4.45), we arrive at

$$\lambda_0 v = Tv.$$

Since $\|v\| = \lim_{n \to \infty} \|v_n\| = 1$, v is an eigenvector of T associated with the eigenvalue λ_0. $\qquad\square$

Maybe, the most important sufficient condition is the following immediate consequence of Theorem 4.6 and Lemma 4.4.

Theorem 4.7 (Krasnoselski). *Let X be a Banach space, Ω an open set in $X \times \mathbb{R}$ and let $f \colon \Omega \to X$ be of the form*

$$f(x, \lambda) = \lambda x - Tx + G(x, \lambda) \tag{4.46}$$

for all $(x, \lambda) \in \Omega$. Here

$$T \text{ is a linear compact operator from } X \text{ into } X, \tag{4.47}$$

$$G \colon \Omega \to X \text{ is a compact (nonlinear) operator,} \tag{4.48}$$

$$G(x, \lambda) = o(\|x\|), \qquad \|x\| \to 0$$
$$\text{uniformly for all } \lambda \in \mathcal{J} \text{ for any bounded set } \mathcal{J} \subset \mathbb{R}, \tag{4.49}$$

$$(o, \lambda_0) \in \Omega, \tag{4.50}$$

$$\lambda_0 \text{ is a nonzero eigenvalue of } T$$
$$\text{with odd (algebraic) multiplicity.} \tag{4.51}$$

Then λ_0 is a point of bifurcation of

$$f(x, \lambda) = o.$$

Example 4.7. (Application of the Krasnoselski Local Non-potential Bifurcation Theorem) Let us consider the Dirichlet boundary value problem

$$\begin{cases} u''(x) + \lambda u(x) + g(x, u(x), \lambda) = 0 & \text{in} \quad (0, \pi), \\ u(0) = u(\pi) = 0. \end{cases} \tag{4.52}$$

We shall summarize first the results about the linear problem

$$\begin{cases} u''(x) + \lambda u(x) = -f(x) & \text{in } (0, \pi), \\ u(0) = u(\pi) = 0 \end{cases} \tag{4.53}$$

where $f \in L^1(0,\pi)$. By a *weak solution* of (4.53) we understand a function $u \in W_0^{1,2}(0,\pi)$ for which the integral identity

$$\int_0^\pi \left(u'(x)w'(x) - \lambda u(x)w(x)\right)\,\mathrm{d}x = \int_0^\pi f(x)w(x)\,\mathrm{d}x$$

holds for all $w \in W_0^{1,2}(0,\pi)$. In Appendix D we prove that if $f \in C[0,\pi]$, then any weak solution u of (4.53) satisfies $u \in C^2[0,\pi]$, $u(0) = u(\pi) = 0$, i.e., u is a classical solution.

The weak formulation of (4.53) can be restated as the operator equation

$$u - \lambda Bu = Bf \tag{4.54}$$

where the operator $B\colon L^1(0,\pi) \to W_0^{1,2}(0,\pi)$ is defined by

$$\langle Bu, v\rangle = \int_0^\pi u(x)v(x)\,\mathrm{d}x \tag{4.55}$$

for any $u \in L^1(0,\pi)$, $v \in W_0^{1,2}(0,\pi)$. Here $\langle \cdot, \cdot\rangle$ denotes the scalar product in $W_0^{1,2}(0,\pi)$ defined by

$$\langle u, v\rangle = \int_0^\pi u'(x)v'(x)\,\mathrm{d}x.$$

Then $B\colon L^1(0,\pi) \to W_0^{1,2}(0,\pi)$ is a continuous linear operator due to the continuity of the embedding of $W_0^{1,2}(0,\pi)$ into $C[0,\pi]$.

A number λ is an eigenvalue of the boundary value problem

$$\begin{cases} u''(x) + \lambda u(x) = 0 & \text{in} \quad (0,\pi), \\ u(0) = u(\pi) = 0 \end{cases} \tag{4.56}$$

if there is a nonzero function $u \in C^2[0,\pi]$ which solves (4.56). Since in (4.56) we have $f \equiv 0$, i.e. $f \in C[0,\pi]$, Theorem 4.17 implies that the problem (4.56) is equivalent to the operator equation

$$u - \lambda Bu = o, \tag{4.57}$$

and so λ is an eigenvalue of (4.56) if and only if $\lambda \neq 0$ and $\lambda = \frac{1}{\mu}$ where μ is an eigenvalue of B.

The operator $B\colon W_0^{1,2}(0,\pi) \to W_0^{1,2}(0,\pi)$ is compact. This follows easily from the compact embedding of $W_0^{1,2}(0,\pi)$ into $L^1(0,\pi)$. It is easy to verify that B is a self-adjoint operator from $W_0^{1,2}(0,\pi)$ into $W_0^{1,2}(0,\pi)$, and that the eigenvalues μ_n and the corresponding normalized eigenfunctions $u_n = u_n(x)$ are of the form

$$\mu_n = \frac{1}{n^2}, \qquad u_n(x) = \frac{1}{n}\sqrt{\frac{2}{\pi}}\sin nx, \qquad n \in \mathbb{N}.$$

It then follows from the Fredholm Alternative that

$$Im\,(I - \lambda B) = W_0^{1,2}(0, \pi) \qquad \text{if} \quad \lambda \neq n^2, \quad n \in \mathbb{N},$$

$$Im\,(I - n^2 B) = \{F \in W_0^{1,2}(0, \pi) : \langle F, \sin nx \rangle = 0\}, \qquad n \in \mathbb{N}.$$

Let us prove that

$$Ker\,(I - n^2 B)^2 = Ker\,(I - n^2 B), \qquad n \in \mathbb{N}. \tag{4.58}$$

This would imply that the algebraic multiplicity of the eigenvalue $\mu_n = \frac{1}{n^2}$ is 1.

Let $u \in Ker\,(I - n^2 B)^2$, i.e.

$$(I - n^2 B)^2 u = (I - n^2 B)(I - n^2 B)u = o.$$

Then for some $a \in \mathbb{R}$ we have

$$(I - n^2 B)u = a \sin nx.$$

Thus $a \sin nx \in Im\,(I - n^2 B)$, and so

$$\langle a \sin nx, \sin nx \rangle = a \,\|\sin nx\|^2 = 0.$$

Since $\|\sin nx\| \neq 0$, it must be $a = 0$. Then $(I - n^2 B)u = o$, i.e. $u \in Ker\,(I - n^2 B)$, and (4.58) is proved.

We shall transform the boundary value problem (4.52) to the form

$$f(u, \lambda) := u - \lambda Bu + G(u, \lambda) = o.$$

We set $X = W_0^{1,2}(0, \pi)$,

$$G(u, \lambda) = BN(u, \lambda), \qquad N(u, \lambda)(x) = g(x, u(x), \lambda).$$

If $g = g(x, s, \lambda)$ is a continuous function from $[0, \pi] \times \mathbb{R} \times \mathbb{R}$ into $\mathbb{R}$, then G is a compact operator from $W_0^{1,2}(0, \pi) \times \mathbb{R}$ into $W_0^{1,2}(0, \pi)$. This fact follows easily from the continuity of the Nemytski operator $N\colon C[0, \pi] \times \mathbb{R} \to C[0, \pi]$, and the embeddings $W_0^{1,2}(0, \pi) \subset\subset C[0, \pi]$ and $C[0, \pi] \subset L^1(0, \pi)$.

Let us assume that

$$g(x, s, \lambda) = o(|s|), \qquad |s| \to 0,$$

uniformly with respect to $x \in [0, \pi]$ and λ from any bounded set in $\mathbb{R}$. This assumption implies

$$G(u, \lambda) = o(\|u\|), \qquad \|u\| \to 0,$$

uniformly with respect to λ from any bounded set in $\mathbb{R}$. Now we apply Theorem 4.7 to get the assertion that "$\lambda = n^2$ *is a point of bifurcation of* (4.52)".

Remark 4.7. (Comparison of the Crandall–Rabinowitz Bifurcation Theorem and the Krasnoselski Local Non-potential Bifurcation Theorem) Let us consider f of the form

$$f(x, \mu) = x - \mu T x + G(x, \mu)$$

where

$$T \colon X \to X \text{ is a compact linear operator,} \\ X \text{ is a Banach space,} \tag{4.59}$$

$$G \colon X \times \mathbb{R} \to X \text{ is a compact (nonlinear) operator,} \tag{4.60}$$

$$\lim_{\|x\| \to 0} \frac{G(x, \mu)}{\|x\|} = o \text{ holds uniformly with respect to } \mu \\ \text{from any bounded set in } \mathbb{R}. \tag{4.61}$$

Let λ_0 be a nonzero eigenvalue of T and

$$\dim Ker\,(\lambda_0 I - T) = 1. \tag{4.62}$$

Denote by v, $\|v\| = 1$, an eigenvector of T, associated with λ_0, and set $\mu_0 = \frac{1}{\lambda_0}$.

Let us compare the assumptions of Theorems 4.5 and 4.7. One of the essential differences consists in the smoothness assumptions: while in Theorem 4.5 we must have $G \in C^p$, $p > 2$, Theorem 4.7 requires G compact (and hence continuous) but also "small" at zero. The assumption (4.61) guarantees that

$$A = f_1'(o, \mu_0) = I - \mu_0 T, \qquad f_{1,2}''(o, \mu_0) = -T. \tag{4.63}$$

Theorem 4.5 requires

$$Im\,A = \overline{Im\,A}, \tag{4.64}$$

$$\operatorname{codim} Im\,A = 1, \tag{4.65}$$

$$f_{1,2}''(o, \mu_0)v \notin Im\,A. \tag{4.66}$$

The compactness of T implies that A is a Fredholm operator of index 0, hence (4.64) holds, and also

$$\operatorname{codim} Im\,A = \dim Ker\,A = 1,$$

i.e., (4.65) is true, too. The last assumption (4.66) is closely connected with the algebraic multiplicity of λ_0 as follows from the assertion below.

The assumption (4.66) *is verified if and only if*

$$n_{\lambda_0} = \dim \bigcup_{k=1}^{\infty} Ker\, A^k = 1. \tag{4.67}$$

Let us prove first (4.67) $\Rightarrow$ (4.66). Assume the contrary: $f''_{1,2}(o, \mu_0)v \in Im\, A$. According to (4.63) it means $-Tv \in Im\, A$. Since $v = \mu_0 Tv$, we have $v \in Im\, A$, too. Then there exists $w \in X$ such that $v = Aw$. But $v \in Ker\, A$, i.e. $w \in Ker\, A^2$. At the same time $w \notin Ker\, A$. This implies

$$\dim Ker\, A^2 > \dim Ker\, A, \qquad \text{i.e.} \qquad n_{\lambda_0} > 1,$$

a contradiction.

Let us prove now (4.66) $\Rightarrow$ (4.67). Let $w \in Ker\, A^2$, and set $u = Aw$. Then $Au = A^2 w = o$ what implies $u \in Ker\, A$. Since $Ker\, A$ is generated by v, there exists $a \in \mathbb{R}$ such that $u = av$. At the same time $u = Aw \in Im\, A$. For $a \neq 0$ we would have $-Tv = -\lambda_0 v = -\frac{\lambda_0}{a} u \in Im\, A$, a contradiction with (4.66). Hence $a = 0$ and $u = Aw = o$, i.e. $w \in Ker\, A$. This proves

$$Ker\, A^2 \subset Ker\, A.$$

Since the opposite inclusion is evident, we proved

$$Ker\, A^2 = Ker\, A.$$

By induction we now easily prove that

$$Ker\, A^{n+1} = Ker\, A^n$$

for any $n \in \mathbb{N}$.

Exercise 4.8. Consider the boundary value problem (4.52) and formulate the most general assumptions on g (with g depending also on u') in such a way that the Crandall–Rabinowitz Bifurcation Theorem will apply. Compare then your result with the result of Example 4.7 in this section.

4.6. Global bifurcation theorems

In this section we shall write the bifurcation equation in the form

$$f(x, \mu) := x - \mu Tx + G(x, \mu) = o \tag{4.68}$$

while the assumptions about T and G will be the same as in the previous section. The following result is due to Rabinowitz [6], [7].

Theorem 4.9 (Rabinowitz Global Bifurcation Theorem). *Let X be a Banach space, Ω an open set in $X \times \mathbb{R}$, $(o, \mu_0) \in \Omega$, $\mu_0 \neq 0$. Let us assume:*

$$T \text{ is a compact linear operator from } X \text{ into } X, \qquad (4.69)$$

$$G \text{ is a compact (nonlinear) operator from } \Omega \text{ into } X, \qquad (4.70)$$

$$G(x, \mu) = o(\|x\|), \qquad \|x\| \to 0,$$
$$\text{uniformly for } \mu \text{ from any bounded set in } \mathbb{R}, \qquad (4.71)$$

$$\lambda_0 = \frac{1}{\mu_0} \text{ is an eigenvalue of } T \text{ of odd (algebraic) multiplicity.} \quad (4.72)$$

Denote by $\mathcal{S}$ the closure of all solutions of (4.68) with $x \neq o$, i.e.

$$\mathcal{S} = \overline{\{(x, \mu) \in \Omega : x \neq o, \ f(x, \mu) = o\}},$$

and let $\mathcal{C}$ be a component of $\mathcal{S}$ which contains the point (o, μ_0).
    Then either

(1) $\mathcal{C}$ is not a compact set in Ω,
   or
(2) $\mathcal{C}$ contains an even number of points (o, μ) where $\mu \neq 0$ and $\lambda = \frac{1}{\mu}$ is an eigenvalue of T with odd (algebraic) multiplicity.

Proof. We shall follow the proof of Izé [4]. The idea is following. We will *assume that $\mathcal{C}$ is compact, and prove that it contains an even number of points described in 2.*

Since $\mathcal{C}$ is compact, it contains only a finite number of points (o, μ) where $\mu \neq 0$ and $\lambda = \frac{1}{\mu}$ is an eigenvalue of T (see Figure 4.12): we shall denote them by

$$(o, \mu_0), \ldots, (o, \mu_{k-1}).$$

Since $\mathcal{C}$ is a component of $\mathcal{S}$ in Ω and $\mathcal{S}$ is closed, there exists an open set $\tilde{\Omega} \subset \Omega$ such that $\tilde{\Omega}$ separates $\mathcal{C}$ from $\mathcal{S} \setminus \mathcal{C}$, i.e. $\mathcal{C} \subset \tilde{\Omega}$ and $\tilde{\Omega} \cap (\mathcal{S} \setminus \mathcal{C}) = \emptyset$. We prove that $\tilde{\Omega}$ can be chosen in such a way that (see Figure 4.13)

(a) $\tilde{\Omega}$ is open and bounded,
(b) $\mathcal{C} \subset \tilde{\Omega}$, $\mathcal{S} \cap \partial \tilde{\Omega} = \emptyset$,
(c) $(o, \mu_j) \in \tilde{\Omega}$, $j = 0, \ldots, k-1$, but
    $(o, \mu) \notin \tilde{\Omega}$ for $\mu \neq 0$, $\lambda = \frac{1}{\mu} \in P_\sigma(T)$, $\mu \neq \mu_j$, $j = 0, \ldots, k-1$.

Indeed, let $\mathcal{U}$ be a δ-neighbourhood of $\mathcal{C}$ such that $\mathcal{U} \setminus \mathcal{C}$ does not contain any point (o, μ), $\mu \neq 0$, $\frac{1}{\mu} \in P_\sigma(T)$. The set $\mathcal{K} = \overline{\mathcal{U}} \cap \mathcal{S}$ is then compact,

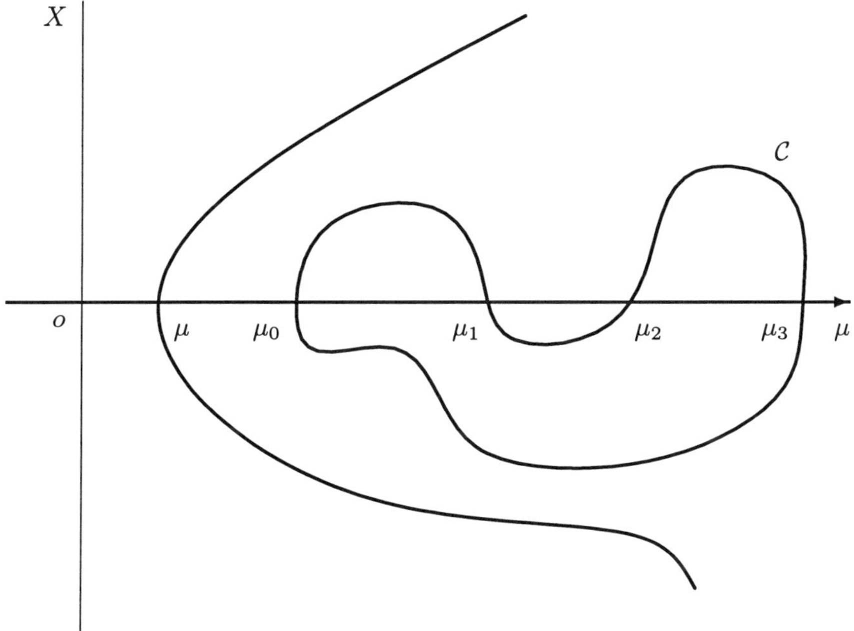

Figure 4.12.

and obviously $C \cap (\partial \mathcal{U} \cap \mathcal{S}) = \emptyset$. Then there exist compact disjoint sets $\mathcal{K}_1, \mathcal{K}_2 \subset \mathcal{K}$ such that

$$\mathcal{K} = \mathcal{K}_1 \cup \mathcal{K}_2, \qquad C \subset \mathcal{K}_1, \qquad \partial \mathcal{U} \cap \mathcal{S} \subset \mathcal{K}_2.$$

Hence $\tilde{\Omega}$ can be chosen as an ε-neighbourhood of $\mathcal{K}_1$ with

$$\varepsilon < \min \{ dist \, (\mathcal{K}_1, \mathcal{K}_2), \delta \}.$$

For any $r > 0$ define $f_r \colon \tilde{\Omega} \to X \times \mathbb{R}$ as follows:

$$f_r(x, \mu) - (f(x, \mu), \|x\|^2 - r^2). \tag{4.73}$$

Then obviously

$$f_r(x, \mu) = o \qquad \Longleftrightarrow \qquad f(x, \mu) = o \quad \text{and} \quad \|x\| = r.$$

(In other words, the function f_r "counts" the solutions of $f(x, \mu) = o$ which belong to the sphere $\|x\| = r$.) Then, thanks to the choice of $\tilde{\Omega}$ and the

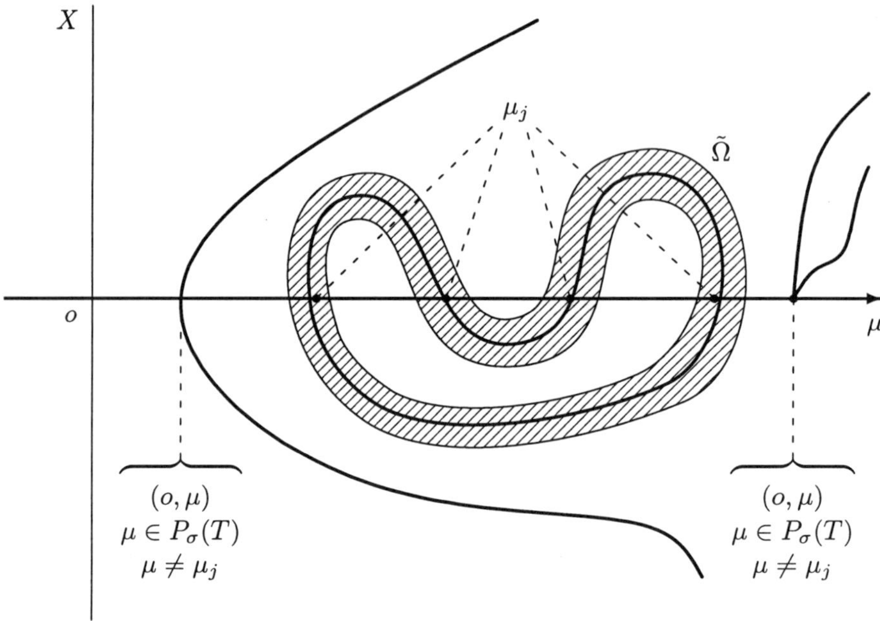

Figure 4.13.

homotopy invariance property of the degree (Appendix C), we have that

$$\deg\left[f_r; \tilde{\Omega}, o\right]$$

is well defined and independent of $r > 0$.

The rest of the proof consists in the calculation of this degree for sufficiently large r and for sufficiently small r.

Step 1: sufficiently large r. The boundedness of $\tilde{\Omega}$ implies that there exists $C > 0$ such that for any $(x, \mu) \in \tilde{\Omega}$ we have $\|x\| < C$. Then for $r > C$ the equation

$$f_r(x, \mu) = o$$

has no solution in $\tilde{\Omega}$, and so, according to 5 of Theorem 4.15, we have

$$\deg[f_r; \tilde{\Omega}, o] = 0.$$

Step 2: sufficiently small r. For $j = 0, \ldots, k - 1$ set

$$\mathcal{U}_j(r, \varepsilon) := \{(x, \mu) : \|x\|^2 + |\mu - \mu_j|^2 < r^2 + \varepsilon^2\},$$

and choose $\varepsilon > 0$ so small that the sets $\overline{\mathcal{U}_j(\varepsilon, \varepsilon)}$ are pairwise disjoint, belong to $\tilde{\Omega}$, and do not contain $(o, 0)$ (see Figure 4.14).

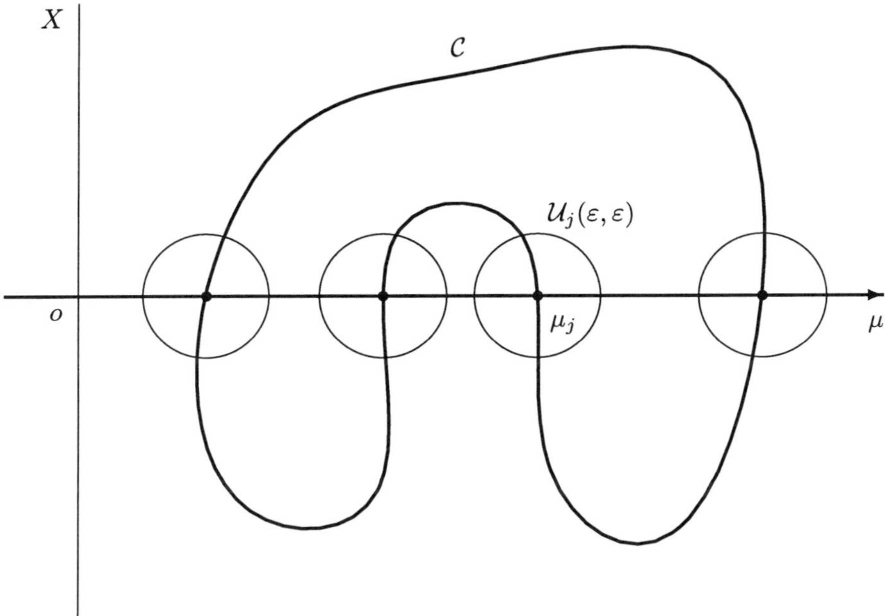

Figure 4.14.

We prove first that there exists $r > 0$ $(r \leq \varepsilon)$ such that

$$x - \mu T x + t G(x, \mu) \neq o \tag{4.74}$$

for all

$$t \in [0, 1], \quad (x, \mu) \in \overline{\tilde{\Omega}}, \quad 0 < \|x\| \leq r, \quad |\mu - \mu_j| \geq \varepsilon, \quad j = 0, \dots, k - 1.$$

Indeed, assume via contradiction that such $r > 0$ does not exist. Then there exist $t_n \in [0, 1]$ and $(x_n, \mu^n) \in \overline{\tilde{\Omega}}$, $n \in \mathbb{N}$, $o \neq x_n \to o$, $|\mu^n - \mu_j| \geq \varepsilon$, $j = 0, \dots, k - 1$, *not* satisfying (4.74), i.e.

$$x_n - \mu^n T x_n + t_n G(x_n, \mu^n) = o. \tag{4.75}$$

We can assume, without loss of generality, that $\mu^n \to \tilde{\mu}$. It follows from the construction of $\tilde{\Omega}$ that $\frac{1}{\tilde{\mu}} \notin P_\sigma(T)$. On the other hand, it follows from

(4.75) that (setting $y_n = \frac{x_n}{\|x_n\|}$)

$$y_n - \mu^n T y_n + t_n \frac{G(x_n, \mu^n)}{\|x_n\|} = o. \qquad (4.76)$$

Now, the compactness of T and (4.71) imply that for some $y \neq o$ ($y_{n_k} \to y$ for some subsequence) we have

$$y - \tilde{\mu} T y = o,$$

a contradiction.

We shall write $\mathcal{U}_j = \mathcal{U}_j(r, \varepsilon)$ for simplicity. It follows from 2 of Theorem 4.15 that

$$\deg\left[f_r; \tilde{\Omega}, o\right] = \sum_{j=0}^{k-1} \deg\left[f_r; \mathcal{U}_j, o\right]. \qquad (4.77)$$

Let μ_j be fixed. It follows from the choice of $\varepsilon > 0$ that for $0 < |\mu - \mu_j| \leq \varepsilon$ we have

$$\frac{1}{\mu} \notin P_\sigma(T).$$

Then for any such μ the degree

$$\deg\left[I - \mu T; B(o; r), o\right]$$

is well defined. Moreover, the homotopy invariance property of the degree implies that it is locally constant with respect to μ. Denote

$$i_-^j = \deg\left[I - (\mu_j - \varepsilon)T; B(o; r), o\right],$$
$$i_+^j = \deg\left[I - (\mu_j + \varepsilon)T; B(o; r), o\right].$$

It follows from Lemma 4.9 that

$$\deg\left[f_r; \mathcal{U}_j, o\right] = i_-^j - i_+^j. \qquad (4.78)$$

If n_j is the (algebraic) multiplicity of $\lambda_j = \frac{1}{\mu_j}$, then it follows from Theorem 4.16 that

$$i_+^j = (-1)^{n_j} i_-^j.$$

Hence for n_j even we obtain

$$\deg\left[f_r; \mathcal{U}_j, o\right] = i_-^j - i_+^j = 0, \qquad (4.79)$$

while for n_j odd we have

$$\deg\left[f_r; \mathcal{U}_j, o\right] = 2 i_-^j. \qquad (4.80)$$

It follows from (4.77)–(4.80) that

$$\deg\left[f_r; \tilde{\Omega}, o\right] = 2 \sum_{\substack{j=0 \\ n_j \text{ odd}}}^{k-1} i_-^j.$$

Since this degree must be equal to zero (see Step 1 of this proof), there must be an even number of eigenvalues of odd algebraic multiplicity among $\mu_0, \ldots, \mu_{k-1}$. $\qquad\square$

Corollary 4.3. *If $\Omega = X \times \mathbb{R}$ in Theorem 4.9, then the first alternative 1 reduce to $\mathcal{C}$ is unbounded in $X \times \mathbb{R}$ and 2 remains unchanged.*

Proof. Let $(x, \mu) \in \mathcal{C}$. Then

$$x = \mu T x - G(x, \mu).$$

This implies that if $\mathcal{C}$ is bounded in $X \times \mathbb{R}$, it is also relatively compact because

$$A(x, \mu) = \mu T x - G(x, \mu)$$

is a compact operator. But $\mathcal{C}$ is closed in the complete space $X \times \mathbb{R}$, and so it is compact. We thus proved that if $\mathcal{C}$ is bounded in $X \times \mathbb{R}$, it is also compact. $\qquad\square$

We shall now pay special attention to the bifurcation from *simple eigenvalues*. Let us assume that X is a Hilbert space and T is a self-adjoint operator (these assumptions are technical and can be dropped). Let λ_0 be an eigenvalue of T, $n_{\lambda_0} = 1$, and let v, $\|v\| = 1$, be the eigenvector associated with λ_0. For $\varepsilon \in (0, 1)$ define (see Figure 4.15)

$$\mathcal{K}_\varepsilon = \{(x, \mu) \in X \times \mathbb{R} : |\langle x, v \rangle| > \varepsilon \|x\|\},$$

$$\mathcal{K}_\varepsilon^+ = \{(x, \mu) \in X \times \mathbb{R} : \langle x, v \rangle > \varepsilon \|x\|\},$$

$$\mathcal{K}_\varepsilon^- = \{(x, \mu) \in X \times \mathbb{R} : -\langle x, v \rangle > \varepsilon \|x\|\}.$$

Let

$$\mathcal{S} = \overline{\{(x, \mu) \in X \times \mathbb{R} : x \neq o, \ f(x, \mu) = o\}},$$

and $\mathcal{C}$ be the component of $\mathcal{S}$ containing the point (o, μ_0).

It is possible to prove that there exists $t_0 > 0$ such that

$$(\mathcal{S} \setminus \{(o, \mu_0)\}) \cap \overline{B((o, \mu_0); t_0)} \subset \mathcal{K}_\varepsilon$$

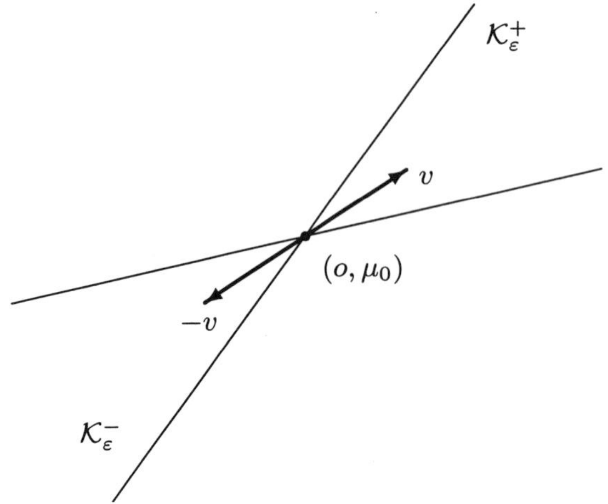

Figure 4.15.

(see Dancer [2]). For $t \in (0, t_0]$ define (see Figure 4.16)

$$\mathcal{D}_t^+ = \{(o, \mu_0)\} \cup (\mathcal{S} \cap \overline{B((o, \mu_0); t)} \cap \mathcal{K}_\varepsilon^+),$$

$$\mathcal{D}_t^- = \{(o, \mu_0)\} \cup (\mathcal{S} \cap \overline{B((o, \mu_0); t)} \cap \mathcal{K}_\varepsilon^-).$$

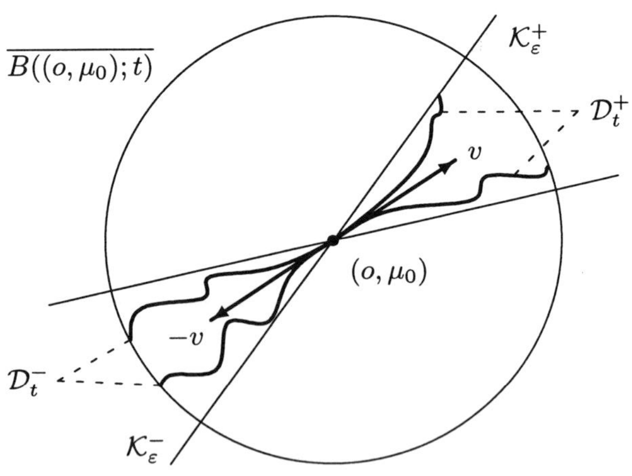

Figure 4.16.

Further, let $\mathcal{C}_t^+$ be the component of $\overline{\mathcal{C} \setminus \mathcal{D}_t^-}$ containing (o, μ_0), and, similarly, $\mathcal{C}_t^-$ be the component of $\overline{\mathcal{C} \setminus \mathcal{D}_t^+}$ containing (o, μ_0) (see Figure 4.17). Finally, we put

$$\mathcal{C}^+ = \overline{\bigcup_{0 < t \leq t_0} \mathcal{C}_t^+}, \qquad \mathcal{C}^- = \overline{\bigcup_{0 < t \leq t_0} \mathcal{C}_t^-}.$$

Then $\mathcal{C}^+$ and $\mathcal{C}^-$ are connected sets and $\mathcal{C}^+ \cup \mathcal{C}^- \subset \mathcal{C}$.

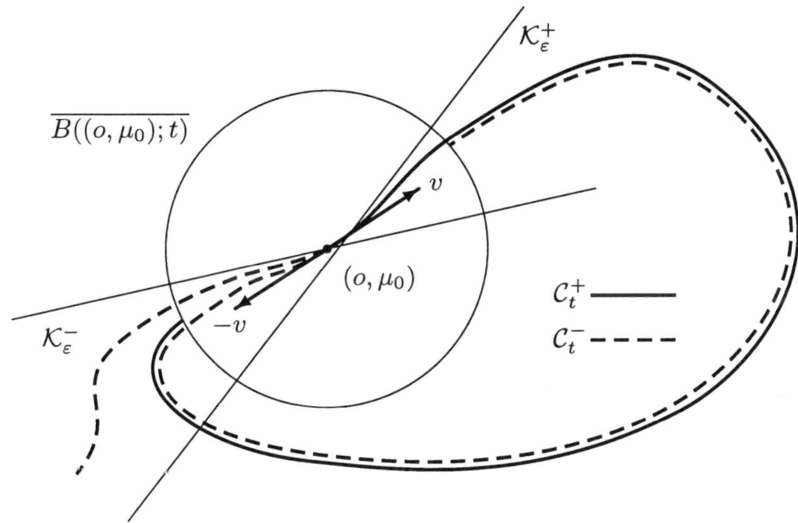

Figure 4.17.

Theorem 4.10 (Dancer Global Bifurcation Theorem). *The sets $\mathcal{C}^+$ and $\mathcal{C}^-$ are either both unbounded, or*

$$\mathcal{C}^+ \cap \mathcal{C}^- \neq \{(o, \mu_0)\}.$$

The proof of this assertion follows similar ideas as that of Theorem 4.9. The reader can found it in [2].

Remark 4.8. The meaning of $\mathcal{C}^\pm$ is the following. The sets $\mathcal{C}^\pm$ describe the branches of nontrivial solutions which bifurcate in the *direction of the eigenvectors* $\pm v$ (see Figure 4.18).

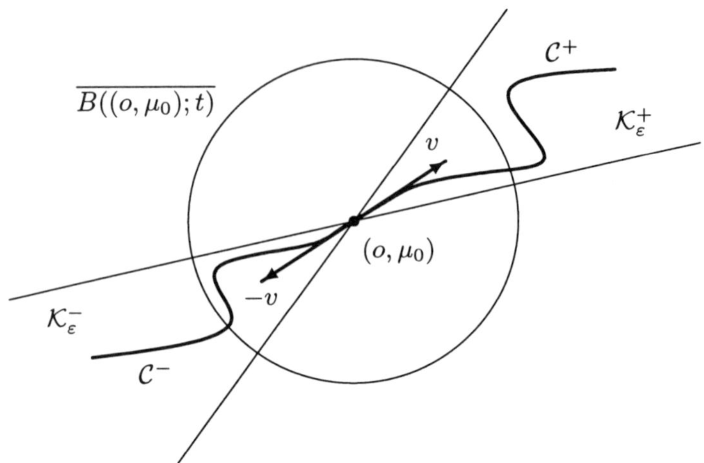

Figure 4.18.

Example 4.8. (Application of the Dancer Global Bifurcation Theorem) Let us consider again the Dirichlet boundary value problem[d]

$$\begin{cases} u''(x) + \lambda u(x) + g(x, u(x), \lambda) = 0 & \text{in} \quad (0, \pi), \\ u(0) = u(\pi) = 0, \end{cases} \tag{4.81}$$

and assume that $g = g(x, s, \lambda)$ is a continuous function from $[0, \pi] \times \mathbb{R} \times \mathbb{R}$ into $\mathbb{R}$ satisfying

$$g(x, s, \lambda) = o(|s|), \qquad |s| \to 0,$$

uniformly with respect to $x \in [0, \pi]$ and λ from any bounded set in $\mathbb{R}$. Since $\lambda = n^2$, $n \in \mathbb{N}$, is a simple eigenvalue of (4.56), we can apply both global bifurcation theorems of this section.

It follows from Theorem 4.10 that there are maximal connected sets $\mathcal{C}^+$ and $\mathcal{C}^-$ of solutions of (4.81) which both contain the point (o, n^2), and either both are unbounded, or $\mathcal{C}^+ \cap \mathcal{C}^- \neq \{(o, n^2)\}$. We show that the latter case cannot occur if $g = g(x, s, \lambda)$ is locally Lipschitz continuous with respect to the variable s.

The function g is locally Lipschitz continuous with respect to s if for any interval $\mathcal{J} \subset \mathbb{R}$ there exists a constant $k > 0$ such that for any $x \in [0, \pi]$ and any $s_1, s_2, \lambda \in \mathcal{J}$ we have

$$|g(x, s_1, \lambda) - g(x, s_2, \lambda)| \leq k|s_1 - s_2|.$$

---

[d]We shall use the same notation as in Section 4.5.

This property of g implies that the initial value problem (with λ fixed)

$$\begin{cases} u''(x) + \lambda u(x) + g(x, u(x), \lambda) = 0, \\ u(x_0) = u_0, \quad u'(x_0) = u_1, \qquad x_0 \in [0, \pi] \end{cases} \tag{4.82}$$

has a unique solution defined in $[0, \pi]$. In particular, (4.82) with $u_0 = u_1 = 0$ has only the trivial solution since $g(x, 0, \lambda) = 0$.

The regularity result (cf. Remark 4.13) implies that for any $(u, \lambda) \in \mathcal{C}^{\pm}$ we have $u \in C^2[0, \pi]$. Also it follows from the Uniqueness Theorem for (4.82) that the number of nodes of u in $(0, \pi)$ is finite for any u such that $(u, \lambda) \in \mathcal{C}^{\pm}$ for some λ.

Let $(u_k, \lambda_k) \in \mathcal{C}^{\pm} \to (u_0, \lambda_0) \in \mathcal{C}^{\pm}$ in $X \times \mathbb{R}$. Then it follows from the embedding $W_0^{1,2}(0, \pi) \subset\subset C[0, \pi]$ that $u_k \to u_0$ in $C^2[0, \pi]$. In particular, this implies that the number of nodes of a nontrivial solution does not change along the branch $\mathcal{C}^{\pm}$. Indeed, consider $(u_k, \lambda_k) \in \mathcal{C}^+$ such that $\|u_k\| \to 0$, $\lambda_k \to n^2$. Then it follows from

$$u_k - \lambda_k B u_k + G(u_k, \lambda_k) = o \tag{4.83}$$

exactly as in the proof of Lemma 4.5 that

$$v_k := \frac{u_k}{\|u_k\|} \to \frac{1}{n}\sqrt{\frac{2}{\pi}} \sin nx \qquad \text{in} \quad W_0^{1,2}(0, \pi).$$

The embedding $W_0^{1,2}(0, \pi) \subset\subset C[0, \pi]$ and (4.83) then implies that the above convergence holds even in $C^2[0, \pi]$. In particular, it means that for any $(u, \lambda) \in \mathcal{C}^+$ we have that

(1) u has exactly $n - 1$ nodes in $(0, \pi)$,
(2) $u'(0) > 0$.

Similarly, for $(u, \lambda) \in \mathcal{C}^-$ we obtain

(1) u has exactly $n - 1$ nodes in $(0, \pi)$,
(2) $u'(0) < 0$.

From here we conclude that

$$\mathcal{C}^+ \cap \mathcal{C}^- = \{(o, n^2)\}.$$

Hence both sets $\mathcal{C}^+$ and $\mathcal{C}^-$ are unbounded in $W_0^{1,2}(0, \pi) \times \mathbb{R}$.

Let us emphasize that this means that $\mathcal{C}^{\pm}$ are unbounded either with respect to $\|u\|$, or with respect to λ (or with respect to both $\|u\|$ and λ!).

Some further properties of g might provide more information about the properties of $\mathcal{C}^{\pm}$ (e.g., boundedness with respect to u — if there are a priori estimates for all solutions — and unboundedness with respect to λ; or vice versa boundedness with respect to λ and unboundedness with respect to $\|u\|$ — so called blow up of the solution for finite λ, etc.).

Exercise 4.11.   Consider the boundary value problem (4.81) and apply Theorem 4.9 to get conclusions about the bifurcation branches. Formulate further assumptions on g which will imply unboundedness of the branches with respect to $\|u\|$ and λ, respectively.

4.7. Potential bifurcation theorems

Let us remind the definition of a potential operator.

Definition 4.2. Let Ω be a subset of a Hilbert space H, $f\colon \Omega \to H$. We say that f has *potential* (in Ω) if there exists a functional $F\colon \Omega \to \mathbb{R}$ which is Fréchet differentiable in Ω, and for any $x \in \Omega$ we have

$$f(x) = F'(x). \tag{4.84}$$

Remark 4.9. Let us explain how to interpret the equality (4.84). Fréchet differential $F'(x)$ is a continuous linear mapping of H into $\mathbb{R}$. It follows from the Riesz Representation Theorem (see, e.g., [9]) that there is a unique point $z \in H$ such that

$$F'(x)y = \langle y, z \rangle, \quad \|z\| = \|F'(x)\|_*,$$

for any $y \in H$ (here $\langle \cdot, \cdot \rangle$ is the scalar product on H, $\| \cdot \| = \sqrt{\langle \cdot, \cdot \rangle}$ is the norm on H and $\| \cdot \|_*$ is the (usual) norm in the space of all continuous linear functionals on H).

In what follows we shall identify $F'(x)$ with $z \in H$ and we will study points of bifurcation of the equation

$$\lambda x - F'(x) = o. \tag{4.85}$$

The main objective of this section is to prove that (under the assumptions $F(o) = o$, $F'(o) = o$ and some assumptions about the smoothness of F) *every nonzero eigenvalue of $F''(o)$ is a point of bifurcation of (4.85)*.

Theorem 4.12 (Krasnoselski Potential Bifurcation Theorem).
Let F be a weakly continuous (nonlinear) functional on a Hilbert space

H. Assume that

$$F \text{ is differentiable in some neighbourhood } \mathcal{U}(o) \\ \text{of the origin in } H, \tag{4.86}$$

$$F' \text{ is compact on } \mathcal{U}(o), \tag{4.87}$$

$$\text{there exists } F''(o) = A, \tag{4.88}$$

$$F(o) = o, \quad F'(o) = o. \tag{4.89}$$

Then $\lambda_0 \neq 0$ is a point of bifurcation of

$$\lambda x - F'(x) = o \tag{4.90}$$

if and only if λ_0 is an eigenvalue of the operator A.

Remark 4.10. Note that the equation (4.90) is a special case of the equation

$$o = \lambda x - Tx + G(x, \lambda)$$

from Section 4.4. Indeed, the left hand side of (4.90) can be written as

$$\lambda x - F''(o)x + [F''(o)x - F'(x)]$$

where $F''(o)$ is a compact linear operator, and

$$F''(o)x - F'(x) = o(\|x\|), \qquad \|x\| \to 0.$$

Proof.   [of Theorem **4.12**] Note first that the implication

$$\text{``If } \lambda_0 \neq 0 \text{ is a point of bifurcation of (4.90),} \\ \text{then } \lambda_0 \text{ is an eigenvalue of } A\text{''}$$

follows from Lemma 4.5.

So we shall concentrate on the proof of the reversed implication. Roughly speaking, we know that the "linearization of (4.90)", i.e., the equation

$$(\lambda I - F''(o))x = o$$

has a nontrivial solution, and we want to show that there is also a nontrivial solution of the "close" but nonlinear equation (4.90).

The basic idea of the proof consists in the fact that (4.90) is a necessary condition for x to be a saddle point of F subject to the sphere

$$S(o; r) := \left\{ x \in H : J(x) = \frac{1}{2}r^2 \right\} \qquad \text{where} \quad J(x) = \frac{1}{2}\|x\|^2.$$

Here we use the fact that identity is differential of the functional J, and the Lagrange Multiplier Method. Later we prove the existence of a sufficiently large number of saddle points of F on $S(o; r)$. If we restrict to spheres with sufficiently small radius, we get saddle points converging to zero. The last part of the proof consists of showing that the corresponding Lagrange multipliers can be chosen close to λ_0.

Let us assume that $\lambda_0 \neq 0$ is an eigenvalue of the operator A. The assumptions (4.87) and (4.88) guarantee that $F''(o)$ is a linear self-adjoint operator (see [9]). We can assume, without loss of generality, that $\lambda_0 > 0$.

Let us start with a geometrical interpretation of the points $x \in S(o; r)$ such that

$$\lambda x = F'(x). \tag{4.91}$$

In this case the differential $F'(x)$ is perpendicular to the sphere $S(o; r)$ at x. Then x can be looked for as a limit of those points of the sphere $S(o; r)$ in which the tangent projections of F' converge to zero. More precisely we have the following

Lemma 4.6. *For $z \in H$, $z \neq o$, set*

$$D(z) = F'(z) - \frac{\langle F'(z), z \rangle}{\langle z, z \rangle} z \tag{4.92}$$

($D(z)$ is the projection of $F'(z)$ to the tangent space of $S(o; \|z\|)$ at z). Let $x_n \in S(o; r)$, $x_n \rightharpoonup x$, F' be continuouous in x, and

$$\lim_{n \to \infty} F'(x_n) = y \neq o, \qquad \lim_{n \to \infty} D(x_n) = o. \tag{4.93}$$

Then $x_n \to x$, $y = F'(x)$, $x \neq o$, and

$$\lambda x - F'(x) = o \qquad where \quad \lambda = \frac{1}{r^2} \langle F'(x), x \rangle. \tag{4.94}$$

Proof. [of Lemma **4.6**] From the weak convergence $x_n \rightharpoonup x$ and from (4.93) we obtain

$$\langle F'(x_n), x_n \rangle \to \langle y, x \rangle,$$

and hence

$$\frac{\langle F'(x_n), x_n \rangle}{r^2} x_n \rightharpoonup \frac{\langle y, x \rangle}{r^2} x.$$

At the same time, from the definition of $D(x_n)$ and (4.93) we have

$$\frac{\langle F'(x_n), x_n \rangle}{r^2} x_n = F'(x_n) - D(x_n) \to y.$$

Hence

$$y = \frac{1}{r^2} \langle y, x \rangle x.$$

Since $y \neq o$, we have $x \neq o$, and also $\langle y, x \rangle \neq 0$. The definition of $D(x_n)$ and $D(x_n) \to o$ yields

$$x_n = r^2 \frac{F'(x_n) - D(x_n)}{\langle F'(x_n), x_n \rangle} \to r^2 \frac{y}{\langle y, x \rangle} = x.$$

Continuity of F' at x then implies

$$y = F'(x), \qquad \text{i.e.} \qquad F'(x) = \frac{\langle y, x \rangle}{r^2} x. \qquad \square$$

We shall *continue in the proof of Theorem* 4.12 in the following way: we will look for a curve on the sphere $S(o; r)$ which starts at a fixed point x, the values of F along this curve do not decrease, and after some finite time (even if large) we shall reach "almost" saddle point of F. More precisely, we are looking for a curve $k = k(t, x)$, $t \in [0, \infty)$, $x \in S(o; r)$ such that

$$k(0, x) = x, \tag{4.95}$$

and for all $t \in (0, \infty)$ we require

$$k(t, x) \in S(o; r), \qquad \text{i.e.} \qquad \|k(t, x)\|^2 = r^2.$$

The last expression implies

$$\frac{\mathrm{d}}{\mathrm{d}t} \|k(t, x)\|^2 = 0$$

what is equivalent to

$$\left\langle \frac{\mathrm{d}}{\mathrm{d}t} k(t, x), k(t, x) \right\rangle = 0 \tag{4.96}$$

for all $t \in (0, \infty)$.

The equality (4.96) states that for all $t \in (0, \infty)$ the element $\frac{\mathrm{d}}{\mathrm{d}t} k(t, x)$ is perpendicular to $k(t, x)$. This will be satisfied if we look for the solution of the initial value problem

$$\begin{cases} \dfrac{\mathrm{d}}{\mathrm{d}t} k(t, x) = D(k(t, x)), & t \in (0, \infty), \\[2mm] k(0, x) = x. \end{cases} \tag{4.97}$$

We shall skip the detailed discussion about the solvability of (4.97). We restrict ourselves only to the statement that if the right hand side D of the equation in (4.97) is a bounded and Lipschitz function of k, then for arbitrary $x \in S(o; r)$ there exists a unique solution of (4.97) which is defined on the whole interval $(0, \infty)$ (see, e.g., [3]). This would force us to assume that F' is Lipschitz continuous in some neighbourhood of o. Continuity of D with respect to k is not enough to get this result in general. However, the special form of D and the compactness of F' guarantee the existence of a solution of (4.97) which is defined on the whole interval $(0, \infty)$ and which depends continuously on the initial condition $x \in S(o; r)$ (see, e.g., [5]).

Let k be a solution of the initial value problem (4.97). Then it has the following important properties:

(a) For any $t \in (0, \infty)$ we have

$$\frac{d}{dt} F(k(t, x)) = \langle F'(k(t, x)), D(k(t, x)) \rangle = \| D(k(t, x)) \|^2 \geq 0.$$

In other words, the values of the functional F do not decrease along k regardless the choice of $x \in S(o; r)$.

(b) For any $t \in (0, \infty)$ we have

$$F(k(t, x)) = F(x) + \int_0^t \| D(k(\tau, x)) \|^2 \, d\tau.$$

Since F is bounded on $S(o; r)$, there exists a sequence $\{t_i\} \subset (0, \infty)$ such that

$$\lim_{i \to \infty} D(k(t_i, x)) = o.$$

(c) Since $\{k(t_i, x)\}$ is bounded, we can select a weakly convergent subsequence.

To summarize we have the following

Lemma 4.7. *For any $x \in S(o; r)$ there exist a sequence $\{t_i\} \subset (0, \infty)$ and $x_0 \in H$ such that*

$$k(t_i, x) \rightharpoonup x_0, \tag{4.98}$$

$$D(k(t_i, x)) \to o, \tag{4.99}$$

$$\{F(k(t_i, x))\} \text{ is a non-decreasing sequence.} \tag{4.100}$$

If we also prove $F'(k(t_i, x)) \to y \neq o$, then the assumptions of Lemma 4.6 are verified with $x_n = k(t_n, x)$, and so the existence of a solution x of (4.90) with λ described by (4.94) will be proved.

In what follows we show that the convergence above as well as the fact that λ given by (4.94) is sufficiently close to λ_0 can be proved making an appropriate choice of the initial condition $x \in S(o; r)$.

Recall that $A = F''(o)$ is a compact self-adjoint linear operator in the Hilbert space H. Its spectrum consists of a countable set of eigenvalues (with one possible limit point $\lambda = 0$) and the point 0. We shall split the set of all eigenvalues to the parts $\lambda \geq \lambda_0$, and $\lambda < \lambda_0$, respectively. We shall denote by H_1, and H_2, respectively, the corresponding closed linear subspaces generated by the eigenvectors. The eigenspace associated with λ_0 will be denoted by H_0. Let P_1, P_2 be the orthogonal projections of H onto H_1, H_2, respectively (see Figure 4.19).

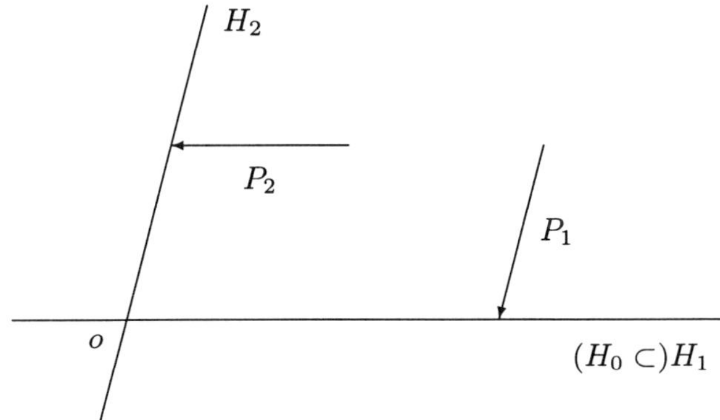

Figure 4.19.

It follows from the *Courant–Weinstein Variational Principle* (see Appendix G) that if we want to get close to the eigenvalue λ_0, we should start from an initial condition $x \in S(o; r) \cap H_0$. At the same time we want to guarantee that along the curve k we shall not get too far from λ_0.

Let us denote

$$S_1(o; r) = \{x \in H_1 : \|x\| = r\}.$$

Lemma 4.8. *There exists $r_0 > 0$ such that $S(o; r_0) \subset \mathcal{U}(o)$, and for all $0 < r < r_0$ we have*

(1) there is no $t \in [0, \infty)$ for which the set $k(t, S_1(o; r))$ is contractible to a point (see Appendix F) in

$$R = \{x \in H : P_1 x \neq o\},$$

(2) for any $t \in [0, \infty)$ there exists $x_t \in S_1(o; r)$ such that

$$P_1 k(t, x_t) \in H_0, \qquad i.e. \qquad k(t, x_t) \in H_0 \oplus H_2.$$

Proof.   [of Lemma **4.8**] Let us realize that the assertion 2 follows directly from 1 and Lemma 4.12. Hence we prove 1. According to Lemma 4.10 it is sufficient to prove that for any t the set $k(t, S_1(o; r))$ is a continuous deformation of the set $S_1(o; r)$. Indeed, the set $S_1(o; r)$ is not contractible to a point in R according to Lemma 4.11. Since k is a continuous function of both variables, it is sufficient to prove that it assumes only values from R: we want to prove that

$$\forall t \in [0, \infty), \ x \in S_1(o; r) \qquad P_1 k(t, x) \neq o. \tag{4.101}$$

We have

$$F(k(0, x)) = F(x) \geq \frac{1}{2}\langle F''(o)x, x\rangle - \varepsilon(\|x\|)\|x\|^2 = \left(\frac{1}{2}\lambda_0 - \varepsilon(\|x\|)\right)\|x\|^2$$

where $\varepsilon(r) \to 0$ as $r \to 0$. Since $F(k(t, x))$ is monotone non-decreasing in t, we conclude from here that

$$F(k(t, x)) \geq \left(\frac{1}{2}\lambda_0 - \varepsilon(r)\right) r^2. \tag{4.102}$$

On the other hand, we have the estimate from above (we write k instead of $k(t, x)$ for the sake of brevity):

$$F(k) = \frac{1}{2}\langle F''(o)k, k\rangle + \left[F(k) - \frac{1}{2}\langle F''(o)k, k\rangle\right] \leq$$

$$\leq \frac{1}{2}\langle F''(o)k, P_1 k\rangle + \frac{1}{2}\langle F''(o)k, P_2 k\rangle + \varepsilon(\|k\|)\|k\|^2$$

(see Appendix H for details).

Denote

$$\mu = \max\{\lambda : \lambda \in P_\sigma(F''(o))\}, \qquad \nu = \sup\{\lambda \in P_\sigma(F''(o)) : \lambda < \lambda_0\}.$$

Then

$$F(k) \leq \frac{\mu}{2}\|P_1 k\|^2 + \frac{\nu}{2}\|P_2 k\|^2 + \varepsilon(\|k\|)\|k\|^2 =$$

$$= \frac{\nu}{2}\|k\|^2 + \frac{\mu - \nu}{2}\|P_1 k\|^2 + \varepsilon(\|k\|)\|k\|^2.$$

Hence, due to the fact $\|k\| = r$, we have

$$F(k) \leq \frac{\nu}{2}r^2 + \frac{\mu - \nu}{2}\|P_1 k\|^2 + \varepsilon(r)r^2. \tag{4.103}$$

It follows from (4.102) and (4.103) that

$$\|P_1 k(t, x)\|^2 \geq \frac{\lambda_0 - \nu}{\mu - \nu} r^2 - \frac{4}{\mu - \nu} \varepsilon(r) r^2.$$

This implies the existence of r_0 such that for any $r \leq r_0$

$$\|P_1 k(t, x)\|^2 \geq a r^2 \tag{4.104}$$

where $a = a(r_0) > 0$. This completes the proof of Lemma 4.8. $\qquad\square$

We shall *finish the proof of Theorem* 4.12.

Step 1. Let $t(n) \to \infty$ be an arbitrary sequence of positive numbers. Let $x(n)$ be such a point from $S_1(o; r)$ for which

$$P_1 k(t(n), x(n)) \in H_0$$

(its existence follows from 2 of Lemma 4.8). Since $S_1(o; r)$ is compact, we can select a strongly convergent subsequence (denoted again by $\{x(n)\}$) such that

$$\lim_{n \to \infty} x(n) = \tilde{x}. \tag{4.105}$$

Step 2. It follows from Lemma 4.7 that there is a sequence $\{t_i\}$ such that

$$k(t_i, \tilde{x}) = x_i \rightharpoonup x_0 \qquad \text{in} \quad H,$$

and at the same time also

$$D(x_i) \to o.$$

Step 3. The compactness of F' implies that (passing again to a subsequence if necessary) there exists $y \in H$ such that

$$\lim_{i \to \infty} F'(x_i) = y.$$

We show that $y \neq o$. Indeed, we have

$$\langle F'(x_i), P_1 x_i \rangle \to \langle y, P_1 x_0 \rangle.$$

Also, for all $i \in \mathbb{N}$, we have the following estimate

$$\langle F'(x_i), P_1 x_i \rangle = \langle F''(o) x_i, P_1 x_i \rangle + \langle F'(x_i) - F''(o) x_i, P_1 x_i \rangle \geq$$

$$\geq \lambda_0 \|P_1 x_i\|^2 - \varepsilon(\|x_i\|) \|x_i\|^2 \geq \frac{1}{2} \lambda_0 a r^2$$

for all $r \leq r_0$ due to (4.104). This immediately implies $\langle y, P_1 x_0 \rangle \neq 0$, and so $y \neq o$.

 P. Drábek

Step 4. We just verified the assumptions of Lemma 4.6. Hence $x_i \to x_0$ in H, and x_0 solves (4.90) with λ given by (4.94):

$$\lambda x_0 - F'(x_0) = o, \qquad \lambda = \frac{1}{r^2}\langle F'(x_0), x_0 \rangle.$$

Step 5. The last step consists in proving the fact that for $r > 0$ small enough λ is arbitrarily close to λ_0. Let us estimate

$$|\lambda - \lambda_0| = \frac{1}{r^2}\left|\langle F'(x_0), x_0 \rangle - \lambda_0\langle x_0, x_0 \rangle\right| \le$$

$$\le \frac{1}{r^2}\left[\left|\langle F'(x_0) - F''(o)x_0, x_0 \rangle\right| + \right.$$

$$\left. + 2\left|\left\langle \frac{1}{2}F''(o)x_0, x_0 \right\rangle - F(x_0)\right| + \left|2F(x_0) - \lambda_0\langle x_0, x_0 \rangle\right|\right] =$$

$$= \frac{1}{r^2}\left|2F(x_0) - \lambda_0\langle x_0, x_0 \rangle\right| + \varepsilon(r).$$

Since $\varepsilon(r) \to 0$ as $r \to 0$, it suffices to estimate

$$\frac{1}{r^2}\left|2F(x_0) - \lambda_0\langle x_0, x_0 \rangle\right|.$$

The weak continuity of F implies

$$F(x_0) = \lim_{i \to \infty} F(x_i). \tag{4.106}$$

Since F is non-decreasing along k, we also have

$$F(x_i) = F(k(t_i, \tilde{x})) \ge F(\tilde{x}). \tag{4.107}$$

Since $\tilde{x} \in S_1(o; r)$,

$$\left\langle \frac{1}{2}F''(o)\tilde{x}, \tilde{x} \right\rangle \ge \frac{\lambda_0}{2}r^2. \tag{4.108}$$

Then (4.106)–(4.108) imply an estimate from below:

$$F(x_0) \ge \left(\frac{\lambda_0}{2} - \varepsilon(r)\right) r^2. \tag{4.109}$$

On the other hand, we shall derive an estimate from above for $F(x_0)$. Since $x(n) \to \tilde{x}$ and k depends continuously on the initial condition, for fixed $i \in \mathbb{N}$

$$k(t_i, x(n)) \to k(t_i, \tilde{x}).$$

The weak continuity of F implies that for fixed $i \in \mathbb{N}$ and $r > 0$ there exists $n_0 \in \mathbb{N}$ such that for all $n \geq n_0$ we have

$$F(k(t_i, \tilde{x})) \leq F(k(t_i, x(n))) + r^3. \tag{4.110}$$

But for any fixed $i \in \mathbb{N}$ we find $n_i \geq n_0$ such that $t(n_i) > t_i$, and the monotonicity of F along k then implies

$$F(k(t_i, x(n_i))) \leq F(k(t(n_i), x(n_i))). \tag{4.111}$$

The choice of $x(n)$ from Step 1 guarantees that $k(t(n_i), x(n_i)) \in H_0 \oplus H_2$, and so (writing k_i instead of $k(t(n_i), x(n_i))$) we have the estimate

$$F(k_i) \leq \frac{1}{2}\langle F''(o)k_i, k_i\rangle + \varepsilon(\|k_i\|)\|k_i\|^2 \leq \left(\frac{\lambda_0}{2} + \varepsilon(r)\right) r^2. \tag{4.112}$$

But (4.106), (4.110) and (4.111) reduce (4.112) to

$$F(x_0) \leq \left(\frac{\lambda_0}{2} + \varepsilon(r)\right) r^2. \tag{4.113}$$

Both estimates (4.109) and (4.113) yield that

$$\frac{1}{r^2}\left|2F(x_0) - \lambda_0\langle x_0, x_0\rangle\right| \to 0$$

as $r \to 0$. This completes the proof of Theorem 4.12. $\qquad\square$

Remark 4.11. It follows from the Krasnoselski Potential Bifurcation Theorem that every nonzero eigenvalue λ_0 of the potential operator is a point of bifurcation. On the other hand, the operator must be of special type, and there is no warranty that there is a curve (or continuum) of nontrivial solutions which emanates from (o, λ_0). In fact, there are counterexamples even in the finite dimension which prove that such a curve need not exist.

Böhme [1] gave an example of a real function of two independent real variables, $F \in C^\infty(\mathbb{R}^2)$, for which $\lambda_0 = 1$ is a point of bifurcation of

$$f(z, \lambda) = \lambda z - F'(z) = o, \qquad z = (x, y) \in \mathbb{R} \times \mathbb{R}, \qquad \lambda \in \mathbb{R}. \tag{4.114}$$

On the other hand, there is no continuous curve of nontrivial solutions of (4.114) which contains the point $((0,0), \lambda_0)$.

Example 4.9. (Application of Krasnoselski Potential Bifurcation Theorem) We shall consider the periodic problem similar to that one studied in Example 4.5:

$$\begin{cases} u''(x) + \lambda u(x) + g(x, u(x), \lambda) = 0 & \text{in} \quad [0, 2\pi], \\ u(0) = u(2\pi), \quad u'(0) = u'(2\pi). \end{cases} \tag{4.115}$$

The difference between (4.115) and (4.34) consists in the fact that here we do not allow g to depend on u'. The reason for this restriction consists in the fact that the boundary value problem (4.34) does not have potential if g depends on u'.

We shall simplify the situation even more and write g in the form

$$g(x, s, \lambda) = (\lambda + 1)\tilde{g}(x, s).$$

Set

$$\tilde{G}(x, s) = \int_0^s \tilde{g}(x, t)\, \mathrm{d}t = \int_0^1 \tilde{g}(x, s\sigma)s\, \mathrm{d}\sigma,$$

i.e., $\tilde{G}$ is the primitive of $\tilde{g}$ with respect to the second variable s.

Put

$$F(u) = \int_0^{2\pi} \left[\frac{1}{2}u^2(x) + \tilde{G}(x, u(x)) \right]\, \mathrm{d}x. \tag{4.116}$$

We shall work in the Hilbert space

$$H := WP = \{u \colon \mathbb{R} \to \mathbb{R} : u \text{ is } 2\pi\text{-periodic}, u \in W^{1,2}(0, 2\pi)\}.$$

Then

$$\langle F'(u), v \rangle = \int_0^{2\pi} \left[u(x)v(x) + \tilde{g}(x, u(x))v(x) \right]\, \mathrm{d}x \tag{4.117}$$

for any $u, v \in H$.

A *weak solution* of the periodic problem is a function $u \in H$ which satisfies the integral identity

$$\int_0^{2\pi} \left[u'(x)v'(x) - \lambda u(x)v(x) - (\lambda + 1)\tilde{g}(x, u(x))v(x) \right]\, \mathrm{d}x = 0 \tag{4.118}$$

for any $v \in H$. The last equality (4.118) can be written as

$$\int_0^{2\pi} \left[u'(x)v'(x) + u(x)v(x) - (\lambda + 1)u(x)v(x) - \right. \\ \left. - (\lambda + 1)\tilde{g}(x, u(x))v(x) \right]\, \mathrm{d}x = 0. \tag{4.119}$$

With respect to the scalar product on H given by

$$\langle u, v \rangle = \int_0^{2\pi} \left[u'(x)v'(x) + u(x)v(x) \right]\, \mathrm{d}x, \qquad u, v \in H,$$

the integral identity (4.119) for $\lambda \neq -1$ can be written as the operator equation

$$\mu u - F'(u) = o$$

where $\mu = \frac{1}{\lambda+1}$. Let us define the operator $B\colon H \to H$ by

$$\langle B(u), v \rangle = \int_0^{2\pi} u(x)v(x)\,\mathrm{d}x.$$

It follows easily from here that B is a bounded linear operator and the compact embedding $H \subset\subset CP$ (see Example 4.5 for definition of CP) yields that B is compact. Then $\mu = \frac{1}{n^2+1}$ is an eigenvalue of B if and only if $\lambda = n^2$ is an eigenvalue of

$$\begin{cases} u''(x) + \lambda u(x) = 0 & \text{in} \quad [0, 2\pi], \\ u(0) = u(2\pi), \quad u'(0) = u'(2\pi). \end{cases}$$

We shall make the following assumptions:

$$\tilde{g}\colon \mathbb{R} \times \mathbb{R} \to \mathbb{R} \text{ is a continuous function and also}$$
$$\frac{\partial \tilde{g}}{\partial u}\colon \mathbb{R} \times \mathbb{R} \to \mathbb{R} \text{ is continuous,} \tag{4.120}$$

$$\tilde{g}(x, 0) = 0, \quad \frac{\partial \tilde{g}}{\partial u}(x, 0) = 0 \qquad \forall x \in \mathbb{R}. \tag{4.121}$$

We shall prove now that F verifies the assumptions of Theorem 4.12:

(1) *Weak continuity* of F. Realize first that

$$F(u) = \frac{1}{2} \int_0^{2\pi} u^2(x)\,\mathrm{d}x + \int_0^{2\pi} \left[\int_0^1 \tilde{g}(x, tu(x))u(x)\,\mathrm{d}t \right] \mathrm{d}x. \tag{4.122}$$

Let $u_n \rightharpoonup u$ in H. The compactness of the embedding $H \subset\subset CP$ yields that we can pass to a subsequence (denoted again by $\{u_n\}$) for which $u_n \to u$ in CP. The continuity of $\tilde{g}$ and (4.122) then imply $F(u_n) \to F(u)$. If F was not weakly continuous at u, then for some sequence we would have $u_n \rightharpoonup u$ and $|F(u_n) - F(u)| \geq \varepsilon_0$. This contradicts the above considerations.

(2) $F(o) = 0$ is an immediate consequence of (4.121).

(3) *Differentiability of* F follows directly from (4.116) and (4.120).

(4) *Compactness* of $F'(u)$. This is a consequence of the compactness of the embedding $H \subset\subset CP$.

(5) $F'(o) = o$ is also a consequence of (4.121).

(6) $F''(o) = B$.

Theorem 4.12 now implies that every eigenvalue $\frac{1}{n^2+1}$, $n = 0, 1, \ldots$, of the operator $B = F''(o)$ is a point of bifurcation of the equation

$$\mu u - F'(u) = o.$$

In other words, for any $n = 0, 1, \ldots$ we have the following assertion:

For arbitrarily small neighbourhood $\mathcal{U}$ of the point $(o, n^2) \in WP \times \mathbb{R}$ there exists $(u, \lambda) \in \mathcal{U}$ such that $u \neq o$ is a weak solution of the periodic problem

$$\begin{cases} u''(x) + \lambda u(x) + (\lambda + 1)\tilde{g}(x, u(x)) = 0 & \text{in} \quad [0, 2\pi], \\ u(0) = u(2\pi), \quad u'(0) = u'(2\pi). \end{cases}$$

Note that the continuity of $\tilde{g}$ and a similar regularity argument as that used in Appendix D imply that every such nontrivial solution satisfies $u \in C^2 P$.

Exercise 4.13. Apply Theorem 4.12 to the Dirichlet and Neumann boundary value problem, respectively.

4.8. Comparison of previous results

To summarize the methods presented in these notes we shall distinguish between three basic methods to deal with the bifurcation phenomenon:

- **Implicit Function Theorem Method (IM)**
  (see Section 4.4)
- **Degree Theory Method (DM)**
  (see Sections 4.5 and 4.6)
- **Variational Method (VM)**
  (see Section 4.7)

It is impossible to say that one method is more general or better than one another. It always depends on the particular situation which method is usable and which provides more interesting information. Nevertheless, it is useful to know what are the differences among the above mentioned basic approaches to the bifurcation problems. However, one has to be aware of the fact that a "disadvantage" in one situation may appear to be an "advantage" in another situation and vice versa. So, the following discussion deserves some abstraction.

"Advantages" of **IM**

- very precise information about the structure of the solution set near the point of bifurcation
- differentiability properties of the solution set near the point of bifurcation, transversality
- possibility to deal with non-potential equations

"Disadvantages" of **IM**

- smoothness requirements on the equation
- local character of the information about the solution set

"Advantages" of **DM**

- less smoothness of the equation is required
- possibility to deal with non-potential equations
- global information about the branches of nontrivial solutions

"Disadvantages" of **DM**

- the solution set need not be a curve even in a small neighbourhood of the point of bifurcation
- global structure of the solution set may be unclear if there is no additional information about "higher order terms"
- bifurcation only from eigenvalues of odd (algebraic) multiplicity

"Advantages" of **VM**

- bifurcation from any eigenvalue
- not much differentiability required

"Disadvantages" of **VM**

- the equation must be of special form
- only local information about nontrivial solutions
- the structure of the solution set might be very wild even very close to the point of bifurcation (it need not form a continuum as shown in Böhme [1]).

A. Implicit Function Theorem

In this appendix we formulate an abstract *Implicit Function Theorem*. Its proof is based on the Banach Contraction Principle, and can be found in various books on the nonlinear functional analysis (see, e.g., [9]).

Theorem 4.14. *Let X, Y, Z be Banach spaces and let $f\colon X \times Y \to Z$ be continuous in $(x_0, y_0) \in X \times Y$, $f(x_0, y_0) = o$, and the partial Fréchet derivative*

$$f_2'(x_0, y_0)\colon Y \to Z$$

be a continuous isomorphism of Y onto Z. Assume, moreover, that

$$f_2'\colon X \times Y \to \mathscr{L}(Y, Z)$$

is a continuous mapping at the point (x_0, y_0). Then there exist $\varepsilon, \delta > 0$ such that for any $x \in X$, $\|x - x_0\| < \delta$, there exists a unique $u(x) \in Y$ such that

$$\|u(x) - y_0\| < \varepsilon, \qquad f(x, u(x)) = o.$$

Moreover, the mapping

$$x \mapsto u(x)$$

is continuous at the point x_0.

Remark 4.12. If we assume that f is of the class C^p, $p \geq 1$, in some neighbourhood of (x_0, y_0), then u is also of the class C^p, $p \geq 1$, in some neighbourhood of x_0.

B.   Proof of Morse Lemma

In this appendix we *prove Lemma* 4.2. Let us look for a mapping $\boldsymbol{\xi}$ in the form

$$\boldsymbol{\xi}(\boldsymbol{x}) = \boldsymbol{R}(\boldsymbol{x})\boldsymbol{x}$$

where $\boldsymbol{R}(\boldsymbol{o}) = \boldsymbol{I}$, $\boldsymbol{R}(\boldsymbol{x})$ is, for any $\boldsymbol{x}$, a matrix $m \times m$.

Denote

$$\boldsymbol{Q} = \frac{1}{2} F''(\boldsymbol{o}).$$

Integrating by parts and using the properties of F we obtain

$$F(\boldsymbol{x}) = F(\boldsymbol{x}) - F(\boldsymbol{o}) = \int_0^1 \frac{\mathrm{d}}{\mathrm{d}t} F(t\boldsymbol{x}) \, \mathrm{d}t =$$

$$= \left[(t - 1) \frac{\mathrm{d}}{\mathrm{d}t} F(t\boldsymbol{x}) \right]_0^1 - \int_0^1 (t - 1) \frac{\mathrm{d}^2}{\mathrm{d}t^2} F(t\boldsymbol{x}) \, \mathrm{d}t =$$

$$= \left\langle \left[\int_0^1 (1 - t) F''(t\boldsymbol{x}) \, \mathrm{d}t \right] \boldsymbol{x}, \boldsymbol{x} \right\rangle = \langle \boldsymbol{B}(\boldsymbol{x})\boldsymbol{x}, \boldsymbol{x} \rangle.$$

Hence we want to find a matrix $\boldsymbol{R} = \boldsymbol{R}(\boldsymbol{x})$ for which

$$\langle \boldsymbol{Q}\boldsymbol{R}(\boldsymbol{x})\boldsymbol{x}, \boldsymbol{R}(\boldsymbol{x})\boldsymbol{x} \rangle = \langle \boldsymbol{B}(\boldsymbol{x})\boldsymbol{x}, \boldsymbol{x} \rangle,$$

or, equivalently,

$$\langle \boldsymbol{R}^*(\boldsymbol{x})\boldsymbol{Q}\boldsymbol{R}(\boldsymbol{x})\boldsymbol{x}, \boldsymbol{x} \rangle = \langle \boldsymbol{B}(\boldsymbol{x})\boldsymbol{x}, \boldsymbol{x} \rangle$$

(here R^* is the adjoint matrix to R). The last identity holds if

$$R^*(x)QR(x) = B(x) \qquad \text{(B.123)}$$

in some neighbourhood of zero in $\mathbb{R}^m$. The matrices Q and B are known, and $B \in C^{p-2}$. The identity (B.123) can be viewed as the equation

$$H(x, R) = R^*QR - B = O.$$

Then H is a map from $\mathcal{W} \times L \subset \mathbb{R}^m \times L$ into S where $\mathcal{W}$ is a neighbourhood of the origin in $\mathbb{R}^m$, L is a subspace of the space K of all $m \times m$ real matrices — L will be determined later — and S is a normed space of all $m \times m$ real symmetric matrices. Since $B(x)$ and Q are symmetric, H is also symmetric. Obviously, $H: \mathcal{W} \times L \to S$ is of the class C^{p-2}, and also

$$H(o, I) = O, \qquad H_2'(o, I)T = (R^*QT + T^*QR)|_{R=I} = QT + T^*Q.$$

Since Q is regular and symmetric, for any $P \in S$, there exists T such that

$$H_2'(o, I)T = P$$

(here $T = \frac{1}{2}Q^{-1}P$).

Let us split K to $K_1 = Ker\, H_2'(o, I)$ and K_2 in such a way that $I \in K_2$. Then $H_2'(o, I)$ is an isomorphism of K_2 onto S. Then we set $L = K_2$ and apply the Implicit Function Theorem. Then there exist a neighbourhood $\mathcal{V}_1$ of zero in $\mathbb{R}^m$ and a neighbourhood $\mathcal{V}_2$ of identity I in L such that for any $x \in \mathcal{V}_1$ there exists a unique $R(x) \in \mathcal{V}_2$ satisfying

$$R^*(x)QR(x) - B(x) = O.$$

The map ξ defined as above then satisfies all relations from Lemma 4.2.

∎

C. Leray–Schauder Degree Theory

In this appendix we shall formulate some useful properties of the *Leray–Schauder degree of the mapping*.

Theorem 4.15. *Let X be a Banach space, Ω a bounded and open set in X, T a compact operator from $\overline{\Omega}$ into X. Denote*

$$f = I - T.$$

Then for any $p \in X \setminus f(\partial\Omega)$ there exists a unique integer

$$\deg[f; \Omega, p]$$

which has the following properties:

(1) (Normalizing property). *For any $p \in \Omega$ we have*

$$\deg\,[I; \Omega, p] = 1.$$

(2) (Additivity property with respect to Ω). *Let Ω_1, Ω_2 be open disjoint subsets of Ω, $p \notin f(\overline{\Omega} \setminus (\Omega_1 \cup \Omega_2))$. Then*

$$\deg\,[f; \Omega, p] = \deg\,[f; \Omega_1, p] + \deg\,[f; \Omega_2, p].$$

(3) (Continuity with respect to f). *For any T as above there exists $\varepsilon > 0$ such that for all compact operators $S\colon \overline{\Omega} \to X$ satisfying*

$$\sup\,\{\|T(x) - S(x)\| : x \in \Omega\} < \varepsilon$$

we have

$$\deg\,[I - T; \Omega, p] = \deg\,[I - S; \Omega, p].$$

(4) (Invariance with respect to translations).

$$\deg\,[f; \Omega, p] = \deg\,[f - p; \Omega, o].$$

(5) (Solvability of the equation).

$$\deg\,[f; \Omega, p] \neq 0 \qquad \Longrightarrow \qquad \exists x \in \Omega \quad f(x) = p.$$

(6) (Continuity with respect to p). *For given f and Ω the degree is constant in components of $X \setminus f(\partial\Omega)$.*

(7) (Dependence on the boundary conditions). *Let S be a compact mapping from $\overline{\Omega}$ into X. If $T|_{\partial\Omega} = S|_{\partial\Omega}$, $g = I - S$, then for any $p \in X \setminus f(\partial\Omega)$ we have*

$$\deg\,[f; \Omega, p] = \deg\,[g; \Omega, p].$$

(8) (Degree of the Cartesian product). *Let Ω_i, $i = 1, 2$, be open bounded subsets of Banach spaces X_i, T_i compact operators from $\overline{\Omega_i}$ into X_i, $f_i = I - T_i$, $p_i \in X_i \setminus f_i(\partial\Omega_i)$. Set*

$$X = X_1 \times X_2, \qquad \Omega = \Omega_1 \times \Omega_2, \qquad f = (f_1, f_2), \qquad p = (p_1, p_2).$$

Then

$$\deg\,[f; \Omega, p] = \deg\,[f_1; \Omega_1, p_1]\,\deg\,[f_2; \Omega_2, p_2].$$

(9) (Homotopy invariance property). *Let Ω be an open bounded set in X, $S\colon [0,1] \times \overline{\Omega} \to X$ a continuous operator, $H(t, x) = x - S(t, x)$. Assume that for all $t \in [0, 1]$, $x \in \partial\Omega$ we have*

$$H(t, x) \neq p.$$

Let either

$$S([0,1] \times \overline{\Omega}) \ \text{is relatively compact,}$$

or

(a) *for any* $t \in [0,1]$ *the operator* $S(t,\cdot)$ *is compact,*
(b) *for any* $\eta > 0$ *there exists* $\delta > 0$ *such that for all* $t_1, t_2 \in [0,1]$:

$$|t_1 - t_2| < \delta \quad \Longrightarrow \quad \sup\left\{\|S(t_1,x) - S(t_2,x)\| : x \in \overline{\Omega}\right\} < \eta.$$

Then

$$\deg\left[H(0,\cdot); \Omega, p\right] = \deg\left[H(1,\cdot); \Omega, p\right].$$

In the Bifurcation Theory we need the notion of the *index of an isolated solution* x_0 of the equation

$$f(x) = p \quad \text{in} \quad \Omega.$$

We still assume that f is of the form $f = I - T$ where T is a compact operator. In this case, for any $\eta > 0$ sufficiently small

$$\deg\left[f; B(x_0; \eta), p\right]$$

is independent of η (we use the property 2 of Theorem 4.15). Hence the following limit is well defined:

$$i(x_0) = \lim_{\eta \to 0_+} \deg\left[f; B(x_0; \eta), p\right].$$

Then $i(x_0)$ is called the *Leray–Schauder index of the point* x_0 *(with respect to the mapping f and the point p)*.

If $f(x) = p$ has a finite number of isolated solutions in Ω, then (using again 2 of Theorem 4.15) we obtain

$$\deg\left[f; \Omega, p\right] = \sum_{x \in \Omega : f(x) = p} i(x).$$

We use several times the following *Leray–Schauder Index Formula*.

Theorem 4.16. *Let* Ω *be a bounded subset of a Banach space* X, $T\colon \overline{\Omega} \to X$ *a compact operator,* $f = I - T$, *and* $x_0 \in \Omega$ *an isolated solution of* $f(x) = p$. *Assume that* T *has the Fréchet differential at* x_0 *(denoted by* $T'(x_0)$*), and*

$$1 \notin P_\sigma(T'(x_0))$$

where $P_\sigma(T'(x_0))$ is the point spectrum of $T'(x_0)$. Then

$$i(x_0) = (-1)^\beta \qquad where \quad \beta = \sum_{\substack{\lambda \in P_\sigma(T'(x_0)) \cap \mathbb{R} \\ \lambda > 1}} n_\lambda$$

and n_λ is the algebraic multiplicity of the eigenvalue λ:

$$n_\lambda = \dim \bigcup_{k=1}^{\infty} Ker\,(\lambda I - T'(x_0))^k.$$

Let us point out that the sum above is always finite due to the compactness of $T'(x_0)$.

The proofs of Theorems 4.15 and 4.16 can be found, e.g., in [9].

D. Regularity of weak solution

Let us consider the Dirichlet boundary value problem

$$\begin{cases} -u''(x) + \varphi(u(x)) = f(x) & \text{in} \quad (0,\pi), \\ u(0) = u(\pi) = 0. \end{cases}$$

Assume that φ is continuous on $\mathbb{R}$. If $f \in C(0,\pi)$, then the *classical solution* is defined as a function $u \in C^2(0,\pi) \cap C[0,\pi]$ which satisfies the equation and the boundary conditions point-wise.

If $f \in L^1(0,\pi)$, then the *weak solution* is defined as a function $u \in W_0^{1,2}(0,\pi)$ for which the integral identity

$$\int_0^\pi \left(u'(x)w'(x) + \varphi(u(x))w(x) \right) \mathrm{d}x = \int_0^\pi f(x)w(x)\,\mathrm{d}x$$

is satisfied for any $w \in W_0^{1,2}(0,\pi)$.

It follows from here that for the weak solution the second derivative of u in the sense of distributions belongs to $L^1(0,\pi)$, and the equation holds almost everywhere in $(0,\pi)$. So, in general, every classical solution is also a weak solution, and the converse is not always true. However, we have the following *regularity result*:

Theorem 4.17. *Let $f \in C[0,\pi]$ and u be a weak solution. Then $u \in C^2[0,\pi]$ and u is a classical solution.*

Proof. It is clear that the boundary conditions are satisfied because u belongs to $W_0^{1,2}(0,\pi) \hookrightarrow C[0,\pi]$. Using the integration by parts, we can rewrite the integral identity above as follows

$$\int_0^\pi \left[u'(x) - \int_0^x \left(\varphi(u(t)) - f(t) \right) dt \right] w'(x)\, dx = 0$$

for any $w \in W_0^{1,2}(0,\pi)$. Denote

$$F(x) = u'(x) - \int_0^x \left(\varphi(u(t)) - f(t) \right) dt,$$

and set

$$w(x) := \int_0^x \left(F(t) - C \right) dt, \qquad C = \frac{1}{\pi} \int_0^\pi F(x)\, dx.$$

Because $w \in W_0^{1,2}(0,\pi)$, it follows then from the integral identity that

$$0 = \int_0^\pi F(x)(F(x) - C)\, dx = \int_0^\pi (F(x) - C)^2\, dx.$$

Hence $F(x) = C$, i.e.,

$$u'(x) = \int_0^x \left(\varphi(u(t)) - f(t) \right) dt + C$$

holds almost everywhere in $(0,\pi)$. This implies $u' \in C^1[0,\pi]$. Taking the derivative we get

$$u''(x) = \varphi(u(x)) - f(x)$$

which proves the assertion. $\qquad\square$

Remark 4.13. It is obvious that the same result holds true if φ is replaced by a continuous function $g = g(x, s, \lambda)$.

E. Analogue of the Leray–Schauder Index Formula

In this appendix we prove an *analogue of the Leray–Schauder Index Formula*.

Lemma 4.9. *Let f_r, $\mathcal{U}_j$, i_-^j, i_+^j be as in the proof of Theorem 4.9. Then*

$$\deg\left[f_r; \mathcal{U}_j, o \right] = i_-^j - i_+^j.$$

Proof. We shall connect f_r with a more simple mapping using a suitable homotopy. Let us define this homotopy in the following way:

$$\forall t \in [0,1] \quad f_{r,t} : \mathcal{U}_j \to X \times \mathbb{R} : f_{r,t}(x, \mu) = (y_t, \tau_t),$$

$$y_t = x - \mu T x + t g(x, \mu),$$

$$\tau_t = t(\|x\|^2 - r^2) + (1-t)(\varepsilon^2 - (\mu - \mu_j)^2).$$

We prove that for any $t \in [0, 1]$

$$o \notin f_{r,t}(\partial \mathcal{U}_j).$$

Assume the contrary, i.e., there exist $t \in [0, 1]$ and $(x, \mu) \in \partial \mathcal{U}_j$ such that

$$f_{r,t}(x, \mu) = o.$$

The fact that $(x, \mu) \in \partial \mathcal{U}_j$ implies

$$\|x\|^2 + (\mu - \mu_j)^2 = r^2 + \varepsilon^2.$$

At the same time, from

$$0 = \tau_t = t(\|x\|^2 + (\mu - \mu_j)^2) - t(r^2 + \varepsilon^2) + \varepsilon^2 - (\mu - \mu_j)^2,$$

we obtain $\mu = \mu_j \pm \varepsilon$, and so $\|x\| = r$. This together with $y_t = o$ contradicts (4.74). The homotopy invariance property of the degree then implies

$$\deg\left[f_r; \mathcal{U}_j, o\right] = \deg\left[f_{r,0}; \mathcal{U}_j, o\right].$$

The mapping $f_{r,0}$ is now easier to deal with. Indeed, the point o has two preimages: $(o, \mu_j - \varepsilon)$ and $(o, \mu_j + \varepsilon)$ with respect to the mapping

$$f_{r,0}(x, \mu) = (x - \mu T x, \varepsilon^2 - (\mu - \mu_j)^2).$$

In both points the Fréchet differential $f'_{r,0}$ is injective:

$$f'_{r,0}(0, \mu)(u, \overline{\mu}) = (u - \mu T u, -2(\mu - \mu_j)\overline{\mu}).$$

Let us choose sufficiently small neighbourhoods of points $(o, \mu_j \pm \varepsilon)$ in the following way:

$$\mathcal{U}^{\pm} = \mathcal{U}^1 \times ((\mu_j \pm \varepsilon) \cup \mathcal{U}^2)$$

where $\mathcal{U}^1$ is a neighbourhood of the origin in X and $\mathcal{U}^2$ is a neighbourhood of the origin in $\mathbb{R}$. Applying now 8 of Theorem 4.15 and Theorem 4.16, we obtain:

$$\deg\left[f_{r,0}; \mathcal{U}_j, o\right] = i(o, \mu_j - \varepsilon) + i(o, \mu_j + \varepsilon) =$$

$$= \deg\left[f_{r,0}; \mathcal{U}^-, o\right] + \deg\left[f_{r,0}; \mathcal{U}^+, o\right] =$$

$$= \deg\left[I - (\mu_j - \varepsilon)T; \mathcal{U}^1, o\right] \deg\left[2\varepsilon\overline{\mu}; \mathcal{U}^2, o\right] +$$

$$+ \deg\left[I - (\mu_j + \varepsilon)T; \mathcal{U}^1, o\right] \deg\left[-2\varepsilon\overline{\mu}; \mathcal{U}^2, o\right] =$$

$$= i_-^j \cdot 1 + i_+^j \cdot (-1) = i_-^j - i_+^j. \qquad \square$$

F. Contractible sets

Definition 4.3. We say that a set $\mathcal{K}$ is a *continuous deformation of a set* $\mathcal{M}$ in a metric space $Q \supset \mathcal{M} \cup \mathcal{K}$ if there exists a continuous mapping $f \colon [0,1] \times \mathcal{M} \to Q$ such that

$$f(0,x) = x \quad \forall x \in \mathcal{M}, \qquad f(1,\mathcal{M}) = \mathcal{K}.$$

In particular, if $\mathcal{K}$ consists of one point, we say that $\mathcal{M}$ is *contractible to a point in Q*.

Next assertions summarize some useful properties of the sets which are contractible to a point.

Lemma 4.10. *Let $\mathcal{F}$ be a subset of a metric space Q and let $\mathcal{G}$ be a continuous deformation of $\mathcal{F}$ in Q. If $\mathcal{F}$ is not contractible to a point in Q, then also $\mathcal{G}$ is not contractible to a point in Q.*

Proof. Assume the contrary, i.e., the set $\mathcal{G}$ is contractible to a point in Q. Following the definition there exists a continuous mapping $f_1 \colon [0,1] \times \mathcal{G} \to Q$ and a point $y \in Q$ such that for all $x \in \mathcal{G}$ we have

$$f_1(0,x) = x, \qquad f_1(1,x) = y.$$

On the other hand, $\mathcal{G}$ is a continuous deformation of $\mathcal{F}$, i.e., there exists a continuous mapping $f_2 \colon [0,1] \times \mathcal{F} \to Q$ such that

$$f_2(0,x) = x \quad \forall x \in \mathcal{F}, \qquad f_2(1,\mathcal{F}) = \mathcal{G}.$$

Let us define a mapping $f \colon [0,1] \times \mathcal{F} \to Q$ as follows:

$$f(t,x) = \begin{cases} f_2(2t,x) & \text{for} \quad t \in \left[0,\tfrac{1}{2}\right], \, x \in \mathcal{F}, \\ f_1(2t-1, f_2(1,x)) & \text{for} \quad t \in \left(\tfrac{1}{2},1\right], \, x \in \mathcal{F}. \end{cases}$$

We easily verify that f is a continuous map which contracts $\mathcal{F}$ to a point y in Q, a contradiction. $\qquad\square$

Let H_1 and H_2 be two closed subsets of a Hilbert space H such that

$$H = H_1 \oplus H_2.$$

Let $P_i \colon H \to H_i$, $i = 1, 2$, be projections, and assume that

$$\dim H_1 < \infty.$$

Set

$$R = \{x \in H : P_1 x \neq o\}.$$

                                    P. Drábek

The set R equipped with the metric induced by the norm in H, is a metric space.

Lemma 4.11. *The set $S_{1,r} = S(o;r) \cap H_1$ is not contractible to a point in R.*

Proof.   It is enough to prove this assertion for the sphere with the radius $r = 1$. Let us denote it by S_1. We proceed in two steps. We prove first that if S_1 is contractible to a point in R, then it must be contractible to a point in S_1. In the second step we show that this fact contradicts the Brouwer Fixed Point Theorem.

Step 1. If S_1 is contractible to a point in R, then there exists a continuous mapping $f\colon [0,1] \times S_1 \to R$ and $x_0 \in R$ such that

$$f(0,x) = x, \qquad f(1,x) = x_0 \qquad \forall x \in S_1.$$

For $t \in [0,1]$, $x \in S_1$ set

$$g(t,x) = \frac{P_1 f(t,x)}{\|P_1 f(t,x)\|}.$$

Then g deforms continuously the set S_1 to the point $\frac{P_1 x_0}{\|P_1 x_0\|}$ in S_1.

Step 2. In this part we can restrict ourselves to a finite dimensional space H_1. Let the unit sphere $S_1 \subset H_1$ be contractible to a point in S_1, i.e., there exists a continuous map $g\colon [0,1] \times S_1 \to S_1$ and a point $x_0 \in S_1$ such that

$$g(0,x) = x, \qquad g(1,x) = x_0 \qquad \forall x \in S_1.$$

Now, we shall define $h\colon \overline{B(o;1)} \to \overline{B(o;1)}$ by

$$h\colon x \mapsto \begin{cases} -g\left(1 - \|x\|, \frac{x}{\|x\|}\right) & \text{for} \quad x \neq o, \\ -x_0 & \text{for} \quad x = o. \end{cases}$$

Then h is continuous. Since $\dim H_1 < \infty$, the Brouwer Fixed Point Theorem implies that there exists $y \in \overline{B(o;1)}$ such that

$$h(y) = y.$$

Since h assumes only values from S_1, we have $y \in S_1$, $\|y\| = 1$. On the other hand,

$$h(y) = -g(0,y) = -y$$

that is a contradiction. $\qquad\square$

Lemma 4.12. *Let $\mathcal{F}$ be a subset of a metric space R. If there exists $x_0 \in H_1$, $\|x_0\| = 1$, such that*

$$P_1(\mathcal{F}) \cap \{y \in H_1 : y = ax_0, \ a \in \mathbb{R}\} = \emptyset,$$

then $\mathcal{F}$ is contractible to a point in R.

Proof. Define $f: [0,1] \times \mathcal{F} \to R$ as follows

$$f(t,x) = \begin{cases} x + 2tx_0(1 - \langle x, x_0 \rangle) & \text{for} \quad t \in \left[0, \tfrac{1}{2}\right], \ x \in \mathcal{F}, \\ x_0 + 2(1-t)(x - \langle x, x_0 \rangle x_0) & \text{for} \quad t \in \left(\tfrac{1}{2}, 1\right], \ x \in \mathcal{F} \end{cases}$$

(see Figure 4.20).

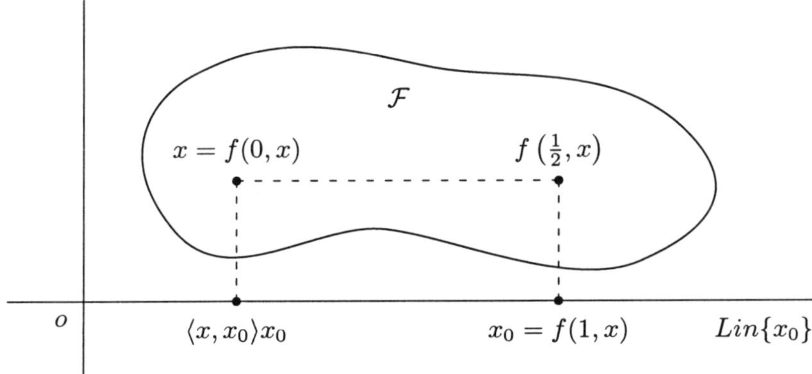

Figure 4.20.

The mapping f is continuous and deforms $\mathcal{F}$ to the point x_0. It is sufficient to verify that for any $t \in [0,1]$, $x \in \mathcal{F}$ we have

$$P_1 f(t,x) \neq o.$$

Indeed, for any $t \in \left[0, \tfrac{1}{2}\right]$ we have

$$P_1 f(t,x) = 2t(1 - \langle x, x_0 \rangle)x_0 + P_1 x,$$

for $t \in \left(\tfrac{1}{2}, 1\right]$ we have

$$P_1 f(t,x) = (1 - 2(1-t)\langle x, x_0 \rangle)x_0 + 2(1-t)P_1 x.$$

For $t \in [0,1)$ we have then $P_1 f(t,x) \neq o$ due to the assumption $P_1(\mathcal{F}) \cap Lin\{x_0\} = \emptyset$. For $t = 1$ we have

$$P_1 f(t,x) = x_0 \neq o. \qquad \square$$

G. Courant–Weinstein Variational Principle

We shall formulate and prove the *Courant–Weinstein Variational Principle*.

Theorem 4.18. *Let H be a real Hilbert space, $A\colon H \to H$ a compact self-adjoint linear operator. Let A be nonnegative, i.e., for any $x \in H$ we have*

$$\langle Ax, x \rangle \geq 0.$$

Assume that positive eigenvalues λ_n of A form the non-increasing sequence

$$\lambda_1 \geq \lambda_2 \geq \lambda_3 \geq \cdots \geq \lambda_n \geq \cdots,$$

and the multiplicity of an eigenvalue λ states how many times this λ repeats in the above sequence. Then for any $n \in \mathbb{N}$,

$$\lambda_n = \sup_{\substack{X \subset H \\ \dim X = n}} \ \min_{\substack{x \in X \\ \|x\| = 1}} \ \langle Ax, x \rangle.$$

(Here X is an arbitrary linear subspace of H of the dimension equal to n.)

Proof. Let $\{u_k\}$ be eigenvectors of A, and assume that u_k is associated with λ_k. We can assume, without loss of generality, that $\{u_k\}$ forms an orthonormal basis of H (this fact follows from the Schmidt Orthonormalizing Process). For $n \in \mathbb{N}$ fixed we denote

$$\mu_n = \sup_{\substack{X \subset H \\ \dim X = n}} \ \min_{\substack{x \in X \\ \|x\| = 1}} \ \langle Ax, x \rangle.$$

Our aim is to prove $\mu_n = \lambda_n$.

Step 1. We prove that $\mu_n \geq \lambda_n$. Set

$$X_0 = Lin\{u_1, \ldots, u_n\}.$$

Then X_0 is a linear subspace of H, $\dim X_0 = n$, and clearly

$$\mu_n \geq \min_{\substack{x \in X_0 \\ \|x\| = 1}} \ \langle Ax, x \rangle.$$

But we can calculate the minimum of the quadratic form on the right hand side. For $x \in X_0$, $\|x\| = 1$ we have

$$x = \sum_{i=1}^{n} x_i u_i, \qquad \sum_{i=1}^{n} x_i^2 = 1.$$

Then

$$\langle Ax, x \rangle = \left\langle \sum_{i=1}^{n} x_i \lambda_i u_i, \sum_{j=1}^{n} x_j u_j \right\rangle = \sum_{i=1}^{n} \lambda_i x_i^2 \geq \lambda_n \sum_{i=1}^{n} x_i^2.$$

Hence

$$\mu_n \geq \lambda_n \sum_{i=1}^{n} x_i^2 = \lambda_n.$$

Step 2. We prove $\mu_n \leq \lambda_n$. Set

$$Y = \overline{Lin\{u_i\}_{i=n}^{\infty}}.$$

Then $\operatorname{codim} Y = n - 1$. Let X be an arbitrary linear subspace of H, $\dim X = n$. Then necessarily

$$\dim (X \cap Y) > 0,$$

and the space $X \cap Y$ must contain some element $x \neq o$. We can assume $\|x\| = 1$. Since $x \in Y$, we have

$$x = \sum_{i=n}^{\infty} x_i u_i, \qquad \sum_{i=n}^{\infty} x_i^2 = 1.$$

The estimate of the quadratic form $\langle Ay, y \rangle$ on the unit sphere in X yields:

$$\min_{\substack{y \in X \\ \|y\|=1}} \langle Ay, y \rangle \leq \langle Ax, x \rangle = \sum_{i=n}^{\infty} \lambda_i x_i^2 \leq \lambda_n \sum_{i=n}^{\infty} x_i^2 = \lambda_n.$$

Since X is arbitrary, we have

$$\mu_n = \sup_{\substack{X \subset H \\ \dim X = n}} \min_{\substack{y \in X \\ \|y\|=1}} \langle Ay, y \rangle \leq \lambda_n. \qquad \square$$

H. Asymptotics for C^2-functionals

Let X be a Banach space and $F \colon X \to \mathbb{R}$ a functional which is twice Fréchet differentiable in $o \in X$ (i.e., there exists $F''(o)$). Then there exists $\varepsilon \colon \mathbb{R} \to \mathbb{R}$ such that

$$\left| \left\langle \frac{1}{2} F''(o)x, x \right\rangle + F'(o)x + F(o) - F(x) \right| \leq \varepsilon(\|x\|)\|x\|^2$$

and

$$\|F''(o)x + F'(o) - F'(x)\| \leq \varepsilon(\|x\|)\|x\|$$

where

$$\lim_{r \to 0_+} \varepsilon(r) = 0.$$

Note that the estimates above are immediate consequences of the definition of the Fréchet derivative of F.

References

[1] R. Böhme, *Die Lösung der Verzweigungsproblemen für Nichtlineare Eigenwertprobleme*, Math. Z. **127** (1972), 105–126.

[2] E. N. Dancer, *On the structure of solutions of nonlinear eigenvalue problems*, Indiana Univ. Math. J. **23** (1974), 1069–1076.

[3] J. Diendonné, *Foundations of Modern Analysis*, Academic Press, New York–-London, 1960.

[4] J. A. Izé, *Bifurcation Theory for Fredholm Operators*, Memoirs of American Mathematical Society, no. 174, American Mathematical Society, 1976.

[5] J. Naumann, *An abstract differential equation and the potential bifurcation theorems by Krasnoselski*, Comment. Math. Univ. Carolinæ **28** (1987), no. 2, 261–276.

[6] P. H. Rabinowitz, *A Global Theorem for Nonlinear Eigenvalue Problem and Applications. Contribution of Nonlinear Functional Analysis*, Academic Press, New York, 1971.

[7] P. H. Rabinowitz, *Some global results for nonlinear eigenvalue problems*, J. Funct. Anal. **7** (1971), 487–513.

[8] J. Stará and O. John, *Funkcionální analýza. Nelineární úlohy*, Státní pedagogické nakladatelství, Praha, 1986 (in Czech).

[9] E. Zeidler, *Nonlinear Functional Analysis and Its Applications*, vol. I, II/A, II/B, III and IV, Springer Verlag, Inc., New York, 1986.

Chapter 5

A nonlinear eigenvalue problem

Peter Lindqvist

Department of Mathematics
Norwegian University of Science and Technology
N-7491 Trondheim, Norway

Contents

5.1. Introduction

My lectures at the *Minicorsi di Analisi Matematica* at Padova in June 2000 are written up in these notes[a]. They are an updated and extended version of my lectures [37] at Jyväskylä in October 1994. In particular, an account of the exciting recent development of the asymptotic case is included, which is called the ∞-eigenvalue problem. I wish to thank the University of Padova for financial support. I am especially grateful to Massimo Lanza de Cristoforis for his kind assistance. I thank Harald Hanche-Olsen for his kind help with final adjustments of the typesetting.

These lectures are about a nonlinear eigenvalue problem that has a serious claim to be the right generalization of the linear case. By now I have lectured on four continents about this theme and my reason for sticking to this seemingly very peculiar problem is twofold. First, one can study the interesting questions without any previous knowledge of spectral

---

[a]A short comment on the uniqueness proof in [11] has been added later.

175

theory. Second, to the best of my knowledge there are many open problems easy to state. The higher eigenvalues are "mysterious".

The leading example of a linear eigenvalue problem is to find all nontrivial solutions of the equation $\Delta u + \lambda u = 0$ with boundary values zero in a given bounded domain in $\mathbf{R}^n$. This is the Dirichlet boundary value problem. (In the Neumann boundary value problem the normal derivative is zero at the boundary.) Needless to say, this has been generalized in numerous ways: to Riemann surfaces and manifolds, to equations $\Delta u + \lambda u + V u = 0$ with a potential V, to more general differential operators than the Laplacian, and so on.

However, when talking about nonlinear eigenvalue problems, there is seldom any eigenvalue at all involved. For example, one just considers the existence of *positive* solutions. The extremely popular and very interesting Emden-Fowler equation

$$\Delta u + |u|^{\alpha - 1} u = 0$$

is of this type. If $\alpha \neq 1$, the parameter λ plays no role in the equation $\Delta u + \lambda |u|^{\alpha - 1} u = 0$, since it can be scaled out: multiply u by a suitable constant to see this. In equations of the type

$$\Delta u + \lambda u + |u|^{\alpha - 1} u = 0$$

the parameter λ is stabilizing, when the exponent α is critical. Though interesting as they are, I will not consider these problems here. I refer to Professor Donato Passaseo's lectures about *Nonlinear Elliptic Equations Involving Critical Sobolev Exponents*.

My objective is the nonlinear eigenvalue problem

$$div(|\nabla u|^{p-2} \nabla u) + \lambda |u|^{p-2} u = 0 \tag{5.1}$$

with $u = 0$ on the boundary of a bounded domain Ω in the n-dimensional Euclidean space. Here $1 < p < \infty$ and for $p = 2$ we are back to the linear case $\Delta u + \lambda u = 0$. Note that

$$\lambda = \frac{\int\limits_{\Omega} |\nabla u|^p \, dx}{\int\limits_{\Omega} |u|^p \, dx}, \tag{5.2}$$

if u is a solution, not identically zero. (Here $dx = dx_1 \, dx_2 \cdots dx_n$ is the Lebesgue measure.) Thus it appears that $\lambda > 0$. Minimizing this so called nonlinear *Rayleigh quotient* among all admissible functions we arrive at Eqn (5.1) as the corresponding Euler-Lagrange equation. The first one to study it in any serious way seems to have been F. de Thélin in 1984, cf. [51]. The so-called p-harmonic operator $div(|\nabla u|^{p-2}\nabla u)$ appears in many contexts in physics: non-Newtonian fluids, reaction-diffusion problems, non-linear elasticity, and glaceology, just to mention a few applications.

The range of p in the p-harmonic operator

$$\Delta_p u = div(|\nabla u|^{p-2}\nabla u)$$
$$= |\nabla u|^{p-4}\left\{|\nabla u|^2\Delta u + (p-2)\sum \frac{\partial u}{\partial x_i}\frac{\partial u}{\partial x_j}\frac{\partial^2 u}{\partial x_i \partial x_j}\right\}$$

is usually $1 \le p \le \infty$. The case $p = 1$ is the mean curvature operator (with a minus sign)

$$H = -\Delta_1 u = -div\left(\frac{\nabla u}{|\nabla u|}\right)$$

and the fascinating asymptotic case $p = \infty$ is related to the operator

$$\Delta_\infty u = \sum_{i,j=1}^{n} \frac{\partial u}{\partial x_i}\frac{\partial u}{\partial x_j}\frac{\partial^2 u}{\partial x_i \partial x_j}.$$

In Section 5.6 the theory of viscosity solutions is used to treat the latter case. An amazing "differential equation" replaces (5.1). Arcane phenomena occur.

Many results are readily extended to equations of the more general form

$$\sum_{i,j=1}^{n} \frac{\partial}{\partial x_i}\left(\left|\sum_{k,m=1}^{n} a_{km}(x)\frac{\partial u}{\partial x_k}\frac{\partial u}{\partial x_m}\right|^{\frac{p-2}{2}} a_{ij}(x)\frac{\partial u}{\partial x_j}\right) + \lambda\rho(x)|u|^{p-2}u = 0$$

where the matrix (a_{km}) satisfies the *ellipticity condition*

$$\sum_{i,j=1}^{n} a_{ij}(x)\xi_i\xi_j \ge |\xi|^2$$

for all $\xi = (\xi_1, \xi_2, \ldots, \xi_n)$, and, by assumption, $\rho(x) \ge \varepsilon > 0$. The weaker restriction $\rho(x) > 0$ leads to considerable technical difficulties, not to mention the case when the density $\rho(x)$ is allowed to change signs. See [54]. It is likely that the theory works, when (a_{km}) is a Muckenhoupt weight. The essential feature here is that *solutions may be multiplied by constants.* Indeed, among all the equations

$$div\left(|\nabla u|^{p-2}\nabla u\right) + \lambda|u|^{s-2}u = 0$$

only the homogeneous case $s = p$ has the proper structure of a "typical eigenvalue problem", to quote an expression in [5].

In passing, I mention that the density $\rho(x)$ in the equation

$$div(|\nabla u|^{p-2}\nabla u) + \lambda\rho(x)|u|^{p-2}u = 0$$

is very decisive. Indeed, if we take $\rho(x)^{-p}$ to be the *distance function* $\delta(x) = \mathrm{dist}(x, \partial\Omega)$ in a *convex* domain Ω, then there is no eigenfunction at all: 0 is the only solution. Moreover, the sharp lower bound in the inequality ("Hardy's inequality").

$$\left(1 - \frac{1}{p}\right)^p < \frac{\int_\Omega |\nabla\varphi|^p dx}{\int_\Omega \left|\frac{\varphi}{\delta}\right|^p dx}, \quad \varphi \in C_0^\infty(\Omega),$$

is not attained for any admissible function, if Ω is convex. It is curious that this sharp bound depends only on p. This phenomenon was observed by S. Agmon. See [40].

The reader who wants to learn this topic does well in reading the first volume of the celebrated book by Courant & Hilbert and, perhaps, the book by Polya & Szegö, before passing on to so called modern expositions like [9] and [50]. The lecture [32] by E. Lieb is very illuminating. See [12] about spectral theory on manifolds. About elliptic partial differential equations we refer to the books [24], [30], and [23]. See also [19]. The book *Metodi diretti nel calculo variazioni* by E. Giusti is an excellent source of information.

NOTE ADDED IN PROOF. The reference E. Lieb [31] has come to my attention. It contains an interesting result about the minimum of the nonlinear Rayleigh quotient. Thus it appears that E. Lieb was the first one to study the nonlinear eigenvalue problem in several variables.

5.2. Preliminary results

Throughout these lectures Ω will denote a *bounded* domain in $\mathbf{R}^n$. For most of the theorems no regularity assumptions are needed about the boundary $\partial\Omega$. The equation will be interpreted in the weak sense.

Definition 5.1. We say that $\mathrm{u} \in W_0^{1,p}(\Omega)$, $u \not\equiv 0$, is an eigenfunction, if

$$\int_\Omega |\nabla u|^{p-2}\nabla u \cdot \nabla\eta \, dx = \lambda \int_\Omega |u|^{p-2}u\eta \, dx \tag{5.3}$$

whenever $\eta \in C_0^\infty(\Omega)$. The corresponding real number λ is called an eigenvalue.

The Sobolev space $W_0^{1,p}(\Omega)$ is the completion of $C_0^\infty(\Omega)$ with respect to the norm

$$\|\varphi\| = \left\{ \int_\Omega (|\varphi|^p + |\nabla\varphi|^p)\, dx \right\}^{\frac{1}{p}}.$$

As usual, $C_0^\infty(\Omega)$ is the class of smooth functions with compact support in Ω. By standard elliptic regularity theory *an eigenfunction is continuous*, i.e., it can be made continuous after a modification in a set of measure zero. See for example [23], [24], [30]. Indeed, even the *gradient ∇u is locally Hölder continuous*, the Hölder exponent depending only on n and p. See [17] or [53] for this deep regularity result, the first proof of which is credited to N. Uraltseva.

In regular domains the boundary value zero is attained in the classical sense. For example, any domain satisfying an exterior cone condition is surely regular enough. As a matter of fact, the regular boundary points can be characterized by a version of the celebrated Wiener criterion, formulated by Mazj'a [39] in a nonlinear setting. See [22] and [29]. It is known that those boundary points ξ where the requirement

$$\lim_{x \to \xi} u(x) = 0$$

fails is a set of p-capacity zero. That is to say that the irregular boundary points form a very small set. If $p > n$, then every boundary point is regular!

It is not difficult to see that *every eigenvalue λ is positive.* Indeed, by approximation, u itself will do as test-function in (5.3). Therefore

$$\lambda = \frac{\int_\Omega |\nabla u|^p\, dx}{\int_\Omega |u|^p\, dx}.$$

It is useful to have an explicit lower bound. The familiar Sobolev inequality $\|u\|_{np/(n-p)} \le C\|\nabla u\|_p$, where $C = C(n,p)$ and $1 < p < n$, implies

$$\lambda \ge \frac{1}{C^p\, |\Omega|^{p/n}}. \tag{5.4}$$

This lower bound for the eigenvalues is valid also for $p \ge n$. It is instructive to prove it directly. Suppose that $\varphi \in C_0^\infty(\Omega)$ where Ω is the parallelepiped

$0 < x_1 < a_1,\ 0 < x_2 < a_2, \ldots, 0 < x_n < a_n.$ Then

$$\varphi(x_1, x_2, \ldots, x_n) = \int_0^{x_1} \frac{d\varphi(t, x_2, \ldots, x_n)}{dt}\, dt,$$

$$|\varphi(x_1, x_2, \ldots, x_n)|^p \le x_1^{p-1} \int_0^{a_1} \left| \frac{d\varphi(t, x_2, \ldots, x_n)}{dt} \right|^p dt,$$

$$\int_0^{a_1} |\varphi(x_1, x_2, \ldots, x_n)|^p\, dx_1 \le \frac{a_1^p}{p} \int_0^{a_1} |D_1\varphi(t, x_2, \ldots, x_n)|^p\, dt$$

and an integration with respect to the remaining variables $x_2, \ldots, x_n$ gives the estimate

$$\int_0^{a_1} \int_0^{a_2} \cdots \int_0^{a_n} |\varphi(x_1, x_2, \ldots, x_n)|^p\, dx_1 dx_2 \cdots dx_n$$

$$\le \frac{a_1^p}{p} \int_0^{a_1} \int_0^{a_2} \cdots \int_0^{a_n} |D_1\varphi(x_1, x_2, \ldots, x_n)|^p\, dx_1 dx_2 \cdots dx_n.$$

Therefore

$$\frac{\displaystyle\int_\Omega |\nabla\varphi|^p\, dx}{\displaystyle\int_\Omega |\varphi|^p\, dx} \ge \frac{p}{a_1^p}. \tag{5.5}$$

We only used the fact that $\varphi(0, x_2, \ldots, x_n) \equiv 0$. (Since also

$$\varphi(a_1, x_2, \ldots, x_n) = 0$$

we can readily improve the lower bound a little, replacing a_1 by $a_1/2$.) Note that we may write $a_2, a_3 \ldots,$ or a_n instead of a_1. The shortest side yields the best estimate.

Essentially the same reasoning can be used to prove the estimate

$$\lambda \ge \frac{\text{Const.}}{R^p}$$

in a *regular domain*, R denoting the radius of the largest inscribed ball in the smallest "box" containing Ω. This means that the eigenvalues are large even in very long, yet narrow domains. See [41] in the linear case.

The Harnack inequality: *If u is a non-negative eigenfunction, then*

$$\max_{B_r} u \leq C \min_{B_r} u$$

whenever $B_{2r} \subset \Omega$. Here B_r and B_{2r} are concentric balls of radii r and $2r$. The constant C depends only on n and p. This result is due to Trudinger [55], who proved it in 1967 using the celebrated Moser iteration. The inequality implies that, if $u \geq 0$ in Ω, then $u > 0$. As we will see in Section 5.4, a positive eigenfunction must correspond to the smallest eigenvalue

$$\lambda_1 = \inf_{\varphi} \frac{\int_\Omega |\nabla\varphi|^p \, dx}{\int_\Omega |\varphi|^p \, dx} \tag{5.6}$$

where the infimum is taken over all $\varphi \in C_0^\infty(\Omega)$, $\varphi \not\equiv 0$. By standard compactness arguments it is easily seen that the infimum is attained for a function u in $W_0^{1,p}(\Omega)$. But, if u is minimizing, so is $|u|$. By the Harnack inequality $|u| > 0$. By continuity, either $u > 0$ in Ω or $u < 0$ in Ω. Hence *a first eigenfunction does not change signs.*

To prove existence, the following slightly simplified version of the Rellich-Kondrachov theorem is useful.

Lemma 5.1 (Rellich–Kondrachov). *Let $p > 1$. Suppose that $u_1, u_2, \ldots$ are functions in $W_0^{1,p}(\Omega)$ and that $\|\nabla u_k\|_{p,\Omega} \leq L < \infty$, when $k = 1, 2, 3, \ldots$. Then there is a function $u \in W_0^{1,p}(\Omega)$ such that $u_{k_j} \to u$ strongly in $L^p(\Omega)$ and $\nabla u_{k_j} \rightharpoonup \nabla u$ weakly in $L^p(\Omega)$ for some subsequence.*

Proof. This is a combination of the weak compactness of L^p and the Rellich-Kondrachov imbedding theorem. See [49, §11, pp. 82-85] or [57, Theorem 2.5.1, p. 62]. As a matter of fact, the convergence is better than we have stated. $\square$

We end this section by proving two results. First, the spectrum is a closed set. This fact would properly belong to Section 5.5. Second, we bound the eigenfunctions. This fact is needed in Section 5.4.

Theorem 5.1. *The spectrum is a closed set.*

Proof. Suppose that a sequence $\lambda_1, \lambda_2, \ldots$ of eigenvalues converges to $\lambda \neq \infty$ and let $u_1, u_2, \ldots$ denote the eigenfunctions, normalized by the condition $\|u_k\|_{p,\Omega} = 1$. We have

$$\int_\Omega |\nabla u_k|^{p-2} \nabla u_k \cdot \nabla \eta \, dx = \lambda_k \int_\Omega |u_k|^{p-2} u_k \eta \, dx \qquad (5.7)$$

for each $\eta \in C_0^\infty(\Omega)$. We claim that λ is an eigenvalue. By the normalization

$$\lambda_k = \int_\Omega |\nabla u_k|^p \, dx.$$

By the Rellich-Kondrachov Theorem there is a subsequence and a function $u \in W_0^{1,p}(\Omega)$ such that $u_{k_j} \to u$ strongly in $L^p(\Omega)$ and $\nabla u_{k_j} \rightharpoonup \nabla u$ weakly in $L^p(\Omega)$. We have to prove that this u is the eigenfunction corresponding to λ. By the equation itself we have

$$\int_\Omega \left[|\nabla u_{k_j}|^{p-2} \nabla u_{k_j} - |\nabla u|^{p-2} \nabla u \right] \cdot \nabla (u_k - u) \, dx$$

$$= \lambda_{k_j} \int_\Omega |u_{k_j}|^{p-2} u_{k_j} (u_{k_j} - u) \, dx - \int_\Omega |\nabla u|^{p-2} \nabla u \cdot (\nabla u_{k_j} - \nabla u) \, dx.$$

The first integral on the right-hand side approaches zero, because of the convergence $\|u_{k_j} - u\|_{p,\Omega} \to 0$, and so does the second integral by the weak convergence of the gradients. Therefore we have obtained that

$$\lim_{j \to \infty} \int_\Omega \left[|\nabla u_{k_j}|^{p-2} \nabla u_{k_j} - |\nabla u|^{p-2} \nabla u \right] \cdot \left[\nabla u_{k_j} - \nabla u \right] dx = 0.$$

The elementary inequality

$$2^{1-p} |w_2 - w_1|^p \leq \left[|w_2|^{p-2} w_2 - |w_1|^{p-2} w_1 \right] \cdot (w_2 - w_1), \qquad p \geq 2,$$

for vectors in $\mathbf{R}^n$ shows that the limit above implies the strong convergence

$$\lim_{j \to \infty} \int_\Omega |\nabla u_{k_j} - \nabla u|^p \, dx = 0.$$

There is a similar inequality for $p < 2$. Thus we can pass to the limit under the integral sign in (5.7) to obtain

$$\int_\Omega |\nabla u|^{p-2} \nabla u \cdot \nabla \eta dx = \lambda \int_\Omega |u|^{p-2} u \eta dx.$$

This shows that λ is an eigenvalue, since the normalization prevents u from being identically zero. $\qquad\square$

It is evident that an eigenfunction is bounded in a regular domain. But there are continuous functions in the Sololev space $W_0^{1,p}(\Omega)$ that are unbounded. Therefore we had better write down a proof of

$$\sup_{x \in \Omega} |u(x)| < \infty.$$

Lemma 5.2. *The bound*

$$\|u\|_{\infty,\Omega} \le 4^n \lambda^{\frac{n}{p}} \|u\|_{1,\Omega} \tag{5.8}$$

holds for the eigenfunction u in any bounded domain Ω in $\mathbf{R}^n$.

Proof. The interesting method in [30, Lemma 5.1, p. 71] yields this estimate. (The constant 4^n is not optimal.) To this end, we may assume that u is positive at some point. The function

$$\eta(x) = \max\{u(x) - k, 0\}$$

will do as test-function in (5.3) and so we obtain

$$\int_{A_k} |\nabla u|^p \, dx = \lambda \int_{A_k} |u|^{p-2} u(u - k) dx \tag{5.9}$$

where

$$A_k = \{x \in \Omega | u(x) > k\}.$$

Clearly $k |A_k| \le \|u\|_{1,\Omega}$ and $|A_k| \to 0$ as $k \to \infty$.

By the elementary inequality $a^{p-1} \le 2^{p-1}(a-k)^{p-1} + 2^{p-1}k^{p-1}$ we have

$$\int_{A_k} u^{p-1}(u - k) dx \le 2^{p-1} \int_{A_k} (u - k)^p dx + 2^{p-1}k^{p-1} \int_{A_k} (u - k) dx. \tag{5.10}$$

The Sobolev inequality yields

$$\int_{A_k} (u - k)^p dx \le \left(2^{-1} |A_k|\right)^{\frac{p}{n}} \int_{A_k} |\nabla u|^p \, dx, \tag{5.11}$$

when applied to each component of the open set A_k. (The constant $\frac{1}{2}$ is not essential.)

Combining (5.9), (5.10), and (11.9) we arrive at

$$\left[1 - 2^{p-2}\lambda |A_k|^{\frac{p}{n}}\right] \int_{A_k} (u - k)^p dx \le 2^{p-2}k^{p-1}\lambda |A_k|^{\frac{p}{n}} \int_{A_k} (u - k) dx.$$

In the first factor $2^{p-2}\lambda|A_k|^{\frac{p}{n}} \leq \frac{1}{2}$, when $k \geq k_1 = 2^{n(p-1)/p}\lambda^{n/p}\|u\|_1$. Using the Hölder inequality and dividing out we finally obtain the estimate

$$\int_{A_k}(u-k)dx \leq 2\lambda^{\frac{1}{p-1}}k\,|A_k|^{1+\frac{p}{(p-1)n}} \tag{5.12}$$

for $k \geq k_1$. This is the inequality needed in [30, Lemma 5.1, p.71] to bound $\operatorname{ess\,sup}u$.

Indeed, writing

$$f(k) = \int_{A_k}(u-k)dx = \int_k^\infty |A_t|\,dt,$$

we have $f'(k) = -|A_k|$ and hence (5.12) can be restated as

$$f(k) \leq 2\lambda^{\frac{1}{p-1}}k\,[-f'(k)]^{1+\frac{p}{(p-1)n}},$$

when $k \geq k_1$. If f is positive in the interval $[k_1, k]$, then an integration of the differential inequality leads to

$$k^{\frac{\varepsilon}{1+\varepsilon}} - k_1^{\frac{\varepsilon}{1+\varepsilon}} \leq \left[2\lambda^{\frac{1}{p-1}}\right]^{\frac{1}{1+\varepsilon}}\left[f(k_1)^{\frac{\varepsilon}{1+\varepsilon}} - f(k)^{\frac{\varepsilon}{1+\varepsilon}}\right]$$

where $\varepsilon = p/(p-1)n$. Since $f(k_1) \leq f(0) = \|u\|_1$ and $f(k) \geq 0$ on the right-hand side, this clearly bounds k and hence $f(k)$ is zero sooner or later. The quantitative bound is seen to be

$$k \leq 2^{1+\frac{2n(p-1)}{p}}\lambda^{\frac{n}{p}}\|u\|_1. \tag{5.13}$$

This means that $f(k) = 0$, if (5.13) is not fulfilled, i.e. $\operatorname{ess\,sup}u$ is never greater than the right-hand side.

To bound $\operatorname{ess\,inf}u$, consider the function $-u$. $\qquad\square$

Let me mention a difficult question. Can an eigenfunction be zero at all the points of an open subset of Ω? This is *the problem of unique continuation*. Except for the first eigenfunction this seems to be an open problem. Zero has a special status. No eigenfunction can have a constant value different from zero in an open subdomain. This is evident from the equation.

5.3. The one-dimensional case

In the case of one independent variable all the eigenvalues are explicitly known. This was first studied by Ôtani in connexion with the determination of the best constant in some Sobolev type inequalities. The equation is

$$\left(|u'|^{p-2}\,u'\right)' + \lambda\,|u|^{p-2}\,u = 0$$

where $u = u(x)$, $a \le x \le b$, and $u(a) = 0$, $u(b) = 0$. The equation is readily integrated and, via the first integral

$$|u'|^p + \frac{\lambda|u|^p}{p-1} = \text{Constant},$$

one arrives at the expression

$$\lambda(p) = (p-1)\left\{\frac{2}{b-a}\int_0^1 \frac{dt}{(1-t^p)^{1/p}}\right\}^p$$

for the first eigenvalue, cf. [42]. This is the minimum of the Rayleigh quotient

$$\frac{\int_a^b |u'(x)|^p\,dx}{\int_a^b |u(x)|^p\,dx}$$

taken among all $u \in C^1[a,b]$ with $u(a) = u(b) = 0$. The expression for $\lambda(p)$ is easily evaluated and the result is

$$\sqrt[p]{\lambda(p)} = \frac{2\pi\sqrt[p]{p-1}}{(b-a)p\sin\frac{\pi}{p}}.$$

The rather striking result

$$\sqrt[p]{\lambda(p)} = \sqrt[q]{\lambda(q)}, \quad \frac{1}{p} + \frac{1}{q} = 1$$

can be read off for conjugated exponents p and q. In terms of Rayleigh quotients

$$\min \frac{\|u'\|_p}{\|u\|_p} = \min \frac{\|v'\|_q}{\|v\|_q}, \quad \frac{1}{p} + \frac{1}{q} = 1.$$

See [34].

The spectrum can be completely determined. The eigenvalues are precisely

$$\lambda(p), 2^p\lambda(p), 3^p\lambda(p), \ldots, k^p\lambda(p), \ldots$$

The eigenfunctions are obtained from the first one. Let u_1 denote the first eigenfunction in $[0,1]$. Extend it as an odd function to $[-1,0]$ and, then, periodically to the whole real axis, i.e., $u_1(x) = -u_1(-x)$, $u_1(x+2) = u_1(x)$. The higher eigenfunctions are

$$u_k(x) = u_1(kx), \qquad k = 1, 2, 3, \ldots$$

In the linear case we have the eigenvalue $k^2\pi^2$ corresponding to the normalized eigenfunction

$$\sqrt{2}\sin(k\pi x), \qquad k = 1, 2, 3, \ldots$$

The spectrum is discrete in the one-dimensional case. The eigenvalues are simple and the kth eigenfunction has $k-1$ nodes (zeros inside the interval) and k nodal intervals of equal length.

An example in [10] shows that the Fredholm alternative does not hold for the equation

$$(|u'|^{p-2}u')' + \lambda|u|^{p-2}u = f(x)$$

in the nonlinear case $p \neq 2$. A solution can exist even if

$$\langle u_1, f \rangle = \int_0^1 u_1(x)f(x)dx \neq 0.$$

Some other orthogonality condition seems to be called for.

5.4. The first eigenfunction

The first eigenfunction (the Ground State) has many special properties. It is the only positive eigenfunction. *The restriction of a higher eigenfunction to a nodal domain is a first eigenfunction* (with respect to this smaller domain).

The first eigenvalue or the principal frequency is

$$\lambda_1 = \inf_\varphi \frac{\int_\Omega |\nabla\varphi|^p \, dx}{\int_\Omega |\varphi|^p \, dx} \tag{5.14}$$

where $\varphi \in C_0^\infty(\Omega)$, $\varphi \not\equiv 0$. By (5.4) $\lambda_1 > 0$. Using a (normalized) minimizing sequence $\varphi_1, \varphi_2, \ldots$ we obtain a function $u_1 \in W_0^{1,p}(\Omega)$ such that

$$\lambda_1 = \frac{\int_\Omega |\nabla u_1|^p \, dx}{\int_\Omega |u_1|^p \, dx}.$$

The compactness argument needed in the existence proof is provided by the Rellich-Kondrachov Theorem. A well-known device due to Lagrange shows that u_1 is a weak solution to the equation

$$div\left(|\nabla u|^{p-2}\nabla u\right) + \lambda_1 |u|^{p-2} u = 0.$$

If u_1 is minimizing, so is $|u_1|$ and therefore also $|u_1|$ satisfies the equation. Since $|u_1| \geq 0$, we must have $|u_1| > 0$ by Harnack's inequality. By continuity either $u_1 > 0$ in Ω or $u_1 < 0$ in Ω. We have established the following result.

Lemma 5.3. *There exists a positive eigenfunction corresponding to the principal frequency λ_1. This eigenfunction minimizes the Rayleigh quotient among all functions in the Sobolev space $W_0^{1,p}(\Omega)$. Moreover, a minimizer is a first eigenfunction and does not change signs.*

Some basic facts can easily be read off from the Rayleigh quotient. First, if $\Omega_1 \subset \Omega_2$, then $\lambda_1(\Omega_1) \geq \lambda_1(\Omega_2)$, since there are more competing functions in Ω_2. Second, the quantity $p\lambda(p)^{1/p}$ increases with p. (The notation $\lambda_1 = \lambda(p)$ indicates the dependence of p.)

Being a solution to Eqn (5.3), the eigenfunction shares many properties with solutions to more general quasilinear eigenvalue problems. But here we would like to emphasize the following specific features:

I **"Isoperimetric" property**. Among all domains with the same volume (area) the ball (the disc) has the smallest principal frequency.[b]

II **Concavity**. For any bounded convex domain $\log u$ is concave, u denoting a positive eigenfunction [48, Theorem 1].

III **Uniqueness**. The first eigenfunctions are essentially unique in any bounded domain: given p, they are merely constant multiples of each other. Moreover, they have no zeros in the domain and they are the only eigenfunctions not changing sign.

IV **Stability**. For any bounded sufficiently regular domain the principal frequency is continuous as a function of p. [35, Theorem 6.1]. [c] In very irregular domains there is some anomaly.

V **Superharmonicity**. In a convex domain the first eigenfunction is superharmonic, for $p \geq 2$. (We mean that $"\Delta u \leq 0"$.)

---

[b] For the second eigenvalue there is a characterization in the linear case, cf [44].
[c] No similar result is known for the second eigenfunction.

VI Asymptotic formula. As $p \to \infty$ we have

$$\lim_{p \to +\infty} \sqrt[p]{\lambda(p)} = \frac{1}{\max\limits_{x \in \Omega} \operatorname{dist}(x, \partial\Omega)}$$

In other words, the reciprocal number of the radius of the largest inscribed ball in the domain gives the principal frequency for the case $p = \infty$!

The uniqueness (III) for *arbitrary* bounded domains was first proved in [33]. A new proof was found in [3]. Recently, an elegant variational proof was found by Belloni and Kawohl, cf. [11]. The radial case has been studied by F. de Thélin [52] and a good reference for C^2-domains is [48, Theorem A.1]. Other references for regular domains are [7], [6] and [2]. As we said, the restriction of a higher eigenfunction to a nodal domain is a first eigenfunction there. Though the original domain is as regular as we please, it is not clear that this is inherited by the nodal domains. Therefore it is important to prove the uniqueness in arbitrary domains. The proof will be discussed below. The logarithmic concavity [d] mentioned in (II) is due to S. Sakaguchi [48], when $p \neq 2$, and the linear case is credited to H. Brascamp & E. Lieb. The proof by Sakaguchi is based on a convexity principle of N. Korevaar. The superharmonicity (V) is a consequence of (II) and the formula

$$(p-2)|\nabla u|^3 \Delta_1 u + \frac{\Delta_p u}{|\nabla u|^{p-4}} = (p-1)|\nabla u|^2 \Delta u,$$

which connects the Laplacian Δu with the p-Laplacian $\Delta_p u = \nabla \cdot (|\nabla u|^{p-2}\nabla u)$ and the *mean curvature operator* $-\Delta_1 u = -\operatorname{div}\left(\frac{\nabla u}{|\nabla u|}\right)$. The formula has to be interpreted in the *viscosity sense*. Property (I) follows by spherical symmetrization (Schwarz symmetrization), cf. [28, p.90]. The ball is (essentially) the only optimal shape, cf [8]. For $p = 2$ this is the celebrated conjecture of Lord Rayleigh, proved by E. Krahn [e] and G. Faber. The asymptotic formula (VI) is postponed to Section 5.6.

Let us begin by discussing the uniqueness (III). The first eigenvalue is simple. That is, all the first eigenfunctions in a fixed domain are merely constant multiples of each other.

---

[d]The reader might find it strange that the property does not depend on p. The corresponding result for the equation $\operatorname{div}\left(|\nabla u|^{p-2}\nabla u\right) = -1$ is that $u^{1-1/p}$ is a concave function.

[e]See "Edgar Krahn 1894-1961", a centenary volume edited by Š. Lumiste & J. Peetre, IOS Press, Amsterdam 1994, pp. 81-106.

Theorem 5.2. *The first eigenvalue is simple in any bounded domain.*

Proof. Suppose that u and v are two first eigenfunctions. So are $|u|$ and $|v|$. Thus the situation is reduced to the case $u > 0$ and $v > 0$. As Anane has observed in [2], the result would follow by certain balanced calculations, if the function $\eta = u - v^p u^{1-p}$ were, a priori, admissible as test-function in

$$\int_\Omega |\nabla u|^{p-2}\, \nabla u \cdot \nabla \eta dx = \lambda_1 \int_\Omega |u|^{p-2}\, u\eta dx$$

and $v - u^p v^{1-p}$ in the similar equation for v. We use the modified test-functions

$$\eta = \frac{(u+\varepsilon)^p - (v+\varepsilon)^p}{(u+\varepsilon)^{p-1}} \quad \text{and} \quad \frac{(v+\varepsilon)^p - (u+\varepsilon)^p}{(v+\varepsilon)^{p-1}},$$

ε being a positive constant. Then

$$\nabla \eta = \left\{ 1 + (p-1) \left(\frac{v+\varepsilon}{u+\varepsilon} \right)^p \right\} \nabla u - p \left(\frac{v+\varepsilon}{u+\varepsilon} \right)^{p-1} \nabla v$$

and, by symmetry, the gradient of the test-function in the corresponding equation for v has a similar expression, yet with u and v interchanged. Using the fact that u and v are bounded (Section 5.2), we easily see that $\eta \in W_0^{1,p}(\Omega)$.

Instead of reproducing the whole proof in [33] we write down the calculations only for $p = 2$, that is, a non-linear proof of the linear case is presented. Inserting the chosen test-functions into their respective equations and adding these, we obtain the simple expression

$$\int_\Omega \left(u_\varepsilon^2 + v_\varepsilon^2 \right) |\nabla \log u_\varepsilon - \nabla \log v_\varepsilon|^2\, dx$$

$$= \lambda_1 \int_\Omega \left[\frac{u}{u_\varepsilon} - \frac{v}{v_\varepsilon} \right] \left(u_\varepsilon^2 - v_\varepsilon^2 \right) dx, \quad (5.15)$$

where we have written $u_\varepsilon = u(x) + \varepsilon$ and $v_\varepsilon = v(x) + \varepsilon$. As ε approaches zero, it is plain that the right hand side tends to zero. By Fatou's lemma

$$\int_\Omega \left(u^2 + v^2 \right) |\nabla \log u - \nabla \log v|^2\, dx = 0.$$

The integrand must be zero. Hence $u\nabla v = v\nabla u$ a.e. Thus $u = Cv$ or $v = Cu$ for some constant. This proves the case $p = 2$.

If $p \geq 2$, then the inequality [f]

$$|w_2|^p \geq |w_1|^p + p\,|w_1|^{p-2}\,w_1 \cdot (w_2 - w_1) + \frac{|w_2 - w_1|^p}{2^{p-1} - 1}$$

should be used. Take $w_2 = \nabla \log v_\varepsilon$ and $w_1 = \nabla \log u_\varepsilon$. There is a counterpart valid when $1 < p < 2$. For the details we refer to [33]. $\qquad\square$

As a byproduct of the proof we can conclude the following

Theorem 5.3. *A positive eigenfunction is always a first eigenfunction.*

Proof. Suppose that $v > 0$ is an eigenfunction with the eigenvalue λ. Let $u > 0$ denote the first eigenfunction. In the case $p = 2$ the previous calculation yields that

$$\int_\Omega \left(u_\varepsilon^2 + v_\varepsilon^2\right) |\nabla \log u_\varepsilon - \nabla \log v_\varepsilon|^2 \, dx = \int_\Omega \left[\lambda_1 \frac{u}{u_\varepsilon} - \lambda \frac{v}{v_\varepsilon}\right] \left(u_\varepsilon^2 - v_\varepsilon^2\right) dx.$$

This exhibits that the right-hand member is non-negative. Hence, letting ε tend to zero, we have

$$(\lambda_1 - \lambda) \int_\Omega \left(u^2 - v^2\right) dx \geq 0.$$

If $\lambda \neq \lambda_1$, then $\lambda > \lambda_1$ and $\int_\Omega \left(u^2 - v^2\right) dx \leq 0$. This is an impossible situation, since u can be replaced by $2u, 3u, \ldots$. We have proved that $\lambda = \lambda_1$. The case $p \neq 2$ is rather similar. $\qquad\square$

A simple proof of the simplicity of λ_1 has recently been given by Belloni and Kawohl, cf [11]. It is based on the admissible function

$$w = \left(\frac{u^p + v^p}{2}\right)^{1/p}$$

in the Rayleigh quotient. A short calculation yields

$$|\nabla w|^p = \frac{u^p + u^p}{2} \left|\frac{u^p \nabla \log u + v^p \nabla \log v}{u^p + v^p}\right|^p.$$

Because the positive quantities $u^p/(u^p + v^p)$ and $v^p/(u^p + v^p)$ add up to 1, we can use Jensen's inequality for convex functions to obtain the estimate

$$\left|\frac{u^p \nabla \log u + v^p \nabla \log v}{u^p + v^p}\right|^p \leq \frac{u^p |\nabla \log u|^p + v^p |\nabla \log v|^p}{u^p + v^p}.$$

---

[f]The inequality seems to be due to L. Evans, see [18, p. 250]. The best constant is not the abovementioned $1/(2^{p-1} - 1)$.

Thus we have

$$|\nabla w|^p \leq \frac{1}{2}|\nabla u|^p + \frac{1}{2}|\nabla v|^p.$$

The inequality is strict at points where $\nabla \log u \neq \nabla \log v$. Now we can conclude that

$$\lambda_1 \leq \frac{\displaystyle\int_\Omega |\nabla w|^p dx}{\displaystyle\int_\Omega w^p dx} \leq \frac{\dfrac{1}{2}\displaystyle\int_\Omega |\nabla u|^p dx + \dfrac{1}{2}\displaystyle\int_\Omega |\nabla v|^p dx}{\dfrac{1}{2}\displaystyle\int_\Omega u^p dx + \dfrac{1}{2}\displaystyle\int_\Omega v^p dx} = \lambda_1.$$

If $\nabla \log u \neq \nabla \log v$ in a set of positive measure, then we would have a strict inequality above, which is a contradiction. This proves that u and v are constant multiples of each other. — This elegant proof is not, as is were, capable of establishing that a positive eigenfunction is a first one.

About the concavity of $\log u$ we refer directly to [48]. It is worth noting that the first eigenfunction u itself is never concave, the one dimensional case beeing an exception. In a ball in $\mathbf{R}^n$ even u^α is concave for some α, $1/n < \alpha < 1$. In [36] I have conjectured that, among all convex domains, the ball has the best concavity exponent. Even the linear case seems to be unsettled.

The **stability of the principal frequency $\lambda_1 = \lambda(p)$, when p varies** is rather intriguing. This topic is discussed in [35]. By the Hölder inequality

$$p\lambda(p)^{\frac{1}{p}} < s\lambda(s)^{\frac{1}{s}}, \text{ when } 1 < p < s < \infty,$$

so that the one-sided limits in

$$\lim_{s \to p-} \lambda(s) \leq \lambda(p) = \lim_{s \to p+} \lambda(s). \tag{5.16}$$

exist. The last equality is almost evident. Normalizing the eigenfunctions so that $\|u_s\|_{s,\Omega} = 1$ we actually have

$$\lim_{s \to p+} \int_\Omega |\nabla u_s - \nabla u_p|^p \, dx = 0$$

as s approaches p from above. When s approaches p from below, even the adjusted version

$$\lim_{s \to p-} \int_\Omega |\nabla u_s - \nabla u_p|^s \, dx = 0 \tag{5.17}$$

is plainly false in irregular domains, when $p \leq n$. However, (5.17) implies that

$$\lim_{s \to p-} \lambda(s) = \lambda(p). \tag{5.18}$$

We think that (5.18) implies (5.17).

Given any p, $1 < p \leq n$, there is a bounded domain Ω in $\mathbf{R}^n$ such that

$$\lim_{s \to p+} \lambda(s) < \lambda(p),$$

and, *a fortiori*, (5.17) cannot hold for the normalized eigenfunctions. The explanation is a rather interesting phenomenon. A sequence of eigenfunctions u_s will converge to a function $u \in W^{1,p}(\Omega)$. One has $u \in W_0^{1,s}(\Omega)$ for every $s < p$. This u is a weak solution to the equation $div\left(|\nabla u|^{p-2}\nabla u\right) + \lambda |u|^{p-2} = 0$ in Ω, except that it fails to be in the right space $W_0^{1,p}(\Omega)$. To cause such a delicate effect, one needs a closed set Ξ_p such that $\mathrm{cap}_s(\Xi_p) = 0$, when $s < p$, yet $\mathrm{cap}_p(\Xi_p) > 0$. It is known how to construct such sets as generalized Cantor sets. The final domain Ω will be of the form $B \setminus \Xi_p$, where B is a sufficiently large ball containing Ξ_p in its interior. For a complete discussion of the "p-stability" we refer to our fairly technical paper in "Potential Analysis". See also [26].

The question about **variations of the domain**, instead of of the exponent p, is relatively simple. Let $\Omega_1 \subset \Omega_2 \subset \Omega_3 \subset \cdots$ be an exhaustion of Ω,

$$\Omega = \bigcup_{j=1}^{\infty} \Omega_j.$$

Then

$$\lim_{j \to \infty} \lambda_1^{(p)}(\Omega_j) = \lambda_1^{(p)}(\Omega), \tag{5.19}$$

where the notation is evident. (By a remark in the book by Courant-Hilbert, this is not true for the corresponding Neumann problem, when $p = 2$. One has to define the admissible variations of the domain in a careful way, when the normal derivative at the boundary is involved.)

To prove (5.19), we note that

$$\lambda_1^{(p)}(\Omega_1) \geq \lambda_1^{(p)}(\Omega_2) \geq \cdots \geq \lambda_1^{(p)}(\Omega)$$

Given $\varepsilon > 0$, there is a function $\varphi \in C_0^{\infty}(\Omega)$ such that

$$\lambda_1^{(p)}(\Omega) > \frac{\displaystyle\int_{\Omega} |\nabla \varphi|^p \, dx}{\displaystyle\int_{\Omega} |\varphi|^p \, dx} - \varepsilon,$$

since $\lambda_1^{(p)}(\Omega)$ is the infimum. Being a compact set, the support of φ is covered by a finite number of the sets $\Omega_1, \Omega_2, \ldots$. Hence $\operatorname{supp}\varphi \subset \Omega_j$ for j large enough. Thus

$$\lambda_1^{(p)}(\Omega_j) \leq \frac{\displaystyle\int_{\Omega_j} |\nabla\varphi|^p \, dx}{\displaystyle\int_{\Omega_j} |\varphi|^p \, dx} = \frac{\displaystyle\int_{\Omega} |\nabla\varphi|^p \, dx}{\displaystyle\int_{\Omega} |\varphi|^p \, dx}$$

so that

$$\lambda_1^{(p)}(\Omega) > \lambda_1^{(p)}(\Omega_j) - \varepsilon$$

for all large j. It is plain that $\lambda_1^{(p)}(\Omega) \geq \lim \lambda_1^{(p)}(\Omega_j)$. This proves the desired result.

Indeed, extending the eigenfunctions $u_j \in W_0^{1,p}(\Omega_j)$ as zero in $\Omega \setminus \Omega_j$ so that $u_j \in W_0^{1,p}(\Omega)$, the strong convergence $\|\nabla u - \nabla u_j\|_{p,\Omega} \longrightarrow 0$ holds for the normalized eigenfunctions (that is, $\|u_j\|_{p,\Omega} = 1$). Here u is the first eigenfunction in Ω. The proof is not difficult.

5.5. Higher eigenvalues

The operator $-\Delta$ has a discrete spectrum $\lambda_1 < \lambda_2 \leq \lambda_3 \leq \cdots$ and $\lambda_k \longrightarrow \infty$ as $k \longrightarrow \infty$. Each eigenvalue is repeated according to its multiplicity. Weyl's theorem about the asymptotic behaviour of the eigenvalues states that

$$\lim_{k \to \infty} \frac{\lambda_k^{n/2}}{k} = \frac{\text{Const.}}{|\Omega|}.$$

The corresponding eigenfunctions $u_1, u_2, u_3, \ldots$ can be chosen to satisfy

$$\langle u_k, u_j \rangle = \int_{\Omega} u_k u_j \, dx = \delta_{ij}.$$

This orthogonality is the key to the linear case $\Delta u + \lambda u = 0$. We recommend the classical book by Courant & Hilbert.

It is more difficult to prove that also the equation

$$\operatorname{div}\left(|\nabla u|^{p-2} \nabla u\right) + \lambda |u|^{p-2} u = 0$$

has infinitely many eigenvalues. There are several methods that work. However, the main open problem is quite the opposite. Are there more eigenvalues than the chosen method produces? If so, how can one exhaust the spectrum? Can all the eigenvalues be numerated? To the best of my knowledge the nonlinear spectrum has not been proved to be discrete, not even when the domain Ω is a ball or a cube.

In order to describe how higher eigenvalues are produced we have to introduce an auxiliary concept, the genus of Krasnoselskij. The proof will be skipped. About the method we refer to [9], [46], and [50].

If A is a symmetric[g] and closed subset of a Banach space, then its *genus* $\gamma(A)$ is defined as the smallest integer k for which there exists a continuous odd mapping $\varphi\colon A \longrightarrow \mathbf{R}^k \setminus \{0\}$. Thus $\varphi(v) = -\varphi(-v)$, when $v \in A$. If no such integer exists, then we define $\gamma(A) = \infty$. Especially, $\gamma(A) = \infty$, if A contains the origin, since $\varphi(0) = 0$ for odd mappings. See [46] and [St, Chapter II] about this concept.

Let $\sum_k$ denote the collection of all symmetric subsets A of $W_0^{1,p}(\Omega)$ such that $\gamma(A) \geq k$ and the set $\{v \in A|\, \|v\|_{p,\Omega} = 1\}$ is compact. *The numbers*

$$\lambda_k = \inf_{A \in \sum_k} \max_{v \in A} \frac{\displaystyle\int_\Omega |\nabla v|^p \, dx}{\displaystyle\int_\Omega |v|^p \, dx} \tag{5.20}$$

are eigenvalues and there are infinitely many of them, cf. [21] and [6]. The fact that this minimax procedure yields eigenvalues is often explained through the Palais-Smale condition.

These "minimax eigenvalues" $\lambda_1 < \lambda_2 \leq \lambda_3 \cdots$ satisfy an estimate of the type encountered in Weyl's theorem. According to [21] there are two positive constants depending only on n and p such that

$$\frac{c_1}{|\Omega|} \leq \frac{\lambda_k^{\frac{n}{p}}}{k} \leq \frac{c_2}{|\Omega|}$$

as $k \to \infty$. See also [6]. Unfortunately, *it is not known whether the described procedure exhausts the spectrum.* Are there other eigenvalues than those listed in (5.20)? Therefore the asymptotic result is of limited interest, so far.

As the notation in (5.20) indicates, λ_1 is the first eigenvalue. As we will see, λ_1 is isolated. It is possible to show that λ_2 is the second one. An unpublished manuscript [4] of A. Anane and M. Tsouli contains a minimax

---

[g]Symmetric means that $-v \in A$, if $v \in A$.

characterization of the second eigenvalue in terms of the functional

$$I(v) = \left(\int_\Omega |\nabla v|^p \, dx \right)^2 - \int_\Omega |v|^p \, dx.$$

Their proof is easily adapted to the Rayleigh quotient: for $k = 2$ (5.20) yields *the second eigenvalue* λ_2, that is $\lambda_2 = \min_{\lambda < \lambda_1} \lambda$. No such identification is yet known for eigenvalues higher than the second. The second eigenvalue is not known to be isolated, when its multiplicity is ignored.

The *nodal domains* are defined as the connected components of the sets $\{u > 0\}$ and $\{u < 0\}$. See [14] and [1].

Theorem 5.4. *Any eigenfunction has only a finite number of nodal domains.*

Proof. Let u be an eigenfunction corresponding to λ. If N_j denotes a component of one of the sets $\{x \in \Omega | u(x) > 0\}$ and $\{x \in \Omega | u(x) < 0\}$, then $u \in W_0^{1,p}(N_j)$. By the Sobolev inequality

$$|N_j| \geq C(n,p) \lambda^{-\frac{n}{p}}.$$

Summing up, we have

$$|\Omega| \geq \sum_j |N_j| > C(n,p) \lambda^{-\frac{n}{p}} \sum_j 1$$

so that the number of nodal domains is bounded by a constant times $\lambda^{n/p} |\Omega|$. $\qquad\square$

Theorem 5.5. *The first eigenvalue is isolated.*[h]

Proof. Suppose that there is a sequence of eigenvalues λ'_k tending to λ_1 (these are not supposed to be minimax eigenvalues). If u_k denotes the corresponding normalized eigenfunction, then

$$\int_\Omega |\nabla u_k|^p \, dx = \lambda'_k, \quad \int_\Omega |u_k|^p \, dx = 1.$$

By compactness arguments there are a subsequence and a function $u \in W_0^{1,p}(\Omega)$ such that $\nabla u_{k_j} \rightharpoonup \nabla u$ weakly and $u_{k_j} \to u$ strongly in $L^p(\Omega)$.

---

[h] For *smooth* domains this is credited to Anane, cf. [2].

By weak lower semicontinuity

$$\frac{\displaystyle\int_\Omega |\nabla u|^p \, dx}{\displaystyle\int_\Omega |u|^p \, dx} \le \lim_{j\to\infty} \lambda'_{k_j} = \lambda_1$$

so that u is the first eigenfunction. Since u does not change signs, we may take $u > 0$.

If $\lambda'_k \neq \lambda_1$, then u_k must change signs in Ω. Both sets $\Omega_k^+ = \{u_k > 0\}$ and $\Omega_k^- = \{u_k < 0\}$ are non-empty and their measures cannot tend to zero, since

$$\left|\Omega_k^+\right| \ge C(n,p)(\lambda'_k)^{-\frac{n}{p}}, \quad \left|\Omega_k^-\right| \ge C(n,p)(\lambda'_k)^{-\frac{n}{p}}.$$

This prevents u_{k_j} from converging to a positive function in $L^p(\Omega)$. Indeed, the sets

$$\Omega^+ = \limsup \Omega_{k_j}^+, \quad \Omega^- = \limsup \Omega_{k_j}^-$$

have positive measure by a well-known "Selection Lemma". We may assume that $u = \lim u_{k_j}$ a.e. in Ω. Passing to suitable subsequences we can show that $u \ge 0$ a.e. in Ω^+ and $u \le 0$ a.e. in Ω^-. This is a contradiction. $\qquad\square$

There are many more open problems about the spectrum of the p-Laplacian than those that have been mentioned here, in passing. To mention two more: Is every eigenvalue of finite multiplicity? What about multiplicity in the situation with general boundary values? Consider the equation $div\left(|\nabla u|^{p-2}\nabla u\right) + \lambda |u|^{p-2} u = 0$ with given boundary values, say φ. This has always at least one solution. Does it have several solutions, if λ happens to be an eigenvalue? In the linear case one just adds solutions to see this.

5.6. The asymptotic case

It is instructive to see what happens when $p \to \infty$. Arcane phenomena occur in this fascinating case. Let

$$\lambda(p) = \inf_\varphi \frac{\displaystyle\int_\Omega |\nabla\varphi|^p dx}{\displaystyle\int_\Omega |\varphi|^p dx} = \inf_\varphi \frac{\|\nabla\varphi\|_p^p}{\|\varphi\|_p^p}$$

denote the principal frequency and write

$$\Lambda_\infty = \inf_\varphi \frac{\|\nabla\varphi\|_\infty}{\|\varphi\|_\infty}, \tag{5.21}$$

where $\varphi \in C_0^\infty(\Omega)$. It turns out that the distance function

$$\delta(x) = \operatorname{dist}(x, \partial\Omega)$$

"solves" the minimization problem:

$$\Lambda_\infty = \frac{\|\nabla\delta\|_\infty}{\|\delta\|_\infty}. \tag{5.22}$$

To see this, notice that

$$|\varphi(x)| \le \|\nabla\varphi\|_\infty \delta(x)$$

by the Mean Value Theorem. Hence

$$\frac{\|\nabla\varphi\|_\infty}{|\varphi(x)|} \ge \frac{1}{\delta(x)} \ge \frac{1}{\|\delta\|_\infty} = \frac{\|\nabla\delta\|_\infty}{\|\delta\|^\infty},$$

since $|\nabla\delta(x)| = 1$ a.e. in Ω. Thus we conclude that

$$\frac{\|\nabla\varphi\|_\infty}{\|\varphi\|_\infty} \ge \frac{\|\nabla\delta\|_\infty}{\|\delta\|_\infty}$$

for each admissible φ. This proves (5.22).

However, the minimization problem often has too many solutions in $W^{1,\infty}(\Omega) \bigcap C(\overline{\Omega})$ with boundary values 0. In order to define the genuine ∞-eigenfunctions, one has to find the limit equation of

$$div(|\nabla u|^{p-2}\nabla u) + \lambda(p)|u|^{p-2}u = 0$$

as $p \to \infty$. It is shown in [27] that the limit equation is

$$\max\left\{\Lambda_\infty - \frac{|\nabla u(x)|}{u(x)}, \Delta_\infty u(x)\right\} = 0 \tag{5.23}$$

for positive solutions. (At each point x in Ω the larger of the two quantities is equal to zero.) Here

$$\Delta_\infty u = \sum_{i,j=1}^n \frac{\partial u}{\partial x_i} \frac{\partial u}{\partial x_j} \frac{\partial^2 u}{\partial x_i \partial x_j} \tag{5.24}$$

is the so called ∞-Laplacian. Unfortunately, the second derivatives of the solutions do not always exist. The above equation has to be interpreted *in the viscosity sense*, because it does not have any weak formulation with test-functions under the integral sign. We refer to [27] about all this.

Definition 5.2. Let $u \geq 0$ and $u \in C(\Omega)$. We say that u is a viscosity solution of the equation (5.23), if

(i) whenever $x_0 \in \Omega$ and $\varphi \in C^2(\Omega)$ are such that $u(x_0) = \varphi(x_0)$ and $u(x) < \varphi(x)$, when $x \neq x_0$, then

$$\Lambda_\infty - \frac{|\nabla\varphi(x_0)|}{\varphi(x_0)} \geq 0 \quad \text{or} \quad \Delta_\infty\varphi(x_0) \geq 0.$$

(ii) whenever $x_0 \in \Omega$ and $\varphi \in C^2(\Omega)$ are such that $u(x_0) = \varphi(x_0)$ and $u(x) > \varphi(x)$, when $x \neq x_0$, then

$$\Lambda_\infty - \frac{|\nabla\varphi(x_0)|}{\varphi(x_0)} \leq 0 \quad \text{and} \quad \Delta_\infty\varphi(x_0) \leq 0.$$

Notice that each point requires its own family of test-functions.

The essential feature is that the difference $u(x) - \varphi(x)$ attains its extremum at the touching point x_0, where the derivatives of the test-function are to be evaluated.

For example, when Ω is the ball $|x| < 1$, the infinity ground state is

$$u(x) = 1 - |x|.$$

We have $\Delta_\infty u(x) = 0$, when $x \neq 0$. The origin is the important point. Here $\Lambda_\infty = 1$ is determined. Since there are no test-functions touching from below at $x_0 = 0$, condition (ii) is automatically regarded as fullfilled. If the function

$$\varphi(x) = 1 + \langle a, x \rangle + 0(|x|^2)$$

touches from above, we must have

$$1 + \langle a, x \rangle \geq 1 - |x|$$

as $x \to 0$. Hence $|a| \leq 1$ and so

$$\Lambda_\infty - \frac{|\nabla\varphi(0)|}{\varphi(0)} = 1 - \frac{|a|}{1} \geq 0,$$

that is, condition (i) holds.

In passing, let me mention that in a square (cube) the distance function δ does not solve the equation. This means that it is not the limit of the ground states u_p, as $p \to \infty$. Recall (5.21) and (5.22).

As we observed

$$\Lambda_\infty = \frac{1}{\max\limits_{x \in \Omega} \text{dist}(x, \partial\Omega)}. \tag{5.25}$$

Thus the principal frequency can be detected from the geometry: it is the reciprocal number of the radius of the largest ball that can be inscribed in the domain Ω. This is an advantage. For example, if Ω is the punctured ball $0 < |x| < 1$, then $\Lambda_\infty = 1/2$. We should point out that *all boundary points are regular in the case $p = \infty$.* The solution is zero even at isolated boundary points! The equation

$$\max\left\{\Lambda - \frac{|\nabla u(x)|}{u(x)}, \Delta_\infty u(x)\right\} = 0$$

has a positive solution with zero boundary values only when $\Lambda = \Lambda_\infty$. No other Λ will do. In this respect we have a typical eigenvalue.

Let us consider the formula in VI, Section 5.4.

Lemma 5.4. $\quad \Lambda_\infty = \lim_{p\to\infty} \sqrt[p]{\lambda(p)}.$

Proof. Using the distance function as test-function in the Rayleigh quotient, we have

$$\sqrt[p]{\lambda(p)} \le \frac{\|\nabla\delta\|_p}{\|\delta\|_p}$$

and hence

$$\limsup_{p\to\infty} \sqrt[p]{\lambda(p)} \le \frac{\|\nabla\delta\|_\infty}{\|\delta\|_\infty} = \Lambda_\infty.$$

To achieve the inequality

$$\liminf_{p\to\infty} \sqrt[p]{\lambda(p))} \ge \Lambda_\infty$$

we use a compactness argument for the eigenfunctions u_p. For p large enough

$$\sqrt[p]{\lambda(p)} = \frac{\|\nabla u_p\|_p}{\|u_p\|_p} < \Lambda_\infty + 1.$$

With the normalization $\|u_p\|_p = 1$ the norms $\|\nabla u_p\|_m$ are uniformly bounded, when $p \ge m$. Using a diagonalization procedure, we can select a subsequence u_{p_j} that converges weakly in each $W^{1,q}(\Omega)$, $q < \infty$, and uniformly in each $C^\alpha(\Omega)$, $\alpha < 1$, to a function denoted by u_∞. By the weak lower semicontinuity

$$\frac{\|\nabla u_\infty\|_q}{\|u_\infty\|_q} \le \liminf_{j\to\infty} \frac{\|\nabla u_{p_j}\|_q}{\|u_{p_j}\|_q} \le \liminf_{j\to\infty} \frac{\|\nabla u_{p_j}\|_{p_j} |\Omega|^{\frac{1}{q}-\frac{1}{p_j}}}{\|u_{p_j}\|_q}$$

200 P. Lindqvist

$$= \liminf_{j \to \infty} \frac{\lambda(p_j)^{1/p_j} |\Omega|^{\frac{1}{q} - \frac{1}{p_j}}}{\|u_{p_j}\|_q} \leq \frac{|\Omega|^{\frac{1}{q}}}{\|u_\infty\|_q} \liminf_{j \to \infty} \lambda(p_j)^{1/p_j}$$

Taking the normalization into account and letting $q \to \infty$, we obtain

$$\frac{\|\nabla u_\infty\|_\infty}{\|u_\infty\|_\infty} \leq \liminf_{j \to \infty} \lambda(p_j)^{1/p_j}.$$

The left-hand side is $\geq \Lambda_\infty$, because u_∞ is admissible in the quotient. The right-hand side can be replaced by $\liminf \lambda(p)^{1/p}$, since we can begin the construction with an arbitrary sequence of p's. $\qquad \Box$

Much more is known but there are also challenging open problems in the case $p = \infty$. The interested reader can find some pieces of information in P. Juutinen, P. Lindqvist & J. Manfredi: *The infinity Laplacian: examples and observations*, Institut Mitag-Leffler, Report 26, 1999/2000.

References

[1] G. Alessandrini: On Courant's nodal domain theorem, Preprint (Univ. Trieste, Quaderni Matematici, II Serie, **331**) 1994.

[2] A. Anane: Simplicité et isolation de la premiére valeur propre du p-laplacien avec poids, C. R. Acad. Sci. Paris **305**, Sér. I Math., 1987, pp. 725–728.

[3] W. Allegretto & Y. Huang: A Picone's identity for the p-Laplacian and applications, Nonlinear Analysis **32**, 1998, pp. 819–830.

[4] A. Anane & N. Tsouli: Sur la seconde valeur propre du p-Laplacien, Université Mohammed I Oujda, Manuscript 1994.

[5] J. García Azorero & I. Peral Alonso: Existence and nonuniqueness for the p-Laplacian: Nonlinear eigenvalues, Communications in Partial Differential Equations **12**, 1987, pp. 1389–1430.

[6] J. García Azorero & I. Peral Alonso: Comportement asymptotique des valeurs propres du p-laplacien, C.R. Acad. Sci. Paris **307**, Série I, 1988, pp. 75–78.

[7] T. Bhattacharya: Some results concerning the eigenvalue problem for the p-Laplacian, Annales Academiæ Scientarum Fennicæ, Series A.I. Mathematica, **14**, 1989, pp. 325–343.

[8] T. Bhattacharya: A proof of the Faber–Krahn inequality for the first eigenvalue of the p-Laplacian, Annali di Matematica pura ed applicata (IV) **177**, 1999, pp. 225–240.

[9] P. Blanchard & E. Brüning: Variational Methods in Mathematical Physics, Springer-Verlag, New York 1992.

[10] P. Binding & P. Drábek & Y. Huang: On the Fredholm alternative for the p-Laplacian, Proceedings of the American Mathematical Society **125** (1997), pp. 3555–3559.

[11] M. Belloni & B. Kawohl: A direct uniqueness proof for equations involving the p-Laplace operator, Manuscripta Mathematica **109** (2002), pp. 229–231.

[12] I. Chavel: Eigenvalues in Riemannian Geometry, Academis Press, New York 1984.

[13] L. Caffarelli & X. Cabré: Fully Nonlinear Elliptic Equations, Colloquium Publications **43**, American Mathematical Society, Providence 1995.

[14] R. Courant & D. Hilbert: Methoden der Mathematischen Physik, Erster Band, Berlin 1931.

[15] R. Courant & D. Hilbert: Methods of Mathematical Physics, Volume I, New York 1953.

[16] M. Crandall: Viscosity solutions: A primer. In Lecture Notes in Mathematics 1660, Springer-Verlag 1997, pp. 1–43.

[17] E. Dibenedetto: $C^{1+\alpha}$ local regularity of weak solutions of degenerate elliptic equations, Nonlinear Analysis **7**, 1983, pp. 827–850.

[18] L. Evans: Quasiconvexity and partial regularity in the calculus of variations, Archive for Rational Mechanics and Analysis **95**, 1986, pp. 227–252.

[19] L. Evans: Weak Convergence Methods for Nonlinear Partial Differential Equations, CBMS, Regional Conference Series in Mathematics 74, American Mathematical Society 1990.

[20] L. Evans & R. Gariepy: Measure Theory and Fine Properties of Functions, CRC Press, Boca Raton 1992.

[21] L. Friedlander: Asymptotic behavior of the eigenvalues of the p-Laplacian, Communications in Partial Differential Equations **14**, 1989, pp. 1059–1069.

[22] R. Gariepy & W. Ziemer: A regularity condition at the boundary for solutions of quasilinear elliptic equations, Archive for Rational Mechanics and Analysis **67**, 1977, pp. 25–89.

[23] M. Giaquinta: Introduction to Regularity Theory for Nonlinear Elliptic Systems, Birkhäuser Verlag, Basel 1993.

[24] D. Gilbarg & N. Trudinger: Elliptic Partial Differential Equations of Second Order, 2nd Edition, Springer-Verlag, Berlin 1983.

[25] E. Giusti: Metodi Diretti nel Calcolo Variazione, Unione Matematica Italiana, Bologna 1994.

[26] L. Hedberg & T. Kilpeläinen: On the stability of Sobolev spaces with zero boundary values, Mathematica Scandinavica **85**, 1999, pp. 245–258.

[27] P. Juutinen & P. Lindqvist & J. Manfredi: The ∞-eigenvalue problem, Archive for Rational Mechanics and Analysis **148**, 1999, pp. 89–105.

[28] B. Kawohl: Rearrangements and Convexity of Level Sets in PDE (Lecture Notes in Mathematics 1150), Springer-Verlag, Heidelberg 1985.

[29] T. Kilpeläinen & P. Lindqvist: Nonlinear ground states in irregular domains. Indiana University Mathematics Journal **49**, 2000, pp. 325–331.

[30] O. Ladyzhenskaya & N. Ural'tseva: Linear and Quasilinear Elliptic Equations, Academic Press, New York 1968.

[31] E. Lieb: On the lowest eigenvalue of the Laplacian for the intersection of two domains, Inventiones Mathematicae **74**, 1983, pp. 441-448.

[32] E. Lieb: The stability of matter: from atoms to stars, Bulletin of the American Mathematical Society **22**, no. 1, 1990, pp. 1–49.

[33] P. Lindqvist: On the equation $div\left(|\nabla u|^{p-2}\nabla u\right) + \lambda |u|^{p-2} u = 0$, Proceedings of the American Mathematical Society **109**, Number 1, 1990, pp. 157–

164. Addendum, ibidem **116**, Number 2, 1992, pp. 583–584.

[34] P. Lindqvist: Note on a nonlinear eigenvalue problem, Rocky Mountain Journal of Mathematics **23**, 1993, pp. 281–288.

[35] P. Lindqvist: On non-linear Rayleigh quotients, Potential Analysis **2**, 1993, pp. 199–218.

[36] P. Lindqvist: A note on the nonlinear Rayleigh quotient, Proceedings of the twenty-first Nordic congress of mathematicians, Lecture notes in pure and applied mathematics, Marcel Dekker, Inc., New York 1994.

[37] P. Lindqvist: On a nonlinear eigenvalue problem, Berichte Univ. Jyväskylä Math. Inst. **68** (1995), pp. 33–54.

[38] P. Lindqvist & J. Manfredi & E. Saksman: Superharmonicity of nonlinear ground states. Revista Matematica Iberoamericana **16**, 2000, pp. 17–28.

[39] V. Maz'ja: On the continuity at a boundary point of solutions of quasi-linear elliptic equations, Vestnik Leningradskogo Universiteta **25**, 1970, pp. 42–55 (in Russian). Vestnik Leningrad Univ. Math. **3**, 1976, pp. 225–242 (English translation).

[40] M. Marcus & V. Mizel & Y. Pinchover: On the best constant for Hardy's inequality in $\mathbb{R}^n$, Transactions of the American Mathematical Society **350**, 1998, pp. 3237–3255.

[41] R. Osserman: Isoperimetric inequalities and eigenvalues of the Laplacian, Proceedings of the International Congress of Mathematicians, Helsinki 1978, Volume I, pp. 435–442.

[42] M. Ôtani: A remark on certain nonlinear elliptic equations, Proceedings of the Faculty of Science, Tokai University **19**, 1984, pp. 23–28.

[43] D. Passaseo: Nonlinear elliptic equations involving critical Sobolev exponents. Lecture Notes, Padova 2000.

[44] G. Pólya: On the characteristic frequences of a symmetric membrane, Math. Z. **63**, 1955, pp. 331–337.

[45] G. Pólya & G. Szegö: Isoperimetric Inequalities in Mathematical Physics, Princeton University Press, Princeton 1951.

[46] P. Rabinowitz: Some aspects of nonlinear eigenvalue problems, Rocky Mountain Journal of Mathematics **3**, 1973, pp. 161–202.

[47] F. Riesz & B. SZ.-Nagy: Vorlesungen über Funktionalanalysis, Deutscher Verlag der Wissenschaften, Berlin 1956.

[48] S. Sakaguchi: Concavity properties of solutions to some degenerate quasi-linear elliptic Dirichlet problems, Annali della Scuola Normale Superiore de Pisa, Serie IV (Classe di Scienze) **14**, 1987, pp. 403–421.

[49] S. Sobolev: Applications of Functional Analysis in Mathematical Physics, American Mathematical Society, Providence, RI 1963.

[50] M. Struwe: Variational Methods (Applications to Nonlinear Partial Differential Equations and Hamiltonian Systems), Springer-Verlag, Berlin 1990.

[51] F. de Thélin: Quelques résultats d'existence et de non-existence pour une E.D.P. elliptique non linéaire, C.R. Acad. Sci. Par **299**, Série I. Math. 1984, pp. 911–914.

[52] F. de Thélin: Sur l'espace propre associé à la première valeur propre du pseudo-laplacien, C.R. Acad. Sci. Paris **303**, Série I. Math. 1986, pp. 355–

358.

[53] P. Tolksdorf: Regularity for a more general class of quasi-linear elliptic equations, Journal of Differential Equations **51**, 1984, pp. 126–150.

[54] A. Touzani: Quelques résultats sur le A_p-laplacien avec poids indéfini, Universite Libre de Bruxelles 1991–92. (Thèse.)

[55] N. Trudinger: On Harnack type inequalities and their application to quasilinear elliptic equations, Communications on Pure and Applied Mathematics **20**, 1967, 721–747.

[56] L. Veron & M. Guedda: Quasilinear elliptic equations involving critical Sobolev exponents, Nonlinear Analysis TMA **13**, no 8, 1989, pp. 879–902.

[57] W. Ziemer: Weakly Differentiable Functions, Springer-Verlag, New York 1989.

Chapter 6

Nonlinear elliptic equations with critical and supercritical Sobolev exponents

Donato Passaseo

Università di Lecce
Dipartimento di Matematica "E. De Giorgi"
Prov.le Lecce-Arnesano, P.O. Box 193
73100 - Lecce (Italy)

Contents

6.1. Introduction and statement of the problems

The purpose of these notes is to present a survey of some recent results dealing with existence, nonexistence and multiplicity of nontrivial solutions for semilinear elliptic equations, whose nonlinear term has critical or supercritical growth.

Let us consider, for example, the following Dirichlet problem

$$P(\Omega,p) \quad \begin{cases} \Delta u + |u|^{p-2}u = 0 & \text{in } \Omega \\ u = 0 & \text{on } \partial\Omega \\ u \not\equiv 0 & \text{in } \Omega, \end{cases} \tag{6.1}$$

where Ω is a smooth bounded domain of $\mathbb{R}^n$, $n \geq 3$, and $p \geq \frac{2n}{n-2}$ ($\frac{2n}{n-2} = 2^*$ is the critical Sobolev exponent).

In particular we are interested to find positive solutions of $P(\Omega,p)$ or also sign changing (nodal) solutions with a prescribed number of nodal regions.

This equation is a simplified model of some variational problems, coming from Differential Geometry, Mechanics, Mathematical Phisics, Chemistry, whose common feature is the lack of compactness: for example a well known problem in Differential Geometry, the Yamabe's problem (see [1], [41], [44], [45]), is related to the solvability of a problem like (6.1) with $p = 2^*$; supercritical nonlinearities arise in some combustion models; lack of compactness also occur in Yang–Mills equations, etc...

In our problem the lack of compactness is due to the presence of critical or supercritical exponents: it is well known that $H_0^{1,2}(\Omega)$ is continuously embedded in $L^p(\Omega)$ for $p \leq 2^*$ and that the embedding is compact only for $p < 2^*$.

Since the nonlinear term in (6.1) is homogeneous, one can easily verify that solving problem $P_a(\Omega, p)$ is equivalent to finding critical points for the energy functional

$$f(u) = \int_\Omega |Du|^2 dx,$$

constrained on the manifold

$$M_p(\Omega) = \{ u \in H_0^{1,2}(\Omega) : \int_\Omega |u|^p dx = 1 \}.$$

There is a sharp contrast between the cases $p < 2^*$ and $p \geq 2^*$.

If $p < 2^*$, then the infimum $\inf_{M_p(\Omega)} f$ is achieved by a positive function, giving rise to a positive solution of $P(\Omega, p)$, independently of the shape of the domain Ω (indeed, one can find infinitely many solutions exploiting the symmetry properties of f and $M_p(\Omega)$). On the contrary, if $p \geq 2^*$, the infimum $\inf_{M_p(\Omega)} f$ is not achieved (as we shall see below).

Hence the problem cannot be simply solved by minimization arguments and the solutions (when there exist) correspond to higher critical values. But several difficulties also arise when trying to find critical points by means of the usual topological methods of the Calculus of Variations (like Morse Theory, Ljusternik–Schnirelman category, linking methods, etc ...), since the corresponding functional does not satisfy the Palais–Smale compactness condition when $p \geq 2^*$.

It is not only a problem of methods: there is a deep reason which explains the impossibility of applying these methods in a standard way. In fact every solution of problem

$$\begin{cases} \Delta u + g(u) = 0 & \text{in } \Omega \\ u = 0 & \text{on } \partial\Omega \end{cases} \tag{6.2}$$

must verify the following Pohozaev's identity (see [36]):

$$(1 - \frac{n}{2}) \int_\Omega g(u)u\,dx + n \int_\Omega G(u)\,dx = \frac{1}{2} \int_{\partial\Omega} (x \cdot \nu)(\frac{\partial u}{\partial \nu})^2 d\sigma, \qquad (6.3)$$

where $G(u) = \int_0^u g(t)\,dt$ and ν denotes the outward normal to $\partial\Omega$. As a consequence (for $g(u) = |u|^{p-2}u$) we have the following nonexistence result:

Theorem 6.1 (Pohozaev [36]). *If Ω is a star–shaped domain and $p \geq 2^*$, then the problem $P(\Omega, p)$ has no solution.*

After Pohozaev's Theorem, the researches in this topic followed two directions:

1) exploiting the shape of the domain Ω in order to regain the existence of solutions,
2) modifying the equation by lower–order terms.

The first direction of research is supported by the following observation: assume Ω is an annulus (i.e. $\Omega = \{x \in \mathbb{R}^n : 0 < r_1 < |x| < r_2\}$); then, exploiting the radial symmetry of Ω, it is easy to see (as pointed out by Kazdan and Warner [15]) that $P(\Omega, p)$ has a positive radial solution and infinitely many nodal radial solutions for all p. This leads to a natural question, pointed out by Nirenberg (see [5]): what happens if Ω has the same topology of an annulus, but not the same radial symmetry properties? is there still a positive solution, at least for $p = 2^*$? and if we assume only that Ω is not contractible, in itself, to a point?
This question has been answered by Bahri and Coron in [2], where the following theorem is proved (see also [11] and [38]).

Theorem 6.2 (Bahri–Coron [2]). *Assume Ω is a smooth bounded domain of $\mathbb{R}^n$, having non trivial topology (i.e. there exists an integer $k \geq 1$ such that either $H_{2k-1}(\Omega, \mathbb{Q}) \neq 0$ or $H_k(\Omega, \mathbb{Z}/2\mathbb{Z}) \neq 0$).*
Then problem $P(\Omega, 2^*)$ has at least one positive solution.

Remark 6.1. It is clear that any domain with nontrivial topology is not contractible in itself to a point. When $n = 3$, the converse is also true. On the contrary, when $n \geq 4$ the converse fails (i.e. there exist noncontractible domains with trivial homology groups). Thus, if $n \geq 4$, it is still an open problem whether the conclusion of Theorem 6.2 holds under the sole assumption that Ω is not contractible. However, note that the assumption "Ω has nontrivial topology" covers a large variety of domains.

After the results of Pohozaev and Bahri–Coron, the following two natural questions arise (see [5]):

Question 6.1 (Brezis). *Assume $p = 2^*$. Can one replace in Pohozaev's Theorem the assumption "Ω is star–shaped" by "Ω has trivial topology"? In other words, are there domains Ω with trivial topology on which $P(\Omega, 2^*)$ has a positive solution?*

Question 6.2 (Rabinowitz). *What happens when $p > \frac{2n}{n-2}$? Pohozaev's Theorem still holds. On the other hand, if Ω is an annulus, it is easy to see that $P(\Omega, p)$ has radial solutions for all p. So the question is: assuming Ω is a domain with nontrivial topology, is there still a solution of $P(\Omega, p)$ for all p?*

Now let us consider the effect of lower–order terms: we deal with the problem

$$P_a(\Omega, 2^*) \quad \begin{cases} \Delta u - a(x)u + |u|^{2^*-2}u = 0 & \text{in } \Omega \\ u = 0 & \text{on } \partial\Omega \\ u \not\equiv 0 & \text{in } \Omega, \end{cases} \qquad (6.4)$$

where $a(x) \in L^{n/2}(\Omega)$.

Motivations for the study of this problem come, for example, from this simple observation: assume Ω is any bounded domain and denote by $\lambda_1 < \lambda_2 \leq \lambda_3 \ldots$ the eigenvalues of $-\Delta$ with zero Dirichlet boundary condition; then general bifurcation results (see [4], [18], [37]) guarantee that $P_a(\Omega, 2^*)$ has solution if $a(x) \equiv -\lambda$, where λ is a constant sufficiently close to the eigenvalues λ_i of $-\Delta$ in $H_0^{1,2}(\Omega)$. In particular, if $\lambda < \lambda_1$ and $|\lambda - \lambda_1|$ is small enough, then there exists a positive solution.

On the other hand this equation is related to the solution of Yamabe's problem (see [1], [41] [44], [45]), where the coefficient $a(x)$ represent a scalar curvature.

The first results in this direction have been stated by Brezis and Nirenberg (see [5], [7]).

The energy functional related to problem $P_a(\Omega, 2^*)$ is

$$f_a(u) = \int_\Omega [|Du|^2 + a(x)u^2]dx; \qquad (6.5)$$

the Pohozaev's identity, satisfied by the solutions of $P_a(\Omega, 2^*)$, becomes

$$\int_\Omega [a + \frac{1}{2}(x \cdot Da)]u^2 dx + \frac{1}{2}\int_\Omega (x \cdot \nu)(\frac{\partial u}{\partial \nu})^2 d\sigma = 0. \qquad (6.6)$$

If $a(x) < 0$ somewhere in Ω and $n \geq 4$, then the infimum $\inf\limits_{M_{2^*}(\Omega)} f_a$ is achieved (the situation is more complicated in the case $n = 3$).

On the contrary, if $a(x) \geq 0$ everywhere in Ω, then the infimum is not achieved, but problem $P_a(\Omega, 2^*)$ may still have positive solutions. In fact, as showed in [5], it is easy to construct such an example, where $a(x) > 0$ in Ω and $P_a(\Omega, 2^*)$ has a positive solution (see section 6.3). This leads to the following natural question:

Question 6.3 (Brezis [5]). *Find general conditions on the nonnegative function $a(x)$ which guarantee the existence of solutions for $P_a(\Omega, 2^*)$, independently of the domain's shape (even in star–shaped domains).*

Note that, if Ω is star–shaped and $P_a(\Omega, 2^*)$ has solution, then $a(x)$ cannot be a positive constant, because of Pohozaev's identity (6.6).

6.2. Effect of the domain's shape

In this section we are concerned with the case $a(x) = 0$ and $p \geq 2^*$; Questions 6.1 and 6.2 are answered.

Notice that $\inf\limits_{M_p(\Omega)} f = 0$ for all $p > 2^*$ (so this infimum cannot be achieved).

For $p = 2^*$ we have

$$\inf\limits_{M_{2^*}(\Omega)} f = S, \qquad (6.7)$$

where S is the best constant for the Sobolev embedding $H_0^{1,2}(\Omega) \hookrightarrow L^{2^*}(\Omega)$.

It is well known that S is independent of Ω and depends only on the dimension n: this property is an easy consequence of the fact that the ratio $\|Du\|_2/\|u\|_{2^*}$ is invariant under dilations and translations. Moreover

S cannot be achieved in any bounded domain Ω, otherwise (extending a minimizing function by zero outside Ω) it should be achieved even in any star–shaped domain containing Ω, in contradiction with Pohozaev's nonexistence result.

S is attained only when $\Omega = \mathbb{R}^n$ and the minimizing function is unique, modulo translations and dilations (see [6], [17], [43]).

Answer to Question 6.1

The first attempts to answer Question 6.1 are some results (by Carpio Rodriguez, Comte, Lewandowski, Schaaf) extending Pohozaev's nonexistence theorem to some contractible but non star–shaped domains: in [8], for example, it is proved that $P(\Omega, 2^*)$ has no solution if the domain Ω is obtained removing from a sphere a frustum of cone having vertex outside the sphere (in such a way that the obtained domain is not star–shaped); also, in the case $p > 2^*$, nonexistence results hold in some dumb–bell shaped domains.

However the answer to Question 6.1 is negative since it is possible to prove existence results in some bounded contractible domains Ω: for example, if Ω is an annulus pierced by removing a cylinder thin enough, then $P(\Omega, 2^*)$ has positive solutions (see [12], [13], [23]).

Indeed it is possible to find bounded contractible domains Ω where the number of positive solutions of $P(\Omega, 2^*)$ is arbitrarily large:

Theorem 6.3 (see [23]). *For every positive integer h, there exists a bounded contractible domain Ω_h, such that $P(\Omega_h, 2^*)$ has at least h distinct positive solutions.*

Sketch of the proof. In order to obtain such a domain Ω_h, it suffices to argue as follows. For every positive integer h, let us consider the domain

$$T^h = \{x = (x_1, \ldots, x_n) \in \mathbb{R}^n : \sum_{i=1}^{n-1} x_i^2 < 1,\ 0 < x_n < h+1\}. \tag{6.8}$$

For all $j \in \{1, \ldots, h\}$, put $c_j = (0, \ldots, 0, j) \in \mathbb{R}^n$; fixed $\sigma_1, \ldots, \sigma_h$ such that $0 < \sigma_j < \frac{1}{2}$ for all $j \in \{1, \ldots, h\}$, set

$$D^h = T^h \setminus \bigcup_{j=1}^{h} \overline{B(c_j, \sigma_j)} \tag{6.9}$$

and, for all $\epsilon_1, \ldots, \epsilon_h$ in $]0, 1[$, define

$$\chi_{\epsilon_j}^j = \{x \in \mathbb{R}^n : \sum_{i=1}^{n-1} x_i^2 \leq \epsilon_j^2, \; j \leq x_n \leq j+1\} \tag{6.10}$$

$$\Omega_{\epsilon_1,\ldots,\epsilon_h} = D^h \backslash \bigcup_{j=1}^{h} \chi_{\epsilon_j}^j. \tag{6.11}$$

The assertion of Theorem 6.3 holds with $\Omega = \Omega_{\epsilon_1,\ldots,\epsilon_h}$ when $\epsilon_1, \ldots, \epsilon_h$ are small enough. In fact $\Omega_{\epsilon_1,\ldots,\epsilon_h}$ is a bounded contractible domain and $P(\Omega_{\epsilon_1,\ldots,\epsilon_h}, 2^*)$ has at least h solutions $u_{\epsilon_1}, \ldots, u_{\epsilon_h}$. These solutions are obtained as local minimum points of the energy functional f on $M_{2^*}(\Omega)$, constrained on the subspace of the functions having radial symmetry with respect to the x_n–axis (notice that the infimum $\inf\limits_{M_{2^*}(\Omega)} f$ is not achieved, not even in the subspace of the radial functions).

Moreover, for all $j = 1, \ldots, h$, the method used in the proof shows that, as $\epsilon_j \to 0$, $u_{\epsilon_j} \to 0$ weakly in $H_0^{1,2}(\Omega)$, $f\left(\frac{u_{\epsilon_j}}{\|u_{\epsilon_j}\|_{2^*}}\right) \to S$ and the energy $|Du_{\epsilon_j}|^2$ concentrates like a Dirac mass near a point of the x_n–axis. $\square$

Remark 6.2. In [12] and [13] Dancer and Ding prove that the positive solution one can find in an annulus, persists if the annulus is perturbed removing a subset of small capacity; moreover the solution in the perturbed domain converges to the solution in the annulus, as the capacity of the perturbation tends to zero. Therefore in a pierced annulus, or equivalently in a domain Ω_{ϵ_1} like in the proof of Theorem 6.3, Dancer and Ding prove the existence of a solution $\tilde{u}_{\epsilon_1}$ (see [12], [13]). However, let us point out that the solution $\tilde{u}_{\epsilon_1}$ obtained by Dancer and Ding is distinct from the solution u_{ϵ_1} given by Theorem 6.3, because $u_{\epsilon_1} \to 0$ weakly in $H_0^{1,2}(\Omega)$, while $\tilde{u}_{\epsilon_1}$ converges strongly to a solution in the limit domain (which is nontrivial in the sense of Bahri–Coron).

Thus the existence of the solutions $u_{\epsilon_1,}, \ldots, u_{\epsilon_h}$ of $P(\Omega_{\epsilon_1,\ldots,\epsilon_h}, 2^*)$, which does not depend on the solvability of the limit problem, seems to be related to other new phenomena and suggests that every perturbation of a given domain, which modifies its topological properties and is obtained removing a subset having small capacity, gives rise to solutions vanishing as the capacity of the perturbation tends to zero. Indeed it is also possible to evaluate the number of positive solutions by the topological properties of the perturbation. These results can be summarized as follows (see [26], [34], [35] for more details).

Definition 6.1 (see [14], [34]). *Let X be a topological space and X_1, X_2 two closed subset of X, such that $X_2 \subseteq X_1$.*
We say that the relative category in X of X_1 with respect to X_2 is m (and write $\operatorname{cat}_X[X_1, X_2] = m$) if m is the smallest positive integer such that

$$X_1 = \bigcup_{s=0}^{m} F_s , \qquad X_2 \subseteq F_0,$$

where, for all $s = 0, 1, \ldots, m$, F_s is a closed subset and there exists $h_s \in C^0([0,1] \times F_s, X)$ such that

I) $h_s(0, x) = x \quad \forall x \in F_s, \quad \forall s = 0, 1, 2, \ldots, m$
II) $\forall s \geq 1 \ \exists p_s \in X : h_s(1, x) = p_s \quad \forall x \in F_s$
III) $h_0(1, x) \in X_2 \quad \forall x \in F_0; \ h_0(t, x) \in X_2 \quad \forall x \in F_0 \cap X_2, \quad \forall t \in [0, 1]$.

Note that $\operatorname{cat}_X[X_1, \emptyset]$ is the well known Ljusternik–Schnirelman category.

Proposition 6.1 (see [34]). *Let Ω be a given bounded domain in $\mathbb{R}^n$, $n \geq 3$, and K be a closed subset of Ω.*
Then, if the capacity of K is small enough, problem $P(\Omega \backslash K, 2^*)$ has at least $\operatorname{cat}_{\bar{\Omega}}(\bar{\Omega}, \overline{\Omega \backslash K})$ positive solutions, which converge weakly to zero as the capacity of K tends to zero, and concentrate like a Dirac mass.

In the previous proposition, as well as in Theorem 6.3, a basic tool is given by the concentration–compactness principle of Lions (see [17]) or by a global compactness result of Struwe (see [42]), which allow us to overcome the difficulties given by the lack of compactness and, in particular, to show that the Palais–Smale condition is satisfied in the energy range $]S, 2^{2/n}S[$ (i.e. every sequence $(u_i)_i$ in $M_{2^*}(\Omega)$, such that $f(u_i) \to c \in]S, 2^{2/n}S[$ and $\operatorname{grad}_{M_{2^*}(\Omega)} f(u_i) \to 0$ in $H_0^{1,2}(\Omega)$, is relatively compact).
Notice that the method of the proof can be iterated in order to show that several independent perturbations produce several distinct positive solutions.
It is clear that this result can be applied in a large variety of geometric situations and allows to obtain an arbitrarily large number of positive so-

lutions. In particular one can obtain multiple positive solutions in domains with several small holes like in [38] (without requiring, unlike [38], that the holes are spherical).

On the other hand one can obtain more than one solution even by a unique but topologically complex perturbation:

Example: Let $n = 3$ and set

$$C = \{(x_1, x_2, x_3) \in \mathbb{R}^3 : x_3 = 0, \ x_1^2 + x_2^2 = 1\}$$
$$\Omega = \{x \in \mathbb{R}^3 : \operatorname{dist}(x, C) < \tfrac{1}{2}\}$$

and, for $\epsilon \in \,]0, \tfrac{1}{2}[\,$,

$$\Omega_\epsilon = \{x \in \mathbb{R}^3 : \epsilon < \operatorname{dist}(x, C) < \frac{1}{2}\}.$$

Then $\operatorname{cat}_{\bar{\Omega}}(\bar{\Omega}, \bar{\Omega}_\epsilon) = 2$ and so, for $\epsilon > 0$ small enough, $P(\Omega_\epsilon, 2^*)$ has at least two solutions whose energy concentrates like Dirac mass as $\epsilon \to 0$.

Notice that Ω_ϵ has radial symmetry with respect to x_3–axis and so it is easy to find radial solutions; however, let us point out that the solutions given by Proposition 6.1 cannot have radial symmetry because of their asymptotic behaviour as $\epsilon \to 0$. On the other hand no symmetry assumption is required in Proposition 6.1, which, for example, can guarantee the existence of two positive solutions, for $\epsilon > 0$ small enough, in a domain of the form

$$\tilde{\Omega}_\epsilon - \{x \in \mathbb{R}^3 : \operatorname{dist}(x, C) < \frac{1}{2}, \ \operatorname{dist}(x, \tilde{C}) > c\},$$

where

$$\tilde{C} = \{x = (x_1, x_2, x_3) \in \mathbb{R}^3 : x_3 = 0, \ (x_1 - \frac{1}{3})^2 + x_2^2 = 1\},$$

which does not have any symmetry property.

The supercritical case: answer to Question 6.2

The answer to Question 6.2 is negative since, as we shall see below, it is possible to find pairs (Ω, p), where Ω is a smooth bounded domain of $\mathbb{R}^n$, nontrivial in the sense of Bahri–Coron, $p > \frac{2n}{n-2}$ and problem $P(\Omega, p)$ has no solution.

In the supercritical case a crucial role seems to be played by the critical exponents

$$2^*(n - k) = \frac{2(n - k)}{(n - k) - 2} \qquad k = 1, \ldots, (n - 3), \tag{6.12}$$

corresponding to lower dimensions. In fact, the possibility of finding a nontrivial domain Ω, such that $P(\Omega, p)$ has no solution, is strictly related to the position of p with respect to the critical exponents (6.12), as showed by the following result:

Theorem 6.4 (see [27], [28]). *For every positive integer k there exists a smooth bounded domain Ω in $\mathbb{R}^n$, with $n \geq k+3$, homotopically equivalent to the k–dimensional sphere S_k (hence nontrivial if $k \geq 1$), such that problem $P(\Omega, p)$ has no solution for $p \geq \frac{2(n-k)}{(n-k)-2}$, while, for $2 < p < \frac{2(n-k)}{(n-k)-2}$, it has infinitely many solutions and at least one of them is positive*

(see [27] and [28] for detailed statements concerning more general nonlinear terms).

Sketch of the proof. For all $x = (x_1, \ldots, x_n) \in \mathbb{R}^n$, let us set

$$P_1^k(x) = (x_1, \ldots, x_k, \ x_{k+1}, \ 0, \ldots, 0) \in \mathbb{R}^n, \qquad P_2^k(x) = x - P_1^k(x)$$

and define

$$S_k = \{x \in \mathbb{R}^n : |x| = 1, \ P_2^k(x) = 0\}$$
$$T_k(\rho) = \{x \in \mathbb{R}^n : \text{dist}(x, S_k) < \rho\}.$$

If $0 < \rho < 1$, then the domain $T_k(\rho)$ is homotopically equivalent to S_k and the conclusion of the theorem holds for $\Omega = T_k(\rho)$.

The proof is based on the following generalized Pohozaev's identity: for all $v \in C^1(\bar{\Omega}, \mathbb{R}^n)$, the solutions of problem $P(\Omega, p)$ must satisfy

$$\frac{1}{2} \int_{\partial\Omega} (v \cdot \nu) \left(\frac{\partial u}{\partial \nu}\right)^2 d\sigma =$$

$$\int_\Omega (dv[Du] \cdot Du)dx + \int_\Omega \frac{|u|^p}{p} \, \text{div} \, v dx - \frac{1}{2} \int_\Omega |Du|^2 \, \text{div} \, v dx.$$

A suitable choice of the function v when $\Omega = T_k(\rho)$ implies that $u \equiv 0$ if $p \geq \frac{2(n-k)}{(n-k)-2}$, which is a contradiction.

On the contrary, if $2 < p < \frac{2(n-k)}{(n-k)-2}$, then one can exploit the radial symmetry of the domain $T_k(\rho)$ with respect to the co–ordinates $x_1, \ldots, x_{k+1}$ in order to find nontrivial solutions in the subspace of the radial symmetric functions (which is compactly embedded in $L^p(T_k(\rho))$ if $0 < \rho < 1$ and $p < \frac{2(n-k)}{(n-k)-2}$).

Note that no symmetry assumption is required for the nonexistence result in the case $p \geq \frac{2(n-k)}{(n-k)-2}$. $\qquad\qquad\qquad\qquad\qquad\qquad\qquad$ $\square$

Remark 6.3. The previous proposition allows us to find a nontrivial domain only when $k \geq 1$ (in fact only in this case $T_k(\rho)$ has nontrivial topology in the sense of Bahri–Coron). Therefore, in the case $n \geq 4$ and $\frac{2n}{n-2} < p < 2^*(n-1) = \frac{2(n-1)}{(n-1)-2}$, or also in the case $n = 3$ and $p > 2^*(3) = 6$, Theorem 6.4 does not gives counterexamples and Question 6.2 could have a positive answer: it is still an open problem.

Moreover, let us mention the following question, which seems suggested by the proof of Theorem 6.4: assume Ω has the k–dimensional homology group $H_k(\Omega, \mathbb{Z}/2\,\mathbb{Z}) \neq 0$; does this assumption guarantee that $P(\Omega, p)$ has (positive) solutions for $2 < p < \frac{2(n-k)}{(n-k)-2}$, if $n \geq k + 3$, and for all $p > 2$ if $n < k + 3$?

On the other hand there exist examples of bounded contractible domains Ω (the same ones introduced above), such that $P(\Omega, p)$ has positive or nodal solutions for all $p > 2$:

Theorem 6.5 (see [29], [30], [31]). *Let* $p > \frac{2n}{n-2}$ *and* $\Omega_{\epsilon_1,\dots,\epsilon_h}$ *be the bounded contractible domains above defined (see* (6.11)*). Then there exists* $\bar{\epsilon} > 0$ *such that, for all* $\epsilon_1, \dots, \epsilon_h$ *in* $]0, \bar{\epsilon}[$ *, problem* $P(\Omega_{\epsilon_1,\dots,\epsilon_h}, p)$ *has at least h distinct positive solutions* $u_{\epsilon_1}, \dots, u_{\epsilon_h}$ *and at least* h^2 *nodal solutions* $u_{\epsilon_i,\epsilon_j}(i,j = 1,\dots,h)$, *having exactly two nodal regions (i.e. both* $u^+_{\epsilon_i,\epsilon_j}$ *and* $u^-_{\epsilon_i,\epsilon_j}$ *have connected support).*
Moreover, for all $i = 1,\dots,h$, $u_{\epsilon_i} \to 0$ *strongly in* $H_0^{1,2}$ *as* $\epsilon_i \to 0$ *and* $\frac{u_{\epsilon_i}}{\|u_{\epsilon_i}\|_p}$ *concentrates like a Dirac mass; for all* $i, j = 1,\dots,h$, $u^+_{\epsilon_i} \to 0$ *in* $H_0^{1,2}$ *as* $\epsilon_i \to 0$, $u^-_{\epsilon_j} \to 0$ *in* $H_0^{1,2}$ *as* $\epsilon_j \to 0$, *and* $\frac{u^+_{\epsilon_i}}{\|u^+_{\epsilon_i}\|_p}$, $\frac{u^-_{\epsilon_j}}{\|u^-_{\epsilon_j}\|_p}$ *concentrate like Dirac mass.*

Sketch of the proof. Let us point out that, even if the solutions we find in the supercritical case present qualitative properties and asymptotic behaviour, as $\epsilon_i \to 0$, analogous to the ones obtained in the critical case (see Theorem 6.3), there is a deep difference from the point of view of the variational framework. In fact in the critical case the solutions in Theorem 6.3

correspond to local minimum points, in the subspace of the radial functions, for the energy functional constrained on $M_{2^*}(\Omega)$. On the contrary, when $p > \frac{2n}{n-2}$, the functional f constrained on $M_p(\Omega)$ has no local minimum point, not even in the subspace of the functions having radial symmetry with respect to the x_n–axis.

The solutions in Theorem 6.5 are obtained using a special device: we modify the functional introducing some obstacle in order to avoid some concentration phenomena, related to the lack of compactness, then we obtain the solutions as local minimum points for the modified functional and, finally, we prove that these modifications do not change the Euler–Lagrange equation, when $\epsilon_1, \ldots, \epsilon_h$ are sufficiently small.

Moreover, the method used in the proof shows that, for all $i = 1, \ldots, h$, $\frac{u_{\epsilon_i}}{\|u_{\epsilon_i}\|_p}$ concentrates near the cylinder $\chi^i_{\epsilon_i}$ (see (6.10)) as $\epsilon_i \to 0$, while $\frac{u^+_{\epsilon_i,\epsilon_j}}{\|u^+_{\epsilon_i,\epsilon_j}\|_p}$ and $\frac{u^-_{\epsilon_i,\epsilon_j}}{\|u^-_{\epsilon_i,\epsilon_j}\|_p}$ concentrate near $\chi^i_{\epsilon_i}$ and $\chi^j_{\epsilon_j}$ respectively, as the size of these cylinders tends to zero. $\qquad\qquad\square$

Theorem 6.5 shows that it is possible to find an arbitrarily large number of nontrivial solutions of $P(\Omega, p)$ in correspondence to several perturbations of a given domain. On the other hand, it is possible to obtain many nontrivial solutions even in correspondence to a unique perturbation, as shown by the following theorem (but these solutions may have, conceivably, more than two nodal regions).

Theorem 6.6 (see [32]). *Let $p > \frac{2n}{n-2}$ and, for all $\epsilon > 0$, set*

$$A = \{x \in \mathbb{R}^n : 1 < |x| < 2\}$$
$$C_\epsilon = \{x = (x_1, \ldots, x_n) \in \mathbb{R}^n : \sum_{i=1}^{n-1} x_i^2 \le \epsilon^2, \ x_n > 0\}$$
$$\Omega_\epsilon = A \backslash C_\epsilon.$$

Then, for all positive integer k there exists $\bar{\epsilon}_k > 0$ such that problem $P(\Omega_\epsilon, p)$ has at least $2k$ distinct solutions for all $\epsilon \in \,]0, \bar{\epsilon}_k[$.
Moreover all the solutions tend strongly to zero in $H_0^{1,2}$ as $\epsilon \to 0$ and their energy concentrates like a Dirac mass near a point of the x_n–axis.

Sketch of the proof. The main tool is a truncation method: we modify the nonlinear term near the region where concentration phenomena (and lack of compactness) could occur; then we find infinitely many solutions of the modified problem; finally we analyse the behaviour of these solutions as $\epsilon \to 0$ and show that many solutions of the modified problem (indeed a number which goes to infinity as $\epsilon \to 0$) are solutions of $P(\Omega_\epsilon, p)$ for ϵ sufficiently close to zero, since they tend to concentrate outside the region where the nonlinear term has been modified. $\hfill \square$

In conclusion, we have that in the critical case the nontriviality of the domain Ω is only a sufficient but not necessary condition in order to guarentee the existence of nontrivial solutions for $P\left(\Omega, \frac{2n}{n-2}\right)$ (because of the results of [2], [12], [13], [23], etc...). On the contrary, when $p > \frac{2n}{n-2}$, this condition is neither sufficient, nor necessary (as shown in [27], [28], [29], [30], [31], [32]).

It is a widely open problem to find what kinds of geometrical properties of the domain Ω are related to the solvability of $P(\Omega, p)$ for $p > \frac{2n}{n-2}$ (for more recent results on this subject, see also [19], [20], [21], [22] and references therein).

6.3. Effect of lower–order terms

The first results in this direction have been stated by Brezis and Nirenberg (see [5], [7]). First, let us observe that, in order to have positive solutions of problem $P_a(\Omega, 2^*)$, it is essential that the linear operator $-\Delta + a$ is positive, i.e. the first eigenvalue

$$\mu_1 = \min\{f_a(u) : u \in H_0^{1,2}(\Omega), \int_\Omega u^2 dx = 1\}$$

is positive.

In fact, denoted by $\varphi_1 > 0$ the corresponding eigenfunction, if u is a positive solution of $P_a(\Omega, 2^*)$, we obtain (multiplying by φ_1 and integrating by parts)

$$\mu_1 \int_\Omega u\varphi_1 dx = \int_\Omega u^{2^*-1}\varphi_1 dx,$$

which implies $\mu_1 > 0$.

It follows that

$$I_a \overset{\text{def}}{=} \inf_{M_{2^*}(\Omega)} f_a \geq 0. \tag{6.13}$$

When this infimum I_a is achieved, a minimizing function gives rise to a positive solution of $P_a(\Omega, 2^*)$. Notice that we have always

$$I_a \leq S,$$

as one can easily verify testing the functional f_a on a suitable sequence of functions (see [5], [7]).

A useful tool, in order to prove that I_a is achieved, is given by the following lemma.

Lemma 6.1 (see [5], [7]). *Let $a \in L^{n/2}(\Omega)$ and assume that (see (6.13))*

$$I_a < S. \tag{6.14}$$

Then the infimum I_a is achieved.

The proof is based on the fact that every minimizing function is relatively compact if $I_a < S$ (indeed Palais–Smale condition holds in the energy range $]-\infty, S[$: see [5], [6], [7], [17], [42]).
Thus the problem is to find concrete assumptions which guarantee that (6.14) holds. The cases $n = 3$ and $n \geq 4$ are quite different.
In the case $n \geq 4$, the main result is the following

Theorem 6.7 (see [5], [7]). *Assume Ω is any bounded domain in $\mathbb{R}^n$, with $n \geq 4$. Then, the following properties are equivalent:*

1) $a(x) < 0$ somewhere on Ω (it suffices in a neighbourhood of a point)
2) $I_a < S$ (and so I_a is achieved).

In the special case where $a(x)$ is a constant function, we obtain the following corollary (other existence and multiplicity results, concerning the case where $a(x)$ is a negative constant, can be found for example in [9] and [10])

Corollary 6.1 (see [5], [7]). *Let Ω be a bounded domain in $\mathbb{R}^n$, with $n \geq 4$, and assume $a(x) = -\lambda$ (constant).*
Then problem $P_{(-\lambda)}(\Omega, 2^*)$ has at least one positive solution for $0 < \lambda < \lambda_1$ (λ_1 denotes the first eigenvalue of $-\Delta$ in $H_0^{1,2}(\Omega)$); $P_{(-\lambda)}(\Omega, 2^*)$ has no positive solution for $\lambda \geq \lambda_1$; for $\lambda \leq 0$ $P_{(-\lambda)}(\Omega, 2^*)$ has no solution if Ω is a star–shaped domain.
Moreover the solution u_λ, for $0 < \lambda < \lambda_1$, converges to zero in $H_0^{1,2}(\Omega)$ as $\lambda \to \lambda_1$, while concentrates like a Dirac mass as $\lambda \to 0$.

The situation is more complicated in the case $n = 3$.

Theorem 6.8 (see [5], [7]). *Assume Ω is any bounded domain in $\mathbb{R}^3$ and $a(x) = -\lambda$ (constant) in Ω.*
Then there exists a constant $\lambda^*(\Omega) \in]0, \lambda_1[$, such that

a) $I_{(-\lambda)} = S$ for $\lambda \leq \lambda^*(\Omega)$ and $I_{(-\lambda)}$ is not achieved for $\lambda < \lambda^*(\Omega)$
b) $I_{(-\lambda)} < S$ for $\lambda > \lambda^*(\Omega)$ and so there exists a positive solution for
* $\lambda^*(\Omega) < \lambda < \lambda_1$ (no positive solution can exist for $\lambda > \lambda_1$).*

It is not clear what happens for $\lambda \leq \lambda^*(\Omega)$; a complete solution is given only in the case where Ω is a ball.

Theorem 6.9 (see [5], [7]). *Assume Ω is a ball in $\mathbb{R}^3$. Then $\lambda^*(\Omega) = \frac{1}{4}\lambda_1$ (see Theorem 6.8) and there exists no positive solution of problem $P_{(-\lambda)}(\Omega, 2^*)$ for $\lambda \leq \frac{1}{4}\lambda_1$.*

When $a(x) \geq 0$ everywhere on Ω, then $I_a = S$ and so I_a cannot be achieved in any bounded domain Ω (otherwise S should be achieved too). Moreover $P_a(\Omega, 2^*)$ cannot have solution in star–shaped domains, if $a(x)$ is a nonnegative constant (because of Pohozaev's identity (6.6)).
However, let us emphasize that, even in the case $a(x) \geq 0$ on Ω, problem $P_a(\Omega, 2^*)$ may still have solutions; but these solutions cannot be obtained by minimization arguments and correspond to higher critical values (obtained by topological methods of Calculus of Variations). In fact, as shown by Brezis in [5], in any bounded domain Ω it is easy to construct an example of a nonnegative (noncostant) function $a(x)$ such that $P_a(\Omega, 2^*)$ has solution: fix $\psi \in C_0^\infty(\Omega)$ such that $\psi \geq 0$, $\psi \not\equiv 0$, and let v be the solution of the problem

$$\begin{cases} \Delta v + \psi = 0 & \text{in } \Omega \\ v = 0 & \text{on } \partial\Omega \end{cases}$$

so that $v > 0$ on Ω. If we set

$$a = t^{2^*-1} v^{2^*-1} - \frac{\psi}{v},$$

then $a(x)$ is a smooth function. Moreover it is easy to verify that $a(x) > 0$ everywhere on Ω, if t is a sufficiently large positive constant, and that $u = tv$ solves problem $P_a(\Omega, 2^*)$. This example motivates Question 6.3.

Note that Pohozaev's identity (6.6) shows that an obvious necessary condition for the existence of a solution for $P_a(\Omega, 2^*)$ is that $[a(x) + \frac{1}{2}(x \cdot Da(x))]$ should be negative somewhere on Ω.

An answer to Question 6.3 is given in [24].

Theorem 6.10 (see [24]). *Let Ω be a smooth bounded domain in $\mathbb{R}^n$, with $n \geq 3$, and x_0 be a fixed point in Ω. Let $\bar{\alpha} \in L^{n/2}(\Omega)$ and $\alpha \in L^{n/2}(\mathbb{R}^n)$ be two nonnegative functions and assume that $\|\alpha\|_{L^{n/2}(\mathbb{R}^n)} \neq 0$.*

Then there exists $\bar{\mu} > 0$ such that, for all $\mu > \bar{\mu}$, problem $P_a(\Omega, 2^*)$, with

$$a(x) = \bar{\alpha}(x) + \mu^2 \alpha[\mu(x - x_0)],$$

has at least one solution u_μ. Moreover

$$S < f_a\left(\frac{u_\mu}{\|u_\mu\|_{2^*}}\right) < 2^{2/n} S$$

and

$$\lim_{\mu \to +\infty} f_a\left(\frac{u_\mu}{\|u_\mu\|_{2^*}}\right) = S.$$

If we assume in addition that

$$\|\alpha\|_{L^{n/2}(\mathbb{R}^n)} < S(2^{2/n} - 1), \tag{6.15}$$

then $P_a(\Omega, 2^*)$ has at least another solution $\hat{u}_\mu$. Moreover

$$f_a\left(\frac{u_\mu}{\|u_\mu\|_{2^*}}\right) < f_a\left(\frac{\hat{u}_\mu}{\|\hat{u}_\mu\|_{2^*}}\right) < 2^{2/n} S$$

and

$$\liminf_{\mu \to +\infty} f_a\left(\frac{\hat{u}_\mu}{\|\hat{u}_\mu\|_{2^*}}\right) > S.$$

The proof is obtained using topological methods of Calculus of Variations.

An important tool is given by the results stated in [17] and [42], which give a description of the behaviour of the Palais–Smale sequences and allow us to prove that Palais–Smale condition is satisfied in the energy range $]S, 2^{2/n} S[$.

Remark 6.4. Note that the assumption on the nonnegative function $a(x)$ in Theorem 6.10 seems to be fairly general. In fact, if we assume for example that $x_0 = 0$, $\bar{\alpha} \equiv 0$ and

$$\alpha(x) = \begin{cases} 1 & \text{if } |x| \leq 1 \\ |x|^{-\beta} & \text{if } |x| > 1, \end{cases}$$

then, if $\beta \leq 2$ (i.e. $\alpha \notin L^{n/2}(\mathbb{R}^n)$) and Ω is a bounded domain star-shaped with respect to zero, problem $P_a(\Omega, 2^*)$, with $a(x) = \mu^2 \alpha(\mu x)$, has no solution for any $\mu > 0$, because of Pohozaev's identity (6.6); on the contrary, if $\beta > 2$ (i.e. $\alpha \in L^{n/2}(\mathbb{R}^n)$), Theorem 6.10 guarantees the existence of solutions for μ large enough, without any assumption on the shape of Ω (if μ is small enough and Ω is star–shaped, no solution can exist because $a(x)$ is constant in Ω).

Remark 6.5. The method used in the proof of Theorem 6.10 can be iterated in order to obtain more general multiplicity results concerning functions $a(x)$ of the form

$$a(x) = \bar{\alpha}(x) + \sum_{i=1}^{h} \mu_i^2 \alpha_i[\mu_i(x - x_i)],$$

where $x_1, \ldots, x_h$ are points in Ω, $\bar{\alpha}$ in $L^{n/2}(\Omega)$ and $\alpha_1, \ldots, \alpha_h$ in $L^{n/2}(\mathbb{R}^n)$ are nonnegative functions and $\lambda_1, \ldots, \lambda_h$ are positive parameters. Indeed, there exist at least h distinct positive solutions when the concentration parameters $\mu_1, \ldots, \mu_h$ are large enough, and at least $2h$ positive solutions when, in addition, the functions $\alpha_1, \ldots, \alpha_h$ satisfy condition (6.15).
Note that it is not necessary to choose distinct concentration points in order to obtain h or $2h$ distinct positive solutions: it suffices to choose only the concentration parameters $\mu_1, \ldots, \mu_h$ in a suitable way (some possible choices of these parameters are described in [24]).

Remark 6.6. In [3] and [25] one can find some results concerning the case where $\Omega = \mathbb{R}^n$.

Consider the problem

$$\begin{cases} \Delta u - [\epsilon + \alpha(x)]u + u^{2^*-1} = 0 & \text{in } \mathbb{R}^n \\ u > 0 & \text{in } \mathbb{R}^n \\ \int_{\mathbb{R}^n} |Du|^2 dx < \infty, \end{cases}$$

where α in $L^{n/2}(\mathbb{R}^n)$, with $\alpha \not\equiv 0$, is a given nonnegative function and $\epsilon \geq 0$.

In [3] it is proved that there exists at least one solution, under the assumption that $\epsilon = 0$ and α satisfies condition (6.15).

In [21] we show that there exist, for $\epsilon > 0$ small enough, at least two solutions, u_ϵ and $\hat{u}_\epsilon$, such that

$$S < f_{a_\epsilon}\left(\frac{u_\epsilon}{\|u_\epsilon\|_{2^*}}\right) < f_{a_\epsilon}\left(\frac{\hat{u}_\epsilon}{\|\hat{u}_\epsilon\|_{2^*}}\right) < 2^{2/n}S,$$

where $a_\epsilon(x) = [\epsilon + \alpha(x)]$.

Moreover u_ϵ vanishes as $\epsilon \to 0$, while $\hat{u}_\epsilon$ converges to a solution of the limit problem.

Finally, let us mention a result (stated in [33]), where we exploit the combined effect of both the lower–order terms and the domain shape.

Notice that the positive solutions u_λ given by Corollary 6.1 concentrate as $\lambda \to 0$ like Dirac mass near points of Ω. Indeed, for every family $(\tilde{u}_\lambda)_{\lambda>0}$ of functions in $M_{2^*}(\Omega)$, such that $f_{(-\lambda)}(\tilde{u}_\lambda) \leq S \ \forall \lambda > 0$, there exists a family of points $(x_\lambda)_{\lambda>0}$ in Ω such that the functions $\tilde{u}_\lambda(x - x_\lambda)$ tend, as $\lambda \to 0$, to δ_0, the Dirac mass in zero.

This fact enable us to relate the topological properties of the sublevels of the energy functional $f_{(-\lambda)}$ to the shape of the domain Ω. Thus, taking also into account the behaviour of the Palais–Smale sequences described in [5], [17], [42], it is possible to evaluate the number of positive solutions of $P_{(-\lambda)}(\Omega, 2^*)$, when λ is a positive constant sufficiently close to zero, by the Ljusternik–Schnirelman category of Ω, or by other topological invariants (see [16], [33], [39]).

Theorem 6.11 (see [33]). *Let Ω be a bounded domain in $\mathbb{R}^n$, with $n \geq 4$, and denote by m its Ljusternik–Schnirelman category.*

Then there exists $\bar{\lambda} \in]0, \lambda_1[$ such that, for all $\lambda \in]0, \bar{\lambda}[$, problem $P_{(-\lambda)}(\Omega,$

2^*) *has at least* m *distinct positive solutions,* $u_{1,\lambda}, \ldots, u_{m,\lambda}$, *such that*

$$f_{(-\lambda)}\left(\frac{u_{i,\lambda}}{\|u_{i,\lambda}\|_{2^*}}\right) < S \quad \forall i = 1, \ldots, m.$$

If we assume in addition that Ω *is not contractible in itself (i.e.* $m > 1$*), then there exists another solution* $u_{m+1,\lambda}$ *such that*

$$S < f_{(-\lambda)}\left(\frac{u_{m+1,\lambda}}{\|u_{m+1,\lambda}\|_{2^*}}\right) < 2^{2/n} S.$$

Remark 6.7. The lower energy solutions $u_{1,\lambda}, \ldots, u_{m,\lambda}$ converge weakly to zero in $H_0^{1,2}(\Omega)$ as $\lambda \to 0$ and concentrate like Dirac mass near some points of Ω. Moreover (see [40]) the concentration points are the critical points of the regular part of the Green's function for Laplace operator with zero Dirichlet boundary condition.

The higher energy solution $u_{m+1,\lambda}$ either converges as $\lambda \to 0$ to a solution of the limit problem, or converges weakly to zero and can be decomposed as sum of at most two functions whose energy concentrates like Dirac mass near two points of Ω.

Finally, let us point out that, if Ω has some symmetry properties, then the number of solutions may increase considerably: for example, if Ω is a domain homotopically equivalent to the $(k-1)$–dimensional sphere S_{k-1} and is symmetric with respect to a point $x_0 \notin \Omega$, then problem $P_{(-\lambda)}(\Omega, 2^*)$ has, for $\lambda > 0$ small enough, at least $2k+1$ solutions, even if the Ljusternik–Schnirelman category of Ω in itself is only 2.

References

[1] T. Aubin. "Equations différentielles nonlinéaires et problème de Yamabe concernant la courbure scalaire". J. Math. Pures et Appl. 55, (1976), 269–293.

[2] A. Bahri – J.M. Coron. "On a nonlinear elliptic equation involving the Sobolev exponent: the effect of the topology of the domain". Comm. Pure Appl. Math. Vol. 41, (1988), 253–294.

[3] V. Benci – G. Cerami. "Existence of positive solutions of the equation $-\Delta u + a(x)u = u^{\frac{N+2}{N-2}}$ in $\mathbb{R}^N$". J. Funct. Anal., Vol. 88, n. 1, (1989), 90–117.

[4] R. Böhme. "Die Lösung der Verzweigungsgleichungen für nichtlineare Eigenwertprobleme". Math. Z., 127, (1972), 105–126.

[5] H. Brezis. "Elliptic equations with limiting Sobolev exponents. The impact of topology" in "Proceedings 50th Anniv. Courant Inst.", Comm. Pure Appl. Math. vol. 39, (1986), 517–539.

[6] H. Brezis – E. Lieb. "A relation between pointwise convergence of functions and convergence of functionals". Proc. Amer. Math. Soc. 88, (1983), 486–490.

[7] H. Brezis – L. Nirenberg. "Positive solutions of nonlinear elliptic equations involving critical Sobolev exponents" Comm. Pure Appl., Math., 36, (1983), 437–477.

[8] A. Carpio Rodriguez – M. Comte – R. Levandowski. "A nonexistence result for a nonlinear equation involving critical Sobolev exponent". Ann. Inst. H. Poicaré, Analyse Non Linéaire, vol. 9, n. 3 (1992), 243–261.

[9] G. Cerami – D. Fortunato –M. Struwe. "Bifurcation and multiplicity results for nonlinear elliptic problems involving critical Sobolev exponents". Ann. Inst. H. Poincaré – Anal. Non Linéaire, 1 (1984), 341–350.

[10] G. Cerami – S. Solimini – M. Struwe. "Some existence results for superlinear elliptic boundary value problems involving critical exponents". J. Funct. Anal. 69, (1986), 289–306.

[11] J.M. Coron. "Topologie et cas limite des injections de Sobolev". C.R. Acad. Sci. Paris, Sér I, 299, (1984), 209–212.

[12] E.N.Dancer. "A note on an equation with critical exponent". Bull. London Math. Soc. 20, (1988), 600–602.

[13] W. Ding. "Positive solutions of $\Delta u + u^{(n+2)/(n-2)} = 0$ on contractible domains". J. Partial Differential Equations, 2 (1989), 83–88.

[14] E.N. Fadell. "Lectures in cohomological index theories of G–spaces, with applications to critical points theory". Rend. Sem. Dip. Mat. Univ. Calabria, 6 (1985).

[15] J. Kazdan – F. Warner. "Remark on some quasilinear elliptic equations". Comm. Pure Appl. Math., 28 (1975), 567–597.

[16] M. Lazzo. "Multiple positive solutions of nonlinear elliptic equations involving critical Sobolev exponents". C.R. Acad. Sci. Sér. I, 314 (1992), Paris, 61–64.

[17] P.L. Lions. "The concentration–compactness principle in the Calculus of Variations: the limit case". Rev. Mat. Iberoamericana, 11 (1985), 145–201 and 12 (1985), 45–121.

[18] A. Marino. "La biforcazione nel caso variazionale". Confer. Sem. Mat. Univ. Bari, n. 132, (1973).

[19] R. Molle – D. Passaseo. "Positive solutions for slightly super-critical elliptic equations in contractible domains". C.R. Acad. Sci. Sér. I, 335 (2002), Paris, no. 5, 459–462.

[20] R. Molle – D. Passaseo. "Nonlinear elliptic equations with critical Sobolev exponent in nearly starshaped domains". C.R. Acad. Sci. Sér. I, 335 (2002), Paris, no. 12, 1029–1032.

[21] R. Molle – D. Passaseo. "Positive solutions of slightly supercritical elliptic equations in symmetric domains". Ann. Inst. H. Poicaré, Analyse Non Linéaire (to appear).

[22] R. Molle – D. Passaseo. "On the existence of positive solutions of slightly supercritical elliptic equations" Preprint Dip. Mat. Univ. Roma "Tor Vergata" (2003).

[23] D. Passaseo. "Multiplicity of positive solutions of nonlinear elliptic equations with critical Sobolev exponent in some contractible domains". Manuscripta Math., 65 (1989), 147–166.

[24] D. Passaseo. "Some sufficient conditions for the existence of positive solutions to the equation $-\Delta u + a(x)u = u^{2^*-1}$ in bounded domains". Ann. Inst. H. Poincaré, vol. 13, n. 2, (1996), 185–227.

[25] D. Passaseo. "Esistenza e molteplicità di soluzioni positive per l'equazione $-\Delta u + (\alpha(x) + \lambda)u = u^{\frac{n+2}{n-2}}$ in $\mathbb{R}^n$." Preprint Dip. Mat. Univ. Pisa (1990).

[26] D. Passaseo. "Su alcune successioni di soluzioni positive di problemi ellittici con esponente critico". Rend. Mat. Acc. Lincei s. 9, vol. 3, (1992), 15–21.

[27] D. Passaseo. "Nonexistence results for elliptic problems with supercritical nonlinearity in nontrivial domains". J. Funct. Anal. 114, n. 1 (1993), 97–105.

[28] D. Passaseo. "New nonexistence results for elliptic equations with supercritical nonlinearity". Diff. Int. Equations, vol. 8, n. 3 (1995), 577–586.

[29] D. Passaseo. "Esistenza e molteplicità di soluzioni positive per equazioni ellittiche con nonlinearità supercritica in aperti contrattili". Rend. Accad. Naz. Sci. detta dei XL, Memorie di Mat. 110, vol. XVI, n. 6 (1992), 77–98.

[30] D. Passaseo. "Multiplicity of nodal solutions for elliptic equations with supercritical exponent in contractible domains" Topological Methods in Nonlinear Analysis vol.8 (1996), 245–262.

[31] D. Passaseo. "Nontrivial solutions of elliptic equations with supercritical exponent in contractible domains" Duke Mathematical Journal, vol.92, n.2 (1998), 429–457.

[32] D. Passaseo. "Multiple solutions of elliptic problems with supercritical growth in contractible domains". (in preparation).

[33] D. Passaseo. "Multiplicity of positive solutions for the equation $\Delta u + \lambda u + u^{2^*-1} = 0$ in non contractible domains". Topological Methods in Nonlinear Analysis, vol. 2 (1993), 343–366.

[34] D. Passaseo. "Relative category and multiplicity of positive solutions for the equation $\Delta u + u^{2^*-1} = 0$." Nonlinear Analysis, vol. 33 (1998), 509–517.

[35] D. Passaseo. "The effect of the domain shape on the existence of positive solutions of the equation $\Delta u + u^{2^*-1} = 0$." Topological Methods in Nonlinear Analysis, vol. 3 (1994), 27–54.

[36] S. Pohozaev. "Eigenfunctions of the equation $\Delta u + \lambda f(u) = 0$." Soviet Math. Dokl. 6 (1965), 1408–1411.

[37] P. Rabinowitz. "Some aspects of nonlinear eigenvalues problems". Rocky Mount. J. Math. 3 (1973), 161–202.

[38] O. Rey. "Sur un problème variationnel non compact: l'effect de petits trous dans le domaine". C.R. Acad. Sci. Paris, t. 308, Série I (1989), 349–352.

[39] O. Rey. "A multiplicity result for a variational problem with lack of compactness". J. Nonlinear Anal. T.M.A., 133 (1989), 1241–1249.

[40] O. Rey. "The role of the Green's function in a nonlinear elliptic equation involving the critical Sobolev exponent." J. Funct. Anal. 89 (1990), 1–52.

[41] R. Schoen. "Conformal deformation of a Riemannian metric to a constant

scalar curvature". J. Diff. Geom. 20 (1984), 479–495.

[42] M. Struwe. "A global compactness result for elliptic boundary value problems involving limiting nonlinearities". Math. Z., 187 (1984), 511–517.

[43] G. Talenti. "Best constants in Sobolev inequality". Ann. Mat. Pura Appl. 110 (1976), 353–372.

[44] N. Trudinger. "Remarks concerning the conformal deformations of Riemannian structures on compact manifolds". Ann. Sc. Norm. Sup. – Pisa, 22 (1968), 265–274.

[45] H. Yamabe. "On a deformation of Riemannian structures on compact manifolds". Osaka Math. J. 12 (1960), 21–37.

Chapter 7

Eigenvalue analysis of elliptic operators

Grigori Rozenblum

*Department of Mathematics,
Chalmers University of Technology and Göteborg University,
Göteborg, Sweden.
e-mail: grigori@math.chalmers.se*

Contents

7.1. Lecture 1. Introduction. Basics in Operator Theory

In 1910 in his lecture on Wolfskehl Congress on pressing problems in Physics, Hendrik Anton Lorentz stated a mathematical problem inspired by the radiation theory created just a few years earlier by James Jeans: to prove that the asymptotic behavior of eigenvalues of the Dirichlet Laplacian in a three-dimensional body depends only on the volume of the body, and not on its form. This, considered as hopeless, problem was almost immediately solved by Herman Weyl [28], who not only proved the above conjecture but also created a number of mathematical tools which turned out to be extremely useful for other problems as well. After Weyl, up to now, the study of asymptotic behavior of eigenvalues of elliptic operators has been one of the most actively developing fields of analysis, having deep relations also with physics, geometry, topology, algebra and number theory.

In these lectures we give an introduction just to one direction in this big field, the asymptotic properties of eigenvalues of operators where singularities of various kind are present, such as unbounded domains, non-smooth boundaries and coefficients, singular weight functions. It is for this kind of problems, the original approach of Weyl proves to be most efficient. The main ingredients in this approach are the variational setting of the eigenvalue problem, a perturbational reasoning, and general eigenvalue estimates.

We are going to present some, now classical, as well and some recent results in this direction. We address the interested reader to the book [23], where other sides of the problem are discussed, as well as to more recent publications [13].

The author wishes to express his gratitude to Prof. Massimo Lanza de Cristoforis for creating the possibility to give these lectures at Minicorsi in Padova, June 2003.

7.1.1. *Variational principle*

We consider a complex Hilbert space $\mathcal{H}$ (look [27], [8], [20] for details). Let A be a self-adjoint operator in $\mathcal{H}$, with domain $\mathcal{D}(A)$. To the operator A we associated its quadratic form $A[u] = (Au, u)$. If the operator is lower semi-bounded, i.e.,

$$A[u] \geq c||u||^2, u \in \mathcal{D}(A),$$

the domain of the quadratic form, $d[A]$, is the closure of $\mathcal{D}(A)$ with respect to the norm $|u|_A^2 = A[u] - (c-1)||u||^2$, which is called A-norm. Obvious changes apply to upper semi-bounded operators.

As it is explained, e.g., in [20], [8], the operator itself can be reconstructed once the quadratic form, together with $d[A]$, is given. In singular situations it is usually convenient to define operators by means of their quadratic forms: the description of the domain of a singular operator is often very implicit, while the domain of the quadratic form can be described explicitly.

It is known from linear algebra that a Hermitian matrix has an orthogonal eigenvector basis. A nontrivial analogy of this fact for operators in a Hilbert space is given by the Spectral Theorem.

Theorem 7.1. *Let A be a self-adjoint operator. Then there exists an operator-valued function $E_t = E_t(A), t \in (-\infty, \infty)$ commuting with A, such that $E_t^* = E_t = E_t^2$, i.e. E_t are orthogonal projections; $E_t E_s = E_{\min(s,t)}, E(-\infty) = 0, E(\infty) = I, s - \lim_{\tau \to t-0} E_\tau = E_t, (Au, u) = \int_{-\infty}^{\infty} t\, d(E_t u, u), u \in \mathcal{D}(A)$ and if A is semi-bounded, $A[u] = \int_{-\infty}^{\infty} t\, d(E_t u, u),$ $u \in d[A]$. E_t is called the spectral function of A.*

The Spectral Theorem leads to the variational principle for the eigenvalue counting function. There are many different formulations of this principle, the one most convenient for our purposes is called *Glazman Lemma*.

In order to count eigenvalues, we denote by $N(\lambda) = N_-(\lambda) = N_-(\lambda, A)$ the total multiplicity of the spectrum of A in the interval $(-\infty, \lambda)$, in other words, $N(\lambda) = \dim E_\lambda(A)\mathcal{H}$. Thus, $N(\lambda)$ equals the number of eigenvalues below λ, counting with multiplicities, provided the spectrum below λ is discrete, and infinity otherwise. For upper semi-bounded operators set $N_+(\lambda, A) = N_-(-\lambda, -A)$; this function counts eigenvalues above λ.

Lemma 7.1 (Glazman Lemma). *Let A be a lower (upper) semi-bounded self-adjoint operator and $\mathcal{M} \subset d[A]$ be a linear subset, dense in A-metric. Then*

$$N_\pm(\lambda, A) = \max \dim\{\mathcal{L} \subset \mathcal{M}, \pm(A[u] - \lambda||u||^2) < 0, u \in \mathcal{L} \setminus \{0\}\}. \quad (7.1)$$

$$N_\pm(\lambda, A) = \min \operatorname{codim}\{\mathcal{L} \subset d[A], \pm(A[u] - \lambda||u||^2) \geq 0, u \in \mathcal{L}\}. \quad (7.2)$$

Here $\operatorname{codim} \mathcal{L}$ *for a subspace* $\mathcal{L} \subset d[A]$ *denotes the minimal number of orthogonality conditions which determine* $\mathcal{L}$ [a]. *Minus and plus signs correspond to upper resp. lower semi-bounded operators.*

---

[a] To save space we will omit $u \in \mathcal{L}$ in expressions like (7.1), (7.2) in the future.

We explain the *proof* of (7.1), with 'minus' sign; (7.2) is proved in a similar way. Denote by $\mathcal{K}$ the subspace $E_\lambda(A)\mathcal{H}$ so $N(\lambda) = \dim \mathcal{K}$. Denote by $\tilde{N}(\lambda)$ the quantity in (7.1). Assume that $\tilde{N}(\lambda) > N(\lambda)$. This means that there exist some $\mathcal{L}$, $\dim \mathcal{L} > \dim \mathcal{K}$, for which the inequality in (7.1) holds. Then there must exist an $u_0 \in \mathcal{L}$, which is orthogonal to $\mathcal{K}$, and therefore, $E_\tau u_0 = 0, t < 0$. Thus

$$A[u_0] = \int\limits_{-\infty}^{\infty} td(E_t u_0, u_0) = \int\limits_{\lambda}^{\infty} td(E_t u_0, u_0) \geq \lambda ||u_0||^2,$$

and this contradicts (7.1).

On the other hand, take some finite integer $N \leq \dim \mathcal{K}$ and an N-dimensional subspace $\mathcal{K}_N \subset \mathcal{K}$. On $\mathcal{K}_N \setminus \{0\}$, we have $A[u] < \lambda ||u||$. Using compactness reasoning and density of $\mathcal{M}$, we can approximate $\mathcal{K}_N$ by a subspace $\mathcal{L}_N$ of the same dimension N so that the above inequality still holds. This gives $N_+(\lambda) \geq N$, and therefore $\tilde{N}(\lambda) \geq N(\lambda)$.$\square$

An important consequence of Glazman Lemma, which we will use systematically, is that the counting function depends on the quadratic form in a monotone way. Let, for example, two operators A_1 and A_2 correspond to quadratic forms $A_1[u], A_2[u]$, so that $d[A_1] \subset d[A_2]$ and $A_1[u] \geq A_2[u], u \in d[A_1]$. Then in (7.1), the set of subspaces over which we maximize, is larger for A_2 than for A_1, and therefore, $N(\lambda, A_2) \geq N(\lambda, A_1)$ for any λ.

7.1.2. *Compact operators*

For a compact operator K, Glazman Lemma can be used with both signs, and one can take the whole Hilbert space $\mathcal{H}$ as $\mathcal{M}$. It is also useful to consider compact operators which are not necessarily self-adjoint. Singular numbers (or s-numbers) of K are defined as square roots of eigenvalues of the compact non-negative operator K^*K. By $n(t, K)$ we denote the distribution function of these s-numbers, $n(t, K) = N_+(t^2, K^*K)$, $t > 0$, i.e., the quantity of s-numbers above t. For $t > 0$, set $n_\pm(t, K) = N_\pm(\pm t, K)$, the number of eigenvalues of proper sign, with absolute value larger than t. If a self-adjoint operator A has discrete spectrum (and is invertible), then eigenvalues of A are inverse quantities to eigenvalues of the compact operator A^{-1}. Thus the study of eigenvalues of elliptic differential operators can be reduced to the study of eigenvalues of their resolvents, which turns out to be much more convenient. In particular, if the operator A is positive, then $N(\lambda, A) = N_+(\lambda^{-1}, A^{-1})$. For a non-semi-bounded A, if we denote by

$N^{(\pm)}(\lambda, A), \lambda > 0$, the number of eigenvalues of A between 0 and $\pm\lambda$, one has $N^{(\pm)}(\lambda, A) = n_{\pm}(\lambda^{-1}, A^{-1})$.

The spectral analysis of singular operators is based on perturbation ideas. First one obtains the required formulas for a certain regular case, and then studies what happens under a singular perturbation. The two crucial inequalities here are the one by Herman Weyl,

$$n(t_1 + t_2, K_1 + K_2) \le n(t_1, K_1) + n(t_2, K_2)$$

having the form

$$n_{\pm}(t_1 + t_2, K_1 + K_2) \le n_{\pm}(t_1, K_1) + n_{\pm}(t_2, K_2) \tag{7.3}$$

for self-adjoint compact operators K_1, K_2, and the one by Ky Fan,

$$n(t_1 t_2, K_1 K_2) \le n(t_1, K_1) + n(t_2, K_2), t_1, t_2 > 0.$$

Complete proofs and generalizations can be found in [11], [25]; here we just explain how (7.3) follows from Glazman Lemma. According to (7.2), we can find subspaces $\mathcal{L}_1, \mathcal{L}_2$ such that

$$\text{codim } \mathcal{L}_j = n_j = N_+(t_j, K_j); (K_j u, u) \le t_j \|u\|^2, u \in \mathcal{L}_j.$$

If we set $\mathcal{L} = \mathcal{L}_1 \cap \mathcal{L}_2, \text{codim } \mathcal{L} \le \text{codim } \mathcal{L}_1 + \text{codim } \mathcal{L}_2 = n_1 + n_2$, then on $\mathcal{L}$ we have $((K_1 + K_2)u, u) \le (t_1 + t_2)\|u\|^2$. Again, applying (7.2), we obtain $N_+(t_1 + t_2, K_1 + K_2) \le \text{codim } \mathcal{L} \le n_1 + n_2$. $\square$

7.1.3. *Asymptotic perturbation lemma*

Generally, one should expect that if one perturbs an operator with a weaker one, the main properties must not change. The lemma we give here (established first in [5]) assigns concrete meaning to this vague statement, as it concerns asymptotics of the spectrum.

Lemma 7.2. *Let K be a compact self-adjoint operator, and for some $q > 0$ and any $\varepsilon > 0$; K may be represented a sum, $K = K_\varepsilon + K'_e$, where*

$$\lim_{t \to +0} n_{\pm}(t, K_\varepsilon)t^q = c_{\pm}(\varepsilon), \limsup_{t \to +0} n_{\pm}(t, K'_\varepsilon)t^q \le \varepsilon.$$

Then there exist limits $\lim_{\varepsilon \to 0} c_{\pm}(\varepsilon) = c_{\pm}$ and $\lim_{t \to +0} n_{\pm}(t, K)t^q = c_{\pm}$.

Proof. Fix some $\delta > 0$. The Weyl inequality gives $n_+(t, K) \le n_+(t(1 - \delta), K_\varepsilon) + n_+(t\delta, K'_\varepsilon)$. Passing to $\limsup$, we obtain

$$c_+^{(+)} = \limsup_{t \to 0} n_+(t, K)t^q \le c_+(\varepsilon)(1 - \delta)^q + \delta^{-q}\varepsilon.$$

On the other hand, applying Weyl inequality to $K_\varepsilon = K + (-K'_\varepsilon)$, we obtain $n_+(t, K) \geq n_+(t(1 + \delta), K_\varepsilon) - n_-(t\delta, K'_\varepsilon)$. Passing here to $\liminf$, we get for $c_+^{(-)} = \liminf_{t \to 0} N_+(t, K)t^q$:

$$c_+^{(-)} \geq \lim_{t \to 0} n_+(t(1 + \delta), K_\varepsilon)t^q - \limsup_{t \to 0} n_-(t\delta, K'_\varepsilon)t^q$$
$$\geq c_+(\varepsilon)(1 + \delta)^{-q} - \delta^{-q}\varepsilon.$$

Thus

$$c_+(\varepsilon)(1 + \delta)^{-q} - \delta^{-q}\varepsilon \leq c_+^{(-)} \leq c_+^{(+)} \leq c_+(\varepsilon)(1 - \delta)^q + \delta^{-q}\varepsilon. \tag{7.4}$$

We set here $\delta = \varepsilon^{1/(q+1)}$ so that $\delta^{-q}\varepsilon \to 0$. Then (7.4) gives $c_+^{(-)} = c_+^{(-)} = \lim c_+(\varepsilon)$. $\qquad\square$

The original Weyl result dealt with the case when $K = K_0 + K'$, for K_0 the spectral asymptotics holds, and for K' the eigenvalues decay faster, $N_\pm(t, K') = o(t^{-q})$. This fact will be also used here.

7.2. Lecture 2. Dirichlet Problem for the Weighted Laplacian in Arbitrary Domains

7.2.1. *The asymptotic formula*

We present in this lecture the proof of the asymptotic formula for eigenvalues for the Dirichlet Laplacian in an arbitrary domain. As it often happens in Mathematics, it turns out to be easier to prove a more general fact, concerning the *weighted* Laplace operator.

Theorem 7.2. *Let $\Omega \in \mathbb{R}^d$, $d \geq 3$, be an arbitrary open set and $p(x)$ be a real function in $L_{\frac{d}{2}}$. Denote by $-\Delta_D$ the (minus) Laplace operator in Ω with Dirichlet boundary conditions. By $N^{(\pm)}(\lambda, -\Delta_D, p)$, $\lambda > 0$, we denote the number of eigenvalues of the spectral problem*

$$-\Delta_D u(x) = \mu p(x) u(x) \tag{7.5}$$

between 0 and $\pm\lambda$. Then

$$\lim_{\lambda \to +\infty} (N^{(\pm)}(\lambda, -\Delta_D, p)\lambda^{-\frac{d}{2}}) = c_d \int_\Omega p_\pm(x)^{\frac{d}{2}} dx, \tag{7.6}$$

where $p_\pm$ is the positive resp. negative part of p and the constant c_d depends only on the dimension, $c_d = (2\pi)^{-d}\omega_d$, and ω_d is the volume of the unit ball in $\mathbb{R}^d$.

Remark 7.1. Note here that the only condition for the asymptotic formulas to hold is finiteness of the integral in the asymptotic coefficient. In low dimensions, $d < 3$, a similar result does not hold, and the proof of a weaker fact requires somewhat different technics.

7.2.2. *Variational statement of spectral problems*

Before explaining the proof of the theorem, we have to assign an exact meaning to the spectral problem (7.5) since in a domain with 'bad' boundary the direct definition of the Dirichlet boundary conditions is impossible. So, we start with a 'nice', bounded domain, and let the function p be smooth and bounded. Then the operator $-\Delta_D$ is well defined, and the domain of its quadratic form, coinciding with the domain of its square root $L = (-\Delta_D)^{1/2}$, is the Sobolev space $H_0^1(\Omega)$. The spectral problem (7.5) by setting $Lu = v$ is transformed to

$$S_p v = sv, v \in L_2(\Omega), S_p = L^{-1}pL^{-1}, s = \mu^{-1}.$$

To the eigenvalue distribution of the operator S_p in the Hilbert space $\mathcal{H} = L_2(\Omega)$ with norm $||v||_{L_2}^2 = ||Lu||_{L_2}^2 = \int_\Omega |\nabla u|^2 dx$, we apply Glazman Lemma, taking as $\mathcal{M}$ the subspace $C_0^\infty(\Omega)$, which is dense in $\mathcal{H}$, thus obtaining

$$n_\pm(s, S_p) = \max \dim\{\mathcal{L} \subset \mathcal{M}, \pm(S_p v, v) > s||v||^2)\}.$$

Returning here to $u = L^{-1}v$, we come to

$$n_\pm(s, S_p) = \max \dim\{\mathcal{L} \subset \mathcal{M}, \pm(pu, u) > s||Lu||^2)\}. \tag{7.7}$$

or, in codimension terms,

$$n_\pm(s, S_p) = \min \operatorname{codim}\{\mathcal{L} \subset H_0^1(\Omega), \pm(pu, u) \leq s||Lu||^2)\}. \tag{7.8}$$

Thus, we obtain that the distribution of *inverse* eigenvalues $s_j = \lambda_j^{-1}$ of our problem is the same as distribution of eigenvalues of the operator T_p defined in the Hilbert space $H_1^0(\Omega)$ with norm $\int_\Omega |\nabla u|^2 dx$ by the quadratic form (pu, u).

We extend this definition to arbitrary domains and arbitrary weight functions. Consider the space $H_0^1(\Omega)$, the closure of $C_0^\infty(\Omega)$ in the norm of the Dirichlet integral,

$$J[u] = J_\Omega[u] = J_{1,\Omega}[u] = \int_\Omega |\nabla u|^2 dx. \tag{7.9}$$

By $I_p[u]$ we denote the quadratic form $\int_\Omega p(x)|u|^2 dx$. Under the conditions of our theorem, the form I_p is bounded in $H_0^1(\Omega)$: the Hölder inequality gives $I_p[u] \leq ||p||_{L_{\frac{d}{2}}} ||u||_q^2$, $q = \frac{2d}{d-2}$, and the latter norm of u is estimated by $J[u]$ according to *Sobolev inequality*, $||u||_q^2 \leq CJ[u]$. Thus the form I_p defines a bounded self-adjoint operator T_p in $H_0^1(\Omega)$, and we say that eigenvalues of this operator are inverse values to eigenvalues of the problem (7.5). For eigenvalues of T_p the variational formulas (7.7), (7.8) are valid.

More generally, if we have a linear space $\mathcal{M}$ and two quadratic forms $\mathfrak{A}, \mathfrak{B}$, $\mathfrak{B} \geq 0$ on $\mathcal{M}$, we denote by $n_\pm(t, \mathfrak{A}, \mathfrak{B}, \mathcal{M})$ the quantity

$$n_\pm(t, \mathfrak{A}, \mathfrak{B}, \mathcal{M}) = \max \dim\{\mathcal{L} \subset \mathcal{M}, \pm\mathfrak{A}[u] > t\mathfrak{B}[u]\}, t > 0, \qquad (7.10)$$

and, similarly, formulas with codimension.

In particular, for a nice domain, we can handle by means of (7.10) the Neumann problem, for which the domain of the quadratic form is the space $H^1(\Omega)$, where $C^\infty(\overline{\Omega})$ is dense. So one has to take $\mathfrak{A} = I_p, \mathfrak{B} = J, \mathcal{M} = C^\infty(\overline{\Omega})$ or $\mathcal{M} = H^1(\Omega)$ in (7.10) to get the variational description of the spectrum of the Neumann Laplacian with weight. Moreover, we can set Dirichlet conditions only on a part Γ of the boundary of Ω and Neumann conditions on the rest of the boundary, by taking as $\mathcal{M}$ in (7.10) the space $C_0^\infty(\overline{\Omega}, \Gamma)$ consisting of functions which vanish near Γ.

The monotonicity of $n_\pm(t, \mathfrak{A}, \mathfrak{B}, \mathcal{M})$, discussed in Sect. 1, holds also in this general situation: the quantity $n_+(t, \mathfrak{A}, \mathfrak{B}, \mathcal{M})(N_-(t, \mathfrak{A}, \mathfrak{B}, \mathcal{M}))$ grows when $\mathfrak{A}$ grows (decreases), $\mathfrak{B}$ decreases, or $\mathcal{M}$ extends. Also, if the space $\mathcal{M}$ is the direct sum of, say, two subspaces, $\mathcal{M} = \mathcal{M}_1 \oplus \mathcal{M}_2$, and the forms $\mathfrak{A}, \mathfrak{B}$ are split into the sum of two forms, $\mathfrak{A} = \mathfrak{A}_1 + \mathfrak{A}_2$, $\mathfrak{B} = \mathfrak{B}_1 + \mathfrak{B}_2$ so that the forms with subscript j vanish on the subspace $\mathcal{M}_{2-j}$, then the *decomposition* formula for eigenvalues holds,

$$n_\pm(t, \mathfrak{A}, \mathfrak{B}, \mathcal{M}) = n_\pm(t, \mathfrak{A}_1, \mathfrak{B}_1, \mathcal{M}_1) + n_\pm(t, \mathfrak{A}_2, \mathfrak{B}_2, \mathcal{M}_2).$$

All this follows directly from (7.10).

One more property will be used to establish stability of asymptotics of the spectrum under weak perturbations.

Proposition 7.1. *Let $\mathfrak{B}[u] = \mathfrak{B}_0[u] + \mathfrak{B}_1[u]$ and the form $\mathfrak{B}_1$ is weak with respect to $\mathfrak{B}_1$ in the following sense: for any $\varepsilon > 0$ there exists a subspace $\mathcal{M}_\varepsilon \subset \mathcal{M}$ with finite codimension $n(\varepsilon)$ such that $|\mathfrak{B}_1[u]| \leq \varepsilon\mathfrak{B}_0[u], u \in \mathcal{M}_\varepsilon$. Then for any q,*

$$\lim_{\tau \to 0} t^q n_\pm(t, \mathfrak{A}, \mathfrak{B}, \mathcal{M}) = \lim_{t \to 0} t^q n_\pm(t, \mathfrak{A}, \mathfrak{B}_0, \mathcal{M}),$$

provided that the latter limit exists.

To establish the above relation it is sufficient to note that on the subspace $\mathcal{M}_\varepsilon$, we have $(1 - \varepsilon)\mathfrak{B}_0[u] \leq \mathfrak{B}[u] \leq (1 + \varepsilon)\mathfrak{B}_0[u]$, therefore if we have some subspace $\mathcal{L}_t$ in $\mathcal{M}$ where $\mathfrak{A}[u] \leq t\mathfrak{B}[u]$ then on the subspace $\mathcal{L}_t \cap \mathcal{M}_\varepsilon$, one has $\mathfrak{A}[u] \leq (1 + \varepsilon)t\mathfrak{B}[u]$, and the codimension of $\mathcal{L}_t$ differs from the codimension of $\mathcal{L}_t \cap \mathcal{M}_\varepsilon$ by no more than $n(\varepsilon)$. This gives the estimate from below,

$$\limsup_{\tau \to 0} t^q n_\pm(t, \mathfrak{A}, \mathfrak{B}, \mathcal{M}) \geq (1 + \varepsilon)^{-q} \limsup_{\tau \to 0} t^q n_\pm(t, \mathfrak{A}, \mathfrak{B}_0, \mathcal{M}),$$

and then we use arbitrariness of ε. The inequalities from the other side follows from the similar reasoning with dimensions. $\square$

7.2.3. *Proof of the asymptotic formula*

Now we can prove our Theorem on asymptotics. The proof goes in four steps, where more and more general situations are handled.

Step 1. The domain Ω is a cube, the weight function p is constant. Formula (7.6) holds both for Dirichlet and Neumann boundary conditions. This case was considered in many textbooks in PDE (see, e.g., [9], [20]). The eigenvalue equation admits separation of variables, eigenvalues are found explicitly and the asymptotic formula is obtained directly.

Step 2. The domain Ω can be cut into a finite set of (open) cubes Q_j, and p is constant, $p = p_j$ on each cube Q_j. Formula (7.6) again holds both for Dirichlet and Neumann boundary conditions. This is explained in the following way. The distribution of Neumann eigenvalues in Ω is described by $N_\pm(t, I_p, J, C^\infty(\overline{\Omega}))$. Now extend here $C^\infty(\overline{\Omega})$ to $\oplus(C^\infty(\overline{Q_j}))$ and replace the forms I_p, J by the sums of the forms I_{p,Q_j}, J_{Q_j} corresponding to integration over Q_j. According to monotonicity property, the number of eigenvalues increases. The decomposition rule, after this, gives the inequality

$$n_\pm(t, I_p, J, C^\infty(\overline{\Omega})) \leq \sum n_\pm(t, I_{p,Q_j}, J_{Q_j}, C^\infty(\overline{Q_j})).$$

For each term in the latter sum the asymptotics is found in Step 1; adding up the asymptotic coefficients, we obtain the required estimate from above for the eigenvalue asymptotics for the Neumann problem. For Dirichlet problem we establish the estimate from below, by similar reasoning. Replace in $n_\pm(t, I_p, J, C_0^\infty(\Omega))$ the space $C_0^\infty(\Omega)$ by a smaller space $\oplus C_0^\infty(Q_j)$, and split the forms I_p, J as above. Then, due to the monotonicity property, and decomposition rule we obtain

$$n_\pm(t, I_p, J, C_0^\infty(\Omega)) \geq \sum n_\pm(t, I_{p,Q_j}, J_{Q_j}, C_0^\infty(Q_j)).$$

Again, for the terms on the right-hand side, the asymptotics is found in Step 1. Adding the asymptotic coefficients, we obtain the estimate from above for the eigenvalue asymptotics for the Dirichlet problem, by the same quantity.

Step 3. Ω is arbitrary, the weight p has support in a smaller bounded domain Ω_0, so that Ω_0 and p satisfy conditions of Step 2. Here the asymptotic formula is established in the following way. Consider $\Omega_1 = \Omega \setminus \overline{\Omega_0}$. We replace in $n_\pm(t, I_p, J, C_0^\infty(\Omega))$ the space $C_0^\infty(\Omega)$ by a smaller space $C_0^\infty(\Omega_0) \oplus C_0^\infty(\Omega_1)$, in other words, require that functions vanish not only near the boundary of Ω, but also at the boundary on Ω_0. Correspondingly split the forms I_p and J into the terms containing integration over Ω_0 and Ω_1. Therefore, following monotonicity and decomposition rules, we get the estimate from below,

$$n_\pm(t, I_p, J, C_0^\infty(\Omega)) \geq n_\pm(t, I_p, J, C_0^\infty(\Omega_0)) + n_\pm(t, I_p, J, C_0^\infty(\Omega_1)).$$

However the last term vanishes, since $I_p = 0$ on Ω_1, and for the first term to the right the asymptotics is established on Step 2. To get an estimate from above, replace $C_0^\infty(\Omega)$ by a larger space $C^\infty(\overline{\Omega_0}) \oplus C^\infty(\overline{\Omega_1}, \partial\Omega_0)$, i.e., admitting a discontinuity at the boundary of Ω_0. Again, we use monotonicity and decomposition, and the term corresponding to Ω_1 vanishes, while the term corresponding to Ω_0 is already taken care of on Step 2.

Step 4. The general case. For a fixed $\epsilon > 0$ find a weight function p_ϵ satisfying condition of Step 3, so that the function $p_\epsilon' = p - p_\epsilon$ has small norm in L_p, $\|p_\epsilon'\|_{L_{\frac{d}{2}}} \leq \varepsilon$. Correspondingly, the operator T_p splits into the sum $T_p = T_{p_\epsilon} + T_{p_\epsilon'}$. For the operator T_{p_ϵ} the asymptotic formula is established on Step 3. As for the operator $T_{p_\epsilon'}$, there is an estimate for its eigenvalues, we formulate now.

Theorem 7.3 (CLR-estimate.). *Under the conditions of Theorem 7.2, the estimate holds*

$$n_\pm(t, T_p) = n_\pm(t, I_p, J, C_0^\infty(\Omega)) \leq Ct^{-\frac{d}{2}} \int p_\pm^{\frac{d}{2}} dx \qquad (7.11)$$

with constant C depending only on the dimension d.

We will discuss this estimate and its generalizations in Lecture 4.

But now, the proof of the asymptotic formula for eigenvalues is concluded by applying the asymptotic perturbation Lemma 7.2. In fact, we have for the operator T_{p_ε} the asymptotic formula $\lim n_\pm(T_{p_\varepsilon})t^{-\frac{d}{2}}$ $= c_d \int (p_\varepsilon)_\pm^{\frac{d}{2}} dx$. Together with (7.11) for T'_{p_ε}, Lemma 7.2 gives that

$$\lim n_\pm(t, T_p)t^{\frac{d}{2}} = \lim_{\varepsilon \to 0} c_d \int (p_\varepsilon)_\pm^{\frac{d}{2}} dx = c_d \int p_\pm^{\frac{d}{2}} dx.$$

7.3. Lecture 3. Second Order Elliptic Operators With Singularities

The approach described in the previous lecture enables one to prove asymptotic formulas for a very general class of elliptic operators, with various types of singularities. In this lecture we show how it works for second order elliptic operators with non-smooth, unbounded coefficients.

7.3.1. *Statement of the problem*

Again we start with the classical setting, and then state the eigenvalue problem in the quadratic form language.

So, let A be an operator in divergence form, $A = -\sum \partial_j a_{jk} \partial_k$ in a domain $\Omega \subset \mathbb{R}^d$. The Hermitian coefficient matrix $\mathbf{a}(x) = (a_{jk}(x)) \in L_{d,\text{loc}}$ consists of measurable functions so that

$$\sum a_{jk}(x)\xi_j \overline{\xi_k} \geq \nu |\xi|^2, \nu > 0, \xi \in \mathbb{C}^d.$$

Thus we admit unbounded coefficients, without any smoothness conditions. We also introduce the weight function $p(x)$. The spectral problem we are going to consider is

$$Au(x) = \lambda p(x)u(x) \tag{7.12}$$

with Dirichlet, and for certain nice domains, Neumann boundary conditions. Of course, in such general conditions, it is impossible to describe explicitly the domain of operator A, so we pass to the quadratic forms formulation, using our reasoning in Lecture 2 as a pattern.

Define the quadratic form $J_{\mathbf{a}}[u] = \int (\mathbf{a}\nabla u, \nabla u)dx$ on functions $u \in C_0^\infty(\Omega)$, and denote by $H_0^1(\mathbf{a})$ the closure of $C_0^\infty(\Omega)$ in this metric. In the space $H_0^1(\mathbf{a}) \subset H_0^1(\Omega)$ consider the operator $T_{\mathbf{a},p}$ generated in $H_0^1(\mathbf{a})$ by the quadratic form $I_p[u] = \int_\Omega p|u|^2 dx$. The eigenvalues of this operator are, in the 'regular' case (everything is smooth), the inverse quantities to eigenvalues of (7.12). In the 'non-regular' case we *define* eigenvalues of (7.12) as the inverse to eigenvalues of $T_{\mathbf{a},p}$. According to the Glazman lemma, for their distribution function $n_\pm(t, T_{\mathbf{a},p}) = n_\pm(t, I_p, J_\mathbf{a}, H_0^1(\mathbf{a}))$ the variational formula holds,

$$n_\pm(t, T_{\mathbf{a},p}) = \max \dim\{\mathcal{L} \subset H_0^1(\mathbf{a}), \pm I_p[u] > t J_\mathbf{a}[u]\}. \tag{7.13}$$

For domains with nice boundaries we can also consider the Neumann boundary conditions. The corresponding eigenvalue counting function is defined in the same way. Indeed, one just has to replace the space $C_0^\infty(\Omega)$ by $C^\infty(\overline{\Omega})$ as in Lecture 2. Now we formulate the spectral asymptotics result.

Theorem 7.4. *Let* $\mathbf{a} \in L_{d,\mathrm{loc}}(\Omega), p \in L_{\frac{d}{2}}(\Omega), d > 2$. *Then for the eigenvalues of the problem* (7.12) *the asymptotic formula holds*

$$\lim_{t \to \infty} (n_\pm(t, T_{\mathbf{a},p}) t^{\frac{d}{2}}) = c_d \int_\Omega p_\pm(x)^{\frac{d}{2}} \det \mathbf{a}(x)^{-\frac{1}{2}} dx. \tag{7.14}$$

We remark that the conditions of the theorem imply, via Hölder inequality, that the coefficient in (7.14) is finite. Compared with results for the weighted Laplacian, the former conditions are, in fact, more restrictive than this finiteness. To be more precise, the coefficient in (7.14) may be finite, due to cancellation of singularities in p and in $\det \mathbf{a}$ but the theorem cannot take into account this cancellation. Up to now, this cancellation property is not yet completely studied. We will also need the same asymptotic result for the Neumann problem in a cube.

We are going to prove our theorem along the lines of the proof of Theorem 7.2. Analyzing this proof, we can see that already on Step 1 we encounter an obstacle, since for variable coefficients one cannot separate variables and find eigenvalues in a cube explicitly. Skipping this for a moment, we see that Steps 2 and 3 go through without any changes, as well as Step 4, as soon as we obtain a variable coefficients version of the CLR estimate. We will see in a moment, that this estimate is needed in Step 1 as well.

So we turn our attention to Step 1: the asymptotic formula (7.14) holds, for Dirichlet and Neumann cases, for Ω being a cube Q and p being a constant. To establish this, we consider the case of a smooth, non-degenerate operator first.

Proposition 7.2. *The asymptotic formula* (7.14) *holds for* $\Omega = Q, p = \mathrm{const}, \mathbf{a} \in C^\infty$.

This result is a classical one, see, e.g., [1].

7.3.2. *Step 1, perturbation*

We pass now to the non-regular case. Let $\mathbf{a}(x)$ be a matrix satisfying the conditions of theorem, and for fixed $\varepsilon > 0$ $\mathbf{a}_\varepsilon(x) \geq \mathbb{N}_0$ be a smooth non-degenerate matrix such that $||\mathbf{a} - \mathbf{a}_\varepsilon||_{L_d} \leq \varepsilon$. It is easy to show that such approximation exists. Denote by A, A_ε self-adjoint operators in $L_2(Q)$ defined by the quadratic forms $J_{\mathbf{a}}, J_{\mathbf{a},\varepsilon}$. To compare the spectral asymptotics of these operators, having in mind applying our asymptotic perturbation lemma (Lemma 7.2), we consider inverse (compact, as it turns out) operators, and study the difference $R_\varepsilon = A^{-1} - A_\varepsilon^{-1}$. Here one has to be rather cautious in writing the expression for this difference. In particular, one cannot apply directly the formula well known from the linear algebra, $A^{-1} - B^{-1} = A^{-1}(B - A)B^{-1}$ due to the fact that the operators $A, B = A_\varepsilon$ have different domains. One comes to an absolutely paradoxical result if one tries to apply the latter 'identity' to A, B being operators corresponding to different sets of boundary conditions for one and the same differential operator[b].

So we apply a different approach, proposed initially by M. Birman in [4]. We consider Dirichlet problem at the moment.

Proposition 7.3 (Resolvent identity).

$$A^{-1} - A_\varepsilon^{-1} = \sum_{j,k} L_{jk}^* K_j, \quad L_{jk} = (a_{jk} - a_{\varepsilon,jk})\partial_k A_\varepsilon^{-1}; \quad K_j = \partial_j A^{-1}. \quad (7.15)$$

To *prove* the formula, we take arbitrary functions $u, v \in L_2$ and consider the sesquilinear form of the operator R_ε :, $(R_\varepsilon u, v) = (u, A^{-1}v) - (A_\varepsilon^{-1} u, v)$. Set $f = A^{-1}u, g = A_\varepsilon^{-1}v, f, g = 0$ on ∂Q. Then we have $(R_\varepsilon u, v) = (Af, g) - (f, A_\varepsilon g)$. Integration by parts gives

$$(R_\varepsilon u, v) = \sum_{j,k} \int_\Omega (\partial_j f)(a_{jk} - a_{jk,\varepsilon})\partial_k g dx = \sum_{j,k}(\partial_j f, (a_{jk} - a_{jk,\varepsilon})\partial_k g).$$

Setting again $f = A^{-1}u, g = A_\varepsilon^{-1}g$, we come to $(R_\varepsilon u, v) = \sum(K_j u, L_{jk}v) = \sum(L_{jk}^* K_j u, v)$, and this proves our proposition.

Now we formulate two more types of CLR estimate - we prove them in Lecture 4.

Proposition 7.4. *For operators K_j and L_{kj} the following estimates hold*

$$n(t, K_j) \leq Ct^{-d}, \quad (7.16)$$

---

[b]try to find correct formula here.

$$\limsup_{t\to 0} n(t, L_{jk}) t^d \le C \int_D |a_{jk} - a_{jk,\varepsilon}|^d dx. \qquad (7.17)$$

The constants in (7.16), (7.17) depend only on the dimension d and the ellipticity constant ν_0.

Note that in (7.17) the estimate holds only in the limit; (7.16) is uniform in t.

Now we apply the Ky Fan inequality to the product of operators $L_{jk}^* K_j$, $n(t_1 t_2, L_{jk}^* K_j) \le n(t_1, L_{jk}) + n(t_2, K_j)$ and set, for given t, $t_1 = \varepsilon^{\frac{1}{2d}} t^{\frac{1}{2}}, t_2 = \varepsilon^{-\frac{1}{2d}} t^{\frac{1}{2}}$. Passing then to $t \to 0$ and using (7.16), (7.17), we come to $\limsup t^{\frac{d}{2}} n(t, L_{jk}^* K_j) \le C \varepsilon^{\frac{d}{2}}$.

There are only a finite, no more than d^2, terms in the representation (7.15) of the difference $A^{-1} - A_\varepsilon^{-1}$, therefore Weyl's inequality for s-numbers gives us

$$\limsup t^{\frac{d}{2}} n_\pm(t, A^{-1} - A_\varepsilon^{-1}) \le C \varepsilon^{1/2}.$$

As a result, we can see that we are in conditions of the Asymptotic Perturbation Lemma from Lecture 1. Applying this Lemma, we arrive at the asymptotic formula (7.14), with constant p, for eigenvalues of the Dirichlet problem in the cube.

The proof of asymptotics for the Neumann problem is somewhat more involved. Shortly, the reasoning goes as follows. Consider a smaller concentric cube Q_ε, with dilation coefficient $1 - \varepsilon$. If $1 = p_\varepsilon + p_\varepsilon'$, with p_ε having support in Q_ε and p_ε' in $S_\varepsilon = Q \setminus Q_\varepsilon$, a CLR-estimate gives that by replacing 1 by p_ε one changes the asymptotical coefficient just slightly, controlled by ε. Next, consider the operator $\mathcal{S} : H^1(\mathbf{a}) \to H_0^1(\mathbf{a})$, multiplication by a compactly supported smooth function having value 1 on Q_ε. One can show that this operator maps any subspace in $H^1(\mathbf{a})$ where the inequality $I_{p_\varepsilon}[u] \ge t J_\mathbf{a}[u]$ holds to a subspace in $H_0^1(\mathbf{a})$ where a similar inequality holds, $I_{p_\varepsilon}[Su] \ge t((1 - \varepsilon) J_\mathbf{a}[Su] - C_\varepsilon ||u||^2)$ thus returning to the Dirichlet problem. The last term does not influence the leading term of asymptotics, as it follows from Lemma 7.3 below and Proposition 7.1, and can be omitted. Finally, one can again, by the price of slightly changing the asymptotics, return the previously omitted term with p_ε' and arrive at Dirichlet problem, for which asymptotics is just established. The details can be found in [6].

7.3.3. *Conclusion of the proof*

Following the lines of the reasoning in Lecture 2, we perform Steps 2 and 3 – the manipulations are made only with domains and the weight function p, and they do not change when passing to variable coefficients.

To finish the proof, we have to make Step 4, passing from the operator with a piecewise constant weight p_ε to a general one. To accomplish it, we need one more CLR-type estimate.

Proposition 7.5. *Under conditions of Theorem 7.4, the estimate holds*

$$n_\pm(t, T_{\mathbf{a},p}) \leq C\|p_\pm\|_{L_{\frac{d}{2}}}^{\frac{d}{2}} t^{-\frac{d}{2}}, \tag{7.18}$$

with a constant depending only on the dimension d and the ellipticity constant ν_0 .

Provided this estimate is established, the proof is concluded exactly as it was done in Lecture 2, by approximating p by a piecewise constant function and applying the perturbation lemma.

Comparing the proofs of Theorems 7.2 and 7.4, we can see two principal ingredients one needs: the first one, the asymptotics of the problem in a regular case, and the second one, estimates of eigenvalues in the general case, involving integral characteristics of coefficients in the problem. Even in more general situations, the same two ingredients are decisive. For regular cases, asymptotic formulas were mostly obtained quite long ago. In the next lecture we are going to discuss the CLR-type estimates.

7.4. Lecture 4. CLR-type estimates

It was, probably, in [20], that the name 'CLR-estimate' has appeared. The notation is deciphered as Cwikel-Lieb-Rozenblum after the names of the persons who gave three first, independent, proofs of the estimate (see [10], [17], [21]). The need for this kind of eigenvalue estimates was felt quite long ago, and it was explicitly formulated by B. Simon in 1973, as a conjecture, without knowledge that it had been already proved in [21]. In the narrow sense, the CLR estimate is the estimate (7.11), however both in [10], [17], [21] as well as in the later publications [16], [15], [24], more general situations are covered, the scope of generalization depending in each case on the method of the proof, since the existing proofs use quite different machinery. Generally, CLR-type estimates are estimates of the spectrum of differential eigenvalue problems, having correct order and expressed in the terms of integral norms of coefficients.

Among existing proofs, each can handle (7.11), (7.16) and (7.18). However (7.17) can be proved (at the present state of knowledge) only by the method of the original paper [21]. Most publications on this matter are rather obscure and/or are only in Russian (see [21], [6], [7]). Probably, the most accessible source where this method is presented is [18], however the forms containing first derivatives only, like our $J_{\mathbf{a}}$, are considered there. The estimate (7.17) requires forms of the second order. Therefore to give this reference is not sufficient, and we have to describe more details here.

We start by explaining some results from geometry and function theory on which the proof is based.

7.4.1. *Covering lemma*

We are going to consider coverings of domains (mostly, cubes) by closed cubes or *bricks* $\overline{\Delta}$, parallelepipeds in $\mathbb{R}^d$ with ratio of longest and shortest edges not larger than 2. Corresponding open sets Δ do not necessarily form a covering, but the part which is not covered has measure zero and therefore is not essential. We say that the covering Ξ of a domain U has multiplicity not greater than $\varkappa$ if each point of U is interior to no more that $\varkappa$ bricks in Ξ. Covering with multiplicity 1 is called *partition*.

Proposition 7.6 (Covering lemma). *Let J be a measure, absolutely continuous with respect to Lebesgue measure on the cube $Q \subset \mathbb{R}^d$. Then, for any $n \geq 1$, there exists a covering Ξ of Q by no more than n bricks $\overline{\Delta} \subset Q$ with multiplicity not greater than 2^d, and for any $\Delta \in \Xi$*

$$J(\Delta) \leq \varkappa n^{-1} J(Q), \quad \varkappa = 2^{d+1}. \tag{7.19}$$

Remark 7.2. This is a simple version of the covering lemma in [21]. A more complicated lemma is needed in the proof of estimates of non-uniformly elliptic equations, see [18].

Proof. There are several ways of establishing this lemma. It may be derived from the general covering principle by M. de Guzman [12], or from a covering theorem due to Besicovitch [2]. We give a somewhat simplified original proof from [21](see also [18]). Set $J(Q) = \rho$. We call the set $G \subset Q$ "poor" if $J(G) \leq \rho \varkappa n^{-1}$ and "rich" if $J(G) \geq \rho \varkappa n^{-1}$ (in the case of equality, G is both rich and poor). Our aim is to cover Q by not more than n poor bricks.

Let Δ be some rich *cube* with edge l. Then at least one of the following possibilities takes place:

i) Δ can be covered by not more than 2^d poor bricks $\Delta_j \subset \Delta$;

ii) Δ can be cut into m_1 poor cubes and m_2 rich cubes, where $m_1 < 2^d$, $m_2 \geq 2$;

iii) there exists a rich cube $\Delta' \subset \Delta$ such that the set $S = \Delta \backslash \Delta'$ is also rich but can be covered by not more than d poor bricks.

To justify this assertion, we cut Δ into 2^d equal cubes with edge $l/2$. If they are all poor, we have case i; if at least two of them are rich, we have the case ii. Hence, suppose that only one of small cubes, namely Δ_1, is rich and all the others are poor. If the set $S_1 = \Delta \backslash \Delta_1$ is rich, we have the case iii. Otherwise, introduce the family of cubes Δ_t, with edge $tl/2$, which are cut out of the same corner of Δ, as Δ_1; $S_t = \Delta \backslash \Delta_t$. For any $0 < t \leq 1$, as one can easily see, S_t can be covered by d bricks, lying in S_t. When t decreases from 1 to 0, the function $\mathcal{J}(\Delta_t)$ decreases continuously to 0 and the function $\mathcal{J}(S_t)$ grows continuously[c]. Set $t_1 = \sup\{t, \Delta_t$ is poor$\}$, $t_2 = \inf\{t, S_t$ is poor $\}$. If $t_1 \geq t_2$, then Δ_{t_1}, S_{t_1} are both poor, and we have the case i, since all bricks covering S_{t_1} are also poor. If $t_1 \leq t_2$ then Δ_{t_2} and S_{t_2} are both rich and we have the case iii.

The construction of the covering in question goes inductively. The *continuation* procedure consists in the following. Suppose that we already have a *partition* of Q into cubes (both rich and poor) and rich sets of the form S, as in iii. We apply the previously described construction to those rich cubes for which the cases ii or iii take place, getting a new partition.

We start with the trivial covering of Q by itself. We assume $n \geq \varkappa\rho$, otherwise the trivial covering of Q is the one we need. We apply the continuation procedure successively. At each step, the number of rich sets of the partition increases, however it cannot exceed $\varkappa^{-1}n$. This means that after several steps, in our partition there will be only poor cubes (n_1 in number), n_2 rich cubes satisfying i, and n_3 rich sets S, as in iii.

We have $n_1 \leq (2^d - 1)n_2$, since in our procedure, poor cubes are produced only in the case ii, and one new rich cube produces no more than $2^d - 2$ poor ones. So, the inequality $n_2 \leq \varkappa^{-1}n$ implies $n_1 \leq (2^d - 2)\varkappa^{-1}n$. Finally, we cover all remaining rich cubes and sets S_k by poor bricks, as in cases i and iii, which produces no more than $2^d(n_2 + n_3) \leq 2^d\varkappa^{-1}n$ poor bricks; there are no more than $2^{d+1}\varkappa^{-1}n = n$ bricks altogether, giving us the desired covering. Our bricks start overlapping only on the final stage

---

[c]It is in this place only that absolute continuity of the measure is needed. This condition can be relaxed, see [6].

(we dealt only with partitions before), so the multiplicity of the covering is not greater than $\varkappa$. $\qquad\square$

7.4.2. *Functional inequalities*

Another ingredient in the proof of CLR estimates is functional inequalities in Sobolev space on cubes and bricks. The sense of these inequalities is that on a subspace of a certain finite codimension in the Sobolev space, the standard Sobolev norm is equivalent to its leading term, containing only highest order derivatives. Such inequalities were obtained by S. Sobolev on early stages of development of the theory of spaces; we however can give a simple proof based on the variational principle described in Lecture 1.

We denote by $||\nabla^l u||^2_\Delta$ the leading term in the norm in the Sobolev space,

$$||\nabla^l u||^2_{L_2(\Delta)} = \sum_{|\alpha|=l} \int_\Delta |D^\alpha u|^2 dx,$$

where sum spreads over multi-indices α of height l. The usual Sobolev norm is given by

$$||u||^2_{H^l(\Delta)} = ||\nabla^l u||^2_\Delta + ||u||^2_{L_2(\Delta)} = J_l[u] + I[u]. \tag{7.20}$$

Lemma 7.3. *For any $\tau > 0$ and brick Δ there exists a subspace $\mathcal{L} \subset H^l(\Delta)$ and a constant $c(d,l)$ such that* codim $\mathcal{L} \leq N = N(d,l,\tau)$ *and*

$$\tau I[u] \leq c(d,l)(\text{meas } \Delta)^{-\frac{2l}{d}} J_l[u]. \tag{7.21}$$

We will use (7.21) both for a fixed $\tau = 1$, where the constant and codimension of the space depend only on d, l, and for arbitrary τ where we do not control codimension.

Proof. First let Δ be the unit cube. Consider the operator T (a compact one, due to the Friedrichs-Sobolev embedding theorem) generated in the Sobolev space $H^l(\Delta)$ by the form $I[u]$. For any t the number $n(t, T)$ is finite. Take $t = \frac{1}{2\tau}$. Due to the variational principle (7.2), there exist a subspace $\mathcal{L}$ of a finite codimension $N = N(d,l,\tau)$ such that $I[u] \leq \frac{1}{2\tau}(J_l[u] + I[u]), u \in \mathcal{L}$, The latter inequality implies

$$\tau I[u] \leq J_l[u], u \in \mathcal{L}, \tag{7.22}$$

i.e. (7.21) for the unit cube. Next, perform a linear transformation of the unit cube to a brick Δ_0, with largest edge having unit length. Each term in J_l will multiply by a factor, lying in a fixed interval, $[1, 2^{2l-d}]$; the form

I gets a factor lying in the interval $[1, 2^{-d}]$. Therefore, for the transformed subspace $\mathcal{L}$, the inequality (7.21) holds, with some other constant c, lying in a fixed interval, not depending on the brick. Finally, perform the dilation, with some coefficient κ of the Δ_0 to some other brick Δ_1. Due to the homogeneity properties of I, J_l, these forms get coefficients, respectively, $\kappa^{-d}, \kappa^{2l-d}$ which are of the order, respectively meas Δ_1^{-1}, meas $\Delta_1^{\frac{2l}{d}-1}$. Setting this in (7.22), we get (7.21). $\qquad\square$

Lemma 7.4. *Let Δ be a brick, $0 < 2(l-r) < d$, $p \in L_q$, $q = \frac{d}{2(l-r)}$ and β be some multi-index of length r. Then there exists a subspace $\mathcal{L} \subset H^l(\Delta)$ of codimension less than $N(d,l,r)$ and constant $c(d,l,r)$ such that*

$$I_{p,\beta}[u] = \int\limits_{\Delta} p(x)|D^{\beta}u|^2 dx \leq c(d,l,r)\|p\|_q J_l[u], u \in \mathcal{L}. \tag{7.23}$$

Proof. First, let Δ_0 be a brick with a unit longest edge. We apply the Hölder inequality to $I_{p,\beta}$, $I_{p,\beta}[u] \leq \|p\|_q \|D^{\beta}u\|^2_{\frac{2q}{q-2}}$. Then from the Sobolev embedding theorem it follows that $I_{p,\beta}[u] \leq C\|p\|_q \|u\|^2_{H^l}$. Now, if u belongs to the subspace $\mathcal{L}$, the H^l norm can be replaced by its leading term, J_l, with a controlled worsening of the constant. This establishes (7.23) for Δ_0. In order to pass to an arbitrary brick, we make a dilation and note that both parts in (7.23) have the same homogeneity order[d]. $\qquad\square$

We use one more inequality of this sort when proving Proposition 7.4.

Lemma 7.5. *Let A be a second order self-adjoint elliptic differential operator with smooth coefficients in a brick Δ, with Dirichlet or Neumann boundary conditions. Then there exists a subspace $\mathcal{L}_A$ in the domain of A, having a finite codimension , such that*

$$\|Au\|^2 \geq \nu_0^{-2}C\|\nabla_2 u\|^2_2, u \in \mathcal{L}_A,$$

with constant C depending only on the dimension.

Proof. For a smooth elliptic operator, the coercive inequality holds, $\|Au\|^2 \geq C\nu_0^{-2}\|\nabla^2 u\|^2_2 - C'\|u\|^2_2, u \in \mathcal{D}(A)$. We apply Lemma 7.3 to find a subspace $\mathcal{L}$ where the second term can be omitted. $\qquad\square$

---

[d]It is this statement that fails in low dimensions, $d \leq 2(l-r)$ and this prevents one from having good eigenvalue estimates in such dimensions.

7.4.3. *Proof of CLR-type estimates*

We have two kinds of estimates. Some of them hold for t small enough only, others are uniform in t. The first kind of estimates is proved by the following reasoning.

Let $Q \subset \mathbb{R}^d$ be a cube. We study the quantity $n_\pm(t, I_{p,\beta},\, J_l, H_l(Q))$, $0 \leq |\beta| = r < l$, $2(l - r) < d$. Consider the functions of sets $\mathcal{J}_\pm(G) = \int_G p_\pm(x)^q dx$, $q = \frac{d}{2(l-r)}$ (separately for plus and minus). We apply to $\mathcal{J}_\pm$ our covering lemma, getting, for given n a covering Ξ with controlled multiplicity, by no more than n bricks, so that $\mathcal{J}_\pm(\Delta) \leq Cn^{-1}\mathcal{J}_\pm(Q)$ for any brick Δ. According to Lemma 7.3, for each brick Δ there exists a subspace $\mathcal{L}_\Delta \subset H^l(\Delta)$ where the inequality

$$I_{p,\beta,\Delta}[u] \leq Cn^{-q^{-1}}\mathcal{J}_\pm(Q)^{q^{-1}}J_{l,\Delta}[u] \tag{7.24}$$

holds, the codimension of $\mathcal{L}_\Delta$ bounded from above by some constant κ_1. Consider the subspace $\mathcal{L}_n$ in $H^l(Q)$ consisting of functions for which restriction to any brick Δ in the covering belongs to $\mathcal{L}(\Delta)$. Since each $\mathcal{L}(\Delta)$ is defined by no more than κ_1 orthogonality conditions, the subspace $\mathcal{L}_n$ is defined by no more than $n\kappa_1$ orthogonality conditions, so it has codimension no larger than $n\kappa_1$. Summing the inequalities of the form (7.24) over all bricks and taking into account the bound for multiplicity, we get $I_{p_\pm,\beta,Q}[u] \leq Cn^{-q^{-1}} \left(\int p_\pm(x)^q dx\right)^{q^{-1}} J_{l,Q}[u]$. Now, for a given $t \leq \mathcal{J}_\pm(Q)^{q^{-1}}$, we take $n \in [n_t, 2n_t], n_t = t^{-q}\mathcal{J}_\pm(Q)$. Then the construction above performed for this n produces a subspace $\mathcal{L}$ in $H^l(Q)$ of codimension not greater that $Cn \leq Ct^{-q}\mathcal{J}_\pm(Q)$ on which $I_{p_\pm,\beta,Q}[u] \leq CtJ_{l,Q}[u]$, and this, according to the variational principle, proves the CLR estimate

$$n_\pm(t, I_{p_\pm,\beta}, J_l, H^l(Q)) \leq Ct^{-q}\int_Q p_\pm^q dx, \tag{7.25}$$

for $t \leq \mathcal{J}_\pm(Q)^{q^{-1}}$.

All estimates we used in Lectures 2,3 follow from (7.25). We start with Theorem 7.3.

Suppose first that Ω is bounded. Then $n_\pm$ increases if we replace Ω by a cube $Q \supset \Omega$, with the weight p extended by zero. For $t \leq \mathcal{J}(Q)^{q^{-1}}$, $\beta = 0, l = 1$, (7.25) coincides with (7.11). In order to take care of large t, just note that from the Hölder and Sobolev inequalities it follows that $I_{p_\pm}[u] \leq C_q\mathcal{J}(Q)^{q^{-1}}J_1[u], u \in H_0^1(Q)$, thus for $t > C_q\mathcal{J}(Q)^{q^{-1}}$ we have $n_\pm(t, I_{p_\pm}, J_1, H_0^1(Q)) = 0$. Thus, by certain increasing the constant, we justify (7.11) for all t.

As for an unbounded domain, note first that the monotonicity property allows one to consider only the case $\Omega = \mathbb{R}^d$. We use the variational principle in the form (7.1). Suppose that for some fixed t, on some finite-dimensional subspace $\mathcal{L} \subset C_0^\infty(\mathbb{R}^d)$, the inequality in (7.1) holds. Then supports of all functions in $\mathcal{L}$ are contained in some cube Q, therefore, the above uniform estimate for this cube can be applied, which gives the required bound.

The estimate (7.18) follows from (7.25) due to monotonicity and the inequality $J_{\mathbf{a}}[u] \geq \nu_0 J_1[u]$.

Next, we pass to (7.16). We represent $K_j = (\partial_j A^{-\frac{1}{2}})A^{-\frac{1}{2}}$. The operator $\partial_j A^{-\frac{1}{2}}$ is bounded due to the inequality $||\partial_j u||^2 \leq \nu_0^{-1} J_{\mathbf{a}}[u]$, and therefore $n(t, K_j) \leq n(t\nu_0, A^{-\frac{1}{2}}) = n(t^2\nu_0^2, A^{-1})$. For the latter quantity we have the variational description as $n(t^2\nu_0^2, I_1, J_{\mathbf{a}}), H_0(\mathbf{a})$, and due to ellipticity condition $J_{\mathbf{a}}[u] \geq \nu_0 J_1[u]$. Now monotonicity and the estimate (7.11) give (7.16)

Next the quadratic form of the operator $L_{jk}^* L_{jk}$ in (7.15) can be written as $||b\partial_k A_\varepsilon^{-1} u||^2$, with $b = a_{jk} - a_{jk,\varepsilon}$. Setting here $u = A_\varepsilon v$, one can describe the distribution function of s-numbers of L_{jk} as $n(t, I_{|b|^2, \beta_k}, ||A_\varepsilon v||^2, H_0^2(Q))$ where β_k is the multi-index having 1 on the place k and zeros on other places. According to Lemma 7.5, on some subspace $\mathcal{L}_0 \subset H_0^2(Q)$ with finite codimension, the norm $||A_\varepsilon v||^2$ can be replaced by $C||\nabla^2 v||^2$. Now, on this subspace we apply (7.25) with $r = 1, l = 2, \beta = \beta_j$ and $b = p^2$. This gives the required estimate (7.17).

7.5. Lecture 5. Magnetic Schrödinger Operators and Semi-Group Domination

7.5.1. *Schrödinger operator and eigenvalue estimates*

One of the most important applications of the CLR-estimate was the bound for eigenvalues of the Schrödinger operator in $\mathbb{R}^d, d \geq 3$. Let $V \in L_{1,\mathrm{loc}}$ be a real function satisfying condition $V_+ \in L_{\frac{d}{2}}$. We consider the quadratic form $H[u] = H_V[u] = \int |\nabla u|^2 dx - \int V|u|^2 dx = J[u] - I_V[u]$. This form, as it follows from Hölder and Sobolev inequalities, is lower semi-bounded and defines the operator $H = H_V$ in $L_2(\mathbb{R}^d)$, the Schrödinger operator corresponding to the differential expression $Hu = -\Delta u - Vu$. We are interested in the number of negative eigenvalues of H. According to the variational principle, $N_-(0, H) = \max\{\dim \mathcal{L} \subset C_0^\infty, J[u] - I_V[u] < 0\}$.

Moving $I_V[u]$ to the left-hand side in the latter inequality, we get

$$N(0, H) = \max\{\dim \mathcal{L} \subset C_0^\infty, J[u] < I_V[u]\} = n_+(1, I_V, J, C_0^\infty). \quad (7.26)$$

Thus, the number of negative eigenvalues of the Schrödinger operator equals the number of eigenvalues in $(0, 1)$ of the weighted Laplacian . Applying here the CLR-estimate (7.11), we get

$$N(0, H_V) \leq C \int V_+(x)^{\frac{d}{2}} dx, \quad (7.27)$$

which is called *the* CLR-estimate in the literature. The easy transformation which led us to (7.26) is the most simple case of Birman-Schwinger principle [3], the powerful tool in the spectral analysis of Schrödinger-like operators. Replacing in (7.26) V by sV, $s \to \infty$, i.e., introducing the coupling constant, we get the equality $N(0, H_{sV}) = n_+(-1, sI_V, J, C_0^\infty) = n_+(s^{-1}, I_V, J, C_0^\infty)$, and Theorem 7.2 gives us the large coupling constant asymptotics of $N(0, H_{sV})$,

$$N(0, H_{sV}) = c_d s^{\frac{d}{2}} \int V_+^{\frac{d}{2}} dx + o(s^{\frac{d}{2}}). \quad (7.28)$$

7.5.2. *Semi-groups, Positivity*

Another object related to the Schrödinger operator is the heat semi-group.

If Z is a non-negative self-adjoint operator in the Hilbert space $\mathcal{H}$ then $U(t) = \exp(-Zt), t \geq 0$ is called the semi-group generated by Z. The semi-group property is the identity $U(t_1+t_2) = U(t_1)U(t_2), t_1, t_2 \geq 0$. Moreover, the semi-group is *contracting*, $||U(t)|| \leq 1$ and *strongly continuous*, $U(t)f \to U(t_0)f$ as $t \to 0$. The semi-group $U(t)$ solves the operator differential equation $\frac{d}{dt}U(t) + ZU(t) = 0, U(0) = I$. If the operator Z has discrete spectrum consisting of eigenvalues λ_j with eigenvectors φ_j, then $U(t)$ has eigenvalues $\exp(-t\lambda_j)$, with the same eigenvectors.

Semigroups are closely related to the resolvent of operators by the following identities,

$$\exp(-tZ) \quad (7.29)$$

$$= s - \lim((I + \frac{t}{n}Z)^{-1})^n, (\lambda + Z)^{-1} = \int_0^\infty \exp(-tZ)\exp(-t\lambda)dt.$$

Semigroups generated by second order elliptic operators with *real* coefficients have one additional property. The space $L_2(\Omega)$ has, in addition to the abstract Hilbert space structure, one more structure, the one of the space of functions, where the notion of *positivity* is present. So we call an operator K *positivity preserving* if it maps any (almost everywhere, a.e.) non-negative function into an a.e. non-negative function. If the operator is an integral one, this property is equivalent to the integral kernel being positive almost everywhere. It follows from (7.29) that for any operator $Z \geq 0$, the positivity preserving properties of the resolvent $(Z + \lambda)^{-1}$ for all $\lambda > 0$ and of the semi-group $\exp(-tZ), t > 0$ are equivalent.

Proposition 7.7. *If Z is an elliptic self-adjoint second order operator with real coefficients, with Dirichlet conditions, then the semi-group $U(t) = \exp(-tZ)$ is positive.*

Proof. The proof follows easily from the minimum principle for parabolic equations. In fact, if $f_0(x) \geq 0$ then $f(.,t) = U(t)f_0$ is the solution of the parabolic equation $\partial_t f + Zf = 0, f(.,0) = f_0$, and, according to the maximum principle, for any given T, the smallest value of f in the cylinder $\Omega \times [0,T]$ is attained on the boundary of Ω or for $t = 0$, and in both cases this is non-negative. $\qquad\square$

We are interested in the special case $Zu = -h\Delta(hu)$ where Δ is the Laplacian in $\mathbb{R}^d$ and h is a real function.

Let us show the relation of this operator to the eigenvalue problem for the weighted Laplacian, discussed in Lecture 2. Recall (7.7),(7.8) where the weighted Laplacian eigenvalues were related to the eigenvalues of the operator $S_p = L^{-1}pL^{-1}$, where $L = (-\Delta)^{\frac{1}{2}}$. For a positive, invertible [e] p, we represent S_p as the product M^*M where $M = p^{\frac{1}{2}}L^{-1}$. Nonzero eigenvalues of the operator M^*M coincide with eigenvalues of $K_p = MM^* = p^{\frac{1}{2}}(-\Delta)^{-1}p^{\frac{1}{2}}$. Therefore the eigenvalues of the original eigenvalue problem, which are inverses of the eigenvalues of K_p, coincide with eigenvalues of the operator $Z_p = K_p^{-1} = h(-\Delta)h$, with $h = p^{-\frac{1}{2}}$.

This relation enables one to derive estimates for the trace of the semi-group $U_p(t) = \exp(-Z_pt)$ from the CLR-estimate. Denote by μ_j the eigenvalues of Z_p and by $N(\lambda, Z_p)$ their distribution function. Integration by parts gives

$$\text{Tr } U_p(t) = \sum e^{-t\mu_j} = \int_0^\infty e^{-t\lambda}dN(\lambda, Z_p) = t\int_0^\infty e^{-t\lambda}N(\lambda, Z_p)d\lambda.$$

---

[e]We will consider only such p here. In [22] it is explained how to treat the general case.

Setting here the estimate from (7.11), we get

$$\text{Tr } U_p(t) \leq Ct \int_0^\infty e^{-t\lambda} \lambda^{\frac{d}{2}} d\lambda \int p^{\frac{d}{2}} dx = Ct^{-\frac{d}{2}} \int p^{\frac{d}{2}} dx. \tag{7.30}$$

With a certain worsening of the constant, one can invert the reasoning and derive (7.11) from (7.30). In fact, for given λ, set $t = \mu_n^{-1}$ in (7.30), where μ_n is the largest eigenvalue of Z below λ, $n = N(\lambda, Z_p)$. Replace the terms in the sum $\sum e^{-t\mu_j}$ with $j \leq n$ by smaller quantities $e^{-t\mu_n} = e^{-1}$ and delete the remaining terms – thus decreasing the sum. This gives $e^{-1}N(\lambda, Z_p) \leq C\lambda^{\frac{d}{2}} \int p^{\frac{d}{2}} dx$, i.e. the CLR-estimate (7.11). This relation was used in [16] to prove (7.11), and was extended to more general operators generating positivity preserving semigroups in [15].

7.5.3. *Semi-group Domination and Eigenvalue Estimates*

We are going to use this relation in order to prove the CLR-estimate for the magnetic Schrödinger operator. This estimate was first established by E. Lieb in [17] by means of rather complicated and fairly specific machinery involving path integration. We present here a general, at the same time rather elementary, approach to obtaining eigenvalue estimates, where the magnetic Schrödinger operator, serves just as a particular case.

Having two operators K, L in $L_2(\Omega)$, Ω being a space with measure, such that K is positivity preserving, we say that L is *dominated* by K if

$$|Lf|(x) \leq K|f|(x), f \in L_2(\Omega) \tag{7.31}$$

almost everywhere. If K, L are integral operators this is equivalent to the kernel domination $|L(x,y)| \leq K(x,y)$ for almost all $(x,y) \in \Omega \times \Omega$. Some properties of K are inherited by L, such as boundedness, compactness, Hilbert-Schmidt property, some others are not, in particular, from domination it does not follow that all eigenvalues of L are smaller than corresponding eigenvalues of K (just consider the matrix example $L = \begin{pmatrix} 1 & 0 \\ 0 & 1 \end{pmatrix}$, $K = \begin{pmatrix} 1 & 1 \\ 1 & 1 \end{pmatrix}$ where K has eigenvalues $\sqrt{2}, 0$ and L has eigenvalues $1, 1$). However if operators in question are semigroups $K = \exp(-tZ), L = \exp(-tY)$, then sufficiently regular eigenvalue estimates for Z imply similar eigenvalue estimates for Y. We present here a result from [22], where one can find it in a considerably more general setting.

Proposition 7.8 (Domination). *Let Z, Y be self-adjoint positive operators in $L_2(\Omega)$, the semi-group $U(t) = \exp(-tZ)$ is positivity preserving and dominates the semi-group $W(t) = \exp(-tY)$. Suppose also that for some $q > 0$, the eigenvalue estimate $N(\lambda, Z) \leq K\lambda^q$ holds for all $\lambda > 0$. Then the eigenvalues of the operator Y satisfy $N(\lambda, Y) \leq C(q)K\lambda^q$ for all $\lambda > 0$ with constant $C(q)$ depending only on q.*

Proof. As it is shown in (7.30), the eigenvalue estimate for Z implies $\mathrm{Tr}\, U(2t) \leq CKt^{-q}, t \in (0, \infty)$. One can also write this as $\|U(t)\|_{HS}^2 = \sum e^{-2t\mu_j(Z)} \leq CKt^{-q}$, where $\|.\|_{HS}$ is the Hilbert-Schmidt norm of the operator (see, e.g., [11]). Now recall that the square of the Hilbert-Schmidt norm of the operator equals the integral of the square of the absolute value of its kernel. Thus, due to domination, $W(t)$ is also a Hilbert-Schmidt operator, moreover $\|W(t)\|_{HS}^2 \leq \|U(t)\|_{HS}^2$ which gives $\mathrm{Tr}\, W(t) \leq CKt^{-q}$. Using the reasoning as in the end of the previous subsection, we obtain the required estimate for eigenvalues of Y. $\qquad\square$

7.5.4. *Magnetic Schrödinger Operator*

The Schrödinger operator describing the motion of an electron in the magnetic field is defined in the following way. Let $\mathbf{a}$ be a real vector-function in $\mathbb{R}^d$ with d components. Formally, the operator corresponds to the differential expression $-\Delta_{\mathbf{a}} u = -(\nabla + i\mathbf{a})^2 = -\sum(\partial_j + ia_j)^2$. The vector field $\mathbf{a}$ is called the magnetic potential and the matrix $\mathbf{b}, b_{jk} = \partial_j a_k - \partial_k a_j$ is the magnetic field itself. The definition of $H_{\mathbf{a}}$ as a self-adjoint operator in the Hilbert space requires certain conditions imposed on $\mathbf{a}$. If $\mathbf{a} \in L_{2,\mathrm{loc}}(\mathbb{R}^d)$, one can consider the quadratic form $-\Delta_{\mathbf{a}}[u] = \sum\|\partial_j u + ia_j u\|^2$ first on $C_0^\infty(\mathbb{R}^d)$, an then on $H_{\mathbf{a}}$, the closure of C_0^∞ in the norm $-\Delta_{\mathbf{a}}[u]$. This form defines the operator $-\Delta_{\mathbf{a}}$. The following fact, fundamental for physics, is called *the diamagnetic inequality*.

Proposition 7.9 (Diamagnetic Inequality). *If $\mathbf{a} \in L_{2,\mathrm{loc}}$ then the semi-group $\exp(t\Delta)$ dominates the semi-group $\exp(t\Delta_{\mathbf{a}})$, $t > 0$*

There are several ways to establish the diamagnetic inequality. The proof we give here (first proposed in [26]) is not the simplest one but, probably, the most enlightening. It is based on an important abstract result from Operator Theory [14].

Theorem 7.5 (Trotter-Kato-Masuda formula). *Let $A_j \geq 0, j = 1, ..., k$ be self-adjoint operators so that the sum $A = A_1 + ... + A_k$ is defined* *in the sense of quadratic forms. Then*

$$\exp(-tA) = s - \lim_{n\to\infty} (\exp(-\frac{t}{n}A_1)\exp(-\frac{t}{n}A_2)....\exp(-\frac{t}{n}A_k))^n$$

Now we are able to prove Proposition 7.9. For j fixed, define $\psi_j(x) = \psi_j(x_1,...,x_j,...,x_d) = \int_0^{x_j} a_j(x_1,...,\xi,...,x_d)d\xi$. Then

$$e^{-i\psi_j}\partial_j e^{i\psi_j} = \tilde{\partial}_j = \partial_j + ia_j. \tag{7.32}$$

Thus the operator $-\Delta_{\mathbf{a}}$ can be represented as a sum (in the sense of quadratic forms) of operators $-\tilde{\partial}_j^2$. We apply Theorem 7.5 to express the semi-group generated by $-\Delta_{\mathbf{a}}$.

$$e^{t\Delta_{\mathbf{a}}} = \lim_{n\to\infty} \left(e^{-i\psi_1}U_1(\frac{t}{n})e^{i(\psi_1-\psi_2)}U_2(\frac{t}{n})...U_d(\frac{t}{n})e^{i\psi_d} \right)^n, \tag{7.33}$$

where $U_j(t) = e^{t\partial_j^2}$. Now note that each $U_j(t)$ is an integral operator with positive kernel, and the expression in (7.33) is a composition of several (rather many) such operators and multiplications by functions having absolute value 1. If we replace these exponents by their absolute value, in other words delete them, the value of the integral can only increase, and this gives $|e^{t\Delta_{\mathbf{a}}}(x,y)| \le e^{t\Delta}(x,y)$ as we need.

The same reasoning gives us a somewhat more general result.

Corollary 7.1. *Let $h > 0$ be a function in $L_{\infty,\mathrm{loc}}(\mathbb{R}^d)$. Consider operators $Z_h = -h\Delta h$, $Y_h = -h\Delta_{\mathbf{a}}h$ (defined by means of quadratic forms). Then the semi-group $\exp(-Z_h)$ dominates $\exp -Y_h$.*

In fact, we can apply Theorem 7.5 to operators $h\partial_j^2 h$ and $h\tilde{\partial}_j^2 h$, for which relation similar to (7.32) also holds.

7.5.5. *The Spectrum of the Magnetic Schrödinger Operator*

Now we can derive results on eigenvalue estimates and asymptotics for the magnetic Schrödinger operator.

Theorem 7.6 (CLR estimate). *For $\mathbf{a} \in L_{2,\mathrm{loc}}$ and $V_+ \in L_{\frac{d}{2}}$,*

$$N(0, -\Delta_{\mathbf{a}} - V) \le C_d \int V_+^{\frac{d}{2}}\,dx \tag{7.34}$$

with constant C_d depending only on the dimension.

Proof. First, note that it is sufficient to consider the case of non-negative V, since, due to the monotonicity in the variational principle, $N(0, -\Delta_{\mathbf{a}} - V) \leq N(0, -\Delta_{\mathbf{a}} - V_+)$. Next consider a function $\tilde{V} \geq V$, $\tilde{V} \in L_{\frac{d}{2}}, (\tilde{V})^{-1} \in L_{\infty,\text{loc}}, \int \tilde{V}^{\frac{d}{2}} \leq 2 \int V^{\frac{d}{2}}$. Then, again due to monotonicity, $N(0, -\Delta_{\mathbf{a}} - V) \leq N(0, -\Delta_{\mathbf{a}} - \tilde{V})$. Set $h(x) = \tilde{V}(x)^{-\frac{1}{2}}$. Then, in notations of Corollary 7.1, the semi-group $\exp(-Z_h)$ dominates the semi-group $\exp(-Y_h)$. For Z_h, according to (7.11), we have the CLR-estimate $N(\lambda, Z_h) \leq C(d)\lambda^{\frac{d}{2}} \int \tilde{V}^{\frac{d}{2}} dx \leq 2C(d)\lambda^{\frac{d}{2}} \int V^{\frac{d}{2}} dx$. Now we can apply Proposition 7.8 which gives us the same estimate, just with a worse constant, for eigenvalues of Y_h. In particular, it holds for $\lambda = 1$, $N(1, Y_h) \leq C \int V^{\frac{d}{2}} dx$. At last we pass from eigenvalue estimates for Y_h to estimates for $-\Delta_{\mathbf{a}} - V$, by means of Birman-Schwinger principle. $\square$

Note here that, generally, one can't say directly that $N(0, -\Delta_{\mathbf{a}} - V) \leq CN(0, -\Delta - V)$. It is the eigenvalue estimate for $-\Delta - sV$ for all s that gives us the estimate for the magnetic Schrödinger operator.

Having the magnetic version of the CLR-estimate, we can now establish the asymptotic formula for eigenvalues.

Theorem 7.7. *Let* $\mathbf{a} \in L_{d,\text{loc}}(\mathbb{R}^d), d \geq 3, p \in L_{\frac{d}{2}}(\mathbb{R}^d)$. *Then for the eigenvalues of the problem*

$$-\Delta_{\mathbf{a}} u = \lambda p(x) u \tag{7.35}$$

in $\mathbb{R}^d$ *the asymptotic formula*

$$\lim_{\lambda \to \infty} N^{\pm}(\lambda, -\Delta_{\mathbf{a}}, p)\lambda^{-\frac{d}{2}} = c_d \int p_{\pm}(x)^{\frac{d}{2}} dx,$$

where c_d *is the constant in (5). Thus, for the weighted magnetic Laplacian the spectral asymptotics does not depend on the magnetic field in* $L_{d,\text{loc}}$.

Note that, due to the Birman-Schwinger principle, the last result gives the strong coupling asymptotics for the negative spectrum of the magnetic Schrödinger operator $H_{\mathbf{a},sV} = -\Delta_{\mathbf{a}} - sV$, see (7.28),

$$N(0, H_{\mathbf{a},sV}) = c_d s^{\frac{d}{2}} \int V_+^{\frac{d}{2}} dx + o(s^{\frac{d}{2}}).$$

Proof. We follow the pattern of the proof of Theorem 7.2. The spectrum of (7.35) is described by the distribution function $n_{\pm}(t, I_p, -\Delta_{\mathbf{a}}, C_0^{\infty}(\mathbb{R}^d))$, where $-\Delta_{\mathbf{a}}[u]$ is the quadratic form of the magnetic Laplacian, which, via integration by parts, equals

$$-\Delta_{\mathbf{a}}[u] = (-\Delta_{\mathbf{a}} u, u) = \int (|\nabla u|^2 + 2\Im(\overline{u}(\mathbf{a}\nabla u)) + |\mathbf{a}|^2 |u|^2) dx. \tag{7.36}$$

Since we have the CLR estimate for the magnetic Laplacian, we can perform Steps 2,3,4 in the proof of the Theorem 7.2, as soon as we have made Step 1, i.e., established the asymptotic formulas for Dirichlet and Neumann problems in a cube Q with a constant (unit) weight p. To do this, it is sufficient to show that the second and third terms in the quadratic form $J_{\mathbf{a}}$, see (7.36), in the cube are weak with respect to the first, leading term, in the sense of Proposition 7.1. The second term is reduced to the third one, since

$$| \int_{\mathbf{Q}} 2\Im(\overline{u}(\mathbf{a}\nabla u))dx| \leq \varepsilon \int_{Q} |\nabla u|^2 dx + \varepsilon^{-1} \int_{Q} |\mathbf{a}|^2 |u|^2 dx.$$

As for the third term, we have $|\mathbf{a}|^2 \in L_{\frac{d}{2}}(Q)$, and from the CLR estimate in the cube follows, in particular, that $n_{\pm}(\varepsilon, I_{|\mathbf{a}|^2}, J_1, H^1(Q))$ is finite for any $\varepsilon > 0$, in other words, due to the variational principle, there exists a subspace $\mathcal{L}_{\varepsilon} \in H^1(Q)$ where $I_{|\mathbf{a}|^2}[u] \leq \varepsilon J_1[u]$. Now it remains to apply Proposition 7.1. $\qquad\qquad\square$

Notes on the Literature

There is a huge literature on estimates and asymptotics of the spectrum of elliptic operators. Probably, the most complete presentation of results obtained before 1989 and respective methods, with exhaustive bibliography, can be found in [23]. The variational approach to the study of eigenvalues is the only one which can handle very singular problems. It, however, cannot give precise remainder estimates in asymptotic formulas or next terms in asymptotics. The advanced technic for this kind of problems is presented in [13].

The results of Lecture 2 are obtained in [5] (for bounded domains) and in [21] for the general case. The results in Lecture 3 admit very far generalizations, see [6]. One can consider elliptic systems of any order, with possible degeneration of ellipticity, and any self-adjoint differential operator of a lower order standing in the place of the weight function. In certain cases it is even possible to take into account cancellation of singularities in asymptotic formulas.

The proof of CLR type estimates in Lecture 4 follows [21] and is an extended version of the presentation in [18].

The semigroup domination approach to the study of the magnetic Schrödinger operator presented in Lecture 5 originates from [18] and was extended to an abstract situation in [22]. The condition of the magnetic potential belonging to $L_{2,\text{loc}}$ is not very restrictive but still it excludes some important examples, in particular, the Aharonov-Bohm magnetic field. Recently the CLR estimate and eigenvalue asymptotics for this case was found in [19].

A comprehensive exposition of the discrete spectrum analysis of the Schrödinger operator is given in [20]. One is directed to [11] and [25] for the theory of compact operators, with a lot of useful inequalities for eigenvalues and singular numbers, of which the Weyl and Fan inequalities are just the first examples.

References

[1] S. Agmon *On kernels, eigenvalues and eigenfunctions of operators related to elliptic problems* Comm. Pure Appl. Math., 18(4) (1965), 627-663

[2] A. Besikovitch *A general form of the covering principle* Proc. Cambr. Phil. Soc., 41 (1945) 103-110

[3] M. Birman *On the spectrum of singular boundary value problems* Am. Math. Soc. Transl.,II. Ser., 53 (1966), 23-60

[4] M. Birman, *Scattering problems for differential operators with constant coefficients* Func. Anal. and Appl., 3(3) (1969), 1-16

[5] M. Birman, M. Solomyak *On the leading term of the spectral asymptotics for non-smooth elliptic problems.* Func. Anal. and Appl., 4 (1971), 265-275

[6] M. Birman, M. Solomyak *Spectral asymptotics of non-smooth elliptic operators* Trans. Moscow Math. Soc. 27 (1975), 1-52; 28 (1975), 1-32

[7] M.Birman, M. Solomyak, *Quantitative analysis in Sobolev imbeding theorems and applications to spectral theory* AMS Transl. Vol 114, AMS, (1980)

[8] M. Birman, M. Solomyak *Spectral Theory of Self-Adjoint operators in Hilbert Space*, Kluver, 1987

[9] R. Courant, D. Hilbert *Methods of Mathematical Physics* Interscience, 1953

[10] M. Cwikel *Weak type estimates for singular values and the number of bound states of Schrödinger operator* Ann. Math., 206 (1977), 93-100

[11] I. Gohberg, M. Krein *Introduction to the Theory of Linear Nonselfadjoint Operators* AMS Transl. Math. Monogr. 18, AMS, 1969

[12] M. de Guzmán *Differentiation of Integrals in R^n*. Lecture Notes in Math., Vol. 481. Springer, 1975.

[13] V. Ivrii *Microlocal Analysis and Precise Spectral Asymptotics* Springer, 1998

[14] T. Kato, K. Masuda *Trotter's product formula for nonlinear semigroups generated by the subdifferentials of convex functionals.* J. Math. Soc. Japan 30 (1978), 169–178.

[15] D. Levin, M.Solomyak, *The Rozenblum-Lieb-Cwikel inequality for Markov generators* J. d'Analyse Math. 71 (1997) 173-193

[16] P.Li, S.T. Yau *On the Schrödinger operator and eigenvalue problem* Comm. Math. Phys, 88 (1983), 309-318

[17] E. Lieb, *Bounds for eigenvalues of Laplace and Schrödinger operators* Bull. AMS, 82(5) (1976), 751-753

[18] M. Melgaard, G. Rozenblum *Spectral estimates for magnetic operators* Math. Scand., 79 (1996), 237-254

[19] M. Melgaard, E. Ouhabaz, G. Rozenblum *Negative discrete spectrum of perturbed multivortex Aharonov-Bohm Hamiltonians* Ann. Henri Poincaré, 5 (2004), 979-1012.

[20] M. Reed, B. Simon *Methods of Modern Mathematical Physics* Voll1, 1972; vol 4, 1979, Academic Press

[21] G. Rozenblum *The distribution of discrete spectrum of singular differential operators* Sov. Math. Dokl., 13 (1972), 245-249; Soviet Mathematics, Izv. VUZ 20(1) (1976), 63-71

[22] G. Rozenblum *Domination of semigroups and estimates for eigenvalues.* St. Petersburg Math. J. 12 (2001), no. 5, 831–845

[23] G. Rozenblum, M. Shubin, M. Solomyak *Spectral Theory for Differential Operators.* Partial Differential Equations, 7. Encyclopedia of Mathematical Sciences, 64, Springer, 1994

[24] G. Rozenblum, M. Solomyak *The Cwikel-Lieb-Rozenblum estimate for generators of positive semigroups and semigroups dominated by positive semigroups* St. Petersburg Math. J. 9 (1998), no. 6, 1195–1211

[25] B. Simon *Trace Ideals and its Applications,* Cambridge Univ., 1979

[26] B. Simon *Maximal and minimal Schrödinger forms,* J. Oper. Theory, 1 (1979), 37-47

[27] J. Weidmann *Linear operators in Hilbert spaces* Graduate Texts in Mathematics, 68. Springer-Verlag, New York-Berlin, 1980.

[28] H. Weyl *Das asymptotische Verteilungsgesetz der Eigenverte linearer partieller Differentialgleichungen* Math. Ann. 71 (1912) 441-479

Chapter 8

A glimpse of the theory of nonlinear semigroups

Edoardo Vesentini

Dipartimento di Matematica
Politecnico di Torino
Corso Duca degli Abruzzi 24
10129 Torino, Italy

Contents

Before acquiring the status of an autonomous geometrical chapter of mathematics, the theory of one-parameter linear semigroups found some of its most relevant motivations[a] in problems related to linear differential and integral operators.

Deep questions in a variety of fields of pure and applied research are at the ouset of a theory of non-linear semigroups that should eventually become an autonomous and self-contained mathematical theory. However, this goal seems now to be still far away, at least in its most general context[b].

The main purpose of these notes (which are a concise résumé of a few lectures delivered in the University of Padua in the Spring of 2002) is to sketch, following essentially the current literature, an approach to one-parameter semigroups of non-linear operators along the same lines followed in the linear case.

---

[a] As exposed, for example, in [6], [8], [5].

[b] With the relevant exception of non-linear semigroups in real Hilbert spaces, [1]. See also [2].

257

8.1. Introduction to the spectral theory for nonlinear operators

Let $\mathcal{E}$ and $\mathcal{F}$ be two complex Banach spaces with norms $\|\ \|_{\mathcal{E}}$ and $\|\ \|_{\mathcal{F}}$, and let K be a subset of $\mathcal{E}$.

The set $\mathrm{Lip}(K, \mathcal{F})$ of all maps $f : K \to \mathcal{F}$ such that

$$p_L(f) := \sup \left\{ \frac{\|f(x) - f(y)\|_{\mathcal{F}}}{\|x - y\|_{\mathcal{E}}} : x, y \in K, x \neq y \right\} < \infty$$

is a complex vector space: the space of all *lipschitz maps* $K \to \mathcal{F}$; p_L is a seminorm on $\mathrm{Lip}(K, \mathcal{F})$.

Chosen any $x_0 \in K$, the map

$$\mathrm{Lip}(K, \mathcal{F}) \ni f \mapsto \|f(x_0)\|_{\mathcal{F}} + p_L(f)$$

is a norm on $\mathrm{Lip}(K, \mathcal{F})$ with respect to which $\mathrm{Lip}(K, \mathcal{F})$ is a Banach space.

Any $f \in \mathrm{Lip}(K, \mathcal{F})$ is uniformly continuous on K.

If $A \in \mathcal{L}(\mathcal{E}, \mathcal{F})$ (the Banach space of all bounded linear maps $\mathcal{E} \to \mathcal{F}$), then $A \in \mathrm{Lip}(K, \mathcal{F})$ and

$$p_L(A) = \sup \left\{ \frac{\|Ax\|_{\mathcal{F}}}{\|x\|_{\mathcal{E}}} : x \in \mathcal{E} \backslash \{0\} \right\} = \|A\|.$$

If $\mathcal{E} = \mathcal{F}$, $\|\ \|_{\mathcal{E}}$, $\|\ \|_{\mathcal{F}}$ and $\mathrm{Lip}(K, \mathcal{F})$ will be replaced by $\|\ \|$ and by $\mathrm{Lip}(K)$.

If $f \in \mathrm{Lip}(K)$ and $g \in \mathrm{Lip}(f(K))$, then $g \circ f \in \mathrm{Lip}(K)$ and

$$p_L(g \circ f) \leq p_L(f) p_L(g).$$

Let $\{f_\nu\}$ be a sequence of maps $K \to \mathcal{E}$ all of whose elements are contained in $\mathrm{Lip}(K)$.

Lemma 8.1. *If*

$$f(x) = \lim_{\nu \to \infty} f_\nu(x)$$

exists for all $x \in K$, and if there is a constant $M > 0$ such that

$$p_L(f_\nu) \leq M$$

for all ν, then $f \in \mathrm{Lip}(K)$ and $p_L(f) \leq M$.

Proof. Let $x \neq y$ in K, and, for any $\epsilon > 0$, let ν be such that

$$\|f(x) - f_\nu(x)\| < \frac{\epsilon}{2}, \quad \|f(y) - f_\nu(y)\| < \frac{\epsilon}{2}.$$

Then,

$$\frac{\|f(x) - f(y)\|}{\|x - y\|} \leq \frac{\|(f(x) - f_\nu(x)) - (f(y) - f_\nu(y))\|}{\|x - y\|} +$$
$$\frac{\|f_\nu(x) - f_\nu(y)\|}{\|x - y\|} < \epsilon + M.$$

$\square$

Proposition 8.1. *If* $p_L(f) = k < 1$, *then:*
1) $I - f$ *is injective, and*

$$p_L(I - f) \leq \frac{1}{1 - p_L(f)}; \tag{8.1}$$

2) if $\overline{B(x_o, r)} \subset K$, *then*

$$(I - f)(\overline{B(x_o, r)}) \supset \overline{B(x_o - f(x_o), r(1 - k))};$$

3) if $K = \mathcal{E}$, $I - f$ *is bijective.*

Proof.   1) For $x_1, x_2 \in K$,

$$\|x_1 - x_2\| \leq \|(x_1 - f(x_1)) - (x_2 - f(x_2))\| + \|f(x_1) - f(x_2)\|$$
$$\leq \|(x_1 - f(x_1)) - (x_2 - f(x_2))\| + k\|x_1 - x_2\|,$$

i.e.,

$$\|x_1 - x_2\| \leq \frac{1}{1 - k}\|(x_1 - f(x_1)) - (x_2 - f(x_2))\|.$$

2) Choose any $y \in \overline{B(x_o - f(x_o), r(1 - k))}$, and let $g : \overline{B(x_o, r)} \to \mathcal{E}$ be defined by

$$g(x) = f(x) + y.$$

Then, $g \in \mathrm{Lip}(\overline{B(x_o, r)})$ and $p_L(g) = k$.
Since moreover

$$\|g(x) - x_o\| \leq \|f(x) - f(x_o)\| + \|y - x_o + f(x_o)\|$$
$$\leq kr + (1 - k)r = r,$$

then

$$g(\overline{B(x_o, r)}) \subset \overline{B(x_o, r)}.$$

3) Since $K = \mathcal{E}$, 1) and 2) yield 3).

$\square$

In the following $2^{\mathcal{E}}$ and $2^{\mathcal{F}}$ will stand for the sets of all subsets of the complex Banach spaces $\mathcal{E}$ and $\mathcal{F}$.

Any subset Γ of $\mathcal{E} \times \mathcal{F}$ defines two subsets

$$\mathcal{D}(\hat{\Gamma}) \subset \mathcal{E} : \{x \in \mathcal{E} : \exists y \in \mathcal{F} \text{ such that } (x,y) \in \Gamma\},$$

$$\mathcal{R}(\hat{\Gamma}) \subset \mathcal{F} : \{y \in \mathcal{F} : \exists x \in \mathcal{E} \text{ such that } (x,y) \in \Gamma\},$$

and a set-valued map

$$\hat{\Gamma} : \mathcal{D}(\hat{\Gamma}) \to 2^{\mathcal{F}} \backslash \{\emptyset\}$$

defined on $x \in \mathcal{D}(\hat{\Gamma})$ by

$$x \mapsto \hat{\Gamma}(x) = \{y \in \mathcal{F} : (x,y) \in \Gamma\}.$$

Vice versa, given a subset E of $\mathcal{E}$ and a set-valued map

$$A : E \to 2^{\mathcal{F}} \backslash \{\emptyset\},$$

the graph of A, *i.e.* the subset of $\mathcal{E} \times \mathcal{F}$ defined by

$$G_A : \{(x,y) : x \in E, y \in A(x)\},$$

is such that

$$A = \widehat{G_A}.$$

In the following, we will be dealing with the case $\mathcal{E} = \mathcal{F}$.

Let $\Upsilon(\mathcal{E})$ be the set of all set-valued maps

$$\mathcal{E} \supset E \to 2^{\mathcal{E}} \backslash \{\emptyset\}.$$

For any $A \in \Upsilon(\mathcal{E})$, the *resolvent set* $r(A)$ of A is, by definition, the set of all $\zeta \in \mathbf{C}$ such that:

$\zeta I - A$ has a dense image in $\mathcal{E}$. That is to say, for any $y \in \mathcal{E}$ and any $\epsilon > 0$ there are $x \in \mathcal{D}(A)$ and $z \in A(x)$ such that

$$\|(\zeta x - z) - y\| < \epsilon;$$

$\zeta I - A$ is injective;

$$(\zeta I - A)^{-1} \in \mathrm{Lip}(\mathcal{M}_\zeta), \tag{8.2}$$

where

$$\mathcal{M}_\zeta = \cup\{\zeta I - A(x) : x \in \mathcal{D}(A)\}.$$

The set $\sigma(A) := \mathbf{C}\backslash r(A)$ is called the *spectrum of A*.

If A is a linear operator $\mathcal{E} \to \mathcal{E}$, $\sigma(A)$ is the spectrum of A as defined in [10] (see also [9]).

Let $\zeta \in r(A)$. In view of (8.2), there exists $M > 0$ such that

$$\left\| (\zeta I - A)^{-1}(y_1) - (\zeta I - A)^{-1}(y_2) \right\| \le M \left\| y_1 - y_2 \right\| \quad \forall \, y_1, y_2 \in \mathcal{M}_\zeta.$$

Letting

$$x_j = (\zeta I - A)^{-1}(y_j) \in \mathcal{D}(A),$$

there exist $z_j \in A(x_j)$ for which

$$y_j = \zeta x_j - z_j.$$

For every $y \in \mathcal{E}$ there are sequences $\{x_\nu\}$ and $\{z_\nu\}$, with $x_\nu \in \mathcal{D}(A)$ and $z_\nu \in A(x_\nu)$, such that

$$\lim_{\nu \to \infty} (\zeta x_\nu - z_\nu) = y. \tag{8.3}$$

Since

$$\left\| x_\nu - x_\mu \right\| \le M \left\| (\zeta x_\nu - z_\nu) - (\zeta x_\mu - z_\mu) \right\|,$$

$\{x_\nu\}$ is a Cauchy sequence. By (8.3), also $\{z_\nu\}$ is a Cauchy sequence.

If

$$x = \lim_{\nu \to \infty} x_\nu, \quad z = \lim_{\nu \to \infty} z_\nu,$$

then

$$y = \zeta x - z.$$

If the graph $G_{\zeta I - A}$ of $\zeta I - A$ is closed, then $(x, z) \in G_{\zeta I - A}$, that is, $x \in \mathcal{D}(A)$ and $z \in A(x)$.

Since, if G_A is closed, $G_{\zeta I - A}$ is closed, that proves

Lemma 8.2. *If A is closed (i.e., G_A is closed), then $r(A)$ consists of all $\zeta \in \mathbf{C}$ for which: $\mathcal{M}_\zeta = \mathcal{E}$, $\zeta I - A$ is injective, and*

$$(\zeta I - A)^{-1} \in \mathrm{Lip}(\mathcal{E}).$$

The following lemma is a further consequence of the above arguments.

Lemma 8.3. *If $\overline{\mathcal{M}_\zeta} = \mathcal{E}$ for some $\zeta \in r(A)$, then $\zeta I - \overline{A}$ is surjective.*

Theorem 8.1. *For any $A \in \Upsilon(\mathcal{E})$,*

$$r(A) \subset r(\overline{A}),$$

or equivalently,

$$\sigma(\overline{A}) \subset \sigma(A).$$

Proof. In view of Lemma 8.3, to show that, if $\zeta \in r(A)$, $\zeta I - \overline{A}$ is bijective, we need only prove that $\zeta I - \overline{A}$ is injective.

Let $x_1, x_2 \in \mathcal{D}(\overline{A})$ and let $z_1 \in \overline{A}(x_1)$, $z_2 \in \overline{A}(x_2)$ be such that

$$\zeta x_1 - z_1 = \zeta x_2 - z_2. \tag{8.4}$$

Since

$$G_{\zeta I - \overline{A}} = \overline{G_{\zeta I - A}},$$

there exist sequences $\{x_1^\nu\}$, $\{x_2^\nu\}$ in $\mathcal{D}(A)$ and sequences $\{z_1^\nu\}$, $\{z_2^\nu\}$ with $z_1^\nu \in A(x_1^\nu)$, $z_2^\nu \in A(x_2^\nu)$, converging respectively to x_1, x_2 and z_1, z_2.

The inequality

$$\|x_1^\nu - x_2^\nu\| \leq M\|(\zeta x_1^\nu - z_1^\nu) - (\zeta x_2^\nu - z_2^\nu)\| \tag{8.5}$$

and (8.4) imply that $x_1 = x_2$ and $z_1 = z_2$.

Thus, for any $\zeta \in r(A)$, $\zeta I - \overline{A}$ is bijective.

We show now that the map $(\zeta I - \overline{A})^{-1}$ is lipschitz.

For any choice of $x_1, x_2 \in \mathcal{D}(\overline{A})$ and $z_1 \in \overline{A}(x_1)$, $z_2 \in \overline{A}(x_2)$, let $\{x_1^\nu\}$, $\{x_2^\nu\}$ in $\mathcal{D}(A)$ and $\{z_1^\nu\}$, $\{z_2^\nu\}$ with $z_1^\nu \in A(x_1^\nu)$, $z_2^\nu \in A(x_2^\nu)$, be sequences converging respectively to x_1, x_2 and z_1, z_2. The equation (8.5), implies that

$$\|x_1 - x_2\| \leq M\|(\zeta x_1 - z_1) - (\zeta x_2 - z_2)\|,$$

proving thereby that

$$(\zeta I - \overline{A})^{-1} \in \mathrm{Lip}(\mathcal{E}).$$

Summing up, $\zeta \in r(\overline{A})$. $\qquad\qquad\qquad\qquad\qquad\qquad\qquad\qquad\square$

Let $\zeta \in r(\overline{A})$. For every $y \in \mathcal{E}$ there exist a unique $x \in \mathcal{D}(\overline{A})$ and a unique $z \in \overline{A}(x)$ such that $\zeta x - z = y$, and the map $y \mapsto x$ is lipschitz. Hence, if $x \in \mathcal{D}(A)$ for every y in the range of $\zeta I - A$, then $\zeta \in r(A)$.

This latter hypothesis is satisfied if the following condition holds:
i) *for every $x \in \mathcal{D}(\overline{A})$ and for every sequence $\{x_\nu\}$ in $\mathcal{D}(A)$ converging to x and such that for every ν there is $y_\nu \in A(x_\nu)$ for which $\{y_\nu\}$ converges to some $y \in \mathcal{E}$, then $x \in \mathcal{D}(A)$ and $A(x) \ni y$.*

If A is a linear operator, this hypothesis is equivalent to requiring that A is *closable*, *i.e.* such that, for every sequence $\{x_\nu\}$ in $\mathcal{D}(A)$ converging to 0 and for which $\{A(x_\nu)\}$ converges to $y \in \mathcal{E}$, then $y = 0$.

Extending this definition to the non-linear case, any operator $A \in \Upsilon(\mathcal{E})$ satisfying condition i) will be called *closable*, and the following theorem holds, which extends a well known fact in the linear case[c]

Theorem 8.2. *If $A \in \Upsilon(\mathcal{E})$ is closable, then $\sigma(A) = \sigma(\overline{A})$.*

For any $\zeta_o \in r(\overline{A})$ let

$$s_o = \frac{1}{p_L((\zeta_o I - \overline{A})^{-1})}$$

and let $\Delta(\zeta_o, s_o)$ be the open disc in $\mathbb{C}$ with center ζ_0 and radius s_0. The following proposition implies that $r(\overline{A})$ is open, or, equivalently, the spectrum of $\overline{A}$ is closed.

Proposition 8.2. *For any $\zeta_0 \in r(\overline{A})$,*

$$\Delta(\zeta_o, s_o) \subset r(\overline{A}),$$

and for any $\zeta \in \Delta(\zeta_o, s_o)$,

$$(\zeta I - \overline{A})^{-1} = (\zeta_o I - \overline{A})^{-1} \circ \left(I + (\zeta - \zeta_o)(\zeta_o I - \overline{A})^{-1} \right)^{-1}$$

for any $\zeta \in \Delta(\zeta_o, s_o)$.

Proof. Being

$$|\zeta - \zeta_o| \, p_L((\zeta_o I - \overline{A})^{-1}) < 1, \tag{8.6}$$

by Proposition 8.1

$$\left(I + (\zeta - \zeta_o)(\zeta_o I - \overline{A})^{-1} \right)^{-1} \in \mathrm{Lip}(\mathcal{E}).$$

Letting

$$F = (\zeta_o I - \overline{A})^{-1} \circ \left(I + (\zeta - \zeta_o)(\zeta_o I - \overline{A})^{-1} \right)^{-1} \in \mathrm{Lip}(\mathcal{E}),$$

we have to show that $F = (\zeta I - \overline{A})^{-1}$; that is to say, for all $x \in \mathcal{E}$, the equation

$$\zeta u - x \in \overline{A}(u) \tag{8.7}$$

---

[c]See, e.g., [9], Theorem 1.9.8., pp. 55-56. The hypothesis whereby $A \in \Upsilon(\mathcal{E})$ is closable was omitted in [9]. Hence, Theorem 1.18.3 of [9], must be replaced by Theorems 8.1 and 8.2.

has the unique solution $u = \overline{A}(x)$.

To establish uniqueness, assume that $\zeta u - x \in \overline{A}(u)$, and let $v = \zeta u - x$. Because $v \in \overline{A}(u)$, then

$$(\zeta_o I - \overline{A})^{-1}(\zeta_o u - v) = u,$$

i.e.,

$$(\zeta_o I - \overline{A})^{-1}(x + (\zeta_o - \zeta)u) = u.$$

Thus u is fixed by the map $C : \mathcal{E} \to \mathcal{E}$ defined by

$$y \mapsto C(y) = (\zeta_o I - \overline{A})^{-1}(x + (\zeta_o - \zeta)y).$$

Since, by (8.6), $p_L(C) < 1$,

$$p_L(C^n) \leq (p_L(C))^n < 1$$

for $n = 1, 2, \ldots$. The classical Banach fixed point theorem for strict contractions of a complete metric space implies that u is the unique fixed point of C (see, *e.g.*, [4]).

To show that $u = F(x)$, let

$$z = \left(I + (\zeta - \zeta_o)(\zeta_o I - \overline{A})^{-1}\right)^{-1}(x),$$

so that

$$(\zeta_o I - \overline{A})^{-1}(z) = F(x),$$

then

$$\zeta F(x) - x = \zeta(\zeta_o I - \overline{A})^{-1}(z) - \left(I + (\zeta - \zeta_o)(\zeta_o I - \overline{A})^{-1}\right)(z)$$
$$= \zeta_o(\zeta_o I - \overline{A})^{-1}(z) \in \overline{A}\left((\zeta_o I - \overline{A})^{-1}(z)\right) = \overline{A}(F(x)).$$

Thus, $u = F(x)$ solves the equation (8.7). $\qquad\qquad\square$

Theorem 8.3. *For any $x \in \mathcal{E}$ the function*

$$r(\overline{A}) \ni \zeta \mapsto (\zeta I - \overline{A})^{-1}(x)$$

is locally lipschitz.

Proof. For $\zeta, \zeta_o \in r(\overline{A})$, $\zeta \neq \zeta_o$,

$$(\zeta I - \overline{A})^{-1}(x) - (\zeta_o I - \overline{A})^{-1}(x) = (\zeta I - \overline{A})^{-1}(x) -$$
$$-(\zeta I - \overline{A})^{-1}(x + (\zeta - \zeta_o)(\zeta_o I - \overline{A})^{-1}(x)).$$

Hence

$$\|(\zeta I - \overline{A})^{-1}(x) - (\zeta_o I - \overline{A})^{-1}(x)\| \le |\zeta - \zeta_o| p_L((\zeta I - \overline{A})^{-1})\|(\zeta I - \overline{A})^{-1}\|.$$

On the other hand,

$$p_L((\zeta I - \overline{A})^{-1}) = \sup \left\{ \frac{1}{\|x - y\|} \|(\zeta I - \overline{A})^{-1}(x) - (\zeta I - \overline{A})^{-1}(y)\| : \right.$$

$$x, y \in \mathcal{E}, x \ne y \} = \sup \left\{ \frac{1}{\|x - y\|} \|(\zeta_o I - \overline{A})^{-1} \times \right.$$

$$\times (I + (\zeta - \zeta_o)(\zeta_o I - \overline{A})^{-1})^{-1}(x) -$$

$$(\zeta_o I - \overline{A})^{-1}(I + (\zeta - \zeta_o)(\zeta_o I - \overline{A})^{-1})^{-1}(y)\| :$$

$$x, y \in \mathcal{E}, x \ne y \}$$

$$\le \|(\zeta_o I - \overline{A})^{-1}\| p_L(I + (\zeta - \zeta_o)(\zeta_o I - \overline{A})^{-1})^{-1}$$

$$\le \frac{\|(\zeta_o I - \overline{A})^{-1}\|}{1 - |\zeta - \zeta_o| p_L(\zeta_o I - \overline{A})^{-1})},$$

because of Proposition 8.1, of the fact that $(\zeta_o I - \overline{A})^{-1}$ is injective and because

$$|\zeta - \zeta_o| p_L(\zeta_o I - \overline{A})^{-1}) < 1.$$

Hence,

$$\|(\zeta I - \overline{A})^{-1}(x) - (\zeta_o I - \overline{A})^{-1}(x)\| \le$$

$$\frac{|\zeta - \zeta_o| \|(\zeta_o I - \overline{A})^{-1}\|}{1 - |\zeta - \zeta_o| p_L(\zeta_o I - \overline{A})^{-1})} \|(\zeta_o I - \overline{A})^{-1}(x)\|.$$

That proves the theorem. $\qquad\square$

Let $f : \mathcal{E} \to \mathcal{E}$ and $g : \mathcal{E} \to \mathcal{E}$ be such that $r(f) \cap r(g) \ne \emptyset$, and let $\zeta \in r(f) \cap r(g)$.

For all $x \in \mathcal{E}$

$$(\zeta I - g)(x) = (\zeta I - f)(x) + (f - g)(x).$$

Hence, if $x = (\zeta I - f)^{-1}(y)$,

$$(\zeta I - g)((\zeta I - f)^{-1}(y)) = y + (f - g)((\zeta I - f)^{-1}(y)),$$

i.e.

$$(\zeta I - f)^{-1} = (\zeta I - g)^{-1} + (\zeta I - g)^{-1} \circ (f - g) \circ (\zeta I - f)^{-1}. \qquad (8.8)$$

For $\zeta \in r(A)$ and $x \in \mathcal{E}$, let

$$H(\zeta, x) = (\zeta I - \overline{A})^{-1}(x).$$

As is well known, if A is a closable linear operator, the function $\zeta \mapsto H(\zeta, x)$ is holomorphic on $r(A)$. Here is an extension of this fact to the non-linear case.

Proposition 8.3. *If H is Fréchet differentiable at $x \in \mathcal{E}$ for all $x \in \mathcal{E}$ and all $\zeta \in r(\overline{A})$, $H(\bullet, x)$ is holomorphic in $r(\overline{A})$. Furthermore,*

$$\frac{d}{d\zeta} H(\zeta, x) = -d_x H(\zeta, x)(H(\zeta, x))$$

for any $x \in \mathcal{E}$ and any $\zeta \in r(\overline{A})$.

Proof. For $x, y \in \mathcal{E}$, with $y \neq 0$, and $\zeta \in r(\overline{A})$, let

$$h(y) := H(\zeta, x + y) - H(\zeta, x) - d_x H(\zeta, x)(y).$$

Then

$$\lim_{y \to 0} \frac{\|h(y)\|}{\|y\|} = 0.$$

If $\kappa \in \mathbf{C}$ is such that $\zeta + \kappa \in r(\overline{A})$, then, by (8.8),

$$\begin{aligned}
H(\zeta + \kappa, x) &= ((\zeta + \kappa)I - \overline{A})^{-1}(x) \\
&= (\zeta - \overline{A})^{-1}(x + (\zeta - \zeta - \kappa)(\zeta + \kappa)I - \overline{A})^{-1}(x)) \\
&= (\zeta - \overline{A})^{-1}(x - \kappa)((\zeta + \kappa)I - \overline{A})^{-1}(x)) \\
&= H(\zeta, x - \kappa H(\zeta + \kappa, x)),
\end{aligned}$$

and therefore

$$\begin{aligned}
H(\zeta + \kappa, x) &- H(\zeta, x) \\
&= H(\zeta, x - \kappa H(\zeta + \kappa, x)) - H(\zeta, x) \\
&= -\kappa d_x H(\zeta, x)(H(\zeta + \kappa, x)) + h(-\kappa H(\zeta + \kappa, x)).
\end{aligned}$$

Since $H(\bullet, x)$ is locally lipshitz, the conclusion follows. $\square$

Theorem 8.4. *Let K be a non-empty, closed and connected subset of $\mathbb{C}$. Let $A \in \Upsilon(\mathcal{E})$, let $\zeta_o \in K$ and let $\phi : K \to \mathbb{R}_+$ be continuous and such that:*

$$\|x_1 - x_2\| \leq \phi(\zeta)\|\zeta(x_1 - x_2) - (z_1 - z_2)\| \tag{8.9}$$

for all $x_1, x_2 \in \mathcal{D}(\overline{A})$, all $z_1 \in \overline{A}(x_1)$, $z_2 \in \overline{A}(x_2)$ and all $\zeta \in K$, and that

$$\overline{(\zeta_o I - \overline{A})(\mathcal{D}(\overline{A}))} = \mathcal{E}. \tag{8.10}$$

Then:

$$K \subset r(\overline{A}); \tag{8.11}$$

$$p_L((\zeta I - \overline{A})^{-1}) \leq \phi(\zeta) \ \forall \, \zeta \in K, \tag{8.12}$$

and

$$\overline{(\zeta I - \overline{A})(\mathcal{D}(\overline{A}))} = \mathcal{E} \ \forall \, \zeta \in K. \tag{8.13}$$

Proof. We begin by showing that, for all $y \in \mathcal{E}$, there exists a unique $x \in \mathcal{D}(\overline{A})$ solving the equation

$$\zeta_o x - y \in \overline{A}(x). \tag{8.14}$$

By (8.9), there exist sequences $\{x_\nu\}$, with $x_\nu \in \mathcal{D}(\overline{A})$, and $\{z_\nu\}$, with $z_\nu \in \overline{A}(x_\nu)$, such that, letting

$$\zeta_o x_\nu - z_\nu = y_\nu,$$

then

$$\lim_{\nu \to \infty} y_\nu = y.$$

It follows from (8.9) that

$$\|x_\nu - x_\mu\| \leq \phi(\zeta_o)\|y_\nu - y_\mu\|$$

for all indices ν and μ.

Since $\mathcal{E}$ is complete, $\{x_\nu\}$ converges to some $x \in \mathcal{D}(\overline{A})$. Hence, $\{z_\nu\}$ converges to some $z \in \overline{A}(x)$ for which

$$\zeta_o x - z = y.$$

Thus, (8.14) holds.

Let $x' \in \mathcal{D}(\overline{A})$ and $z' \in \overline{A}(x')$ be such that

$$\zeta_o x' - z' = y.$$

Since

$$\|x - x'\| \leq \phi(\zeta_o)\|(\zeta_o x - z) - (\zeta_o x' - z')\|$$
$$= \|y - y\| = 0,$$

then $x = x'$, *i.e.*, the solution of the equation (8.14) is unique.

Since, furthermore,

$$p_L(\zeta_o I - \overline{A})^{-1}) \leq \phi(\zeta_o) \tag{8.15}$$

then

$$\zeta_o \in r(\overline{A}) \cap K.$$

The intersection $K_o = K \cap r(\overline{A})$ is non-empty and open in K.

To establish (8.11), we prove now that K_o is also closed.

Let $\zeta \in \overline{K_o}$, and let $\{\zeta_\nu\}$, with $\zeta_\nu \in K_o$, be a sequence converging to ζ. By Proposition 8.2 and by (8.9),

$$\Delta\left(\zeta_\nu, \frac{1}{\phi(\zeta_\nu)}\right) \cap K \subset K_o.$$

If $\epsilon > 0$ is such that $\phi(\zeta_\nu) < \frac{1}{\epsilon}$, then

$$\Delta(\zeta_\nu, \epsilon) \cap K \subset K_o,$$

and therefore $\zeta \in K_o$. Since K_o is open in K, closed and non-empty, and since K is connected, then $K = K_o \subset r(\overline{A})$.

Thus

$$p_L((\zeta I - \overline{A})^{-1}) \le \phi(\zeta) \ \forall\, \zeta \in K.$$

We complete the proof of the theorem showing that (8.13) holds. Indeed, if λ is a continuous linear form on $\mathcal{E}$ such that

$$\langle \zeta x - y, \lambda \rangle = 0 \ \forall\, x \in \mathcal{D}(\overline{A}), \ \ \forall y \in \overline{A}(x),$$

then $\lambda = 0$ because $\zeta \in r(\overline{A})$. $\qquad\qquad\qquad\qquad\qquad\qquad$ $\square$

Note Many of the results established in this section can be found in [4].

8.2. Accretive operators in Banach spaces

A linear operator $X : \mathcal{D}(X) \subset \mathcal{E} \to \mathcal{E}$ is called accretive when

$$\|(sI + X)x\| \ge s\|x\| \ \forall\, x \in \mathcal{D}(X), \ \forall\, s \in \mathbb{R}_+^*.$$

This definition has been extended to the non-linear case in the following way. A map $A \in \Upsilon(\mathcal{E})$ is said to be *accretive* if

$$\|x_1 - x_2\| \le \|x_1 - x_2 + s(y_1 - y_2)\|$$

for all $x_1, x_2 \in \mathcal{D}(A), y_1 \in A(x_1), y_2 \in A(x_2), s \in \mathbb{R}_+^*$.

This section is devoted to sketching some elementary aspects of this extension. Many deeper results can be found, *e.g.*, in [4] and [3].

Clearly, if A is accretive, $\overline{A}$ is accretive, and $\alpha I + A$ is injective for all $\alpha \in \mathbb{R}_+^*$.

Lemma 8.4. *If $\overline{A}$ is accretive and if*

$$r(\overline{A}) \cap \mathbb{R}_-^* \neq \emptyset, \tag{8.16}$$

then

$$\mathbb{R}_-^* \subset r(\overline{A}) \tag{8.17}$$

and

$$p_L(\zeta I - \overline{A}) \leq \frac{1}{\zeta} \ \forall \, \zeta \in \mathbb{R}_+^*. \tag{8.18}$$

Proof. If $\zeta \in \mathbb{R}_-^*$,

$$\|x_1 - x_2\| \leq \frac{1}{|\zeta|}\|\zeta(x_1 - x_2) - (y_1 - y_2)\|$$

for all $x_1, x_2 \in \mathcal{D}(A), y_1 \in A(x_1), y_2 \in A(x_2)$.

Because of (8.16),

$$(\zeta_0 I - \overline{A})(\mathcal{D}(\overline{A})) = \mathcal{E}$$

for some $\zeta_0 \in \mathbb{R}_-^*$. By Theorem 8.4, (8.18) holds, and $[c, +\infty) \subset -r(A)$ for all $c > 0$. $\qquad\square$

If A is accretive and (8.17) holds, A is called *m-accretive*.
If

$$(I + A)(\mathcal{D}(A)) = \mathcal{E}, \tag{8.19}$$

$I + A$ is closed (and therefore A is closed) because otherwise $I + \overline{A}$ could not be injective. Thus, $-1 \in r(\overline{A})$ and therefore A is *m*-accretive.

At this point, one easily concludes that an accretive operator A is *m*-accretive if, and only if, (8.19) holds. This fact can be improved showing (exercise) that, if A is accretive and there is some $\alpha > 0$ for which

$$\overline{(I + \alpha A)(\mathcal{D}(A))} = \mathcal{E},$$

then A is *m*-accretive.

Ordering the operators by inclusion of their graphs, an accretive operator is said to be a maximal accretive operator if every accretive operator C such that $G_A \subset G_C$ coincides with A. By the Zorn lemma, every accretive operator is contained in a maximal accretive operator.

Theorem 8.5. *If A is m-accretive, A is maximal accretive.*

Proof. Let C be accretive and such that $G_A \subset G_C$. If $x_0 \in \mathcal{D}(C)$ and $y_0 \in C(x_0)$, then $x_0 + y_0 \in (I + C)(x_0)$. As a consequence of (8.19), there is some $x_1 \in \mathcal{D}(A)$ for which

$$x_0 + y_0 \in x_1 + A(x_1) \subset x_1 + C(x_1). \tag{8.20}$$

Since C is accretive, and therefore $I + C$ is injective, then $x_0 = x_1$, so that, by (8.20),

$$x_0 + y_0 \in x_0 + C(x_0),$$

i.e., $y_0 \in C(x_0)$. Thus, $A = C$. $\qquad\qquad\square$

Corollary 8.1. *If A is m-accretive, A is closed.*

Theorem 8.6. *Let $A \in \Upsilon(\mathcal{E})$. If there exists $\zeta_o < 0$ such that $(-\infty, \zeta_o) \subset r(\overline{A})$ and*

$$p_L((\zeta I - A)^{-1}) \leq \frac{1}{|\zeta|} \; \forall \, \zeta \in (-\infty, \zeta_o),$$

then A is m-accretive.

Proof. Let: $\zeta < \zeta_o$, $x_1, x_2 \in \mathcal{D}(A)$, $z_1 \in A(x_1)$, $z_2 \in A(x_2)$, $y_1 = \zeta x_1 - z_1$, $y_2 = \zeta x_2 - z_2$.

Then,

$$(\zeta I - A)(x_1) \ni \zeta x_1 - z_1 = y_1,$$

$$(\zeta I - A)(x_2) \ni \zeta x_2 - z_2 = y_2,$$

and therefore

$$x_1 = (\zeta I - A)^{-1}(y_1), \quad x_2 = (\zeta I - A)^{-1}(y_2).$$

Hence,

$$\|x_1 - x_2\| \leq p_L((\zeta I - A)^{-1})\|y_1 - y_2\| \leq \frac{1}{-\zeta}\|y_1 - y_2\|$$

$$= \frac{1}{-\zeta}\|\zeta(x_1 - x_2) - (z_1 - z_2)\|,$$

that is,

$$-\zeta\|x_1 - x_2\| \leq \|\zeta(x_1 - x_2) - (z_1 - z_2)\|$$

for all $\zeta < \zeta_o$.

Letting $\zeta_o - \zeta = \tau$, then $\tau > 0$ and the latter inequality yields

$$
\begin{aligned}
(\tau - \zeta_o)\|x_1 - x_2\| &\leq \|(\zeta_o - \tau)(x_1 - x_2) - (z_1 - z_2)\| \\
&= \| - \tau(x_1 - x_2) - (z_1 - z_2) + \zeta_o(x_1 - x_2)\| \\
&\leq \|\tau(x_1 - x_2) + z_1 - z_2\| - \zeta_o\|x_1 - x_2\|,
\end{aligned}
$$

whence

$$
\tau\|x_1 - x_2\| \leq \|\tau(x_1 - x_2) + z_1 - z_2\|
$$

for all $\tau > 0$.

That shows that A is accretive. Since $r(\overline{A}) \cap \mathbf{R}_-^* \neq \emptyset$, A is m-accretive.

Examples show, [6], that being maximal accretive does not imply being m-accretive. However this implication does hold in the case of Hilbert spaces.

8.3. Monotone operators

Let $\mathcal{H}$ be a real Hilbert space with inner product $(\ |\ )$. Accretive operators on $\mathcal{H}$ are also called *monotone operators*.

Let $A \in \Upsilon(\mathcal{H})$.

Theorem 8.7. *The operator A is monotone if, and only if,*

$$
(x - \xi|y - \eta) \geq 0 \tag{8.21}
$$

for all $x, \xi \subset \mathcal{D}(\Lambda)$ and all $y \in \Lambda(x), \eta \in A(\xi)$.

Proof. First of all, for any $s \in \mathbb{R}$,

$$
\|x - \xi + s(y - \eta)\|^2 = \|x - \xi\|^2 + s^2\|y - \eta\|^2 + 2s(x - \xi|y - \eta).
$$

As a consequence of (8.21),

$$
\|x - \xi + s(y - \eta)\|^2 \geq \|x - \xi\|^2
$$

for all $x, \xi \in \mathcal{D}(A)$, all $y \in A(x), \eta \in A(\xi)$ and all $s \geq 0$. That shows that, if (8.21) holds, A is accretive.

Viceversa, if A is accretive,

$$
\|x - \xi\|^2 \leq \|x - \xi\|^2 + s^2\|y - \eta\|^2 + 2s(x - \xi|y - \eta),
$$

that is

$$
s\|y - \eta\|^2 + 2(x - \xi|y - \eta) \geq 0
$$

for all $s > 0$. That yields (8.21) for all $x, \xi \in \mathcal{D}(A)$, all $y \in A(x), \eta \in A(\xi)$ and all $s > 0$. $\qquad\square$

As in the case of accretive operators, a monotone operator A on $\mathcal{H}$ is said to be *maximal monotone* if G_A is maximal in the set of the graphs of all monotone operators on $\mathcal{H}$, ordered by inclusion. By the Zorn lemma, every monotone operator is contained in a maximal monotone operator.

An immediate consequence of Theorem 8.7 is the following characterization of maximal monotone operators.

Theorem 8.8. *The operator A is a maximal monotone operator if, and only if, A is monotone and the fact that (8.21) holds for all $x, y \in \mathcal{H}$ and all $\xi, \eta \in G_A$ implies that $x \in \mathcal{D}(A)$ and $y \in A(x)$.*

Theorem 8.9. *The operator A is maximal monotone if, and only if, it is m-accretive.*

Proof. The "if" part is Theorem 8.5. For a proof of the "only if" part, see, *e.g.*, [9], pp.249-252 and p.257. $\qquad\square$

Here are some examples of monotone operators.

Lemma 8.5. *Let $K \subset \mathcal{H}$ and let $f : K \to \mathcal{H}$ be a contraction. Then, $A = I - f$ is monotone.*

Proof. If $x_1, x_2 \in K, y_j = x_j - f(x_j), (j = 1, 2)$,

$$
\begin{aligned}
(x_1 - x_2 | y_1 - y_2) &= (x_1 - x_2 | x_1 - x_2 - (f(x_1) - f(x_2))) \\
&= \|x_1 - x_2\|^2 - (x_1 - x_2 | f(x_1) - f(x_2)) \\
&\geq \|x_1 - x_2\|^2 - \|x_1 - x_2\| \, \|x_1 - x_2\| = 0.
\end{aligned}
$$
$\qquad\square$

Let $A \in \Upsilon(\mathcal{H})$ and, for $x \in \mathcal{D}(A)$, let $\text{convex}(A(x))$ be the convex hull of $A(x)$.

Lemma 8.6. *If A is monotone, $x \mapsto \overline{\text{conv}(A(x))}$ is monotone.*

Proof. Let $x_1, x_2 \in \mathcal{D}(A), \ j = 1, 2, \ n \geq 1, \ y_{j,1}, \ldots, y_{j,n} \in A(x_j)$, $t_1, \ldots, t_n \in \mathbf{R}_+$, with $t_1 + \cdots + t_n = 1$.

Then

$$
\left(x_2 - x_1 \Big| \sum_{\alpha=1}^{n} t_\alpha(y_{2,\alpha} - y_{1,\alpha})\right) = \sum_{\alpha=1}^{n} t_\alpha(x_1 - x_2 | y_{2,\alpha} - y_{1,\alpha}) \geq 0.
$$
$\qquad\square$

As a consequence, if A is maximal monotone, $A(x) = \text{convex}(A(x))$ is closed and convex in $\mathcal{H}$ for all $x \in \mathcal{D}(A)$.

Let M be a connected, C^∞ Riemannian manifold of dimension N. Let Ω^q be the vector space of all real, exterior q-forms of class C^∞ on M, and $\Omega_c^q \subset \Omega^q$ be the subspace consisting of all compactly supported elements of Ω^q.

The Riemannian metric defines the Hodge isomorphism

$$* : \Omega^q \xrightarrow{\sim} \Omega^{N-q}.$$

If $d\omega$ is the volume element of the Riemannian metric, for $\phi, \psi \in \Omega^q$, and for any $x \in M$, let $\langle \phi, \psi \rangle_x$ be the real scalar defined by

$$(\phi \wedge *\psi)_x = \langle \phi, \psi \rangle_x \, d\omega(x);$$

the map $\phi, \psi \mapsto \langle \phi, \psi \rangle_x \in \mathbf{R}$ is a positive-definite bilinear form which is a continuous function of $x \in M$ for every choice of ϕ and ψ. Both $\langle \phi, \psi \rangle_x$ and

$$|\phi|_x = \sqrt{\langle \phi, \phi \rangle_x}$$

(which is a norm in the space of anti-symmetric q-vectors in $\mathbf{R}^N$) do not depend on the local representations of ϕ and ψ in any coordinate neighbourhood of x.

Setting, for $\phi, \psi \in \Omega_c^q$,

$$(\phi|\psi) = \int_M \langle \phi, \psi \rangle \, d\omega \ \text{ and } \ \|\phi\| = (\phi|\phi)^{\frac{1}{2}},$$

we define a positive-definite inner product and a norm on Ω_c^q, which thus acquires the structure of a real pre-Hilbert space. Let L^q be its completion.

If $d : \Omega^q \to \Omega^{q+1}$ is the exterior differential operator, the operator $\delta : \Omega^q \to \Omega^{q-1}$ defined by

$$\delta = (-1)^q *^{-1} \circ d \circ *$$

is the formal adjoint of d, in the sense that

$$(d\phi|\chi) = (\phi|\delta\chi)$$

for all $\phi \in \Omega_c^q$, $\chi \in \Omega_c^{q+1}$.

The Laplace-Beltrami operator

$$\Delta = d\delta + \delta d : \Omega_c^q \to \Omega_c^q$$

is linear, symmetric and positive on the domain Ω_c^q:

$$(\Delta\phi|\psi) = (d\phi|d\psi) + (\delta\phi|\delta\psi) = (\phi|\Delta\psi),$$

$$(\Delta\phi|\phi) = \|d\phi\|^2 + \|\delta\phi\|^2 \geq 0$$

for all $\phi, \psi \in \Omega_c^q$.

For $p \geq 2$, the *p-Laplacian*, or *p-Laplace-Beltrami operator*,

$$\Delta_p : \Omega_c^q \to \Omega_c^q$$

is defined on any $\phi \in \Omega_c^q$ by

$$\Delta_p(\phi) = \delta\left(|d\phi|^{p-2}d\phi\right) + d\left(|\delta\phi|^{p-2}\delta\phi\right).$$

For $\phi, \psi \in \Omega_c^q$,

$$(\Delta_p(\phi)|\psi) = \left(|d\phi|^{p-2}d\phi|d\psi\right) + \left(|\delta\phi|^{p-2}\delta\phi|\delta\psi\right)$$

$$= \int_M \left[|d\phi|^{p-2}\langle d\phi|d\psi\rangle + |\delta\phi|^{p-2}\langle\delta\phi|\delta\psi\rangle\right] \, d\omega,$$

and, in particular,

$$(\Delta_p(\phi)|\phi) = \|d\phi\|^p + \|\delta\phi\|^p,$$

where

$$\| \bullet \| = \left(\int_M | \bullet |^p d\omega\right).$$

These identities, yield now

Theorem 8.10. *If $p \geq 2$, Δ_p, with domain $\mathcal{D}\left(\Delta_p\right) = \Omega_c^q$, is monotone.*

Proof. If $\phi_1, \phi_2 \in \Omega_c^q$,

$$(\Delta_p(\phi_1) - \Delta_p(\phi_2)|\phi_1 - \phi_2) =$$

$$\|d\phi_1\|^p + \|\delta\phi_1\|^p + \|d\phi_2\|^p + \|\delta\phi_2\|^p -$$
$$(|d\phi_1|^{p-2}d\phi_1|d\phi_2) - (|\delta\phi_1|^{p-2}\delta\phi_1|\delta\phi_2) -$$
$$(|d\phi_2|^{p-2}d\phi_2|d\phi_1) - (|\delta\phi_2|^{p-2}\delta\phi_2|\delta\phi_1)$$
$$= \|d\phi_1\|^p + \|\delta\phi_1\|^p + \|d\phi_2\|^p + \|\delta\phi_2\|^p -$$
$$\int_M \left[\left(|d\phi_1|^{p-2} + |d\phi_2|^{p-2} \right) \langle d\phi_1|d\phi_2\rangle + \right.$$
$$\left. \left(|\delta\phi_1|^{p-2} + |\delta\phi_2|^{p-2} \right) \langle \delta\phi_1|\delta\phi_2\rangle \right] d\omega$$
$$\geq \int_M \left[|d\phi_1|^p + |d\phi_2|^p - \frac{1}{2} \left(|d\phi_1|^{p-2} + |d\phi_2|^{p-2} \right) \times \right.$$
$$\left(|d\phi_1|^2 + |d\phi_2|^2 \right) + |\delta\phi_1|^p + |\delta\phi_2|^p -$$
$$\left. \frac{1}{2} \left(|\delta\phi_1|^{p-2} + |\delta\phi_2|^{p-2} \right) \times \left(|\delta\phi_1|^2 + |\delta\phi_2|^2 \right) \right] d\omega$$
$$= \frac{1}{2} \int_M \left[|d\phi_1|^p + |d\phi_2|^p - |d\phi_1|^{p-2}|d\phi_2|^2 - |d\phi_2|^{p-2}|d\phi_1|^2 + \right.$$
$$\left. |\delta\phi_1|^p + |\delta\phi_2|^p - |\delta\phi_1|^{p-2}|\delta\phi_2|^2 - |\delta\phi_2|^{p-2}|\delta\phi_1|^2 \right] d\omega.$$

Since, for $p \geq 2, a \geq 0, b \geq 0$,

$$a^p + b^p - a^{p-2}b^2 - a^2b^{p-2} = \left(a^{p-2} - b^{p-2} \right) \left(a^2 - b^2 \right) \geq 0,$$

then

$$(\Delta_p(\phi_1) - \Delta_p(\phi_2)|\phi_1 - \phi_2) \geq 0.$$

$\square$

Lemma 8.7. *If A is maximal monotone, $\overline{\mathcal{D}(A)}$ is convex.*

Proof. See, *e.g.*, [9], Theorem 3.8.7, pp. 259-260. $\square$

8.4. Nonlinear semigroups

Let $A \in \Upsilon(\mathcal{H})$ be a maximal monotone operator. For $x \in \mathcal{D}(A)$, let $\Pi_{A(x)}$ be the orthogonal projector of $\mathcal{H}$ onto the closed affine subspace of $\mathcal{H}$ spanned by $\overline{\text{conv}(A(x))} = A(x)$, and let $A^o(x) = \Pi_{A(x)}(0)$.

The central role of the single valued operator $A^o : \mathcal{D}(A) \to \mathcal{H}$ is clarified by the following theorem.

Theorem 8.11. *If A is maximal monotone, and if $(x, y) \in \overline{\mathcal{D}(A)} \times \mathcal{H}$ is such that*

$$(A^o(\xi) - y|\xi - x) \geq 0 \ \ \forall \xi \in \mathcal{D}(A),$$

then

$$(\eta - y|\xi - x) \geq 0 \ \ (\xi, \eta) \in G_A,$$

i.e., $x \in \mathcal{D}(A)$ and $y \in A(x)$.

As a consequence, A^o determines A in the sense that, if A_1 and A_2 are maximal monotone, if $\mathcal{D}(A_1) = \mathcal{D}(A_2)$ and if $A_1{}^o = A_2{}^o$, then $A_1 = A_2$.

The operator A^o establishes a link between maximal monotone operators and continuous semigroups, which can be viewed as a non-linear extension of the Lumer-Phillips theory in the linear case.

Let $A \in \Upsilon(\mathcal{H})$ be a maximal monotone operator.

Theorem 8.12. *For every $x_o \in \mathcal{D}(A)$ there exists a unique function $x :$ $[0, +\infty) \to \mathcal{H}$ which satisfies the following conditions:*

$x(t) \in \mathcal{D}(A)$ for all $t \in (0, +\infty)$;

x is lipschitz on $(0, +\infty)$ (that is to say

$$\frac{dx}{dt} \in L^\infty((0, +\infty), \mathcal{H})$$

in the sense of distributions) and

$$\left\| \frac{dx}{dt} \right\|_{L^\infty((0,+\infty),\mathcal{H})} \leq \|A^o(x_o)\| ;$$

$-\frac{dx}{dt} \in A(x(t))$ a.e. on $(0, +\infty)$).

The map $t \mapsto x(t)$ is a contraction $\mathcal{D}(A) \to \mathcal{D}(A)$ that can be extended, in a unique way, to a contraction $T(t) : K \to K$ of

$$K = \overline{\mathcal{D}(A)}. \tag{8.22}$$

The map $\mathbb{R}_+^* \ni t \mapsto T(t)$ is a semigroup of contractions of K which is continuous, *i.e.,*

$$\lim_{t \downarrow 0} \|T(t)(x) - x\| = 0 \ \forall \, x \in K,$$

and such that

$$\|T(t)(x_1) - T(t)(x_2)\| \le \|x_1 - x_2\|$$

for all $x_1, x_2 \in K$ and all $t \in \mathbb{R}_+^*$.

Furthermore

$$\lim_{t\downarrow 0} \frac{1}{t}(x - T(t)(x)) = A^o(x) \ \forall \ x \in \mathcal{D}(A). \tag{8.23}$$

Viceversa, the following theorem holds.

Theorem 8.13. *Let K be a closed, convex subset of $\mathcal{H}$ and let $\mathbb{R}_+^* \ni t \mapsto T(t)$ be a continuous semigroups of contractions of K. There exists a unique maximal monotone operator A such that $K = \overline{\mathcal{D}(A)}$ and (8.23) holds.*

Proofs of these facts are given in [1].

References

[1] H.Brézis, Opérateurs maximaux monotones et semi-groupes de contraction dans les espaces de Hilbert, Math. Studies, n. 5, North Holland, Amsterdam, 1973.

[2] Ph. Clement, H.J.A.M. Heijmans, S. Angenent, C.J. van Duijn, B. de Bagter, One parameter semigroups, North Holland, Amsterdam/New York/Oxford/Tokyo, 1987.

[3] M.G.Crandall. *Non linear semigroups and evolution governed by accretive operators*, Proc. Symp. Pure Math., 45, Part I, American Mathematical Society, Providence, R.I., 1986, 305-337.

[4] G.Da Prato, Applications croissantes et équations d'évolution dans les espaces de Banach, Institutiones Mathematicae, Academic Press, London/New York, 1976.

[5] K.-J.Engel and R.Nagel, One-parameter semigroups for linear evolution equations, Springer-Verlag, New York, 2001.

[6] J.A.Goldstein, Semigroups of operators and applications, Oxford University Press, Oxford, 1985.

[7] J.W.Neuberger, *Existence of a spectrum for nonlinear transformations*, Pacific J. of Math., 31 (1969), 157-159.

[8] A.Pazy, Semigroups of linear operators and applications to partial differential equations, Springer-Verlag, New York, 1983.

[9] E.Vesentini, Introduction to continuous semigroups, Second Edition, Scuola Normale Superiore, Pisa, 2002.

[10] K.Yosida, Functional Analysis, Second Edition, Springer-Verlag, Berlin/Heidelberg/New York, 1968.

PART 3
Harmonic analysis

Chapter 9

Integral geometry and spectral analysis

Mark Agranovsky

Department of Mathematics and Statistics
Bar-Ilan University
Ramat-Gan, 52900, Israel
e-mail: agranovs@math.biu.ac.il

Contents

9.1. Introduction

From the general point of view, integral geometry can be understood as a branch of analysis dealing with reconstruction of functions or distributions from their integrals over families of submanifolds. The celebrated article [R] of J. Radon on reconstruction of functions in $\mathbb{R}^3$ by their integrals over 2-planes, published in 1917, may be considered as a starting point for developing integral geometry.

More precisely, the general problem can be described as follows. Given a Riemannian manifold M and a family of submanifolds $M_\alpha \subset M$, one defines the Radon transform as

$$Rf(M_\alpha) = \int_{M_\alpha} f d\mu_\alpha,$$

where d_{μ_α} is an appropriate measures on M_α and f is an integrable function. Thus, we deal with integral operators of specific form.

The following questions are important:
- What is the kernel of the transform R (when is it injective)?
- What is the image of R?
- What is the inverse operator (inversion formula) to R on its image, in the case of injectivity?

In these notes we will mainly deal with the problem of injectivity for Radon transforms arising in certain analytic problems.

Let us consider, for example, the classical Radon transform. In this case $M = \mathbb{R}^n$ and the family $\{M_\alpha\}$ is the family of all hyperplanes in $\mathbb{R}^n$.

Parametrize the hyperplanes by the unit normal vector and the distance to the origin: $\xi_{\omega,d} = \{x \in \mathbb{R}^n : (\omega, x) = d\}$, where $d \in \mathbb{R}$ and $\omega \in S^{n-1}$, the unit sphere in $\mathbb{R}^n$.

Then the *Radon transform R* can be regarded as a function on $S^{n-1} \times \mathbb{R}$ defined by:

$$Rf(\xi_{\omega,d}) = Rf(\omega, d) = \int_{\omega,x)=d} f(x)dm_{n-1}(x), \qquad (9.1)$$

where dm_{n-1} is the $(n-1)$-dimensional volume.

One of the main properties of the transform is that it commutes with transforms from the group $M(n)$ of all rigid motions (isometries) of $\mathbb{R}^n$: If $\tau \in M(n)$, then $R(f \circ \tau)(\xi) = Rf(\tau\xi)$, where ξ is any hyperplane in $\mathbb{R}^n$ and f belongs to the domain of R. This invariance implies that the transform R is closely related to the Laplace operator $\Delta = \frac{\partial^2}{\partial x_1^2} + \cdots + \frac{\partial^2}{\partial x_n^2}$ which is the only (up to a constant factor) second order $M(n)$-invariant differential operator.

In this course, we will consider several important problems arising in analysis and PDE which lead to such Radon transforms, i.e., to transforms invariant under groups of isometries and where this invariance delivers fascinating interaction between integral geometry and spectral analysis of differential operators naturally related to the transforms. The Radon transforms over planes and spheres are important examples of such kind of transforms.

9.2. Radon Transform over Hyperplanes

This section contains more information about the classical Radon transform defined in Section 9.1.

The regular functional classes on which the Radon transform (9.1) is well-defined are: $C_c(\mathbb{R}^n)$, the space of all continuous compactly supported functions, the Schwartz class $S(\mathbb{R}^n)$ of rapidly decreasing functions, and certain spaces of functions with sufficient rate of decay at ∞.

It is convenient to introduce the *Radon projection* on the direction $\omega \in S^{n-1}$:

$$(P^\omega f)(d) = Rf(\omega, d), \quad d \in \mathbb{R}.$$

In polar coordinates $x = r\omega$, $x \in \mathbb{R}^n$, $\omega \in S^{n-1}$, $r \in \mathbb{R}^+$, the Radon projection is related to the Fourier transform

$$(F_n f)(\lambda) = \int_{\mathbb{R}^n} e^{-i(\lambda, x)} f(x) dx$$

in the following nice way:

Projection Slice Theorem.

$$(F_n f)(r\omega) = (F_1 P^\omega f)(r), \quad \omega \in S^{n-1}, \ r \in \mathbb{R}.$$

Proof.

$$(F_n f)(r\omega) = \int_{\mathbb{R}^n} e^{-i(r\omega, x)} f(x) dx$$

$$= \int_{\mathbb{R}^1} e^{-irs} \left(\int_{(\omega, x) = s} f(x) dm_{n-1}(x) \right) ds \qquad (9.2)$$

$$= (F_1 P^\omega)(r). \qquad \qquad \square$$

Corollary. *Let $f \in C_c(\mathbb{R}^n)$ and $P^\omega f = 0$ for infinitely many directions $\omega \in S^{n-1}$. Then $f = 0$.*

Proof. The Projection Slice Theorem implies $(F_n f)(r\omega) = 0$ for an infinite set of vectors $\omega \in S^{n-1}$ and for arbitrary $r \geq 0$. The set $\{\omega\}$ has a limit point and therefore the real-analysis function $F_n f$ vanishes everywhere in $\mathbb{R}^n$. Then $f = 0$.

Thus, any compactly supported continuous function is uniquely defined by its Radon projection on an infinite set of directions. $\qquad\square$

The hyperplane Radon transform of f can be regarded as a function $\varphi(\xi)$ defined on a set $\mathbb{P}^n$ of all hyperplanes in $\mathbb{R}^n$, equipped with the natural topology, by

$$\varphi(\xi) = \int_\xi f(x) dm_{n-1}(x), \quad \xi \in \mathbb{P}^n.$$

The *dual transform* $\check{\varphi}$ is defined by:

$$\check{\varphi}(x) = \int_{x \in \xi} \varphi(\xi) d\mu(\xi),$$

where μ is the (unique) normalized measure on the set of all hyperplanes $\xi \in \mathbb{P}^n$ through the point x, i.e., $x \in \xi$, that is invariant with respect to rotations around x. The measure μ can be taken as the normalized area measure $dA(\omega)$ on the sphere S^{n-1} where ω is the unit normal vector to $\xi = \{y \in \mathbb{R}^n : (\omega, y)\} = (\omega, x)$, so that

$$\check{\varphi}(x) = \int_{S^{n-1}} \varphi(\omega, (\omega, x)) d\omega.$$

Inversion Formula: Let $f \in S(\mathbb{R}^n)$ Then

$$f_{(x)} = \text{const}\, \Delta^{\frac{n-1}{2}} R f(x). \tag{9.3}$$

For $n = 3$, this formula was first written by J. Radon [R]. It says that $f(x)$ can be computed by averaging the Radon tranform Rf around the point x and applying the iterated Laplacian to the result.

The inversion formula (9.3) needs clarification for the case of even dimensions n. The fractional power of Δ is understood in a consistent way by means of functional calculus, based on the Fourier transform, as the Riesz potential

$$(\Delta^{\frac{n-1}{2}} f)(x) = \int_{\mathbb{R}^n} \frac{f(y) dy}{|x - y|^{2n-1}}.$$

We refer the reader to the rigorous detailed proof of the inversion formula (9.3) to the books of Helgason [He] and Natterer [Na], restricting ourselves here to comments on the idea of the proof for odd n.

When $(n-1)/2$ is an integer, the formula (9.3) is obtained as follows. First, since the transform R commutes with rotations, the averaging $\check{R}f$ of Rf coincides with the Radon transform R applied to the result of averaging of f around x. In other words, it is enough to prove (9.3) for f invariant with respect to rotations around x. Then consequently applying the Laplace operator Δ to Rf using the Euler-Darboux equation for spherical means (see the next section) and integrating by parts finally reduces the left hand side in (9.3) to $f(x)$ times a constant.

Remarks.

1. The inversion formula (9.3) is local for n odd and is nonlocal for n even. This means that in odd dimensions one needs to know integrals of f over the hyperplanes near the points x, to reconstruct $f(x)$. On the other hand, for n even, one needs to know the integrals over all hyperplanes.

   In fact, this distinguishing feature of the character of the inversion formula in even- and odd-dimensional spaces is related to the known Huygen's principle for the wave propagation.

2. The formula (9.3) gives a decomposition of the function f into the plane waves $\varphi_\omega(x) = \varphi(\omega, (\omega, x))$-functions constant on parallel hyperplanes with the same normal vector ω. Later on, we will consider such kinds of decompositions related to spherical waves, that is, functions constant on concentric spheres.

9.3. Spherical Radon Transform

The spherical Radon transform is defined by

$$Mf(x,r) = M^r f(x) = \int_{S(x,r)} f(y)dA(y),$$

where $S(x,r) = \{y \in \mathbb{R}^n : |y-x| = r\}$ and dA is the area measure on the sphere $S(x,r)$.

The transform M is $M(n)$-invariant and can be written in the group-theoretical form:

$$Mf(x,r) = \int_{\mathcal{O}(n)} f(x + ky)dk,$$

where $\mathcal{O}(n)$ is the orthogonal group, dk is the Haar measure on $\mathcal{O}(n)$ and $y \in \mathbb{R}^n$ is any vector $|y| = r$.

The spherical transform commutes with the Laplace operator, i.e., $\Delta M^r = M^r \Delta$, which can be readily checked.

Moreover, for C^∞-functions with sufficient rate of decay at ∞, say for functions in $S(\mathbb{R}^n)$, the transform can be decomposed into series of iterated Laplacians:

$$Mf(x,r) = \frac{1}{2^{\frac{n-2}{2}}} \sum_{k=0}^{\infty} \left(\frac{r}{2}\right)^{2k} \frac{1}{k!\Gamma\left(\frac{n}{2}+k\right)} \Delta^k f(x)$$

(Pizzeti formula).

Formally, the decomposition can be obtained as follows. Write $M^r f$ as the convolution

$$M^r f = \chi_r * f,$$

where χ_r is the delta distribution on the sphere $S(r) = S(0,r)$. Applying the Fourier transform yields

$$\widehat{M^r f} = \hat{\chi}_r \cdot \hat{f}.$$

The Fourier transform $\hat{\chi}_r$ is the Bessel function:

$$\hat{\chi}_r(\lambda) = J_{\frac{n-2}{2}}(|\lambda|r)(|\lambda|r)^{\frac{2-n}{2}},$$

and the Pizzeti formula now can be obtained by decomposition of the Bessel function in the power series and taking the inverse Fourier transform.

The following identity is called the Darboux-Euler equation. Let $F(x,r) = M^r f(x)$, f is twice differentiable. Then

$$\left(\frac{\partial^2}{\partial r^2} + \frac{n-1}{r}\frac{\partial}{\partial r}\right) F(x,r) = \Delta_x F(x,r). \tag{9.4}$$

The left hand side is just the radial part Δ_r of the Laplacian Δ, so the equation says $\Delta_r F = \Delta_x F$.

We now give the proof. Introduce the function $F(x,y) = M^{|y|}f(x)$. Then

$$\Delta_y F(x,y) = \int_{\mathcal{O}(n)} \Delta_y f(x+ky)dk = \int_{\mathcal{O}(n)} \Delta_x f(x+ky)dk = \Delta + xF(x,y)$$

and the equation (9.4) follows, since $\Delta_y F(x,y) = \Delta_r F(x,r)$ because F is radial in the y-variable.

Moreover, the Aisgersson theorem says that any solution $F(x,r)$ of the Darboux-Euler equation is the spherical transform $F(x,r) = Mf(x,r)$ of the initial data $f(x) = F(x,0)$.

9.4. Pompeiu Transform and Pompeiu Problem

Both the plane Radon transform and the spherical transform integrate over families of congruent sets (hyperplanes, spheres), that is, the motion group $M(n)$ acts on families of manifolds of integration.

The Pompeiu transform deals with integration over a family of congruent compacts. Let M be a Riemannian manifold with the isometry group $Iso(M)$. With any compact $K \subset M$ of positive measure, one associates the transform

$$(P_K f)(\tau) = \int_{\tau(K)} f(x)dx, \quad \tau \in Iso(M),$$

defined on $f \in L^1_{\mathrm{loc}}(M)$.

The transform

$$P_K : L^1_{\mathrm{loc}}(M) \to C(Iso(M))$$

is called the *Pompeiu transform* (associated with the compact K).

The spherical transform M^r can be regarded as the Pompeiu transform $P_{S(r)}$ associated with the sphere $S(r) = \{x \in \mathbb{R}^n : |x| = r\}$.

The main problem regarding the Pompeiu transform is its injectivity.

In 1929, the Romanian mathematician, D. Pompeiu, published the article [Po] which contains the proof of injectivity of the transform P_K in $\mathbb{R}^n$ for arbitrary compact $K \subset \mathbb{R}^n$, $mes\, K > 0$. However, the result in such generality appeared to be false. Namely, Chakalov [Ch] poined out that the transform P_K, where K is a ball $K = B(a,r)$ or a sphere $K = S(a,r)$, fails to be injective. Indeed, take

$$f(x) = e^{i(\lambda,x)}.$$

Then

$$P_{B(a,r)} f(\tau) = \int_{\tau(B(a,r))} e^{i(\lambda,x)} dx = \int_{B(\tau a,r)} e^{i(\lambda,x)} dx$$

$$= e^{i(\lambda,\tau a)} \int_{B(a,r)} e^{i(\lambda,y)} dy = e^{i(\lambda,\tau a)} J_{\frac{n-2}{2}}(|\lambda|r)|r\lambda|^{\frac{2-n}{2}}, \tag{9.5}$$

for any rigid motion $\tau \in M(n)$. Therefore, if λ is chosen so that $|\lambda| r$ is a zero for the Bessel function $J_{\frac{n-2}{2}}$, then $P_{B(a,r)} f \equiv 0$ while $f \neq 0$.

The following question is known as the Pompeiu problem and still remains open (we formulate the problem for $\mathbb{R}^n$) :

Pompeiu Problem. *If balls are the only compact domains in $\mathbb{R}^n$ with connected boundary, such that the corresponding Pompeiu transforms fail to be injective?*

The answer is unknown even for $n = 2$. Observe that a generic $M(n)$-invariant family of compacts in $\mathbb{R}^n$ has $\dim M(n) = \frac{n(n+1)}{2}$ parameters, while the family of balls of a fixed radii is n-parametric. Thus, the question is whether the lack of the number of parameters is the only reason of noninjectivity of the transform?

Our immediate aim is to link the Pompeiu problem with the so-called Schiffer problem on solvability of an over-determined boundary value problem for the Laplace operator. To this end, we need tools of spectral analysis and synthesis.

9.5. Spectral Synthesis of Invariant Spaces

We will use the standard notations: $\mathcal{E}(\mathbb{R}^n)$ – the space of all C^∞-functions in $\mathbb{R}^n$ with the topology of uniform convergence on compact sets, $\mathcal{E}'(\mathbb{R}^n)$ – the dual space consisting of distributions with compact support.

Let $T \in \mathcal{E}'(\mathbb{R}^n)$. Then the Fourier transform of G is defined by

$$\hat{T}(\lambda) = \langle T, e^{-i(\lambda, \cdot)} \rangle, \quad \lambda \in \mathbb{R}^n.$$

According to the Paley-Wiener theorem [Ru], the image $\widehat{\mathcal{E}'}(\mathbb{R}^n)$ of the space $\mathcal{E}'(\mathbb{R}^n)$ under Fourier transform coincides with the *Paley-Wiener space* $PW(\mathbb{R}^n)$ of all functions $f(x)$ in $\mathbb{R}^n$ possessing extension to $\mathbb{C}^n = \mathbb{R}^n + i\mathbb{R}^n$ as entire functions $f(z_1, \ldots, z_n)$ satisfying the growth estimates

$$\|f\|_{k,A} = \sup_{z \in \mathbb{C}^n} |f(z_1, \ldots, z_n)| \, \big| \, (1 + |z|)^{-k} e^{-A|\operatorname{Im} z|} < \infty,$$

$$k \in \{0\} \cup \mathbb{N}, \ A > 0.$$

The topology in the space $PW(\mathbb{R}^n)$ is defined by the seminorms $\|f\|_{k,A}$ and the Fourier transform is an isomorphism of the spaces.

We will denote $PW(\mathbb{C}^n)$ the space of entire extensions of functions in $PW(\mathbb{R}^n)$ onto $\mathbb{C}^n$.

Definition. *The subspace $Y \subset \mathcal{E}'(\mathbb{R}^n)$ is called translation-invariant if $f \in Y$, $a \in \mathbb{R}^n$, imply that the translation $f_a(x) = f(x-a)$ is also in Y. We call Y $M(n)$-invariant if for any $f \in Y$ and for any rigid motion $\tau \in M(n)$, the function $f_\tau(x) = f(\tau x)$ belongs to Y.*

Closed translation-invariant spaces $Y \subset \mathcal{E}'(\mathbb{R}^n)$ are characterized by invariance under convolutions: if $f \in Y$ and $g \in \mathcal{E}'(\mathbb{R}^n)$ then $f * g \in Y$. In other words, those subspaces Y are ideals with respect to convolution.

Let Y be a closed translation-invariant subspace in $\mathcal{E}'(\mathbb{R}^n)$. Since the Fourier transform maps convolutions to products, the Fourier image $\hat{Y}$ is a closed ideal in $\widehat{\mathcal{E}'}(\mathbb{R}^n) \cong PW(\mathbb{R}^n)$ with respect to multiplication, i.e., $f \in \hat{Y}$, $g \in \widehat{\mathcal{E}'}(\mathbb{R}^n)$ implies $fg \in \hat{Y}$.

If the subspace $Y \subset \mathcal{E}'(\mathbb{R}^n)$ is $M(n)$-invariant, then the Fourier image $\hat{Y} \subset \widehat{\mathcal{E}'}(\mathbb{R}^n)$ is a *rotation-invariant* ideal, which means that $f \in \hat{Y}$, $k \in \mathcal{O}(n)$ imply $f \circ k \in W$.

The celebrated theorem of L. Schwartz (fundamental theorem of mean-periodic functions) gives necessary and sufficient conditions for an ideal in $PW(\mathbb{C})$ to be dense in $PW(\mathbb{C})$:

Theorem. *[Sch] Let $W \subset PW(\mathbb{C})$ be an ideal (with respect to pointwise mutiplication), whose functions have no common zero in $\mathbb{C}$. Then W is dense in $PW(\mathbb{C})$.*

This theorem can be translated for the space $\widehat{\mathcal{E}'}(\mathbb{R}) \cong PW(\mathbb{R}) = PW(\mathbb{C})\big|_{\mathbb{R}^n}$ as follows: a closed ideal $W \subset PW(\mathbb{R})$ coincides with the whole space $PW(\mathbb{R})$ if and only if extensions of functions in W onto $\mathbb{C}$ have no common zero.

It should be mentioned that the statement is not true for the space $PW(\mathbb{R}^n)$, $n > 1$. This is related to the absence of isolated zeros for entire functions in $\mathbb{C}^n$ when $n > 1$.

However, the spectral synthesis is possible for rotation-invariant ideals, due to the fact that the orbit space $\mathbb{R}^n/\mathcal{O}(n)$ is one-dimensional. Here is the precise statement:

Theorem 9.1. *[BST] Let $W \subset PW(\mathbb{R}^n)$ be a rotation-invariant closed idea. If extensions of functions in W as entire functions in $\mathbb{C}^n$ have no common zero in $\mathbb{C}^n$ then $W = PW(\mathbb{R}^n)$.*

Proof. (Sketch) The proof is based on reducing to the Schwartz theorem by means of radialization of functions in W. Namely, introduce the radialization operator that maps any function in W to the function

$$f^{\#}(x) = \int_{\mathcal{O}(n)} f(kx)\,dk.$$

Clearly $f^{\#}$ is radial, i.e., $f^{\#}$ depends only on $|x|$. If $f(z)$, $z \in \mathbb{C}^n$, is the entire extension of f onto $\mathbb{C}^n$ then the entire extension of $f^{\#}$ is given by

$$f^{\#}(z) = \int_{\mathcal{O}(n)} f(kz)\,dk,$$

where the action $z \to kz$ of the group $\mathcal{O}(n)$ in $\mathbb{C}^n$ is defined naturally: $k(x + iy) = kx + iky$, $x, y \in \mathbb{R}^n$.

The radialization operator maps the space $PW(\mathbb{C}^n)$ into itself. Functions in the image, $[PW(\mathbb{C}^n)]^{\#}$, of this operator can be identified, in a natural way, with functions of one complex variable. Moreover, it appears that they form an ideal in the space of even functions in the one-dimensional Paley-Wiener space. The main part of the proof is to check that functions in the ideal have no common zero. It is done by means of Banach algebras. Then the Schwartz theorem yields that the closure of the above ideal contains the function 1 and therefore $1 \in W$. This finishes the proof. $\qquad\square$

Since $M(n)$-invariant subspaces in $\mathcal{E}'(\mathbb{R}^n)$ and rotation-invariant ideals in $PW(\mathbb{R}^n)$ are isomorphic via the Fourier transform, we have

Corollary. *Let $Y \subset \mathcal{E}'(\mathbb{R}^n)$ be a closed $M(n)$-invariant subspace, Fourier transforms whose functions have no common zero in $\mathbb{C}^n$. Then $Y = \mathcal{E}'(\mathbb{R}^n)$.*

9.6. The Schiffer Conjecture

9.6.1. *Characterization of Pompeiu compacts by zero sets of Fourier transforms of their characteristic functions*

We call a compact $K \subset \mathbb{R}^n$ *Pompeiu compact (domain)* if the Pompeiu transform P_k, defined in Section 9.4, is injective.

Any rigid motion $\tau \in M(n)$ is the composition of translation and rotation, hence a compact K being Pompeiu compact is equivalent to the equation

$$\chi_{\sigma(K)} * f = 0, \quad \text{for all} \quad \sigma \in \mathcal{O}(n), \tag{9.6}$$

$f \in L^1_{\mathrm{loc}}(\mathbb{R}^n)$, has the only trivial solution $f = 0$.

Define Y_K as the translation-invariant closed space in $\mathcal{E}'(\mathbb{R}^n)$, spanned by the functions $\chi_{\sigma(K)}$, $\sigma \in \mathcal{O}(n)$.

The Hahn-Banach theorem and (9.6) yield:

Proposition 9.2. $K \subset \mathbb{R}^n$ *is a Pompeiu compact if and only if* $Y_K = \mathcal{E}'(\mathbb{R}^n)$.

Proposition 9.3. $K \subset \mathbb{R}^n$ *is a Pompeiu compact if and only if the Fourier transforms* $\widehat{\chi_{\sigma^{-1}(K)}} = \hat{\chi}_K \circ \sigma$, $\sigma \in \mathcal{O}(n)$, *have no common zero in* $\mathbb{C}^n$.

The functions $\hat{\chi}_K \circ \sigma$ have a common zero $z_0 \in \mathbb{C}$ means that $\hat{\chi}_K$ vanishes identically on the orbit $\mathcal{O}(n) \circ z_0 = \{\sigma z_0 : \sigma \in \mathcal{O}(n)\}$ of the point z_0 under the group $\mathcal{O}(n)$.

When σ runs the group $\mathcal{O}(n)$, the point $\sigma z_0 = \sigma x_0 + i\sigma y_0$ runs all $z = x + iy \in \mathbb{C}^n$ such that $|x| = |x_0|$, $|y| = |y_0|$ and $(x, y) = (x_0, y_0)$. The latter conditions can be written in complex form as

$$z_1^2 + \cdots + z_n^2 = \alpha + i\beta, \quad \text{where} \quad \alpha = |x_0|^2 - |y_0|^2$$

and $\beta = 2(x_0, y_0)$, with the additional restriction $|x| = |x_0|$.

Thus, the orbit $\mathcal{O}(n) \cdot z_0$ is the intersection of the quadric $Q : z_1^2 + \cdots + z_n^2 = \alpha + i\beta$ with the cylinder $|\operatorname{Re} z| = |x_0|$. This intersection is a $2n - 3$-dimensional real submanifold of the complex hypersurface Q in $\mathbb{C}^n$ and the uniqueness theorem implies that the vanishing of the entire function $\hat{\chi}_K$ on the orbit $\mathcal{O}(n) \cdot z_0$ is equivalent to the vanishing of $\hat{\chi}_K$ on the whole quadric Q.

In view of Proposition 9.3, we have, by denoting $\alpha + i\beta = \lambda^2$:

Theorem 9.4. *[BST] The compact* $K \subset \mathbb{R}^n$ *fails to be Pompeiu if and only if the Fourier transform* $\hat{\chi}_K$ *of its characteristic function vanishes on a quadric* $Q_\lambda = \{z \in \mathbb{C}^n : z_1^2 + \cdots + z_n^2 = \lambda^2\}$ *for some* $\lambda \in \mathbb{C} \setminus \{0\}$ *(*$\hat{\chi}_K$ *cannot vanish at* $\lambda = 0$ *because* $\hat{\chi}_K(0) = \int_K dx = \operatorname{mes} K > 0$*).*

9.6.2. *Overdetermined Dirichlet-Neumann boundary value problem*

Let $K \subset \mathbb{R}^n$ be a non-Pompeiu compact of positive measure. Then, as we already know, the Fourier transform $\hat{\chi}_k$ vanishes on a quadric $z_1^2 + \cdots + z_n^2 - \lambda^2$, $\lambda \in \mathbb{C} \setminus \{0\}$. It is equivalent for the entire function $\hat{\chi}_K$ in $\mathbb{C}^n$ to be divisible by the polynomial $z_1^2 + \cdots + z_n^2 - \lambda^2$:

$$\hat{\chi}_K = (z_1^2 + \cdots + z_n^2 - \lambda^2)v,$$

where v is also an entire function in $\mathbb{C}^n$.

Moreover, $v \in PW(\mathbb{C}^n)$ and therefore v is the Fourier transform, $v = \hat{u}$ of a compactly supported function u. Applying the Fourier transform yields:

$$(\Delta + \lambda^2)u = \chi_K \quad \text{in} \quad \mathbb{R}^n. \tag{9.7}$$

The estimate

$$|v(x)| \leq \frac{|\hat{\chi}_K(x)|}{|x|^2 - |\lambda|^2} \leq \frac{1}{|x|^2 - |\lambda|^2}, \quad x \in \mathbb{R}^n,$$

shows, due to the known relation between the rate of decay of a function and the smoothness of its Fourier transform, that $u = \check{v}$ (the inverse Fourier transform) belongs to $C^1(\mathbb{R}^n)$.

Outside of the compact K, the function u solves the equation $(\Delta + \lambda^2)u(x) = 0$, $x \in \mathbb{R}^n \setminus K$, and therefore u is real analytic in the complement of K. Since u has compact support, we conclude that u vanishes on the unbounded component of $\mathbb{R}^n \setminus K$.

Conversely, if (9.7) is solvable, then applying the inverse Fourier transform leads to the conclusion that $\hat{\chi}_K$ vanishes on the quadric $z_1^2 + \ldots z_n^2 = \lambda^2$ and therefore K fails to be Pompeiu compact.

From now on, we will deal with bounded domains $\Omega \subset \mathbb{R}^n$ with connected boundary. According to (9.7), the domain $\overline{\Omega}$ is not Pompeiu if and only if there exists $u \in C^1(\overline{\Omega})$ and $\lambda \in \mathbb{C} \setminus \{0\}$ such that $(\Delta + \lambda^2)u = 1$ in Ω and $u(x) = 0$, $\nabla u(x) = 0$ for $x \in \partial\Omega$ (because u possesses a C^1-extension in $\mathbb{R}^n$ vanishing on $\mathbb{R}^n \setminus \Omega$).

Replacing u by $-\lambda^2 u + 1$ reduces the equation to $(\Delta + \lambda^2)u = 0$, $u(x) = 1$, $\nabla u(x) = 0$ for $x \in \partial\Omega$, and we arrive at

Theorem 9.5. *[BST] Let Ω be a bounded domain in $\mathbb{C}^n$ with connected boundary. Then Ω fails to be a Pompeiu domain if and only if the overdetermined Dirichlet-Neumann boundary value problem*

$$(\Delta + \lambda^2)u = 0 \quad in \quad \Omega \tag{9.8}$$

$$u\big|_{\partial\Omega} = 1$$

$$\nabla u\big|_{\partial\Omega} = 0$$

has a solution $u \in C^1(\overline{\Omega})$ for some $\lambda \in \mathbb{C} \setminus \{0\}$.

Remark. The condition of vanishing the gradient ∇u on $\partial\Omega$ can be replaced by the Neumann boundary condition $\frac{\partial u}{\partial \nu} = 0$, where ν is the unit normal vector to $\partial\Omega$. Also, since the Laplace operator with the Neumann boundary condition is self-adjoint, the eigenvalue λ^2 is real positive

Now the Pompeiu problem can be reformulated in terms of solvability of the boundary value problem (9.8). This reformulation is known as the Schiffer's problem:

Schiffer's Conjecture. *Euclidean balls are the only bounded domain with connected Lipschitz boundary for which the problem (9.8) has a solution.*

Remark. If $\Omega = B(0, R)$ is a ball in $\mathbb{R}^n$ (centered at 0), then the solution to (9.8) exists. It is radial and is expressed via Bessel functions:

$$u(x) = \mathrm{const}\, \frac{J_{n-2}}{2}(\lambda|x|)(\lambda|x|)^{\frac{2-n}{2}}.$$

The constant is chosen to satisfy the condition $u(R) = 1$, and the spectrum $\{\lambda\}$ is countable and is defined by the conditin $\frac{J_n}{2}(\lambda R) = 0$. This follows from the known identity $\frac{d}{dt} J_m(t)t^{-n} = J_{m+1}t^{-(m+1)}$.

Let us formulate here briefly some known results on Schiffer's Conjecture, For further information we refer the reader to the survey of L. Zalcman [Za2], where the state of the art until 1991 is presented. The articles [Za1] and [Za3] give a lovely introduction to the subject.

Regularity of the boundary. A result of Williams [Wi] states that if the problem (9.8) is solvable and $\partial\Omega$ is Lipschitz then $\partial\Omega$ is real-analytic. This means that domains with non real-analytic boundary of the Lipschitz class, for instance polygons, possess the Pompeiu property, i.e., provide injectivity of the Pompeiu transform. Therefore one can assume the real-analytic boundary $\partial\Omega$ in the Schiffer Conjecture.

Infinite set of eigenvalues. C. Berenstein [Be] proved that if (9.8) is solvable for infinitely many eigenvalues λ, then Ω is a ball.

Domains far from balls. Results of Brown and Kahane [BK] state that if Ω is a convex domain whose length is twice its width (cigar-like domain), then Ω is a Pompeiu domain, i.e., (9.8) is unsolvable.

Schiffer Conjecture for partial classes of domains.

Garofalo and Segala [GS] proved the Schiffer Conjecture for domains in the plane whose boundaries are images of the unit circle under a trigonometric polynomia. Ebenfelt [E] proved the Schiffer Conjecture for so-called quadrature domains – images of the unit disk under rational conformal mappings.

9.7. Characterization of Euclidean Balls by Multiplicity of Dirichlet-Neumann Eigenvalues

Let us write again the over-determined Dirichlet-Neumann problem that we deal with:

$$(\Delta + \lambda^2)u = 0 \quad \text{in} \quad \Omega \tag{DN}$$

$$u\big|_{\partial\Omega} = c, \ \frac{\partial u}{\partial v}\Big|_{\partial\Omega} = 0$$

where c is a constant and Ω is a bounded domain with connected real-analytic boundary.

In a parallel way, we consider the Dirichlet problem:

$$(\Delta + \lambda^2)v = 0 \quad \text{in} \quad \Omega \tag{D}$$

$$u\big|_{\partial\Omega} = 0.$$

By $m(\lambda)$ we denote the multiplicity λ as a Dirichlet eigenvalue, i.e., $m(\lambda)$ is the dimension of the solution space to (D).

Note that if Ω is a ball, $\Omega = B(0, R)$, and λ is an eigenvalue for the problem (DN) then separation of variables in spherical coordinates implies that $u(x) = \text{const} \ \dfrac{J_{\frac{n-2}{2}}(\lambda|x|)}{|x|^{\frac{n-2}{2}}}$, and the relation $\frac{d}{dt}\left(J_{\frac{n-2}{2}}(t)t^{\frac{2-n}{2}}\right) = J_{\frac{n}{2}}(t)t^{-n/2}$, and the condition on normal derivative $\frac{\partial u}{\partial v}\big|_{\partial B(0,R)} = 0$ yields that λ must be chosen so that $J_{\frac{n}{2}}(\lambda R) = 0$.

Again, it follows by separation of variables in the problem (D) that the solution space for the problem (D) is spanned by n linearly indpendent solutions

$$x_i J_{\frac{n}{2}}(\lambda|x|)(|x|)^{-n/2}, \quad i = 1, \ldots, n.$$

Thus, if Ω is a ball, then the D-multiplicity $m(\lambda)$ of any D-eigenvalue λ is precisely n.

We want to prove that the opposite is also true. Namely, the following holds:

Theorem 9.6. *Let Ω be a bounded domain in $\mathbb{R}^n$ with real-analytic boundary, homeomorphic to the Euclidean ball. If there exists a DN-eigenvalue $\lambda \neq 0$ such that its D-multiplicity $m(\lambda)$ satisfies $m(\lambda) \leq n$, then D is a ball (and, in fact, $m(\lambda) = n$).*

Proof. We break the proof into 2 main steps.

Vector fields annihilating solutions.

Let $u \neq 0$ be the (unique) solution for the problem (DN) with the given eigenvalue λ. Since $\lambda \neq 0$ then $u \neq$ const. Now we can generate solutions to (D) in the following way. Let us consider the vector fields

$$X_i = \frac{\partial}{\partial x_i}, \quad i = 1, \ldots, n, \quad \text{and} \quad D_{ij} = x_i \frac{\partial}{\partial x_j} - x_j \frac{\partial}{\partial x_i}, \ 1 \leq i \leq j \leq n$$

which form a basis in the Lie algebra G of the motion group $M(n)$ of $\mathbb{R}^n$. Here X_i are infinitesimal translations and D_{ij} are infinitesimal rotations.

Since transformations from the motion group $M(n)$ commute with the Laplace operator Δ, the infinitesimal generators X_i and D_{ij} commute with Δ as well. Therefore $X_i u$ and $D_{ij} u$ are eigenfunctions corresponding to the eigenvalue λ. The boundary conditions for u in (DN) imply $\nabla u\big|_{\partial\Omega} = 0$ and therefore $X_i u$ and $D_{ij} u$ vanish on $\partial\Omega$.

Thus, $X_i u$ and $D_{ij} u$ solve the problem (D) in the domain Ω. Now observe that the functions $X_1 u, \ldots, X_n u$ are linearly independent. Indeed, $\alpha_1 X_1 u + \cdots + \alpha_n X_n u = 0$ for some $\alpha_i \in \mathbb{R}$ would mean that u is constant on any line orthogonal to the hyperplane $\alpha_1 x_1 + \cdots + \alpha_n x_n = 0$ which in turn, along with $u\big|_{\partial\Omega} = c$ would imply $u(x) = c$ for all $x \in \Omega$, which is not the case.

Therefore, since any $n + 1$ solutions to (D) are linearly dependent we obtain that each solution $D_{ij} u$ is a linear combination of $X_1 u, \ldots, X_n u$:

$$D_{ij} u = \sum_{k=1}^{n} c_{ij}^k X_k u.$$

Denote $T_{ij} = D_{ij} - \sum_{k=1}^{n} c_{ij}^k X_k$ so that $T_{ij} u = 0$.

Group symmetries of the level surfaces of the solution.

Let us consider the subspace $\mathfrak{N} \subset \mathcal{G}$ defined as

$$\mathfrak{N} = \{T \in \mathcal{G} : Tu = 0\}.$$

The subspace $\mathfrak{N}$ is a Lie subalgebra in $\mathcal{G}$ because $T, S \in \mathfrak{N}$ imply that the commutator $[T, S] \in \mathfrak{N}$ $([T, S]u = TSu - STu = 0)$.

Denote by N the closed connected Lie subgroup $N \subset M(n)$ with the Lie algebra $\mathfrak{N}$, i.e., $N = \exp \mathfrak{N}$ and let us prove that N is compact. To this end, pick a point $x_0 \in \Omega$ such that $u(x_0) \neq c$, where c is the boundary value of u and consider the orbit

$$N_{x_0} = \{gx : g \in N\}$$

of the point x_0. By construction, $u\big|_{N_{x_0}} = \text{const}$. Since $u(x_0) \neq c = u\big|_{\partial\Omega}$, the orbit N_{x_0} is disjoint from the boundry $\partial\Omega$. Since N_{x_0} is a connected manifold, it is entirely contained in ω and hence is bounded.

Additionally, $N_{x_0} = N/N \cap \mathcal{O}_{x_0}(n)$, where $\mathcal{O}_{x_0}(n) \subset M(n)$ is the stationary subgroup at x_0, is topologically closed and hence N_{x_0} is compact. It follows that the group N is compact as $N \cap \mathcal{O}_{x_0}(n)$ is compact and the coset space N_{x_0} is compact as well.

Any compact Lie subgroup $N \subset M(n)$ is contained in a maximal compact subgroup K of $M(n)$ that is known to be conjugated to $\mathcal{O}(n)$:

$$K = g\mathcal{O}(n)g^{-1} \quad \text{for some} \quad g \in M(n)$$

The corresponding inclusion for Lie algebras holds: $\mathfrak{N} \subset \mathfrak{K} \cong \mathcal{O}(n)$, $\mathcal{O}(n)$ is the Lie algebra of $\mathcal{O}(n)$. Then $\dim\mathfrak{N} \leq \dim\mathfrak{K} = \dim\mathcal{O}(n)$. On the other hand, the vector fields T_{ij} which were constructed earlier belong to $\mathfrak{N}$ and are linearly independent. Indeed, if $\sum \alpha_{ij}T_{ij} = 0$ then

$$\sum_{ij} \alpha_{ij}D_{ij} - \sum_{k}\left(\sum_{ij}\alpha_{ij}e_{ij}^{k}\right)X_k = 0$$

and $\alpha_{ij} = 0$ because the vector fields D_{ij}, X_i are linearly independent(they constitute a basis in $\mathcal{G}$.)

The number of the vector fields T_{ij} equals the number of the operators D_{ij}, i.e., the dimension of the Lie algebra $\mathcal{O}(n)$. Thus, $\dim\mathfrak{N} \geq \dim\mathcal{O}(n)$ and finally $\dim\mathfrak{N} = \dim\mathcal{O}(n) = \dim\mathfrak{K}$. But then $\mathfrak{N} = \mathfrak{K}$ because $\mathfrak{N}$ is a subspace of the finite dimensional space $\mathfrak{K}$.

Coincidence of the Lie algebras implies local homomorphy of the corresponding Lie groups $N \subset K = \exp\mathfrak{K}$, and the subgroup N is the factor group $N = K/H$ over a discrete, and therefore finite, normal subgroup $H \subset K$.

Then any two orbits N_x and K_x, $x \in \Omega$, have a common open subset $\{kx : k \in U\}$, where $H \cap U = \{e\}$, e is the unit element, and since the orbits are connected real-analytic submanifolds,then they coincide everywhere, $N_x = K_x$.

Observe that the orbit $K_x = \{gkg^{-1}(x) : k \in \mathcal{O}(n)\}$ is the Euclidean space centered at $g(0)$ and passing through x. The function u is constant on orbits of the group N and therefore the level sets of u are spheres K_x, $x \in \Omega$. Then the boundary $\partial\Omega$ must be one of the orbits since otherwise $u\big|_{\partial\Omega} = \text{const}$ would imply $u = \text{const}$ near $\partial\Omega$ and hence $u = \text{const}$ on Ω due to real-analyticity of u.

Thus, $\partial\Omega$ is a sphere and, corresondingly, Ω is a Euclidean ball. $\qquad\square$

9.8. Pompeiu-Schiffer Problem for Domains Close to Balls

Now we will use the result of the previous section to study stability in the Pompeiu-Schiffer problem.

We will show that if a sequence of non-Pompeiu domains with bounded DN-eigenvalues shrink to a ball, then all these domains, except a finite number of them, are balls.

By convergence of a sequence of domains, we understand the following:

Let Ω_n be a sequence of compact domains in $\mathbb{R}^n$, each of which is C^2-diffeomorphic to the unit ball $B \subset \mathbb{R}^n$, *id* stands for the identical mapping.

We say that Ω_n C^2-converge to B if the C^2-diffeomorphisms $F_n : B \to \Omega_n$ can be chosen so that $F_n \to id$ in $C^2(B)$.

Theorem 9.7. *[AS] Let Ω_n be a sequence of compact domains in $\mathbb{R}^n$, C^2-diffeomorphic to the unit ball $B \subset \mathbb{R}^n$. Suppose that each Ω_n is a non-Pompeiu domain, i.e., the problem (DN) is solvable, and assume that $\sup|\lambda_n| < \infty$, where λ_n are the DN-eigenvalues. Then then there exists n_0 such that all Ω_n with $n \geq n_0$ are Euclidean balls.*

We describe the main steps of the proof; for a detailed proof we refer the reader to [AS].

Proof. (Sketch)

Step 1. If the conclusion of the theorem is not true, then there exists a subsequence Ω_{n_k} of domains different from the balls. By replacing the sequence Ω_n by subsequence Ω_{n_k}, one can assume that no domain Ω_n is a ball.

There exists a converging subsequence λ_{n_k} of eigenvalues, $\lambda_{n_k} \to \lambda_0$, and, after one more renumerating, it can be assumed that $\lambda_n \to \lambda_0$.

The boundary value problem (DN) in each domain Ω_n can be written in the weak form

$$\int_{\Omega_n} (u_n - c_n)\Delta v dx = -\lambda_n^2 \int_{\Omega_n} w dx, \qquad (9.9)$$

for arbitrary $v \in L^w(\mathbb{R}^n) \cap C^2(\mathbb{R}^n)$. Here u_n is the solution to $(\Delta + \lambda_n^2)u_n = 0$ in Ω_n with the boundary conditions $u_n\big|_{\partial\Omega_n} = c_n$, $\frac{\partial u_n}{\partial v}\big|_{\partial\Omega_n} = 0$, v is normal vector to $\partial\Omega_n$.

Normalize u_n by the condition $\int_{\Omega_n} |u_n|^2 dx = 0$ and extend u_n to $\mathbb{R}^n$ by $u_n(x) = 0$, $x \in \mathbb{R}^n \setminus \Omega$. Using compactness of the unit sphere in $L^2(\mathbb{R}^n)$ with respect to the weak topology, one can select a weak-converging subsequence $u_{n_k} \xrightarrow{w} u_0$, $k \to \infty$.

Again, we can assume that $u_n \xrightarrow{w} u_0$, $n \to \infty$, by replacing the sequence Ω_n by its subsequence. It can be easily seen from the identity (9.9) that the sequence c_n is bounded (if $c_{n_k} \to \infty$ then dividing both sides by c_n and taking into account boundedness of λ_n and $\|u_n\|_{L^2}$ and letting $k \to \infty$ we would get $\int_{\Omega_n} \Delta v dx = 0$ which is not the case).

Thus there exists a convergent subsequence $c_{n_k} \to c_0$. One more renumeration leads us to the following situation:

$$u_n \xrightarrow{w} u_0, \quad \lambda_n \to \lambda_0, \quad c_n \to c_0.$$

Step 2. The L^2-function u_0 is a weak solution to

$$\int_B (u_0 - c_0)\Delta v dx = -\lambda_0^2 \int_{\Omega_n} u_n v dx, \ v \in L^2(\mathbb{R}^n) \cap C^2(\mathbb{R}^n). \tag{9.10}$$

in the unit ball. The regularity theorem for elliptic equations implies that u_0 is a smooth solution. Moreover, it follows that u_0 is radial and therefore coincides with a Bessel function. In particular, u_0 is in $C^1(B)$ and (9.10) implies $u_0\big|_{\partial B} = c_0$, $\frac{\partial u_0}{\partial v}\big|_{\partial B} = 0$.

To conclude that u_0 is a DN-eigenfunction, it remains to check that $u_0 \neq 0$, or equivalently , $c_0 \neq 0$. This is done by energy estimates [AS].

Step 3. The final argument rests on Theorem 9.6 and the theory of perturbations of operators (see [Ka]). First of all, the diffeomorphism $F_k :$ $B \to \Omega_k$ enables us to transfer the Laplace operator on Ω_n to the differential operator Δ_k defined on a function in the unit ball B by

$$\Delta_k v = \Delta(v \circ F_k^{-1}).$$

Then Δ_k has the form

$$\Delta_k = \sum_{i_j=1}^{n} A_{ij}^k(x) \partial_{x_i x_j}^2$$

and $F_k \to id$ in $C^2(B)$ implies that the matrix $A_{ij}^k(x)$ converges to the unit matrix as $k \to \infty$, i.e., Δ_k converge to Δ.

It was shown in Step 2 that λ_0 is a DN-eigenvalue for the unit ball and therefore the Dirichlet multiplicity $m(\lambda_0) = n$. According to the perturbation theory, small perturbations do not increase the multiplicity and hence the eigenvalues λ_k of the perturbed operators Δ_k, close to λ_0, have D-multiplicities $m(\lambda_k) \leq n$. Theorem 9.6 says that then Ω_k is a ball. This contradicts the assumption that no Ω_k is a ball. $\qquad\square$

For the case n=2, a different proof of stability of discs in the Pompeiu problem was given in [Ag]. There the proof uses conformal mappings and family of domains in the question are images under conformal perturbations of the identical map of the unit disc.

Theorem 8.1 implies that any small C^2-deformation of the unit ball by non-Pompeiu domains with bounded DN-eigenvalues reduces to trivial, affine deformation of the unit ball by balls. In other words, balls constitute a homotopy class in the space of all non-Pompeiu domains equipped with a proper topology.

To confirm the Schiffer Conjecture, it would suffice to prove that any non-Pompeiu domain can be deformed into a within the space of non-Pompeiu domains. Consider the case $n = 2$. Let Ω be a simply-connected bounded doman in $\mathbb{C}$, with real-analytic boundary. A natural deformation Ω into the unit disk can be constructed as follows. Assume that $0 \in \Omega$ and let

$$\omega : \Delta \to \Omega$$

be a Riemannian mapping of the unit disk $\Delta \subset \mathbb{C}$ onto Ω with $\omega(0) = 0$. For any $0 < t < 1$ define

$$\omega_t(z) = \frac{1}{t}\omega(tz)$$

and let $\Omega_t = \omega_t(\Delta)$. Then $\omega_t(z) = c_1 z + tc_2 z + \ldots$, where $\omega(z) = \sum_{n=1}^{\infty} c_n z^n$ and we see that $\omega_t \to id$ as $t \to 0$. Therefore, Ω_t tend to the unit disk Δ, as $t \to 0$.

In view of Theorem 8.1, the positive answer to the following conjecture would lead to solving the entire problem:

Conjecture. *If Ω is a non-Pompeiu domain then for any $0 < t < 1$ the domain Ω_t is also a non-Pompeiu domain.*

It is natural to call the domains Ω_t hyperbolically homothetic to Ω. Therefore the above conjecture is that if the over-determined boundary value problem (DN) in Section 9.7 is solvable for Ω then it is solvable for any hyperbolically homothetic domain Ω_t. Note that this statement is consistent with the Schiffer-Pompeiu conjecture that (DN) is solvable only for disks, because domains hyperbolically homothetic to disks are disks. T.Kobayashi [Ko] obtained a stability result similar to Theorem 8.1 by studying perturbations of zero sets of the Fourier transform of the characteristic functions of the domains Ω_n. The relation of such sets to the Pompeiu property is given in Theorem 6.3.

9.9. Spherical Transform Revisited

9.9.1. *Injectivity of pairs of Pompeiu spherical transforms (two-radii theorems)*

Recall that the spherical transform is defined by

$$Mf(x,r) = M^r f(x) = \int_{S(x,r)} f(y)dA(y),$$

where $dA(y)$ is the normalized surface area on the sphere $S(x,r) = \{y \in \mathbb{R}^n : |y - x| = r\}$.

Clearly, the spherical transform can be regarded as the Pompeiu transform P_K with $K = S(0,r)$, defined in Section 9.6.

As we saw in Section 9.6, the transforms $P_{S(0,r)}$ and $P_{B(0,r)}$ (over spheres or balls) fail to be injective. To get injectivity, one considers the pair of the Pompeiu transforms over spheres:

$$P_{r_1,r_2} f(x) = (Mf(x,r_1), Mf(x,r_2))$$

that defines the operator

$$P_{r_1,r_2} : L^1_{\mathrm{loc}}(\mathbb{R}^n) \to C(\mathbb{R}^n, \mathbb{R}^2).$$

The following result, due to L. Zalcman, gives all values r_1, r_2 providing injectivity of the transform P_{r_1,r_2} :

Theorem. *[Za2] The tranfsform P_{r_1,r_2} $(r_1, r_2 > 0)$ is injective so long as $\frac{r_1}{r_2} \neq \frac{z_1}{z_2}$, where z_i are zeros of the Bessel function $J_{\frac{n-2}{2}}$.*

The proof is inferred from the fundamental thereom of Schwartz (Section 9.4). The condition $P_{r_1,r_2}f = 0$ can be rewritten as the pair of the convolution equations

$$f * \sigma_{r_1} = f * \sigma_{r_2} = 0,$$

where σ_r is the surface measure on the sphere $S(0,r)$. Now, the condition for the radii can be translated as the absence of common zeros for the functions

$$\hat{\sigma}_{r_i}(\lambda) = J_{n/2}(|\lambda|r_i)(|\lambda|r_i)^{\frac{2-n}{2}}, \; i = 1,r$$

which, in turn, implies by the Schwartz theorem that the ideal in $\mathcal{E}'(\mathbb{R}^n)$ spanned by the distributions σ_{r_1} and σ_{r_2} is dense in $\mathcal{E}'(\mathbb{R}^n)$. This implies $f = 0$. Conversely, if $\frac{r_1}{r_2} = \frac{z_1}{z_2}$ with $J_{\frac{n-2}{2}}(z_2) = J_{\frac{n-2}{2}}(z_0) = 0$, then the function $f(x) = e^{i(\lambda, x)}$, $|\lambda| = \frac{z_1}{r_1} = \frac{z_2}{r_2}$, is in the kernel of the transform P_{r_1, r_2}.

C. Berenstein and R. Gay [BY] proved local variant of two-radii theorem where one integrates over pairs of spheres contained in a given ball. Series of refined versions of two-radii theorems were obtained by V.Volchkov, see his book [Vo] and the references there.

9.9.2. *Injectivity problem for the spherical Radon projection*

We will call the function $Mf(x, \cdot)$ the *spherical Radon projection* of a functional at the point $x \in \mathbb{R}^n$.

Let X be a class of functions in $\mathbb{R}^n$ (e.g., $X = C(\mathbb{R}^n)$, $C_c(\mathbb{R}^n)$, $L^p(\mathbb{R}^n)$, $S(\mathbb{R}^n)$).

Definition. *A set $\Gamma \subset \mathbb{R}^n$ is said to be a set of injectivity (on X) if $Mf(x, \cdot) = 0$, for all $x \in \Gamma$, implies $f = 0$, for any $f \in X$.*
In other words, a set Γ is a set of injectivity if vanishing of all spherical means woth centers on Γ implies that the function is identicaly zero.

A trivial example of an injectivity set for $C(\mathbb{R}^n)$ is the whole space $\mathbb{R}^n$. The inversion formula for the spherical Radon projection in this case is also trivial: $f(x) = \lim_{r \to 0} Mf(x, r)$. A simple example of a set of non-injectivity (in any reasonable class of functions) is a hyperplane $\Gamma \subset \mathbb{R}^n$. Indeed, for any function f that is odd with respect to reflections around Γ we have $Mf(x, r) = 0$ if $x \in \Gamma$.

The general problem we are going to discuss is:

Problem. *Given a class of functions X, characterize the injectivity sets of the spherical Radon projection.*

Let us emphasize the difference between the injectivity problems for the Pompeiu spherical transform and for the spherical Radon projection. In the first case, the centers of the spheres of integration are arbitrary and the set of radii is under question, while in the second case, conversely, radii are arbitrary and the set of the centers is to be described.

Let us show that this problem is reduced to the injectivity problem for the plane Radon projection (see Section 9.1) restricted to a certain quadric.

Let $\Gamma \subset \mathbb{R}^n$. Associate with the set Γ the cone $C_\Gamma \subset \mathbb{R}^{n+1}$:

$$C_\Gamma = \{-2xt, t) : x \in \Gamma, \ t \in \mathbb{R}\}.$$

Define the paraboloid $\Pi \subset \mathbb{R}^{n+1}$:

$$\Pi : x_{n+1} = x_1^2 + \cdots + x_n^2.$$

Proposition 9.8. *The set $\Gamma \subset \mathbb{R}^n$ is an injectivity set for the spherical Radon projection on $C(\mathbb{R}^n)$ if and only if the cone C_Γ is a cone of injectivity for the plane Radon transform P^ω on $C(\Pi)$, that is if f is a continuous function on the paraboloid Π and $P^\omega f(d) = Rf(\omega, d) = 0$, for all directions $\omega \in C_\Gamma$, and for all distances $d \in \mathbb{R}$ then $f = 0$. Here f is understood as a distribution supported by the paraboloid Π and the Radon transform Rf is understood in the distributional sense.*

Proof. The cone C_Γ being a cone of injectivity for P^ω on the class $C(\Pi)$ means that if $f \in C(\Pi)$ and $P^\omega f = 0$, $\omega \in C_\Gamma$, then $f = 0$.

Consider a hyperplane $(\omega, x) = d$ in $\mathbb{R}^{n+1}$ with $\omega_{n+1} \neq 0$. The intersection of this hyperplane with the paraboloid Π is given by

$$\sum_{i=1}^n \left(x_i + \frac{1}{2} \frac{\omega_i}{\omega_{n+1}} \right)^2 = \frac{d}{\omega_{n+1}} - \frac{1}{4} \sum_{i=1}^n \frac{\omega_i^2}{\omega_{n+1}^2},$$

and therefore the orthogonal projection

$$\pi : \Pi \to \mathbb{R}^n, \qquad \pi(x_1, \ldots, x_n, x_{n+1}) = (x_1, \ldots, x_n)$$

maps the cross-section $\Pi \cap \{(\omega, x) = d\}$ onto the sphere $S(a, r) \subset \mathbb{R}^n$ with $a = -\frac{1}{2\omega_{n+1}}(\omega_1, \ldots, \omega_n)$ and $r = \frac{d}{\omega_{n+1}} - \frac{1}{4}\sum_{i=1}^n \frac{\omega_i^2}{\omega_{n+1}^2}$.

Denote $t = \omega_{n+1}$, then the vector ω is represented as

$$\omega = (\omega_1, \ldots, \omega_n, \omega_{n+1}) = (-2ta, t)$$

and $\omega \in C_\Gamma$ whenever $a \in \Gamma$. Therefore, the family of spheres $S(a, r)$, $a \in \Gamma$, $r > 0$ is just the projection under π of the family of parallel cross-sections of the paraboloid Π by the hyperplanes $(\omega, x) = d$, $d \in \mathbb{R}$. Then we have the equivalence: $Mf(a, \cdot) = 0$ for all $a \in \Gamma$ $(f \in C(\mathbb{R}^n))$ if and only if $P^\omega(f \circ \pi) = 0$ for all $\omega \in C_\Gamma$ and the proposition follows. $\square$

Remark. Proposition 9.8 can be understood in analytic terms, in the spirit of the Projection Slice Theorem.

Indeed, assume that f decays at ∞, for instance, $f \in C_c(\mathbb{R}^n)$. The condition $Mf(x, r) = 0$ for $x \in \mathcal{B} \subset \mathbb{R}^n$ and $r > 0$ is equivalent to

$$\int_{\mathbb{R}^n} e^{-it|x-y|^2} f(y)dy = 0, \quad \text{for all} \quad x \in \Gamma \quad \text{and} \quad t \in \mathbb{R},$$

which is equivalent to

$$\int_{\mathbb{R}^n} e^{-i(-2t(x,y)+|y|^2)} f(y)dy = 0.$$

The latter can be written as

$$U(-2tx, t) = 0 \quad \text{for} \quad (x, t) \in \Gamma \times \mathbb{R},$$

where

$$U(\lambda_1, \ldots, \lambda_{n+1}) = \int_{\mathbb{R}^n} e^{-i(\lambda_1 y_1 + \ldots \lambda_n y_n + \lambda_{n+1} \| y \|^2)} f(y)dy.$$

The function U is the Fourier transform of the distribution $f \circ \pi$ with the support on the paraboloid P.

Thus we see that vanishing of spherical means $Mf(x, \cdot) = 0$, $x \in \Gamma$ on Γ is equivalent to vanishing of the Fourier transform on the generated cone C_Γ, $U|_{C_\Gamma} = 0$. In turn, by the Projection Slice Theorem vanishing of the Fourier transform is equivalent to vanishing of integrals over all hyperplanes with the normal vectors from C_Γ: $P^\omega(f \circ \pi) = 0$, $\omega \in C_\Gamma$.

9.10. Injectivity of the Spherical Radon Projection on Compactly Supported Functions

9.10.1. *The case $n = 2$*

The following theorem gives a complete characterization of the injectivity set for the spherical Radon projection on the space $C_c(\mathbb{R}^2)$ of compactly supported functions in the plane.

Theorem 9.9. *[AQ1] A set $\Gamma \subset \mathbb{R}^2$ is a set of injectivity on $C_c(\mathbb{R}^2)$ if and only if Γ is contained in no set of the form $\Sigma \cup V$, where V is finite and Σ is a bunch of lines through one point with equal angles between adjacent lines (Coxeter system of lines).*

For the detailed proof, we refer the reader to the article [AQ1], while the idea of the proof will be given later, when we discuss the case of arbitrary dimension. Let us comment here only on the "if" part of Theorem 10.1.

Any Coxeter system Σ of lines fails to be a set of injectivity for the following reason. The group $W(\Sigma)$ generated by reflections around the lines in Σ is finite, and any function that is odd with respect to these lines belongs to the kernel of the spherical transform with centers on Σ. It is a little bit more complicated to show that adding any finite set V preserves the kernel to be nontrivial.

Now, a Coxeter system Σ of lines in $\mathbb{R}^2$ can be described in the complex form by the equation $\operatorname{Re} c(z-a)^k = 0$, $z \in \mathbb{C}$, where $c, a \in \mathbb{C}$, a is the vertex (common point) of Σ and $\arg c$ defines rotation of the configuration of lines. Any harmonic homogeneous polynomial h in $\mathbb{R}^2$ can be written as $h(x,y) = \operatorname{Re} c(x+iy)^k$ and therefore Coxeter systems of lines are just translations of zero sets of harmonic homogeneous polynomials. This rewording provides a bridge to the expected result in higher dimensions.

9.10.2. *The case $n > 2$*

The problem of full characterization of an injectivity set on the space $C_c(\mathbb{R}^n)$ with $n > 2$ is still open. On of the reasons is that not much is known about zero sets of harmonic polynomias of n real variables, when $n > 2$. For $n = 2$, the structure of these sets is well understood and this is one of the ingredients of the proof of Theorem 9.9.

Here are some preliminary observations.

Proposition 9.10. *Let $\Gamma \subset \mathbb{R}^n$ is not a set of injectivity on the space $C_c(\mathbb{R}^n)$. Then there is a polynomial P in $\mathbb{R}^n$ with real coefficients, which divides some nonzero harmonic polynomial (P is an harmonic divisor) and an algebraic variety $V \subset \mathbb{R}^n$ of dimension $\dim V < n - 1$ such that*

$$\Gamma \subset P^{-1}(0) \cup V.$$

Proof. Let $f \in C_c(\mathbb{R}^n)$, $f \neq 0$, such that

$$Mf(x,r) = 0 \quad \text{for all} (x,r) \in \gamma \times \mathbb{R}_+.$$

Define polynomials P_k :

$$P_k(x) = \int_{\mathbb{R}^n} |x - y|^{2k} f(y)dy.$$

Not all P_k are zero since otherwise $f = 0$. It is easy to show that $\int_{|x-y|=r} f(y)dy = 0$, for all $r > 0$, is equivalent to $P_k(x) = 0$, for $k = 0, 1, \ldots$, and therefore

$$\Gamma \subset \bigcap_{k=0}^{\infty} P_k^{-1}(0).$$

Let $\tilde{P}_k$ be the natural extension of P_k to $\mathbb{C}^n$.

The Hilbert Finiteness Theorem [vW] states that the intersection of the complex algebraic varieties $\tilde{P}_k^{-1}(0)$ is defined by the finite subfamily of the polynomials P_m :

$$\bigcap_{k=0}^{\infty} \tilde{P}^{-1}(0) = \bigcap_{k=m}^{M} \tilde{P}_k^{-1}(0).$$

Here m is the minimal index k such that $\tilde{P}_k \neq 0$. If $\tilde{P}$ is the greatest common divisor (over $\mathbb{C}$) of the polynomials $\tilde{P}_k$, $k = m, \ldots, M$, then $\bigcap_{k=m}^{M} \tilde{P}_k^{-1}(0) = \tilde{P}^{-1}(0) \cup W$, where W is an algebraic variety in $\mathbb{C}^n$ with $\dim_{\mathbb{C}} W < n-1$. The decomposition follows from the fact that any two irreducible components of two varieties $\tilde{P}_{k_1}^{-1}(0)$ and $\tilde{P}_{k_2}^{-1}(0)$ either coincide or intersect by algebraic variety of lower dimensions.

Now we have, after restriction to $\mathbb{R}^n$:

$$\bigcap_{k=0}^{\infty} P_k^{-1}(0) = P^{-1}(0) \cup V,$$

where $P = \tilde{P}\big|_{\mathbb{R}^n}$, $V = W \cap \mathbb{R}^n$. Observe that

$$\Delta P_k(x) = \text{const} \int_{\mathbb{R}^n} |x - y|^{2(k-1)} f(y) dy = \text{const } P_{k-1}(x).$$

Then $\Delta P_m = \text{const } P_{m-1} = 0$. Therefore P divides harmonic polynomials $P_m \neq 0$. It follows that P has real coefficients. Indeed, since P_m has real coefficients, we have $\overline{\tilde{P}_m(\bar{z})} = \tilde{P}_m$. The polynomial $\tilde{P}$ divides $\tilde{P}_m$ and if $\tilde{P}^*(z) = \overline{\tilde{P}(\bar{z})} \neq \tilde{P}(z)$, then both $\tilde{P}$ and $\tilde{P}^*$ are divisors of $\tilde{P}_m$. It follows that $P_m = |P|^2 Q$ for some polynomial Q. This contradicts the theorem of Brelot-Choquet [BC] which states that no polynomial in $\mathbb{R}^n$, preserving the sign, can divide a nontrivial harmonic polynomial. Therefore $\tilde{P}^* = \tilde{P}$ which means that $\tilde{P}$ has real coefficients. This completes the proof. $\square$

Remark. The set $P^{-1}(0)$ is a $(n-1)$-dimensional submanifold with singularities at the points where the gradient ∇P vanishes. Indeed, $\mathbb{R}^n \setminus P^{-1}(0)$ must be disconnected since otherwise $P(x) \geq 0$, $x \in \mathbb{R}^n$, or $P(x) \leq 0$, $x \in \mathbb{R}^n$ and in both cases we would get a contradiction with the above-mentioned theorem of Brelot-Choquet.

Conjecture. *The polynomial P in Proposition 9.10 can be chosen homogeneous after suitable translation. In other words, the main, $(n-1)$-dimensional component of any set of noninjectivity is contained in a cone with respect to some point.*

Theorem 9.9 confirms the conjecture for $n = 2$. Here is a partial description of injectivity sets in higher dimensions.

Theorem 9.11. *[AQ2] Let Γ be a smooth hypersurface in $\mathbb{R}^n$ having the property that there are two points $a, b \in \Gamma$, $a \neq b$, such that the segment $[a, b]$ is orthogonal to both tangent planes $T_a(\Gamma)$, $T_b(\Gamma)$. Then Γ is a set of injectivity for the spherical Radon projection on the space $C_c(\mathbb{R}^n)$.*

The proofs of Theorem 9.9 and Theorem 9.11 have microlocal analysis, i.e., analysis of analytic wave front sets, as a key ingredient. This strong analytic tool is the basis for proving the support theorem (see Quinto [Q]). The main point in [Q] is: if f has compact support and $Mf(x, \cdot) = 0$ when x belongs to a real-analytic hypersurface S, then a sphere $S(x, r)$ can touch supp f only at points symmetric with respect to the tangent plane $T_x(S)$.

In turn, it is related to the fact that f is in a kernel of a real-analytic elliptic Fourier integral operator (differential of the operator M) and, due to the regularity theorem, analytic wave front sets must cancel at $T_x(S)$-symmetric points, assuming that one of them is a touching point of $S(x, r)$ to supp f. This implies that the symmetric point must be also in supp f. We are not going into more details of the proof, as it would require many prerequisites, so we refer the reader to [AQ1], [AQ2], [AQ3].

For instance, the hyperboloid $x^2 + y^2 - z^2 = 1$ in $\mathbb{R}^3$ is a set of injectivity on $C_c(\mathbb{R}^3)$ while the cone $x^2 + y^2 - 2z^2 = 0$ is not, as follows from

Theorem 9.12. *The zero set of any nonzero harmonic homogeneous polynomial fails to be a set of injectivity for the spherical Radon projection on $C_c(\mathbb{R}^n)$.*

Proof. Let $P \neq 0$ be a homogeneous harmonic polynomial in $\mathbb{R}^n$, $\Gamma = P^{-1}(0)$. Define the distribution $T = P\delta(|x| - 1)$, δ is the delta-function at the origin, supported on the unit sphere. The distribution T acts on test functions φ by

$$\langle T, \varphi \rangle = \int_{|x|=1} P(x)\varphi(x)dx.$$

Let $\psi \in C_c(\mathbb{R}^n)$ be a radial function $\psi(x) = \psi(|x|)$, such that the convolution $f = T * \psi \neq 0$ (it is easy to show that such a function ψ exists since $T \neq 0$).

The function f is continuous and has compact support since T and ψ do. Let us show that $Mf(x, r) = 0$, $r > 0$, when $x \in P^{-1}(0)$. We have

$$
\begin{aligned}
Mf(x, r) &= \int_{|x-y|=r} f(y) dA(y) \\
&= \int_{|x-y|=r} \left(\int_{|z|=1} \psi(y - z) P(z) dA(z) \right) dA(y) \qquad (9.11) \\
&= \int_{|y|=r} \left(\int_{|z|=1} \psi(y + x - z) P(z) dA(z) \right) dA(y).
\end{aligned}
$$

Denote by $\mathcal{O}_x(n)$ the group of orthogonal transformations k of $\mathbb{R}^n$ such that $k(x) = x$. Since ψ is radial and therefore $\mathcal{O}(n)$-invariant, we have $\psi(y + x - z) = \psi(ky + x - kz)$, $k \in O_x(n)$. The measures $dA(y)$ and $dA(z)$ are invariant under rotations and hence, after changing of variables we arrive to

$$
Mf(x, r) = \int_{|y|=z} \left(\int_{|z|=1} \psi(y + x - z) P(k^{-1}z) dA(z) \right) dA(y).
$$

Integrating against the normalized Haar measure dk yields

$$
\begin{aligned}
&Mf(x, r) \\
&= \int_{|y|=r} \left(\int_{|z|=1} \psi(y + x - z) \left(\int_{\mathcal{O}_x(n)} P(k^{-1}z) dk \right) dA(z) \right) dA(y).
\end{aligned}
$$

The inner integral against dk is proportional to the zonal spherical harmonic $Z_x(z)$ [SW, Theorem 2.12]:

$$
\int_{\mathcal{O}_x(n)} P(k^{-1}z) dk = cP(x) Z_x(z)
$$

and since $P(x) = 0$ the integral vanishes for all z. Thus $Mf(x, r) = 0$. $\quad\square$

Theorem 9.13. *Any algebraic variety $V \subset \mathbb{R}^n$ of $\dim V < n - 1$ fails to be a set of injectivity for the spherical Radon projection on $C_c(\mathbb{R}^n)$.*

Proof. We have to prove existence of a function $f \in C_c(\mathbb{R}^n)$, $f \neq 0$, such that $Mf(x, r) = 0$ for all $x \in V$ and $r > 0$.

As in Theorem 9.12 we first prove existence of a nontrivial compactly supported distribution in the kernel of the transform M and then regularize the distribution by convolution with a smooth radial function of compact support to obtain a continuous (even smooth) nonzero function in the kernel.

To this end, we exploit the construction in Section 9.10. We have proven (Theorem 9.9) that $Mf(x,r) = 0$ for $(x,r) \in \Gamma \times \mathbb{R}_+$, $\Gamma \subset \mathbb{R}^n$, if and only if the nonlinear Fourier transform

$$U(\lambda_1, \ldots, \lambda_{n+1}) = \int_{\mathbb{R}^n} e^{-i(\lambda_1 y_1 + \cdots + \lambda_n y_n + \ldots \lambda_{n+1} \|y\|^2)} f(y) dy \qquad (9.12)$$

vanishes on the cone

$$C_\Gamma \subset \mathbb{R}^{n+1}, \quad C_\Gamma = \{(-2xt, t) : x \in \Gamma, \ t \in \mathbb{R})\}.$$

According to the Paley-Wiener theorem, the function f is of compact support corresponding to the functions U, regarded as the Fourier transform of a distribution on the paraboloid $\Pi : y_1^2 + \cdots + y_n^2 = y_{n+1}$, from the Paley-Wiener class $PW(\mathbb{R}^{n+1})$. Extending the construction to distributions enables us to include the case of polynomials U, which corresponds to distributions f supported at 0.

Observe that the function U satisfies the Schrödinger heat equation

$$\frac{\partial U}{\partial \lambda_{n+1}} = i\Delta U \qquad (9.13)$$

where $\Delta = \sum_{i=1}^{n} \frac{\partial^2}{\partial \lambda_i^2}$. Conversely, any function $U \in PW(\mathbb{R}^{nn})$ satsifying the equation (9.13) is the Fourier transform of a distribution with compact support in the paraboloid Π and therefore we conclude that $\Gamma \subset \mathbb{R}^n$ is a set of injectivity for the spherical Radon projection $Mf(x, \cdot)$ if and only if the equation (9.13) has the only trivial solution $U \in PW(\mathbb{R}^{n+1})$ satisfying the condition

$$U\big|_{C_\Gamma} = 0. \qquad (9.14)$$

Let us look for polynomial solutions to (9.13). All of them are of the form

$$U(\lambda_1, \ldots, \lambda_{n+1}) = e^{i\lambda_{n+1}\Delta} U_0(\lambda_1, \ldots, \lambda_n)$$

where $U_0(\lambda_1 \ldots \lambda_n) = U(\lambda_1, \ldots, \lambda_n, 0)$ is the initial value. Therefore, the dimension of polynomial solutions U of $\deg U = m$ is $\dim P_n^m$, where P_n^m is the space of all polynomials in $\mathbb{R}^n$ of degree m.

In our case $\Gamma = V$ – an algebraic variety in $\mathbb{R}^n$ of $\dim V < n - 1$. Then $\dim C_\Gamma < n$. It is known in algebraic geometry that the dimension of the restriction $P_{n+1}^m\big|_W$, W is an algebraic variety, asymptotically, as $m \to \infty$, depends only on the $\dim V$ and does not depend on the variety itself. This means that

$$\dim \left(P_{n+1}^m\big|_{C_\Gamma} \right) \asymp \dim P_{\dim C_\Gamma}^m, \quad \text{as} \quad m \to \infty.$$

Since $\dim C_\Gamma < n$, then for m large

$$\dim(P_{n+1}^m)_{C_\Gamma}) < \dim P_n^m = \dim\{U : \deg U = m\}.$$

Therefore the reistriction operator

$$U \to U\big|_{C_\Gamma} \in P_{n+1}\big|_{C_\Gamma}$$

has a nontrivial kernel when m is large, because it maps the finite dimensional space of solutions U, $\deg U = m$, in the space of less dimension. Thus there exists a polynoimial solution $U \neq 0$ to (9.13) satisfying the condition (9.14), $U\big|_{C_\Gamma} = 0$. $\qquad\square$

From the discussion above, the variety V is a set of noninjectivity for the transform $Mf(x, \cdot)$.

Remark 9.1. We have proven that for any algebraic variety $V \subset \mathbb{R}^n$, $\dim V < n - 1$, (say, an algebraic curve in $\mathbb{R}^3$), there exists a nonzero function $f \in C^\infty(\mathbb{R}^n)$ with compact support that integrates to zero over any sphere centered on V.

In spite of the fact that the statement sounds like an analytic fact, both its proof and its nature are rather algebraic, because for non-algebraic sets such a function does not exist.

Remark 9.2. The reduction to the Schrödinger heat equation (9.14) delivers also an alternative proof of theorem Theorem 9.12. Indeed, if $P \neq 0$ is a homogeneous harmonic polynomial in $\mathbb{R}^n$, then P can be extended to $\mathbb{R}^{n+1}$ by $P(x_1 \ldots x_{n+1}) = P(x_1 \ldots x_n)$ and $P(-2tx, t) = (-2t)^{\deg P} P(x) = 0$ for any $x \in P^{-1}(0)$, so $P\big|_{C_P^{-1}(0)} = 0$. The equation (9.13) is satisfied in the trivial way: $\frac{\partial P}{\partial \lambda_{n+1}} = 0 = i\Delta P$. Therefore, the Cauchy problem for (9.13) with zero data on the cone $P^{-1}(0)$ has a nontrivial solution and hence $P^{-1}(0)$ fails to abe a set of injectivity.

Theorem 9.14. *[A] Any polynomial solution U to the equation*

$$\frac{\partial U}{\partial \lambda_{n+1}} = i\Delta U, \quad U\big|_{C_{P^{-1}(0)}=0}, \tag{9.15}$$

where P is a harmonic homogeneous polynomial, is of the form

$$U(\lambda', \lambda_{n+1}) = e^{i\lambda_{n+1}\Delta}U_0(\lambda') = \sum_{s=0}^{\left[\frac{m}{2}\right]} i^s \frac{\lambda_{n+1}^s}{s!} \Delta^s U_0(\lambda'),$$

$m = \deg U$, $\lambda' = (\lambda_1, \ldots, \lambda_n)$, *where U_0 is representable as*

$$U_0(\lambda') = \sum_{i=0}^{\left[\frac{m}{2}\right]} |\lambda'|^{2i} h_{m-2i}(\lambda')$$

and h_{m-2i} are harmonic polynomials divisible by P.

Thus to understand the dimension of polynomial solutions to U to v(9.14) one has to understand the dimension of harmonic polynomials divisibe by a given harmonic homogeneous polynomial (spherical harmonic) P which is a very nontrivial problem.

In fact, Theorem 9.14, due to the proved relation between injectivity of the spherical Radon projection and uniqueness for the Cauchy problem (9.13),(9.14), characterizes the kernel of the transform $Mf(x, \cdot)$ where $x \in P^{-1}(0)$ in the space of distributions supported at 0. It follows because ker M and the space of solutions to (9.13),(9.14) are related via the Fourier transform.

Moreover, when the set Γ is a cone, ker M has a similar characterization in the much large space $C(\mathbb{R}^n)$:

Theorem 9.15. *[AVZ] Let $\Gamma \subset \mathbb{R}^n$ be a cone with vertex 0 and $f \in C(\mathbb{R}^n)$. Then $Mf(x,r) = 0$, $x \in \Gamma$, $r > 0$, if and only if the Fourier decomposition in polar coordinates*

$$f(r\theta) = \sum_{k=0}^{\infty} f_k(r)Y_k(\theta)$$

includes only spherical harmonics Y_k that vanish on $\Gamma \cup S^{n-1}$.

We saw that understanding of sets of injectivity is closely related to characterization of *harmonic divisors*, that is polynomials that divide nontrivial harmonic polynomials. Some conditions are known for a long time, for instance the Brelot-Choquet theorem [BC] states that harmonic divisors in $\mathbb{R}^n$ must change sign. Complete characterization of harmonic divisors seems to be a very difficult problem. Nevertheless, for quadratic polynomials the complete answer can be given in algebraic terms. For simplicity, we will formulate the result for the case of quadratic forms with simple eigenvalues:

Theorem 9.16. *[AK] Let Q be a real quadratic polynomial in $\mathbb{R}^n$,*

$$Q(x) = (Ax, x) + (b, x) + c,$$

where A is a nondegenerate real $n \times n$ matrix and b, c are vectors in $\mathbb{R}^n$. Let $a_i = 1/\lambda_i, i = 1, \cdots, n$, where λ_i are the (simple) eigenvalues of the matrix A. Then the polynomial Q is a harmonic divisor if and only if for some natural N and numbers $\varepsilon_i, i = 1, \cdots, n$, each of them is 0 or 1, the nonlinear algebraic system (Niven system of equations):

$$\sum_{j=1}^{n} \frac{2\varepsilon_j + 1}{z_s - a_j} + \sum_{q=1, q \neq s}^{N} \frac{4}{z_s - z_q} = 0, \quad s = 1, \cdots, N$$

has a solution $z_i \in \mathbb{R}^n$ such that at least one z_i is zero.

The proof is based on separation of variables for the Laplace operator in ellipsoidal coordinates associated to the quadratic form Q. In these coordinates, the quadric $Q = 0$ becomes a coordinate hyperplane and separation of variables relates existence of a nonzero harmonic polynomial vanishing on the quadric to the existence of a vanshing at the origin polynomial solution to a Fuchsian ordinary differential equation. This ODE is the result of the separation of variables in the Laplace equation. In turn, polynomail solvability this ODE is equaivalent to solvabiulity of the Niven system.

This method works for quadratic polynomials only as it is known that the separation of variables is possible only in second order coordinates. For more detailed discussion of the problem of harmonic divisors we refer to the articles [A], [AK].

Now we pass to the relation between the spherical Radon transform and the wave and heat equations.

9.11. Stationary Sets for the Wave and Heat Equations

Let us consider the heat equation in $\mathbb{R}^n$:

$$u_t = \Delta u, \quad u = u(x, t), \quad (x, t) \in \mathbb{R}^n \times \mathbb{R}_+ \tag{9.16}$$

with the initial data

$$u(x, 0) = f(x) \tag{9.17}$$

and the wave equation

$$u_{tt} = \Delta u, \quad u = u(x,t), \quad (x,t) \in \mathbb{R}^n \times \mathbb{R}_+ \tag{9.18}$$

with the initial data

$$u(x,0) = 0, \quad u_t(x,0) = f(x) \tag{9.19}$$

For each equation, we call a set $S \subset \mathbb{R}^n$ *stationary* if

$$S = S(f) = \{x \in \mathbb{R}^n : u(x,t) = 0, \ \forall t > 0\},$$

where $u \neq 0$ is a solution to (9.16), (9.17) or (9.18), (9.19), respectively.

In other words, the stationary set for the wave equation is the set of unmovable points, and that stationary set for the heat equation is the set of zero temperature.

The description of stationary sets for the problems (9.16) and (9.18) and, especially, for analogous problems in bounded domainsis of great interest.

If the solution u to (9.16) is time-harmonic, i.e., has the form $u(x,t) = \sin \lambda t \cdot \varphi(x)$, then φ is an eigenfunction, $\Delta \varphi = -\lambda^2 \varphi$ and $S = \varphi^{-1}(0)$. Zero sets $\varphi^{-1}(0)$ of eigenfunctions of the Laplace operator are called *nodal sets*. Extensive literature exists which is devoted to the study of nodal sets, a field where many natural problems are still open.

The following proposition relates stationary sets for the problems (9.16), (9.18) in the whole space $\mathbb{R}^n$ to injectivity sets for the spherical Radon projection.

Proposition 9.17. *Assume $f \in C_c(\mathbb{R}^n)$. For both equations (9.16) and (9.18)*

$$S(f) = \{x \in \mathbb{R}^n : Mf(x,r) = 0, \ \forall r > 0\}.$$

Proof.

1. The solution to (9.16), (9.17) is given by the convolution of the initial data with the heat kernel

$$u(x,t) = \frac{1}{(2\pi t)^{n/2}} \int_{\mathbb{R}^n} e^{-\frac{|x-y|^2}{t}} f(y) dy.$$

The condition $x \in S$, i.e., $u(x,t) = 0$ for all $t > 0$ can be rewritten, by substitution $\lambda = \frac{1}{t}$, as

$$\int_{\mathbb{R}^n} e^{-\lambda |x-y|^2} f(y) dy = 0, \ \forall \lambda > 0$$

or

$$\sum_{n=0}^{\infty} \frac{(-\lambda)^k}{k!} \int_{\mathbb{R}^n} |x-y|^{2k} f(y) dy = 0.$$

This is equivalent to $\int_{\mathbb{R}^n} |x-y|^{2k} f(y) dy = 0$, $k-0,1,\ldots$, and the Weierstrass theorem about uniform approximation by polynomials yields

$$\int_{\mathbb{R}^n} \varphi(|x-y|) f(y) dy$$

for any radial function $\varphi \in C_c(\mathbb{R}^n)$. In turn, this is equivalent to vanishing integrals over spheres

$$M f(x,r) = \int_{S(x,y)} f(y) dA(y),$$

for all $r > 0$.

2. The solution to (9.18), (9.19) is given by Kirchhoff's formula:

$$u(x,t) = \text{const}(\partial_t)^{n-2} F(x,t),$$

where

$$F(x,t) = \int_0^1 (t^2 - r^2)^{(n-3)/2} r M(x,r) dr.$$

Therefore, $u(x,t) = 0$, $t > 0$ (i.e., $x \in S$) is equivalent to $F(x,t)$ being polynomial in the t-variable of degree $< n-2$.

The change of variables $r = ts$ yields

$$F(x,t) = t^{n-1} \int_0^1 (1 - s^2)^{(n-3)/2} s M f(x,ts) ds$$

and hence $F(x,t) = O(t^{n-1})$, $t \to 0$. As $\deg F < n-2$, this is possible only if $F(x,t) = 0$ for all $t > 0$. Thus, Mf satisfies Abel equation

$$\int_0^t (t^2 - r^2)^{(n-3)/2} r M f(x,r) dr = 0, \ t > 0,$$

that is known to have the unique solution $Mf(x,r) = 0$, $r > 0$. The proposition is proved. $\square$

Remark. The Proposition is valid for larger classes of initial data f, e.g., for the function of the Schwartz class. One can also consider evolution equations of the type (9.16) or (9.18), involving higher derivatives in t, but we restrict ourselves by the equations having clear physical meaning.

Now we can immediately translate results on injectivity of the transform M on the language of stationary sets.

Theorem 9.18. *For $n = 2$, the stationary sets for the problem (9.16), (9.17) (the membrame equation), with compactly supported initial data, are of the following types:*

1) finite sets
2) union of finite sets and Coxeter configuration of lines.

This theorem is an immediate consequence of Theorem 9.9. It explains vibration of an infinite membrame, which is flat at the initial moment and when the initial velocity is different from zero only in a bounded region. Two possibilities exist: either only a finite set of points remain unmovable, or the membrame vibrates leaving stationary an equiangular configuration of lines through one point. Moreover, the latter case occurs only if the initial velocity f has odd symmetry with respect to a Coxeter configuration and this odd symmetry preserves in time when the membrame oscillates.

Other results of Section 9.10 can be also translated for stationary sets. For instance, Theorem 10.3 says that no hypersurface $\Gamma \subset \mathbb{R}^n$ having two opposite points a and b (as in Theorem 10.3) can be stationary set of a nontrivial solution of the wave equation with compactly supported initial data.

9.12. Closed Stationary Hypersurfaces

In the last section we discussed stationary sets corresponding to compactly supported initial data f in (9.17) and (9.19).

Now we consider the case $f \in L^p(\mathbb{R}^n)$. In this case one cannot expect conical structure of the stationary sets as it takes place for the compactly supported initial data f, $f \in C_c(\mathbb{R}^n)$ and explicit description of the geometry of stationary sets seems to be too difficult a problem.

Nevertheless, we are able to answer the following question: When can stationary sets contain a closed hypersurface?

This question is of special interest and is one of the central problems in the study of nodal sets. One of the reasons is that, for the wave equation, the energy in a bounded region $\Omega \subset \mathbb{R}^n$

$$E(t) = \int_\Omega (u_t^2 + |\nabla u|^2)dx$$

is constant, $E(x) = \text{const}$, if the boundary $\partial\Omega$ is stationary. This can be easily checked by showing $E'(t) = 0$ with help of the equation (9.16) and Green's formula.

Theorem 9.19. *[ABK] The boundary of no bounded domain $\Omega \subset \mathbb{R}^n$ can be stationary for the equations (9.16) or (9.18) with initial data $f \in L^p(\mathbb{R}^n)$ when $p \leq \frac{2n}{n-1}$. For $p > \frac{2n}{n-1}$, the assertion fails.*

Proof. (Sketch) We will give the idea of the proof. All details can be found in [ABK].

Let Ω be a bounded domain in $\mathbb{R}^n$ and $\Gamma = \partial\Omega$. Suppose that Γ is a stationary set for the problem (9.16), (9.17) with $f \in L^p(\mathbb{R}^n$ and u is the solution such that $u(x,t) = 0$ for $x \in \Gamma$ at any time $t > 0$.

The Laplace operator Δ with Dirichlet conditions on Γ is a self-adjoint operator in $L^2(\Omega)$ with discrete spectrum $\{-\lambda_k^2\}_{k=0}^\infty$. Let $\{\psi_k\}_{k=1}^\infty$ be the corresponding orthonormal basis in $L^2(\Omega)$ of eigenfunctions of Δ. For any fixed t decompose $u(x,t)$ in to an L^2-convergent series

$$u(x,t) = \sum_{k=0}^\infty c_k(t)\psi_k(x).$$

The wave equation (9.18) implies $c_k''(t) = -\lambda_\varphi^2 c_k(t)$ and $c_k(t) = \sin \lambda_k t$.

Convolving u in the t-variable with an appropriate function $\chi(t)$ leads to the function

$$v(x,t) = \sin \lambda_k t \cdot \psi_k(x) \neq 0$$

that still satisfies $v(x,t) = 0$ for $(x,t) \in \Gamma \times \mathbb{R}_+$.

The next step is constructing an eigenfunction radial with respect to a certain point $x_0 \in \Omega$ and vanishing on Γ. To this end, we choose a point $x_0 \in \Omega$ such that $\psi_k(x_0) \neq 0$ and radialize v by

$$v^\#(x,t) = \int_{O(n)} v(x_0 + \sigma(x - x_0), t)d\sigma,$$

where $d\sigma$ is the Haar measure on the orthogonal group $O(n)$. Then $v^\#$ is still an eigenfunction of Δ but the Dirichlet condition for $v^\#$ on Γ may no longer be fulfilled. Note that radialization preserves conditions of the type "v belongs to L^p."

Assume, without loss of generality, that $x_0 = 0$. The radialized function $\psi_k^\#$ is a radial eigenfunction of the Laplace operator in a ball $B(0,\varepsilon) \subset \Omega$ with no singularity at $x = 0$, and hence it is a solution of the corresponding

Bessel equation. Therefore

$$\psi_k^{\#}(x) = \text{const} \, |x|^{1-\frac{n}{2}} J_{\frac{n}{2}-1}(\lambda_k |x|)$$

in $B(0, \varepsilon)$ and thus $\psi_k^{\#}$ extends to a global solution of the equation $\Delta \psi_k^{\#} = 0$ in $\mathbb{R}^n$. Correspondingly, $\sin \lambda_k t \cdot \psi_k^{\#}(x)$ extends to a global solution to the wave equation in $\mathbb{R}^n$.

The uniqueness theorem for the wave equation (Section 17, Ch.VI of [CH]) implies that

$$v^{\#}(x,t) = \sin \lambda_k t \cdot |x|^{1-\frac{n}{2}} \psi_k^{\#}(x)$$

for al $(x,t) \in \mathbb{R}^n \times \mathbb{R}_+$. The known asymptotic for the kernel function yields

$$\psi_k^{\#}(x) \sim \text{const} \cos\left(|x| - \frac{\pi}{2} - \frac{n-2}{2}\right) |x|^{\frac{1-n}{2}}$$

and therefore

$$\psi_k^{\#} \notin L_p(\mathbb{R}^n)$$

when $p \leq \frac{2n}{n-1}$.

It can be proved by estimates that all the transformations under f which led us to $\psi_k^{\#}$ (Kirchhoff transform, convolving in the t-variable and radialization) preserve membership in $L^p(\mathbb{R}^n)$. Therefore we conclude that $f \notin L^p(\mathbb{R}^n)$ if $p \leq \frac{2n}{(n-1)}$.

The second part of the Theorem follows from the fact that the function

$$u(x,t) = \sin t \cdot |x|^{1-\frac{n}{2}} J_{\frac{n}{2}-1}(|x|)$$

solves (9.16), (9.17) with $f(x) = |x|^{1-\frac{n}{2}} J_{\frac{n}{2}-1}(|x|)$, and $u(x,t) = 0$ for all $t > 0$ and all x on the sphere $S(0,R)$ where R is any zero of the Bessel function $J_{\frac{n}{2}-1}(R) = 0$. Thus, $S(0,R)$ is a stationary set, corresponding to the initial data f. It remains to observe that $f \in L^p(\mathbb{R}^n)$ for any $p > \frac{2n}{n-1}$. $\square$

Due to equivalence between stationary sets and noninjectivity sets for the transform M, we have

Theorem 9.20. *Let Ω be a bounded domain $\mathbb{R}^n$ and $\Gamma = \partial\Omega$. Then Γ is a set of injectivity for the spherical Radon projection $M f(x,r)$ in any space $L^p(\mathbb{R}^n)$ with $p \leq \frac{2n}{n-1}$. The assertion is not true for $p > \frac{2n}{n-1}$.*

Remark. The result can be interpeted as follows: if the initial velocity decays too fast at infinity, then the energy can be preserved in no bounded region. The critical rate of decay is given by the asymptotic of the Bessel function and is of const $|x|^{-\frac{n-1}{2}}$.

For further development see [AQ3], [AQ4]. In the first article stationary sets for the wave equation in the whole space $\mathbb{R}^n$ with the initial data which are finitely supported distribution are fully described. The essential, $(n-1)-$ dimensional, component of the stationary sets in this case is a cone associated with a harmonic divisor, similarly to the two-dimensional case desribed by Theorem 11.2. In the second article the case of the wave equation in bounded domains is treated. Namely, stationary sets in crystallographic domains are completely described, under assumption that the initial data vanish near the boundary. It turned out that all $(n-1)-$ dimensional stationary sets are formed by hyperplanes which are faces of crystallic substructures.

In [NT], [AR] results of Section 12 about closed stationary hypersurfaces are transfered to the Heisenberg group. Finally, we should mention the recent work [FP] where, in particular, explicit inversion formulae are obtained for the spherical Radon transform with centers on a sphere, and also further generalizations and development of Theorem 12.1 are given.

9.13. Approximation by Spherical Waves

We saw that the problems on injectivity sets for the spherical Radon projection M and on stationary sets for the wave and heat equations are equivalent.

Now we will translate those problems and corresponding results on the language of approximation theory.

We will call *spherical wave centered at the point $a \in \mathbb{R}^n$* any function f in $\mathbb{R}^n$ of the form $f(x) = \varphi(|x - a|)$. Given a set $S \subset \mathbb{R}^n$ denote $W(S)$ the linear span of all spherical waves centered at points $a \in S$.

Let X be one of the functional spaces: $C(\mathbb{R}^n)$ (equipped with the topology of uniform convergence on compact sets), $L^p(\mathbb{R}^n)$, $1 \leq p < \infty$. The dual spaces are correspondingly $M(\mathbb{R}^n)$–the space of Radon measures in $\mathbb{R}^n$ with compact support, $L^q(\mathbb{R}^n)$, where $\frac{1}{p} + \frac{1}{q} = 1$.

Theorem 9.21. *Given a set $S \subset \mathbb{R}^n$, the space $W(S) \cap X$ is dense in X if and only if S is a set of injectivity for the spherical Radon projection M on the space $X' \cap C(\mathbb{R}^n)$.*

Proof. Let $W^{\perp}(S) \subset X$ be the annihilator of the space $W(S)$, i.e., $W^{\perp}(S) = \{\mu \in X' : \mu(f) = 0, \ f \in W(S)\}$. The functions in $W(S)$ are translates $f(x) = \varphi_a(x) = \varphi(|x - a|)$, $a \in S$ of radial functions φ. Then the convolution $f * \psi$ with any radial function $\psi \in C_c(\mathbb{R}^n)$ is in the closure of

$W(S)$. This implies that $W^\perp(S)$ is invariant with respect to convolutions of its elements with continuous radial functions with compact support. These convolutions of measures $\mu \in M(\mathbb{R}^n)$ or functions $f \in L^q(\mathbb{R}^n)$ are continuous functions, and on the other hand the convolutions are dense in $W^\perp(S)$ because one can construct an approximative unit of radial functions from $C_c(\mathbb{R}^n)$. Therefore, $W^\perp(S) \cap C(\mathbb{R}^n)$ is dense in $W^\perp(S)$.

By the Hahn-Banach theorem $W(S) \cap X$ is dense in X if and only if $W^\perp(S) = 0$. Due to the denseness, $W^\perp(S) = 0$ is equivalent to $W^\perp(S) \cap C(\mathbb{R}^n) = 0$. If $f \in W^\perp(S) \cap C(\mathbb{R}^n)$, then

$$\int_{\mathbb{R}^n} \varphi(|x - a|) f(x) dx = 0$$

for any $\varphi \in C_c(\mathbb{R}^n)$ radial and any $a \in S$. Taking a sequence $\{\psi_k\}$ that tends to a δ-function on a sphere $S(a, r)$, we obtain

$$\int_{S(a,r)} f(x) dA(x) = 0, \quad a \in S,$$

i.e., $Mf(a, r) = 0$ for all $a \in S$, $r > 0$.

Conversely, if $f \in C(\mathbb{R}^n)$, then $Mf(a, r) = 0$, $(a, r) \in S \times \mathbb{R}_+$ implies $f \in W^\perp(S) \cap C(\mathbb{R}^n)$. Therefore, $W(S) \cap X$ is dense in X if and only if $Mf\big|_{S \times \mathbb{R}_+} = 0$, $f \in C(\mathbb{R}^n) \cap X'$ implies $f = 0$, i.e., S is a set of injectivity.

Due to proven duality, Theorem 9.9 now can be translated as follows:

Theorem 9.22. *The system of spherical waves $\varphi(|x - a|)$, $a \in S \subset \mathbb{R}^2$, $\varphi \in C(\mathbb{R})$, is complete in $C(\mathbb{R}^2)$, unless either S is finite or $S = V \cup \Sigma$, where V is a finite set and Σ is a Coxeter system of lines in the plane.*

It is easy to understand the reason why $W(\Sigma)$ is not dense in $C(\mathbb{R}^2)$. Indeed, any function $\varphi(|x - a|)$ is even with respect to reflexion around any straight line through a. Therefore, all functions from $W(\Sigma)$ are invariant under reflections around straight lines constituting the Coxeter configuration Σ, while not all functions in $C(\mathbb{R}^2)$ possess this property.

The corollary of Theorem 9.19 and Theorem 9.20 is

Theorem 9.23. *Let Ω be a bounded domain in $\mathbb{R}^n$, $\Gamma = \partial\Omega$. Then $W(\Gamma) \cap L^q(\mathbb{R}^n)$ is dense in $L^q(\mathbb{R}^n)$ so long as $q \geq \frac{2n}{n+1}$. The statement is not true for $q < \frac{2n}{n+1}$.*

Remark. Since $\frac{2n}{n+1} < 2$, the spherical waves in $L^2(\mathbb{R}^n)$, centered on the boundary of any bounded domain, form a complete system in $L^2(\mathbb{R}^n)$. The statement fails for $\Gamma = S(0,R) = \{x \in \mathbb{R}^n : |x| = R\}$ when $q < \frac{2n}{n+1}$, as it follows from the argument with the Bessel function in the proof of Theorem 9.19.

References

[A] M. Agranovsky. On a problem of injectivity for the Radon transform on a paraboloid. *Contemp. Math.*, **251**, 1–14, 2000.

[Ag] M. L. Agranovsky, The stability of the spectrum in the Pompeiu problem. *J.Math.Anal. Appl.*, **178**,1, 269–279, 1993.

[ABK] M. Agranovsky, C. Berenstein, and Kuchment P. Approximation by spherical waves in L^p- spaces. *J. Geom. Anal.*, **6**, 365–383,1996.

[AK] M. Agranovsky, Ya. Krasnov. Quadratic divisors of harmonic polynomias in $\mathbb{R}^n$. *J. Analyse Math.*, **82**, 379–395, 2000.

[AQ1] M. Agranovsky,E.T. Quinto. Injectivity sets for the Radon transform over circes and complete systems of radial functions. *J. of Funct. Anal.*,**139**, 2, (1996),383–414,1996.

[AQ2] M. Agranovsky, E.T. Quinto. Injectivity of the spherical mean operator and related problems. *in: Complex Analysis, Harmonic Analysis and Applications (R. Deville et al, eds.*, Addison Wesley, London, 1996.

[AQ3] M. Agranovsky, E.T. Quinto. Geometry of stationary sets for the wave equation in $\mathbb{R}^n$.The case of finitely supported initial data. *Duke Math. J.*, **107**,1,57–84, 2001.

[AQ4] M. Agranovsky M., E.T. Quinto. Stationary sets for the wave equation in crystallographic domains. *Trans. Amer. Math. Soc.*, **355**, 6, 24–2451, 2003.

[AR] M. Agranovsky,R. Rawat. Injectivity sets for spherical means on the Heisenberg group. *J. of Fourier Analysis and Appl.*, **5**, 4,363–372,1999).

[AS] M. Agranovsky M., A. Semenov, Deformations of balls in Schiffer's conjecture for Riemannian symmetric spaces. *Israel J. Math. .*, **95**,43–59,1996.

[AVZ] M. Agranovsky, V.V. Volchkov, and L.A. Zalcman. Conical injectivity sets for the spherical Radon transform. *Bul. London Math. Soc.*, **31**,231–236, 1999.

[Be] C.A. Berenstein. An inverse spectral theorem and its relation to the Pompeiu problem. *J. Analyse Math.*, **37**,128–144, 1980.

[BY] C.A. Berenstein, R. Gay. A local version of th two-circles theorem. *Israel J. Math.*, **55**,267-288,1986.

[BC] M. Brelot,G. Choquet. Polynomes harmoniques et polyharmoniques. *Colloque sur les Equations aux Derivees Partielles*, Brussels, 45–46,1954.

[BK] L. Brown L., J.-P. Kahane, A note on the Pompeiu problem for convex domains. *Math. Ann.*, **259**,107–110, 1982.

[BST] L. Brown L., B.M. Schreiber, and B.A. Taylor. Spectral synthesis and the Pompeiu problem. *Ann. Inst. Fourier (Grenoble)*, **23**, 3,125–154 , 1973.

[CH] R. Courant, D. Hilbert. *Methods of Mathematical Physics*, vol. 2, Interscience, New York, 1953.

[Ch]   L. Chakalov. Sur un probl'eme de D.Pompeiu. *Annaire Univ. Sofia Fac. Phys.Math.*, **40**,1–44, 1944.

[E]   P. Ebenfelt. Some results on the Pompeiu problem. Ann. Acad. Sci. Fennicae , **18**, 323–341, 1993.

[FP]   D. Finch, D.S.Patch. Determining a function from its mean values over a family of spheres. Preprint, 2002.

[Ga]   T. Gamelin. Uniform Algebras. *Prentice-Hall*, New York, 1969.

[GS]   N. Garofalo N.,F. Segala. New results on the Pompeiu problem. *Trans. AMS*, **325**,273–286, 1991.

[He]   S. Helgason. The Radon Transform. Birkhäuser, Boston, Basel, 1980.

[Ko]   T. Kobayashi. Bounded domains and the zero sets of Fourier transforms. Conf.Proc. Lecture Notes Math.Phys.,IV. Internat. Press, Cambridge, MA, 223–239,1994.

[Ka]   T. Kato. Perturbation Theory for Linear Operators. *Springer-Verlag*, Berlin, 1966.

[Na]   F. Natterer. The Mathematics of Computerized Tomography. *B.G. Teubner, Stuttgartand J. WIley & Sons Ltd.*, 1980.

[NT]   E.K. Narayanan., S. Thangavelu, Injectivity sets for spherical means on the Heisenberg group. *J. Math.Anal. Appl.*, **263**, 2,565–579, 2001.

[Po]   D. Pompeiu. Sur certains systems d'equations lineaires et sur une propriete integrale de fonctions de plusieurs ariables. *C.R. Acad. Sci. Paris*,**188**,1138–1139, 1929.

[Q]   E.T. Quinto. Pompeiu transforms on geodesic spheres in real analytic manifolds. *Israel J. Math.*,**84**,353–363, 1993.

[R]   J. Radon. Uber die Bestimmung von Funktionen durch ihre Integralwerte längs gewisser Manningfaligkeiten. *Ber. Verh. Sächs. Akad. Wiss. Leipzig Math. Nat. Kl.*, **69**,262–277, 1977.

[Ru]   W. Rudin. Functional Analysis, *McGraw Hill, New York*, 1973.

[Sch] L. Schwartz. Théorie générale des fonctions moyenne-périodiques, *Ann. of Math.*,**42**,857–929, 1974.

[SW]   E. Stein E.,G. Weiss, Introduction to Fourier Analysis on Euclidean Spaces, *Princeton University Press*, 1971.

[vW]   van der Waerden B.L. Algebra. *Springer-Verlag, Berlin-Heidelberg-New York*, 1976.

[Vo]   V.V. Volchkov. Integral Geometry and Convolution Equations, *Kluwer Academic Publishers, Dordrechts-Boston-London*, 2003.

[Wi]   S.A. Williams. Analyticity of the boundary for Lipschitz domains without the Pompeiu property. *Indiana Univ. Math. J.*,**30**, 357–369, 1981.

[Za1] L. Zalcman. Analyticity and the Pompeiu problem. *Arch. Rat. Anal. Mech.*,**47**,237–254, 1972.

[Za2] L. Zalcman. A bibliographic survey of the Pompeiu problem. *in: Approximations by Solutions of PDE, (B. Fuglede et al., eds.), Kluwer Acad. Publ.*, 185–194, 1992.

[Za3] L. Zalcman. Offbeat integral geometry. *Amer. Math. Monthly*, **87**, 3, 161–175, 1980.

Chapter 10

Fourier analysis and geometric combinatorics

Alex Iosevich*

Department of Mathematics
University of Missouri-Columbia
Columbia, MO 65211 USA
e-mail: iosevich@math.missouri.edu

Contents

10.1. Introduction

This paper is based on the lectures the author gave in Padova at the Mini-corsi di Analisi Matematica in June, 2002. The author wishes to thank the organizers, the participants, and the fellow lecturers for many interesting and useful remarks. The author also wishes to thank Georgiy Arutyun-yants, Leonardo Colzani, Julia Garibaldi, Derrick Hart, and Bill McClain for many useful comments and suggestions about the content and style of the paper.

The main theme of this paper is an old and beautiful subject of geometric combinatorics. We will not even attempt to cover anything resembling a

---

*Research supported in part by the NSF grant DMS02-45369

significant slice of this broad and influential discipline. See, for example, [22] for a thorough description of this subject. The purpose of this article is to describe Szekely's [28] beautiful and elementary proof of the Szemeredi-Trotter incidence theorem ([26]), a result that found a tremendous number of applications in combinatorics, analysis, and analytic number theory. We shall describe some of the consequences of this seminal result and its interaction with problems and techniques of Fourier analysis and additive number theory.

Definition 10.1. An incidence of a point and a line is a pair (p, l), where p is a point, l is a line, and p lies on l.

Theorem 10.2 (Szemeredi-Trotter). *Let I denote the number of incidences of a set of n points and m lines (or m strictly convex closed curves). Then*

$$I \lesssim n + m + (nm)^{\frac{2}{3}}, \tag{10.1}$$

where here and throughout the paper, $A \lesssim B$ means that there exists a positive constant C such that $A \leq CB$.

Quite often in applications, one uses the following "weighted" version the the Szemeredi-Trotter theorem due to L. Szekely ([28]).

Theorem 10.2'. *Given a set of n points and m simple (no self-intersections) curves in the plane, such that any two curves intersect in at most α points and any two points belong to at most β curves, the number of incidences is at most $C(\alpha\beta)^{\frac{1}{3}}(nm)^{\frac{2}{3}} + m + 5\beta n$.*

The probabilistic proof of Theorem 10.2 (due to Szekely) given below, can be modified (as it is done in [28], Theorem 8) to yield Theorem 10.2'. We outline these modifications at the end of Section 10.2 below where we also briefly describe some of the applications of weighted incidence theory to the theory of diophantine equations.

Corollary 10.3. *Let S be a subset of $\mathbb{R}^2$ of cardinality n. Let $\Delta(S) = \{|x - y| : x, y \in S\}$, where $|\cdot|$ denotes the Euclidean norm. Then*

$$\#\Delta(S) \gtrsim n^{\frac{2}{3}}. \tag{10.2}$$

This estimate is not sharp. It is conjectured to hold with the exponent 1 in place of $\frac{2}{3}$. For the best known exponents to date (around .86), see [23] and [24]. However, Corollary 10.3 is still quite useful as we shall see in the final section of this paper.

Corollary 10.4. *Let A be a subset of $\mathbb{R}$ of cardinality n. Then either $A + A = \{a + a' : a, a' \in A\}$ or $A \cdot A = \{aa' : a, a' \in A\}$ has cardinality $\gtrsim n^{\frac{5}{4}}$.*

This estimate has been recently improved in a number of ways by several authors. See, for example, [3] and references contained therein.

10.2. Proof of Theorem 10.2, Corollary 10.3 and Corollary 10.4

We shall deduce Theorem 10.2 from the following graph theoretic result due to Ajtai et al ([1]), and, independently, to Leighton. Note that for the purposes of this paper, a pair of vertices in a graph can be connected by at most one edge.

Definition 10.5. The crossing number of a graph, $cr(G)$, is the minimal number of crossings over all the possible drawings of this graph in the plane. A crossing is an intersection of two edges not at a vertex.

Definition 10.6. We say that a graph G is planar if there exists a drawing of G in the plane without any crossings.

Theorem 10.7. *Let G be a graph with n vertices and e edges. Suppose that $e \geq 4n$. Then*

$$cr(G) \gtrsim \frac{e^3}{n^2}. \tag{10.3}$$

Before proving Theorem 10.7, we show how it implies Theorem 10.2. Take the points in the statement of the Theorem as vertices of a graph. Connect two vertices with an edge if the two corresponding points are consecutive on some line. It follows that

$$e = I - m. \tag{10.4}$$

If $e < 4n$ we get $I < 4n + m$, which is fine with us. If $e \geq 4n$, we invoke Theorem 10.7 to see that

$$cr(G) \gtrsim \frac{e^3}{n^2} = \frac{(I - m)^3}{n^2}. \tag{10.5}$$

Combining (10.5) with the obvious estimate $cr(G) \leq m^2$, we complete the proof of Theorem 10.2. Observe that strictly speaking, we have only proved Theorem 10.2 for lines and points. In order to extend the argument to translates of the same strictly convex curve, one needs to replace (10.4) with an (easy) estimate $e \gtrsim I$.

We now turn our attention to the proof of Theorem 10.7. Let G be a planar graph with n vertices, e edges, and f faces. Euler's formula (proved by induction) says that

$$n - e + f = 2. \tag{10.6}$$

Combined with the observation that $3f \leq 2e$, we see that in such a planar graph

$$e \leq 3n - 6. \tag{10.7}$$

It follows that if G is any graph, then

$$cr(G) \geq e - 3n. \tag{10.8}$$

We now convert this linear estimate into the estimate we want by randomization. More precisely, let G be as in the statement of Theorem 10.7 and let H be a random subgraph of G formed by choosing each vertex with probability p to be chosen later. Naturally, we keep an edge if and only if both vertices survive the random selection. Let $\mathbb{E}()$ denote the usual expected value. An easy computation yields

$$\mathbb{E}(vertices) = np, \tag{10.9}$$

$$\mathbb{E}(edges) = ep^2, \tag{10.10}$$

$$\mathbb{E}(crossing\ number\ of\ H) \leq p^4 cr(G). \tag{10.11}$$

Observe that the inequality in (10.11) is due to the fact that the number of avoidable crossings in G may decrease once a smaller random subset is extracted.

It follows by linearity of expectation that

$$cr(G) \geq \frac{e}{p^2} - \frac{3n}{p^3}. \tag{10.12}$$

Choosing $p = \frac{4n}{e}$ we complete the proof of Theorem 10.7, and consequently of Theorem 10.2. Theorem 10.2' can be proved in a similar fashion. First one shows that under the assumption that any two vertices are connected by at most β edges, the conclusion of Theorem 10.7 becomes $cr(G) \gtrsim \frac{e^3}{\beta n^2}$. Then, in the application of this estimate to incidences, the upper bound on $cr(G)$ is no longer m^2, but rather $m^2\alpha$. Combining this with the trivial estimates yields the conclusion of Theorem 10.2'.

Further developments in weighted incidence theory have recently led to the following result proved in the case $d = 2$ by S. Konyagin ([20]), and by Iosevich, Rudnev and Ten ([18]) for $d > 2$.

Theorem 10.8. *Let $\{b_j\}_{j=1}^{N}$ denote a strictly convex sequence of real numbers (in the sense that vectors (j, b_j) lie on a strictly convex curve). Then the number of solutions of the equation*

$$b_{i_1} + \cdots + b_{i_d} = b_{j_1} + \cdots + b_{j_d} \tag{10.13}$$

is

$$\lesssim N^{2d-2+2^{-d}}. \tag{10.14}$$

Taking $b_j = j^2$, for example, shows that $N^{2d-2+2^{-d}}$ in (10.14) cannot be replaced by anything smaller than N^{2d-2}. We conjecture that $O(N^{2d-2})$ is the right estimate, up to logarithms, for any strictly convex sequence $\{b_j\}$.

10.3. Proof of Corollary 10.3

Draw a circle of fixed radius around each point in S. By Theorem 10.2, the number of incidences is $\lesssim n^{\frac{4}{3}}$. This means that a single distance cannot repeat more than $\approx n^{\frac{4}{3}}$ times. It follows that there must be at least $\approx n^{\frac{2}{3}}$ distinct distances since the total number of distances is $\approx n^2$. In other words, we just proved that $\Delta(S) \gtrsim n^{\frac{2}{3}}$ as promised.

10.3.1. *Proof of Corollary 10.4*

The choice of lines and points is less obvious here. Let $P = (A+A)\times(A\cdot A)$. Let L be the set of lines of the form $\{(ax, a' + x) : a, a' \in A\}$. We have

$$\#P = \#(A + A) \times \#(A \cdot A), \tag{10.15}$$

$$\#L = n^2, \tag{10.16}$$

while the number of incidences is clearly $n \times n^2 = n^3$. It follows that

$$n^3 \lesssim (\#P)^{\frac{2}{3}} n^{\frac{4}{3}}, \tag{10.17}$$

which means that

$$\#P \gtrsim n^{\frac{5}{2}}. \tag{10.18}$$

It follows that either $\#(A + A)$ or $\#(A \cdot A)$ exceeds a constant multiple of $n^{\frac{5}{4}}$. This completes the proof of Corollary 10.4.

10.4. Application to Fourier analysis

Definition 10.9. We say that a domain $\Omega \subset \mathbb{R}^d$ is spectral if $L^2(\Omega)$ has an orthogonal basis of the form $\{e^{2\pi i x \cdot a}\}_{a \in A}$.

The following result is due to Fuglede ([8]). It was also proved in higher dimensions by Iosevich, Katz and Pedersen ([12]).

Theorem 10.10. *A disc, $D = \{x \in \mathbb{R}^2 : |x| \le r\}$, is not spectral.*

Proof. [of Theorem 10.10] Let A denote a putative spectrum. We need the following basic lemmas:

Lemma 10.11. *A is separated in the sense that there exists $c > 0$ such that $|a - a'| \ge c$ for all $a, a' \in A$.*

Lemma 10.12. *There exists $s > 0$ such that any square of side-length s contains at least one element of A.*

For a sharper version of Lemma 10.12 see [16].

The proof of Lemma 10.11 is straightforward. Orthogonality implies that

$$\int_D e^{2\pi i x \cdot (a-a')} dx = 0, \tag{10.19}$$

whenever $a \ne a' \in A$. Since $\int_D dx = 2\pi r$ and the function $\int_D e^{2\pi i x \cdot \xi} dx$ is continuous, the left hand side of (10.19) would have to be strictly positive if $|a - a'|$ were small enough. This implies that $|a - a'|$ can never be smaller than a positive constant depending on r.

The proof of Lemma 10.12 is a bit more interesting. By Bessel's inequality we have

$$\sum_A |\widehat{\chi}_D(\xi + a)|^2 \equiv |D|^2, \tag{10.20}$$

for almost every $\xi \in \mathbb{R}^d$, since the left hand side is a sum of squares of Fourier coefficients of the exponential with the frequency ξ with respect to the putative orthogonal basis $\{e^{2\pi i x \cdot a}\}_{a \in A}$. We have

$$\sum_{A_\xi} |\widehat{\chi}_D(a)|^2 = \sum_{A_\xi \cap Q_s} + \sum_{A_\xi \cap Q_s^c} = I + II, \tag{10.21}$$

where $A_\xi = A - \xi$ and Q_s is a square of side-length s centered at the origin.

We invoke the following basic fact. See, for example, [25]. We have

$$|\widehat{\chi}_D(\xi)| \lesssim |\xi|^{-\frac{3}{2}}. \tag{10.22}$$

It follows that

$$II \lesssim \sum_{A_\xi \cap Q_s^c} |a|^{-3} \lesssim s^{-1}. \tag{10.23}$$

Choosing s big enough so that $s^{-1} << |D|^2$, we see that $I \neq 0$, and, consequently, that $A_\xi \cap Q_s$ is not empty. This completes the proof of Lemma 10.12.

We are now ready to complete the proof of Theorem 10.10. Intersect A with a large disc of radius R. By Lemma 10.11 and Lemma 10.12, this disc contains $\approx R^2$ points of A. We need another basic fact about $\widehat{\chi}_D(\xi)$, that it is radial, and in fact equals, up to a constant, to $|\xi|^{-1} J_1(2\pi|\xi|)$, where J_1 is the Bessel function of order 1. We also need to know that zeros of Bessel functions are separated in the sense of Lemma 10.11. This fact is contained in any text on special functions. See also [29].

With this information in tow, recall that orthogonality implies that $|a - a'|$ is a zero of J_1. Since the largest distance in the disc of radius R is $2R$ and zeros of J_1 are separated, we see that the total number of distinct distances between the elements of A in the disc or radius R is at most $\approx R$. This is a contradiction since Corollary 10.3 says that R^2 points determine at least $R^{\frac{4}{3}}$ distinct distances. This completes the proof of Theorem 10.10 □

It turns out that not only does $L^2(D)$ not possess an orthogonal basis of exponentials, the numbers of exponentials orthogonal with respect to D is in fact finite. This is a theorem due to Fuglede ([9]) which was extended to all sufficiently smooth well-curved symmetric convex domains by Iosevich and Rudnev ([17]). The latter paper is based on the generalization of the following beautiful geometric principle due to Erdos ([5]).

Theorem 10.13 (Erdos integer distance principle). *Let S be an infinite subset of $\mathbb{R}^d$ such that the distance between any pair of points in S is an integer. Then S is a subset of a line.*

10.5. Applications to convex geometry

The following result is due to Andrews ([2]).

Theorem 10.14. *Let Q be a convex polygon with n integer vertices. Then $n \lesssim |Q|^{\frac{1}{3}}$.*

10.5.1. *Proof of Theorem 10.14*

Let $\mathcal{C}$ denote a strictly convex curve running through the vertices of Q. Let Ω denote the convex domain bounded by $\mathcal{C}$. Let L denote the set of strictly convex curves obtained by translating $\mathcal{C}$ by every lattice point inside Ω. Let P denote the set of lattice points contained in the union of all those translates. By Theorem 10.2 the number incidences between the elements of P and elements of L is $\lesssim |\Omega|^{\frac{4}{3}}$ since $\#L \approx \#P \approx |\Omega|$. Since each translate of $\mathcal{C}$ contains exactly the same number of lattice points,

$$\#\mathcal{C} \cap \mathbb{Z}^2 \lesssim \frac{|\Omega|^{\frac{4}{3}}}{|\Omega|} = |\Omega|^{\frac{1}{3}}. \tag{10.24}$$

This completes the proof of Theorem 10.14. Observe that proof implies the following (easier) estimate.

Lemma 10.15. *Let Γ be a closed strictly convex curve in the plane. Then*

$$\#\{R\Gamma \cap \mathbb{Z}^2\} \lesssim R^{\frac{2}{3}}. \tag{10.25}$$

What sort of an incidence theorem would be required to prove a more general version of this result?

Definition 10.16. We say that $A \subset \mathbb{R}^d$ is well-distributed if the conclusions of Lemma 10.11 and Lemma 10.12 hold for A.

Let A be a well-distributed set, and let A_R denote the intersection of A and the ball of radius R centered at the origin. Observe that $\#A_R \approx R^d$. Let U be a strictly convex hyper-surface contained in the unit ball. Suppose we had a theorem which said that the number of incidences between A_{5R} and a family of hyper-surfaces $\{RU + x\}_{x \in A_R}$ is $\lesssim R^{d\alpha}$. Repeating the argument above, we would arrive at the conclusion that if P is a convex polyhedron with N lattice vertices, then

$$|P| \gtrsim N^{\frac{1}{\alpha - 1}}. \tag{10.26}$$

However, a higher dimensional version of the aforementioned theorem of Andrews says that

$$|P| \gtrsim N^{\frac{d+1}{d-1}}. \tag{10.27}$$

This leads us to conjecture that the putative incidence theorem described above should hold with $\alpha = 2 - \frac{2}{d+1}$, which was recently proved in [15] under additional smoothness assumptions. This result is sharp in view of (10.21) and the following result due to Barany and Larman ([4]).

Theorem 10.17. *The number of vertices of P_R, the convex hull of the lattice points contained in the ball of radius $R \gg 1$ centered at the origin is $\approx R^{d\frac{d-1}{d+1}}$.*

10.6. Higher dimensions

Theorem 10.18. *If $R > 0$ is sufficiently large, then*

$$\#(\Delta(A \cap [-R, R]^d) \gtrsim R^{2-\frac{1}{d}}. \tag{10.28}$$

This result was recently proved in a more general setting, using different methods, by Solymosi and Vu ([27]).

Corollary 10.19. *The ball $B_d = \{x : |x| \le 1\}$ is not spectral in any dimension greater than 1.*

Corollary 10.19 follows from Theorem 10.18 in the same way as Theorem 10.10 follows from Corollary 10.3. Lemma 10.11 and Lemma 10.12 go through without change except that in $\mathbb{R}^d$,

$$|\widehat{\chi}_{B_d}(\xi)| \lesssim |\xi|^{-\frac{d+1}{2}}, \tag{10.29}$$

$\widehat{\chi}_{B_d}(\xi)$ is a constant multiple of

$$|\xi|^{-\frac{d}{2}} J_{\frac{d}{2}}(2\pi|\xi|), \tag{10.30}$$

and the zeroes of $J_{\frac{d}{2}}$ are still separated.

See [25], [29] and/or any text on special functions for the details.

We are left to prove Theorem 10.18. Since A is well-distributed, there is $s > 0$ such that every cube of side-length s contains at least one point of A. Without loss of generality let $s = 1$. Since A is well-distributed, we can find points $P_1, P_2, \ldots, P_d$, such that $|P_i - P_j| \le 10$ and such that for any sequence $R_1, R_2, \ldots, R_d \gg 1$ with $|R_i - R_j| \le 10$, the intersection of d spheres centered at each P_i or radius R_i is transverse. Let O be the center of mass of the polyhedron with vertices given by the points $P_1, \ldots, P_d$. Construct a system of annuli centered at O of width 10, with the first annulus of radius $\approx R$. Construct $\approx R$ such annuli.

It follows from the assumption that A is well distributed that each constructed annulus $\mathcal{A}$ has $\approx R^{d-1}$ points of A. Let

$$\cup_{i=1}^{d}\{|x - P_i| : x \in \mathcal{A}\} = \{d_1, \ldots, d_k\}. \tag{10.31}$$

Let

$$A_j^l = \{x \in \mathcal{A} \cap A : |x - P_l| = d_j\}. \tag{10.32}$$

It is not hard to see that

$$A_j^l = \cup_{1 \le j_m \le k} \cup_{m=1}^{d-1} A_j^l \cap_{l' \ne l} A_{j_m}^{l'}. \tag{10.33}$$

Taking unions of both sides in j and counting, we see that

$$R^{d-1} \lesssim k^d. \tag{10.34}$$

This follows from the fact, which follows by a direct calculation, that the intersection of d spheres in question consists of at most two points. Taking d'th roots and using the fact that we have $\approx R$ annuli with $\approx R^{d-1}$ point of A, we conclude that

$$\#\Delta(A \cap [-R, R]^d) \gtrsim R^{1 + \frac{d-1}{d}} = R^{2 - \frac{1}{d}}, \tag{10.35}$$

as desired.

Observe that while the intersection claim made above is not difficult to verify for spheres, the situation becomes much more complicated for boundaries of general convex bodies, even under smoothness and curvature assumptions. This issue is partially addressed in [10].

Another point of view on distance set problems was recently pursued by Iosevich and Laba ([13]) and, independently, by Kolountzakis ([21]).

Theorem 10.20. *([13] for $d = 2$ and [21] for $d > 2$). Let A be a well-distributed subset of $\mathbb{R}^d$, $d \ge 2$. Let K be a symmetric bounded convex set. Then $\Delta_K(A)$ is separated only if K is a polyhedron with finitely many vertices.*

The more difficult question of which polyhedra can result in separated distance sets is partially addressed in both aforementioned papers, but the question is, in general, unresolved.

10.7. Some comments on finite fields

In this section we consider incidence theorems in the context of finite fields. More precisely, let F_q denote the finite field of q elements. Let F_q^d denote the d-dimensional vector space over F_q. A line in F_q^d is a set of points $\{x+tv : t \in F_q\}$ where $x \in F_q^d$ and $v \in F_q^d \setminus (0,\ldots,0)$. A hyperplane in F_q^d is a set of points $(x_1,\ldots,x_d)$ satisfying the equation $A_1 x_1 + \cdots + A_d x_d = D$, where $A_1,\ldots,A_d, D \in F_q$ and not all A_j's are 0.

It is clear that without further assumptions, the number of incidences between n hyper-planes and n points is $\approx n^2$ and no better, since we can take all n planes to be rotates of the same plane about a line where all the points are located. We shall remove this "difficulty" by operating under the following non-degeneracy assumption.

Definition 10.21. We say that a family of hyperplanes in F_q^d is non-degenerate if the intersection of any d (or fewer) of the hyper-planes in the family contains at most one point.

The main result of this section is the following:

Theorem 10.22. *Suppose that a family $\mathcal{F}$ of n hyper-planes in F_q^d is non-degenerate. Let $\mathcal{P}$ denote a family of n points in F_q^d. The the number of incidences between the elements of $\mathcal{F}$ and $\mathcal{P}$ is $\lesssim n^{2-\frac{1}{d}}$. Moreover, this estimate is sharp.*

We prove sharpness first. Let $\mathcal{F}$ denote the set of all the hyper-planes in F_q^d and $\mathcal{P}$ denote the set of all the points in F_q^d. It is clear that $\#\mathcal{F} \approx \mathcal{P} \approx q^d$. On the other hand, the number of incidences is simply the number of hyper-planes times the number of points on each hyper-planes, which is $\approx q^{2d-1}$. Since $q^{2d-1} = (q^d)^{2-\frac{1}{d}}$, the sharpness of the Theorem 10.22 is proved.

We now prove the positive result. Consider an n by n matrix whose (i,j) entry if 1 if i'th point lies on j's line, and 0 otherwise. The non-degeneracy condition implies that this matrix does not contain a d by 2 sub-matrix consisting of 1's. Using Holder's inequality we see that the number of incidences,

$$I = \sum_{i,j} I_{ij} \leq \left(\sum_i \left(\sum_j I_{ij} \right)^d \right)^{\frac{1}{d}} \times n^{\frac{d}{d}-\frac{1}{d}} \tag{10.36}$$

$$= \left(\sum_i \sum_{j_1,\ldots,j_d} I_{ij_1} \ldots I_{ij_d} \right)^{\frac{1}{d}} \times n^{\frac{d-1}{d}} \lesssim n \times n^{\frac{d-1}{d}} = n^{2-\frac{1}{d}}, \tag{10.37}$$

because when j_k's are distinct, $I_{ij_1} \ldots I_{ij_d}$ can be non-zero for at most one value of i due to the non-degeneracy assumption. If j_k's are not distinct, we win for the same reason. This completes the proof of Theorem 10.22.

Why should the finite field case be different from the Euclidean case? The proof of Szemeredi-Trotter theorem given above suggests that main difference may be the notion of order. In the proof of Szemeredi-Trotter we used the fact that points on a line may be ordered. However, no such notion exists in a finite field. Nevertheless, Tom Wolff conjectured that if q is a prime, then there exists $\epsilon > 0$ such that the number of incidences between n points and n lines in F_q^2 should not exceed $n^{\frac{3}{2}-\epsilon}$ for $n \approx q$. This fact has recently been proved by Bourgain, Katz, and Tao ([3]).

10.8. A Fourier approach

In this sections we briefly outline how some results in geometric combinatorics can be obtained using Fourier analysis. For a more complete description, see, for example, [11], [14], and [15].

We could take a more direct approach, but we take advantage of this opportunity to introduce the following beautiful problem in geometric measure theory.

Falconer Distance Conjecture. *Let $E \subset [0,1]^d$, $d \geq 2$. Suppose that the Hausdorff dimension of E is greater than $\frac{d}{2}$. Then $\Delta(E) = \{|x - y| : x, y \in E\}$ has positive Lebesgue measure.*

We shall not discuss the history and other particulars of the Falconer Distance Problem in this paper. See, for example, [30] and references contained therein for a thorough description of the problem and related machinery. The main thrust of this section is to show that any non-trivial theorem about the Falconer Distance Conjecture can be used to deduce a corresponding "discrete" result about distance sets of well-distributed subsets of $\mathbb{R}^d$.

Theorem 10.23. *Let K be a bounded convex set in $\mathbb{R}^d$, $d \geq 2$, symmetric with respect to the origin. Suppose that the Lebesgue measure of $\Delta_K(E)$ is positive whenever the Hausdorff dimension of $E \subset [0,1]^d$ is greater than s_0, with $0 < s_0 < d$. Let A be a well-distributed subset of $\mathbb{R}^d$. Then*
$$\#\Delta_K(A \cap [-R, R]^d) \gtrsim R^{\frac{d}{s_0}}.$$

The following result is essentially proved in [7].

Theorem 10.24. *Let $E \subset [0,1]^d$, $d \geq 2$, of Hausdorff dimension greater than $\frac{d+1}{2}$. Suppose that K is a bounded convex set, symmetric with respect to the origin, with a smooth boundary and everywhere non-vanishing Gaussian curvature. Then the Lebesgue measure of $\Delta_K(E)$ is positive.*

Theorem 10.23 and 10.24 combine to yield the following "discrete" theorem.

Theorem 10.25. *Let A be a well-distributed subset of $\mathbb{R}^d$, $d \geq 2$. Suppose that K is a bounded convex set, symmetric with respect to the origin, with a smooth boundary and everywhere non-vanishing Gaussian curvature. Then*
$$\#\Delta_K(A \cap [-R,R]^d) \gtrsim R^{2-\frac{2}{d+1}}.$$

Observe that while this result is not as strong as the one given by Theorem 10.18, it is more flexible since it does not require K to be the Euclidean ball.

Proof. [of Theorem 10.23] Let $q_1 = 2$ and choose integers $q_{i+1} > q_i^i$. Let
$$E_i = \{x \in [0,1]^d : |x_k - p_k/q_i| \leq q_i^{-\frac{d}{s}} \tag{10.38}$$
$$\text{for some } p = (p_1, \ldots, p_d) \in A \cap [0,q_i]^d\}.$$

Let $E = \cap E_i$. It follows from the proof of Theorem 8.15 in [6] that the Hausdorff dimension of E is s. Suppose that there exists an infinite subsequence of q_is such that $\#\Delta_K(A \cap [0,q_i]^d) \lesssim q_i^\beta$ for some $\beta > 0$. Then we can cover $\Delta_K(E_i)$ by $\lesssim q_i^\beta$ intervals of length $\approx q_i^{-\frac{d}{s}}$. If $\beta < \frac{d}{s}$, $|\Delta_K(E_i)| \to 0$ as $i \to \infty$. It follows that $\Delta_K(E)$ has Lebesgue measure 0. However, by assumption, $\Delta_K(E)$ is positive if $s > s_0$. The conclusion follows. $\square$

The proof of Theorem 10.23 suggests that one may be able to make further progress on the Erdos Distance Conjecture for well-distributed sets using Fourier methods by studying the Falconer Distance Conjecture for special sets constructed in the previous paragraph.

References

[1] M. Ajtai, V. Chvatal, M. Newborn, and E. Szemeredi Crossing-free subgraphs North-Holland Math Studies 60 1986

[2] G. Andrews A lower bound for the volume of strictly convex bodies with many boundary lattice points Trans. Amer. Math. Soc. 106 270-279

[3] J. Bourgain, N. Katz, and T. Tao A sum-product estimate in finite fields, and applications (preprint) 2003

[4] I. Barany and D. Larman The convex hull of the integer points in a large ball 1999 Math. Ann. 312 167-181

[5] P. Erdos Integral distances Bull. Amer. Math. Soc. 51 996 1945

[6] K. J. Falconer The geometry of fractal sets Cambridge University Press 1985

[7] K. J. Falconer On the Hausdorff dimensions of distance sets Mathematika 32 206-212 1986

[8] B. Fuglede ;Commuting self-adjoint partial differential operators and a group theoretic problem J. Func. Anal. 16 1974 101-121

[9] B. Fuglede Orthogonal exponentials on the ball Expo. Math. 19 267-272 2001

[10] M. Ganichev, A. Iosevich, and I. Laba Generic intersections of boundaries of convex bodies and Fourier analysis (in preparation) 2004

[11] S. Hofmann and A. Iosevich Circular averages and Falconer/Erdos distance conjecture in the plane for random metrics 2003 Proc. Amer. Math. Soc. (accepted for publication)

[12] A. Iosevich, N. Katz, and S. Pedersen Fourier bases and a distance problem of Erdos Math. Res. Lett. 6 1999 251-255

[13] A. Iosevich and I. Laba Distance sets of well-distributed planar sets (accepted for publication by the Journal of Discrete and Computational Geometry) 2003

[14] A. Iosevich and I. Laba K-distance, Falconer conjecture, and discrete analogs 2003 (submitted for publication)

[15] A. Iosevich, I. Laba, and M. Rudnev Incidence theorems via Fourier analysis (in preparation) 2004

[16] A. Iosevich and S. Pedersen How large are the spectral gaps? Pacific J. Math. 192 2000

[17] A. Iosevich and M. Rudnev Combinatorial approach to orthogonal exponentials Inter. Math. Res. Notices 2003 49 1-12

[18] A. Iosevich, M. Rudnev and V. Ten Combinatorial complexity of convex sequences (submitted for publication) 2003

[19] A. Iosevich, E. Sawyer, and A. Seeger Bounds for the mean square lattice point discrepancy. II: Convex domains in the plane (preprint) 2003

[20] S. Konyagin An estimate of the L^1-norm of an exponential sum (in Russian) The theory of approximation of functions and operators 2000 88-89

[21] M. Kolountzakis Distance sets corresponding to convex bodies GAFA (to appear) 2003

[22] P. Agarwal and J. Pach Combinatorial geometry A Wiley-Interscience publication 1995

[23] J. Solymosi and Cs. D. Toth Distinct distances in the plane Discrete Comp. Geom. (Misha Sharir birthday issue) 25 2001 629-634

[24] J. Solymosi, G. Tardos, and C. D. Toth Distinct distances in the plane 25 2001 629-634

[25] E. M. Stein Harmonic Analysis 1993 Princeton University Press

[26] E. Szemeredi and W. Trotter Extremal problems in discrete geometry Com-

binatorica 3 1983 381-392

[27] J. Solymosi and V. Vu Distinct Distances in High Dimensional Homogeneous Set Contemporary Mathematics Series of the AMS (to appear) 2003

[28] L. Szekely Crossing numbers and hard Erdos problems in discrete geometry 1997 6 353-358 Combinatorics, probability and computing

[29] E. M. Stein and G. Weiss Introduction to Fourier analysis on Euclidean spaces 1971 Princeton University Press

[30] T. Wolff Lectures in Harmonic Analysis AMS University Lecture Series 29 2003

Chapter 11

Lectures on eigenfunctions of the Laplacian

Christopher D. Sogge *

Department of Mathematics, Johns Hopkins University
Baltimore, MD 21218

Contents

11.1. Introduction

In this paper we shall present material from lectures given by the author in a workshop in Padova, Italy in June of 2003. The author is very grateful for the hospitality shown to him, especially that of Massimo Lanza de Cristoforis and Paolo Ciatti.

The main goal of the lectures was to see how "geometry", especially long term dynamics of geodesic flow influences the "size" of eigenfunctions on compact Riemannian manifolds. This is also known as the "quantum correspondence principle". In addition, we went over problems in harmonic analysis related to this, such as the restriction theorem for the Fourier transform and estimates for Riesz means. We also presented recent work on manifolds with boundary.

---

*The author was partially supported by the NSF

11.2. Review: Restriction theorem and wave equation

Much of what we shall present is related to the Stein-Tomas [26] restriction theorem for the Fourier transform. Recall that the Fourier transform of an $L^1(\mathbb{R}^n)$ function is

$$\hat{f}(\xi) = \int_{\mathbb{R}^n} e^{-ix\cdot\xi} f(x)\, dx.$$

Since $\hat{f}$ is continuous if $f \in L^1(\mathbb{R}^n)$ the restriction of $\hat{f}$ to S^{n-1} makes perfect sense in this case. This is not the case for $L^2(\mathbb{R}^n)$ functions. For instance the inverse Fourier transform of $\hat{f}(\xi) = (1 - |\xi|^{2n})^{-1/3}$ is in $L^2(\mathbb{R}^n)$, but clearly $\hat{f}(\xi)$ does not restrict to S^{n-1} as a function or even a distribution.

The Stein-Tomas theorem tells us that if $1 < p < 2$ is close enough to 1 and if $f \in L^p(\mathbb{R}^n)$ then one can define the restriction of $\hat{f}$ to S^{n-1} when $n \geq 2$:

Theorem 11.1 (Stein-Tomas [26]). *Let $n \geq 2$ and suppose that*

$$1 \leq p \leq \frac{2(n+1)}{n+3}. \tag{11.1}$$

Then there is a uniform constant C so that if $f \in \mathcal{S}(\mathbb{R}^n)$ then

$$\left(\int_{S^{n-1}} |\hat{f}(\xi)|^2 \, d\sigma(\xi) \right)^{1/2} \leq C\|f\|_{L^p(\mathbb{R}^n)}, \tag{11.2}$$

if $d\sigma$ is the induced Lebesgue measure on S^{n-1}. Consequently, if p is as in (11.1) then there is a bounded linear map $\mathcal{R} : L^p(\mathbb{R}^n) \to L^2(S^{n-1})$ so that $\mathcal{R}f(\xi) = \hat{f}(\xi)|_{S^{n-1}}$ when $f \in \mathcal{S}(\mathbb{R}^n)$.

It is conjectured that if $1 \leq p < \frac{2n}{n+1}$ then $\hat{f}$ restricts to S^{n-1} as an element of $L^q(S^{n-1})$ with $q = \frac{(n+1)p'}{n-1}$, with

$$p' = \frac{p}{p-1}.$$

This result is known to hold when $n = 2$ ([30]). For $n \geq 3$ only partial results are known. See, e.g. ([5]).

One of our main goals will be to obtain discrete analogs of this result when $\mathbb{R}^n$ is replaced by a compact Riemannian manifold. To motivate what happens in this case, let us first state a couple of results that are equivalent to the Stein-Tomas restriction theorem.

To do this, we first notice that (11.2) holds if and only if

$$\limsup_{\lambda \to +\infty} \lambda \int_{\{\xi \in \mathbb{R}^n : |\xi| \in [1, 1+\lambda^{-1}]\}} |\hat{f}(\xi)|^2 \, d\xi \le C \|f\|_{L^p(\mathbb{R}^n)}^2,$$

$$1 \le p \le \frac{2(n+1)}{n+3}. \quad (11.3)$$

If we let

$$m_\lambda(\xi) = \lambda^{1/2} \chi_{|\xi| \in [1, 1+\lambda^{-1}]},$$

then, by duality, we conclude that (11.2) and (11.3) are equivalent to the statement that the multiplier operators

$$(2\pi)^{-n} \int e^{ix \cdot \xi} m_\lambda(\xi) \hat{f}(\xi) \, d\xi$$

are uniformly bounded from $L^2(\mathbb{R}^n)$ to $L^q(\mathbb{R}^n)$ when $q \ge \frac{2(n+1)}{n-1}$. Finally, if we let

$$\chi_\lambda f(x) = (2\pi)^{-n} \int_{|\xi| \in [\lambda, \lambda+1]} e^{ix \cdot \xi} \hat{f}(\xi) \, d\xi, \quad (11.4)$$

then we conclude by scaling that the above inequalities are equivalent to the one that says there is a uniform constant C so that for $\lambda \ge 1$

$$\|\chi_\lambda f\|_{L^q(\mathbb{R}^n)} \le C \lambda^{n(1/2-1/q)-1/2} \|f\|_{L^2(\mathbb{R}^n)}, \quad q \ge \frac{2(n+1)}{n-1}. \quad (11.5)$$

By Plancherel's theorem the χ_λ operators are uniformly bounded from $L^2(\mathbb{R}^n)$ to $L^2(\mathbb{R}^n)$, and so if we apply the M. Riesz interpolation theorem, we also get the bounds

$$\|\chi_\lambda f\|_{L^q(\mathbb{R}^n)} \le C \lambda^{\frac{n-1}{2}(\frac{1}{2}-\frac{1}{q})} \|f\|_{L^2(\mathbb{R}^n)}, \quad 2 \le q \le \frac{2(n+1)}{n-1}, \quad (11.6)$$

for the remaining range of exponents in $[2, \infty]$. Using the fact that the range of exponents in the Stein-Tomas L^2 restriction theorem is sharp, we see that the favorable bounds in (11.5) cannot hold for a larger range of exponents. Similarly, the counterexamples that verify this fact can be used to show that the bounds in (11.6) for the complimentary range of exponents cannot be improved (see e.g., [22]).

Let us give another equivalent formulation of (11.2) that shows how we can use properties of the wave equation for small values of t to prove (11.2). To do this, we choose an even function $\rho \in \mathcal{S}(\mathbb{R})$ satisfying

$$\rho(\tau) \ge 1, \quad |\tau| \le 1, \quad \text{and} \quad \hat{\rho}(t) = 0, \ |t| \ge 1.$$

We then define a variation of the operators in (11.4) by setting

$$\tilde{\chi}_\lambda f(x) = (2\pi)^{-n} \int_{\mathbb{R}^n} e^{ix\cdot\xi} \rho(|\xi| - \lambda)\hat{f}(\xi)\, d\xi$$

$$+ (2\pi)^{-n} \int_{\mathbb{R}^n} e^{ix\cdot\xi} \rho(|\xi| + \lambda)\hat{f}(\xi)\, d\xi. \quad (11.7)$$

Then, by duality and Plancherel's theorem, (11.5) and hence (11.2) hold if and only if there is a constant C so that when $\lambda \geq 1$

$$\|\tilde{\chi}_\lambda f\|_{L^q(\mathbb{R}^n)} \leq C(1+\lambda)^{n(1/2-1/q)-1/2}\|f\|_{L^2(\mathbb{R}^n)}, \quad q \geq \frac{2(n+1)}{n-1}. \quad (11.8)$$

Let us rewrite the approximate spectral projection operators $\tilde{\chi}_\lambda$ using the wave equation. We notice that since $\hat{\rho}$ is even and vanishes for $|t| > 1$

$$\rho(|\xi| \pm \lambda) = (2\pi)^{-1} \int_{-1}^{1} e^{it(|\xi|\pm\lambda)} \hat{\rho}(t)\, dt$$

$$= (2\pi)^{-1} \int_{-1}^{1} \hat{\rho}(t) \cos(t(|\xi| \pm \lambda))\, dt.$$

Consequently,

$$\rho(|\xi| - \lambda) + \rho(|\xi| + \lambda)$$

$$= (2\pi)^{-1} \int_{-1}^{1} \hat{\rho}(t) \Big[\cos(t(|\xi| - \lambda)) + \cos(t(|\xi| + \lambda)) \Big]\, dt$$

$$= 2(2\pi)^{-1} \int_{-1}^{1} \hat{\rho}(t) \cos(t|\xi|) \cos(t\lambda)\, dt.$$

Therefore,

$$\tilde{\chi}_\lambda f(x) = 2(2\pi)^{-n-1} \int_{\mathbb{R}^n} \left(\int_{-1}^{1} \hat{\rho}(t) \cos t\lambda \cos t|\xi|\, dt \right) e^{ix\cdot\xi} \hat{f}(\xi)\, d\xi$$

$$= 2(2\pi)^{-n-1} \int_{-1}^{1} \hat{\rho}(t) \cos t\lambda \left(\int_{\mathbb{R}^n} e^{ix\cdot\xi} \cos t|\xi|\, \hat{f}(\xi)\, d\xi \right) dt$$

$$= \frac{2}{\pi} \int_{0}^{1} \hat{\rho}(t) \cos t\lambda\, u(t, x)\, dt,$$

where u solves the Cauchy problem in $\mathbb{R}_+ \times \mathbb{R}^n$ with Cauchy data $(f, 0)$, i.e.,

$$\begin{cases} \Box u(t, x) = (\partial^2/\partial t^2 - \Delta)u(t, x) = 0 \\ u(0, \cdot) = f(\cdot), \quad \partial_t u(0, \cdot) = 0. \end{cases} \quad (11.9)$$

Δ is of course the Laplacian for $\mathbb{R}^n$ with the flat metric:

$$\Delta = \sum_{j=1}^{n} \partial^2/\partial x_j^2.$$

We point out that, in order to prove the Stein-Tomas L^2 restriction, we only need to know the behavior of the the solution u to the wave equation for small values of t ($|t| < 1$ in the above formulation). Note further that the kernel of $\tilde{\chi}_\lambda$ is given by

$$\tilde{\chi}_\lambda(x, y) = \frac{2}{\pi} \int_0^1 \hat{\rho}(t) \cos t\lambda \, U(t; x, y) \, dt,$$

where

$$U(t; x, y) = (2\pi)^{-n} \int_{\mathbb{R}^n} e^{i(x-y)\cdot\xi} \cos t|\xi| \, d\xi, \qquad (11.10)$$

is the kernel for the solution of the Cauchy problem (11.9).

Let us make a few important observations about this kernel. First $\tilde{\chi}_\lambda(x, y) = 0$ if $|x - y| > 1$ by the finite propagation speed for $\square$, i.e., $U(t; x, y) = 0$ if $|x - y| > t$. Second, by applying the method of stationary phase one can see that

$$\tilde{\chi}_\lambda(x, y) \approx \begin{cases} \lambda^{(n-1)/2} \frac{\cos(\lambda|x-y|)}{|x-y|^{(n-1)/2}}, & \lambda^{-1} \le |x - y| \le 1 \\ \lambda^{n-1}, & |x - y| \le \lambda^{-1}. \end{cases} \qquad (11.11)$$

To handle the $\tilde{\chi}_\lambda$ operators directly we need to appeal to a special case of Stein's oscillatory integral theorem [25]:

Theorem 11.2. *Let $\eta \in C_0^\infty(\mathbb{R}^n)$ and suppose that $\Phi(x, y) \in C^\infty(\mathbb{R}^n \times \mathbb{R}^n)$ is a real function satisfying*

$$rank\Big(\frac{\partial^2 \Phi}{\partial x_j \partial y_k}\Big) = n - 1, \quad x, y \in supp\ \eta$$

and

$$\Sigma_x = \{\nabla_x \Phi(x, y) : y \in supp\ \eta\}, \quad x \in supp\ \Phi \qquad (11.12)$$

has everywhere nonvanishing principal curvatures. If we then set

$$T_\lambda f(x) = \int e^{i\lambda\Phi(x,y)} \eta(x, y) \, f(y) \, dy, \qquad (11.13)$$

then

$$\|T_\lambda f\|_{L^q(\mathbb{R}^n)} \le C\lambda^{-n/q}\|f\|_{L^2(\mathbb{R}^n)}, \quad q \ge \frac{2(n+1)}{n-1}. \qquad (11.14)$$

Note that the second hypothesis in the theorem makes sense. This is because when the mixed Hessian of Φ has rank $n-1$ then the sets Σ_x defined in (11.12) are smooth (immersed) hypersurfaces in $\mathbb{R}^n$.

To be able to use the theorem in applications to eigenfunctions on Riemannian manifolds, we have stated the oscillatory integral theorem for general phase functions whose Hessians have rank $n-1$ and satisfy the curvature condition that the surfaces Σ_x in (11.12) have nonvanishing principal curvatures. To handle the Euclidean case and prove (11.8) we shall need to apply the special case of the theorem where

$$\Phi(x,y) = |x-y|.$$

In this case the hypotheses are fulfilled away from the diagonal $\{(x,y) : x = y\}$ since clearly the mixed Hessian has rank $n-1$, while, for this phase function for every x the sets Σ_x is a subset of the sphere of radius one centered at the origin. Therefore, let us fix an amplitude $\beta \in C_0^\infty(\mathbb{R})$ satisfying, say, $\beta(s) = 0$, $s \notin [1/4, 2]$. If we then set

$$T_\lambda f(x) = \int e^{i\lambda|x-y|}\beta(|x-y|)\, f(y)\, dy, \tag{11.15}$$

then, by Theorem 11.2, the operators must satisfy (11.14).

We can use these bounds to prove (11.8). If we require additionally that $\sum_{-\infty}^{\infty} \beta(2^j s) = 1$, $s \geq 0$, then, by a scaling argument, the integral operator with kernel $\beta(\lambda 2^{-j}|x-y|)\tilde{\chi}_\lambda(x,y)$, $j \geq 1$, must have an $L^2(\mathbb{R}^n) \to L^q(\mathbb{R}^n)$ norm which is $(\lambda 2^{-j})^{n(1/2-1/q)-n}$ times that of the one with kernel

$$\beta(|x-y|)\tilde{\chi}_\lambda(2^j x/\lambda, 2^j y/\lambda).$$

By Stein's theorem and (11.11) the latter integral operator sends L^2 to L^q with norm $O(\lambda^{n-1}\, 2^{-j(n-1)/2}\, 2^{-nj/q})$, assuming that $q \geq \frac{2(n+1)}{n-1}$. By putting these two steps together, we conclude that for such q,

$$\|\tilde{\chi}_\lambda\|_{L^2 \to L^q} \leq C \sum_{1 \leq 2^j \leq \lambda} (\lambda 2^{-j})^{n(1/2-1/q)-n} \times (\lambda^{n-1}\, 2^{-j(n-1)/2}\, 2^{-nj/q})$$

$$= C\lambda^{-1+n(1/2-1/q)} \sum_{1 \leq 2^j \leq \lambda} 2^{j/2} = C\lambda^{n(1/2-1/q)-1/2},$$

as claimed.

Thus, by using properties of the wave equation for small values of t, stationary phase and Stein's oscillatory integral theorem, we have argued that we can obtain bounds that are equivalent to the sharp ones in

$$\|\chi_\lambda f\|_{L^q(\mathbb{R}^n)} \leq C(1+\lambda)^{\sigma(p)}\|f\|_{L^2(\mathbb{R}^n)}, \quad 2 \leq q \leq \infty, \tag{11.16}$$

if

$$\sigma(p) = \begin{cases} n(1/2 - 1/q) - 1/2, & q \geq 2(n+1)/(n-1), \\ \frac{n-1}{2}(1/2 - 1/q), & 2 \leq q \leq 2(n+1)/(n-1), \end{cases} \tag{11.17}$$

and if, as before, χ_λ are the Euclidean spectral projection operators defined in (11.4). Note that if $\chi(\tau) = \chi_{[0,1]}(\tau)$ is the characteristic function of $[0,1]$, then

$$\chi_\lambda = \chi(\sqrt{-\Delta} - \lambda). \tag{11.18}$$

Later we shall show that we can generalize the bounds in (11.17) to the setting of compact Riemannian manifolds without boundary (M, g) of dimension $n \geq 2$ by showing that we have the analog of (11.17) if χ_λ is the operator $\chi(\lambda - \sqrt{-\Delta_g})$, with Δ_g being the Laplacian on (M, g). As above, these bounds can be seen to be sharp, and this result is the natural extension of the Stein-Tomas restriction theorem to compact Riemannian manifolds without boundary. We shall also go over what happens if M has a boundary. In this case, a counterexample of Grieser [11] shows that less favorable bounds can hold. Recent results of Grieser [12] and the author [23] established the bounds when $q = \infty$, and recently Smith and the author [18] established sharp results in the two-dimensional case for Riemannian manifolds with boundary.

11.3. Eigenfunctions on compact Riemannian manifolds without boundary

We shall now consider the setting of compact boundaryless Riemannian manifolds $M = M^n$ of dimension n and Riemannian metrics $g = \sum g_{jk}(x)dx^j dx^k$. Recall that the associated Laplace-Beltrami operator is given in local coordinates by the formula

$$\Delta_g = |g(x)|^{-1/2} \sum \partial_k g^{jk}(x)\partial_j, \quad |g| = \det g_{jk},$$

where (g^{jk}) is the inverse matrix of (g_{jk}). We shall be interested in proving L^p estimates for eigenfunctions of the Laplacian, i.e., solutions of

$$-\Delta_g \phi_\lambda(x) = \lambda^2 \phi_\lambda(x).$$

Recall that the set of eigenvalues of $-\Delta_g$ is discrete here and tends to $+\infty$. Thus, we can order the eigenvalues with respect to multiplicity

$$0 = \lambda_0^2 \leq \lambda_1^2 \leq \lambda_2^2 \leq \cdots. \tag{11.19}$$

We have normalized things so that λ_j is the jth eigenvalue of the first order operator $\sqrt{-\Delta_g}$. We are interested in sharp estimates for

$$L^p(\lambda_j, g) = \frac{\|\phi_{\lambda_j}\|_{L^q(M)}}{\|\phi_{\lambda_j}\|_{L^2(M)}} \tag{11.20}$$

as $\lambda_j \to \infty$. Here $L^q(M)$ denotes the L^q norm with respect to the volume element for g. We shall also want sharp bounds of the form

$$\|\chi_\lambda f\|_{L^q(M)} \leq C(1 + \lambda)^{\delta(q,n)} \|f\|_{L^2(M)},$$

if χ_λ are the spectral projection operators that are the natural analogs of the ones defined in (11.4) for the Euclidean case:

$$\chi_\lambda f(x) = \sum_{\lambda_j \in [\lambda, \lambda+1]} e_j(f), \tag{11.21}$$

with $e_j(f)$ being the projection of f onto the jth eigenspace for $-\Delta_g$. Thus, if $\{e_j(x)\}$ is an orthonormal basis with e_j being an eigenfunction with eigenvalue λ_j, then

$$e_j(f)(x) = e_j(x) \times \left(\int_M f(y)\overline{e_j(y)}\, dy \right),$$

with dy denoting the volume element. Note also that, by the spectral theorem, we have the analog of (11.18)

$$\chi_\lambda = \chi(\sqrt{-\Delta_g} - \lambda),$$

if as before χ is the characteristic function of $[0, 1]$.

In this section we wish to present the following sharp theorem first proved in [20]:

Theorem 11.3. *Let (M, g) be an n-dimensional Riemannian compact manifold without boundary. Then there is a uniform constant C so that if $q \geq 2$ and if $\sigma(p)$ is as in (11.17), then*

$$\|\chi_\lambda f\|_{L^q(M)} \leq C(1 + \lambda)^{\sigma(p)} \|f\|_{L^2(M)}. \tag{11.22}$$

Note that the bounds in (11.22) automatically imply that the ratio in (11.20) is $O((1 + \lambda_j)^{\sigma(p)})$. As we shall see, that if our Riemannian satisfies certain geometric assumptions these bounds can be improved if q is large enough, and in particular if $q = \infty$.

On the other hand, if (M, g) is that standard sphere S^n with the round metric the bounds

$$\frac{\|\phi_{\lambda_j}\|_{L^q(M)}}{\|\phi_{\lambda_j}\|_{L^2(M)}} \leq C(1 + \lambda_j)^{\sigma(p)} \tag{11.23}$$

cannot be improved for any $q \geq 2$. Recall (see, e.g. [19]) that in this case the distinct eigenvalues of $-\Delta_g$ are $k(k + n - 1)$, and they repeat with multiplicity

$$d_k = \frac{(n + k)}{n!\,k!} - \frac{(n + k - 2)!}{n!\,(k - 2)!} = k^{n-1}(2 + O(1/k))/(n - 1)!. \qquad (11.24)$$

Furthermore, if ϕ_k is the zonal spherical harmonic of degree k, which corresponds to $\tilde{\lambda}_j = \sqrt{k(k + n - 1)}$, then

$$\frac{\|\phi_k\|_{L^q(S^n)}}{\|\phi_k\|_{L^2(S^n)}} \approx k^{n(1/2 - 1/q) - 1/2}, \quad q \geq \frac{2(n + 1)}{n - 1},$$

while if ϕ_k is the highest weight spherical harmonic of degree k

$$\frac{\|\phi_k\|_{L^q(S^n)}}{\|\phi_k\|_{L^2(S^n)}} \approx k^{\frac{n-1}{2}(1/2 - 1/q)}, \quad 2 \leq q \leq \frac{2(n + 1)}{n - 1},$$

which implies that (11.23) is sharp in for this particular Riemannian manifold.

To prove Theorem 11.3 we shall use the template that was presented in the last section. To do this, we choose an even function $\rho \in \mathcal{S}(\mathbb{R})$ satisfying

$$\rho(\tau) = 1, \quad |\tau| \leq 1, \quad \text{and} \quad \hat{\rho}(t) = 0, \ |t| \geq r_0,$$

where now, instead of taking $r_0 = 1$ as we did for the Euclidean case, we take

$$r_0 = \mathrm{Inj}(M)/2,$$

with $\mathrm{Inj}(M)$ being the injectivity radius of (M, g). We then define a variation of the operators in (11.22) by setting

$$\tilde{\chi}_\lambda f(x) = \sum_{j=0}^{\infty} \rho(\lambda - \lambda_j)e_j(f) + \sum_{j=0}^{\infty} \rho(\lambda + \lambda_j)e_j(f). \qquad (11.25)$$

Then, by duality and Plancherel's theorem, (11.22) holds if and only if there is a constant C so that when $\lambda \geq 1$

$$\|\tilde{\chi}_\lambda f\|_{L^q(\mathbb{R}^n)} \leq C(1 + \lambda)^{\sigma(p)}\|f\|_{L^2(\mathbb{R}^n)}, \quad q \geq 2. \qquad (11.26)$$

If we use the Fourier transform then, as before, we conclude that

$$\tilde{\chi}_\lambda f(x) = \frac{2}{\pi} \int_0^{r_0} \hat{\rho}(t) \cos t\lambda\, u(t, x)\, dt, \qquad (11.27)$$

where u solves the Riemannian version of the Cauchy problem (11.9), i.e.,

$$\begin{cases} \Box_g u(t, x) = (\partial^2/\partial t^2 - \Delta_g)u(t, x) = 0 \\ u(0, \cdot) = f(\cdot), \quad \partial_t u(0, \cdot) = 0. \end{cases} \qquad (11.28)$$

Here the kernel of $\tilde{\chi}_\lambda$ is given by

$$\tilde{\chi}_\lambda(x,y) = \frac{2}{\pi} \int_0^{r_0} \hat{\rho}(t) \cos t\lambda\, U(t;x,y)\, dt,$$

where

$$U(t;x,y) = \sum_{j=0}^{\infty} \cos t\lambda_j\, e_j(x)\overline{e_j(y)}$$

is the kernel for the solution of the Cauchy problem (11.28).

As in the Euclidean case, we can make a couple important observations about this kernel. First, if $\mathrm{dist}(x,y)$ is the Riemannian distance between x and y, then $\tilde{\chi}_\lambda(x,y) = 0$ if $\mathrm{dist}(x,y) > r_0$ by the finite propagation speed for $\Box_g$. Second, by applying the method of stationary phase one can see that

$$\tilde{\chi}_\lambda(x,y) \approx \begin{cases} \lambda^{(n-1)/2} \frac{\cos(\lambda \mathrm{dist}(x,y))}{\mathrm{dist}(x,y)^{(n-1)/2}}, & \lambda^{-1} \le \mathrm{dist}(x,y) \le r_0 \\ \lambda^{n-1}, & \mathrm{dist}(x,y) \le \lambda^{-1}. \end{cases} \tag{11.29}$$

This calculation follows from the arguments for the Euclidean case if one uses the Hadamard parametrix for $\Box_g$, which says that, in local coordinates, if $\mathrm{dist}(x,y) \le t \le r_0$,

$$U(t;x,y) = U_{\mathbb{R}^n}(t; \mathrm{dist}(x,y)) + \text{Lower order terms},$$

where $U_{\mathbb{R}^n}(t; |x-y|) = U_{\mathbb{R}^n}(t;x,y)$ is the corresponding kernel for the wave equation in $\mathbb{R}_+ \times \mathbb{R}^n$ defined in (11.10).

Since the kernels of $\tilde{\chi}_\lambda$ satisfy (11.29) we can argue as before and see that (11.26) must hold if we can verify that $\Phi(x,y) = \mathrm{dist}(x,y)$ satisfies the conditions of Theorem 11.2. But this just follows from Gauss' lemma. Indeed, since Φ is the Riemannian distance function we have that the mixed Hessian must have rank $n-1$, and if we work in local coordinates, the hypersurfaces Σ_x defined in (11.12) are just

$$\Sigma_x = \{\xi \in \mathbb{R}^n : \sum g^{jk}(x)\xi_j \xi_k = 1\},$$

which of course have positive principal curvatures due to the fact that $(g^{jk}(x))$ is a positive definite matrix.

This completes the sketch of the proof of (11.26), and hence the bounds (11.22) for the spectral projection operators χ_λ.

11.4. Bochner-Riesz Means

In $\mathbb{R}^n$ the Bochner-Riesz means of index δ are given by the formula

$$S^\delta f(x) = (2\pi)^{-n} \int_{|\xi|<1} e^{ix\cdot\xi}(1 - |\xi|^2)^\delta \hat{f}(\xi)\, d\xi.$$

Just by a dilation argument if $S^\delta : L^p(\mathbb{R}^n) \to L^p(\mathbb{R}^n)$ then the operators

$$S^\delta_\lambda f(x) = (2\pi)^{-n} \int_{|\xi|<\lambda} e^{ix\cdot\xi}(1 - |\xi/\lambda|^2)^\delta \hat{f}(\xi)\, d\xi$$

are uniformly bounded on $L^p(\mathbb{R}^n)$ and $S^\delta_\lambda f \to f$ in $L^p(\mathbb{R}^n)$ if $f \in L^p(\mathbb{R}^n)$.

Recall also the Bochner-Riesz conjecture which says that if

$$\delta > \delta(p) = \max\left(n\left|\tfrac{1}{2} - \tfrac{1}{p}\right| - \tfrac{1}{2}, 0\right) \tag{11.30}$$

then one should have $S^\delta : L^p(\mathbb{R}^n) \to L^p(\mathbb{R}^n)$ if $\delta > \delta(p)$. For this reason, $\delta(p)$ is called the critical index for Bochner-Riesz summation. It follows from C. Fefferman's [10] celebrated negative solution of the ball multiplier and earlier work that $\delta > \delta(p)$ is a necessary condition for $S^\delta : L^p(\mathbb{R}^n) \to L^p(\mathbb{R}^n)$.

Many partial results are known. For instance it was shown by C. Fefferman [9] that if one has the $L^p(\mathbb{R}^n) \to L^2(S^{n-1})$ restriction theorem (11.2), then the Bochner-Riesz conjecture is valid for this value of p. Thus, by the Stein-Tomas restriction theorem and duality, the conjecture holds for $p \in [1, 2(n+1)/(n+3)] \cup [2(n+1)/(n-1), \infty]$. Also, Carleson and Sjölin [7] proved that the Bochner-Riesz conjecture is valid for all p when the dimension n equal to 2. Much later Bourgain [5] made an important breakthrough when he showed that for $n \geq 3$ the conjecture is valid for certain $p < 2(n+1)/(n-1)$. Many authors have worked to improve Bourgain's results (see [28]), but the conjecture is still unresolved in the Euclidean case when $n \geq 3$.

Let us now consider the corresponding problems on n-dimensional compact Riemannian manifolds without boundary, (M, g). In this case, the Bochner-Riesz means are given by the forumla

$$S^\delta_\lambda f(x) = \sum_{\lambda_j \leq \lambda} \left(1 - \left|\frac{\lambda_j^2}{\lambda}\right|^2\right)^\delta e_j(f)(x), \tag{11.31}$$

if, as before, $e_j(f)$ is the projection of f onto the j-th eigenspace with eigenvalue $-\lambda_j^2$ for Δ_g. Since eigenfunctions of the Laplacian are smooth and since the spectrum is discrete, it is automatic that if one fixes λ then

348　　　　　　　　　　　　　　　*C. D. Sogge*

$S_\lambda^\delta : L^p(M) \to L^p(M)$ for every p regardless of the value of δ. Thus, the analog of the Bochner-Riesz conjecture for the Riemannian case must involve uniform bounds for the operators S_λ^δ as $\lambda \to \infty$. Specifically, given M, the conjecture for $L^p(M)$ would be that if $\delta > \delta(p)$ then there must be a uniform constant $C = C_{p,\delta}$ so that

$$\|S_\lambda^\delta f\|_{L^p(M)} \le C\|f\|_{L^p(M)}, \quad \lambda > 0. \tag{11.32}$$

Just as in the euclidean case the condition that $\delta > \delta(p)$ is necessary for these bounds to hold. On the other hand, counterexamples of Bourgain [6] and Minicozzi and the author [16] suggest that these bounds might not hold for all values of p if the dimension n is larger than 2.

Let us see how this might be related to the bounds for eigenfunctions presented in the preceding section. Recall that if χ_λ are the spectral projection operators defined by (11.21) then we have the bounds

$$\|\chi_\lambda f\|_{L^2(M)} \le C(1 + \lambda)^{\sigma(p)}\|f\|_{L^p(M)},$$

with

$$\sigma(p) = \begin{cases} n(1/p - 1/2) - 1/2, & 1 \le p \le 2(n+1)/(n+3) \\ \frac{n-1}{2}(1/p - 1/2), & 2(n+1)/(n+3) \le p \le 2. \end{cases}$$

Thus, in the "good" range $1 \le p \le 2(n+1)/(n+3)$ we have that the sharp exponent for L^p bounds for eigenfunctions agrees exactly with the critical index for Bochner-Riesz summation:

$$\|\chi_\lambda f\|_{L^2(M)} \le C(1 + \lambda)^{\delta(p)}\|f\|_{L^p(M)}, \quad 1 \le p \le \frac{2(n+1)}{n+3}. \tag{11.33}$$

On the other hand, if $2(n + 1)/(n + 3) < p \le 2$, we have that $\sigma(p) > \delta(p)$. Thus, we might expect to be able to prove the sharp bounds for Bochner-Riesz summation using the above estimates for χ_λ only in the favorable range of exponents.

It turns out that we can do this:

Theorem 11.4 ([21]). *If $1 \le p \le \frac{2(n+1)}{n+3}$ or if $\frac{2(n+1)}{n-1} \le p \le \infty$ then whenever $\delta > \delta(p)$ (11.32) holds.*

By duality, we only need to prove (11.32) when

$$1 \le p \le \frac{2(n+1)}{n+3}.$$

To prove these inequalities, as in the proof of (11.33), we shall need to use the wave equation. We start with the exact formula

$$S_\lambda^\delta f = \sum_{\lambda_j \le \lambda} (1 - \lambda_j^2/\lambda^2)^\delta e_j(f) \tag{11.34}$$

$$= \frac{\sqrt{\pi}\,\Gamma(1+\delta)}{2\pi} \int_{-\infty}^\infty \lambda \left(\frac{\lambda|t|}{2}\right)^{-\delta-1/2} J_{\delta+1/2}(\lambda|t|) \sum_{j=0}^\infty \cos t\lambda_j\, e_j(f)\, dt$$

$$= \frac{\sqrt{\pi}\,\Gamma(1+\delta)}{2\pi} \int_{-\infty}^\infty \lambda \left(\frac{\lambda|t|}{2}\right)^{-\delta-1/2} J_{\delta+1/2}(\lambda|t|) \left(\cos t\sqrt{-\Delta_g}\, f\right)(x)\, dt,$$

with J_μ being the Bessel function of order μ, and

$$u(t,x) = \left(\cos t\sqrt{-\Delta_g}\, f\right)(x)$$

being the solution of the Cauchy problem (11.28).

To use this formula, as in the preceding section, we choose an even function $\rho \in \mathcal{S}(\mathbb{R})$ satisfying

$$\rho(\tau) = 1, \quad |\tau| \le 1, \quad \text{and} \quad \hat\rho(t) = 0,\ |t| \ge r_0,$$

where, as before, r_0 is half the injectivity radius. We then split

$$S_\lambda^\delta f = C_\delta \int_{-\infty}^\infty \lambda\rho(t) \left(\frac{\lambda|t|}{2}\right)^{-\delta-1/2} J_{\delta+1/2}(\lambda|t|) \left(\cos t\sqrt{-\Delta_g}\, f\right)(x)\, dt$$

$$\tag{11.35}$$

$$+ C_\delta \int_{-\infty}^\infty \lambda(1 - \rho(t)) \left(\frac{\lambda|t|}{2}\right)^{-\delta-1/2} J_{\delta+1/2}(\lambda|t|) \left(\cos t\sqrt{-\Delta_g}\, f\right)(x)\, dt$$

$$= \tilde{S}_\lambda^\delta f + R_\lambda^\delta f,$$

with $C_\delta = \frac{\sqrt{\pi}\,\Gamma(1+\delta)}{2\pi}$. One thinks here of $\tilde{S}_\lambda^\delta$ as the "main" term and R_λ^δ as the remainder. We shall be able to compute the kernel of $\tilde{S}_\lambda^\delta$ very precisely and then estimate it by appealing to Stein's oscillatory integral theorem. On the other hand, the operators R_λ^δ are not "local" and so one cannot hope to compute their kernels with the necessary degree of precision on general compact Riemannian manifolds. On the other hand, fortunately, for the above range of exponents, the bounds for R_λ^δ are a direct consequence of (11.33).

To verify the last assertion we use asymptotics for Bessel functions of a fixed order to see that we can write

$$R_\lambda^\delta f = \int_\infty^\infty (1 - \rho(t)) \sum_\pm \lambda m_\delta^\pm(\lambda t) e^{\pm i\lambda t} \cos t\sqrt{-\Delta_g}\, f\, dt,$$

where

$$\lambda |t^j \partial_t^j m_\delta^\pm(\lambda t)| \le C_{\delta,j} \lambda^{-\delta} |t|^{-1-\delta}, \quad |t| > r_0/2.$$

Consequently, if we fix δ and let β_λ be the inverse Fourier transform of $t \to m_\delta^\pm(\lambda t)(1 - \rho(t))$ then we have

$$|\beta_\lambda(\tau)| \le C_N \lambda^{-\delta}(1 + |\tau|)^{-N},$$

for every N, with constants independent of λ. Since

$$R_\lambda^\delta f = \beta_\lambda(\lambda - \sqrt{-\Delta_g})f + \beta_\lambda(\lambda + \sqrt{-\Delta_g})f,$$

we can use Hölder's inequality, (11.33) and orthogonality to get

$$\|R_\lambda^\delta f\|_p^2 \le C\|R_\lambda^\delta f\|_2^2 = \sum_j |\beta_\lambda(\lambda - \lambda_j) + \beta_\lambda(\lambda + \lambda_j)|^2 \|e_j(f)\|_2^2$$

$$\le C \sum_k (1 + |\lambda - k|)^{-N} \lambda^{-2\delta} \|\chi_k f\|_2^2$$

$$\le C \Big[\sum_k (1 + |\lambda - k|)^{-N} \lambda^{-2\delta} k^{2\delta(p)}\Big] \|f\|_p^2$$

$$\le C\lambda^{-2\delta} \lambda^{2\delta(p)} \|f\|_p \le C\|f\|_p^2,$$

assuming in the second to last inequality that $N > 1 + \delta(p)$, and using in the last inequality our assumption that $\delta > \delta(p)$.

To handle the "main term" we note that, since it only involves a superposition of wave kernels for small values of t, we can compute its kernel using the Hadamard parametrix and stationary phase. By doing so we conclude that the kernel of $\tilde{S}_\lambda^\delta$ is of the form

$$\tilde{S}_\lambda^\delta(x, y) \approx \begin{cases} \lambda^{\frac{n-1}{2} - \delta} \dfrac{\cos(\lambda \mathrm{dist}\ (x,y))}{(\mathrm{dist}\ (x,y))^{(n+1)/2+\delta}}, & \mathrm{dist}\ (x,y) \ge \lambda^{-1} \\ \lambda^n, & \mathrm{dist}\ (x,y) \le \lambda^{-1}, \end{cases}$$

where, as before, dist $(\cdot, \cdot)$ is the Riemannian distance function. Because the kernel has this form, we can appeal to Stein's oscillatory integral theorem (Theorem 11.2) to see that

$$\tilde{S}_\lambda^\delta : L^p \to L^p$$

uniformly in λ if $\delta > \delta(p)$ and p is as above, which completes the proof of Theorem 11.4.

The argument that we have just given is modelled after the proof of the sharp Weyl formula, which says that if $N(\lambda)$ is the number of eigenvalues λ_j which are $\le \lambda$ (counted with multiplicity), then

$$N(\lambda) = (2\pi)^{-n}\mathrm{Vol}\ B_n\ \mathrm{Vol}\ M\ \lambda^n + O(\lambda^{n-1}), \tag{11.36}$$

where Vol B_n is the volume of the unit ball in $\mathbb{R}^n$, and Vol M is the volume of M with respect to the metric g. By (11.24) the $O(\lambda^{n-1})$ bounds for the error term in the Weyl formula cannot be improved for the sphere with the standard metric.

Let us conclude this section by showing how the special case of (11.33) can be used to establish this result. We first note that

$$N(\lambda) = \int_M S_\lambda(x,x)\,dx,$$

where $S_\lambda(x,x)$ is the restriction to the diagonal of the kernel

$$S_\lambda(x,y) = \sum_{\lambda_j \le \lambda} e_j(x)\overline{e_j(y)} = S_\lambda^0(x,y),$$

with S_λ^0 being the Bochner-Riesz mean of index 0. Thus, if $\tilde{S}_\lambda = \tilde{S}_\lambda^0$ is defined as above then we can use the Hadamard parametrix to see that

$$\tilde{S}_\lambda(x,x) = c(x)\lambda^n + O(\lambda^{n-1}),$$

where $c(x)$ integrates to the appropriate constant for (11.36). Consequently, to finish, we just need to see that

$$R_\lambda(x,x) = O(\lambda^{n-1}). \tag{11.37}$$

To see this, we can argue as above to get that

$$|R_\lambda(x,x)| = \left|\sum_j (\beta_\lambda(\lambda - \lambda_j) + \beta_\lambda(\lambda + \lambda_j))|e_j(x)|^2\right|$$

$$\le C_N \sum_j (1 + |\lambda - \lambda_j|)^{-N} |e_j(x)|^2$$

$$\le C_N \sum_j (1 + |\lambda - \lambda_j|)^{-N} \left(\sum_{\lambda \in [j,j+1]} |e_j(x)|^2\right),$$

which means that we would have (11.37) if

$$\chi_\lambda(x,x) = \sum_{\lambda_j \in [\lambda,\lambda+1]} |e_j(x)|^2 \le C\lambda^{n-1}.$$

But this follows from the dual version of the $p = 1$ special case of (11.33) since

$$\sup_x \left(\sum_{\lambda_j \in [\lambda,\lambda+1]} |e_j(x)|^2\right)^{1/2} = \sup_x \left(\int \Big| \sum_{\lambda_j \in [\lambda,\lambda+1]} e_j(x)\overline{e_j(y)} \Big|^2 dy\right)^{1/2} \tag{11.38}$$

$$= \|\chi_\lambda\|_{L^2 \to L^\infty} \le C\lambda^{\frac{n-1}{2}}.$$

11.5. Manifolds with boundary

Let M be an n-dimensional C^∞ open manifold with compact closure and boundary ∂M. Consider a Riemannian metric $g = g_{jk}dx^j dx^k$ on M and the associated Dirichlet-Laplacian $\Delta = \Delta_{g,\mathcal{D}}$. We shall then be concerned with estimates for the eigenfunctions,

$$\begin{cases} -\Delta\phi_\lambda(x) = \lambda^2\phi_\lambda(x), & x \in M \\ \phi_\lambda(x) = 0, & x \in \partial M. \end{cases} \tag{11.39}$$

The eigenvalues are discrete and tend to $+\infty$. As before, we count them with respect to multiplicity and order them as $0 < \lambda_1^2 \leq \lambda_2^2 \leq \lambda_3^2 \leq \cdots$. Also, as before, we wish to study the behavior of the L^p norms, i.e., (11.20), as the eigenvalue goes to infinity. We are also interested in stronger estimates like (11.22) for the spectral projection operator.

Thus, a goal would be try to extend Theorem 11.3 to the setting of compact Riemannian manifolds with boundary. We immediately encounter two difficulties:

- It is much harder to use the wave operator $\square$ in manifolds with boundary.
- Less is true: Rayleigh whispering gallery modes say that the bounds in (11.22) cannot hold for all values of $2 \leq p \leq \infty$. Specifically in every dimension $n \geq 2$ the favorable bounds (11.33) can only hold for a smaller range of exponents.

Let us give a brief explanation of these two facts. We start with the first one. In order to explain the complicated nature of parametrices for wave operators in manifolds with boundary, we start by reviewing what happens for the very simple case of the Dirichlet-wave equation for the half-plane $(t, x) \in \mathbb{R} \times (\mathbb{R}^{n-1} \times \mathbb{R}_+)$, i.e.,

$$\begin{cases} \square u = 0 \\ u(t, x', 0) = 0 \\ u(0, x) = f(x), \quad \partial_t u(0, x) = 0. \end{cases}$$

Here we are writing $x = (x', x_n)$, with $x' = (x_1, \ldots, x_{n-1})$. The kernel for the solution operator is given by the formula

$$U_{\mathcal{D}}(t; x, y) = (2\pi)^{-n} \int_{\mathbb{R}^n} e^{i(x-y)\cdot\xi} \cos t|\xi|\, d\xi$$

$$- (2\pi)^{-n} \int_{\mathbb{R}^n} e^{i(x_r - y)\cdot\xi} \cos t|\xi|\, d\xi, \quad (11.40)$$

where x_r is the reflection of x across the boundary.

In a manifold with boundary it is difficult to construct the second term in the parametrix for the solution kernel. Here, we work in geodesic normal coordinates about y and then think of $(x_r - y)$ as the reflected geodesic normal coordinates of x about y. Simple examples where M is either the interior or exterior of a euclidean ball show that these coordinates become degenerate as x approaches the boundary in a tangential direction from y. Indeed, these two examples tell us that if y is a geodesic distance $d = d(y) << 1$ from the boundary ∂M, then we cannot hope to construct the phase functions corresponding to the second term in (11.40) for all x if the time variable satisfies $t > cd^{1/2}$ for some fixed constant c.

Let us address the other difficulty that arises in trying to extend Theorem 11.3. This is an observation of D. Grieser [11]. We shall focus on the case where $n = 2$, but similar considerations show that the bounds (11.22) in Theorem 11.3 cannot hold for all $2 \le q \le \infty$ for higher dimensions as well. Grieser observed that if (M, g) is the interior of the unit disk with the euclidean metric and if ϕ_λ is a so-called whispering gallery mode with eigenvalue λ^2 then, for $q \le 8$, f_λ has most of its L^q mass in a $\lambda^{-2/3}$ neighborhood of the boundary. Hence

$$\frac{\|f_\lambda\|_2}{\|f_\lambda\|_q} \ge c\lambda^{\frac{2}{3}(\frac{1}{2} - \frac{1}{q})}, \quad q \le 8.$$

Since f_λ is an eigenvalue, we conclude that, for such exponents, we can have the favorable 2-dimensional estimates

$$\|\chi_\lambda f\|_q \le C\lambda^{2(1/2 - 1/q) - 1/2}\|f\|_2,$$

only when

$$2(1/2 - 1/q) - 1/2 \ge \tfrac{2}{3}(1/2 - 1/q),$$

which means that the best possible analog of (11.22) for manifolds with boundary when $n = 2$ would be

$$\|\chi_\lambda f\|_q \le C\lambda^{2(1/2 - 1/q) - 1/2}\|f\|_2, \quad q \ge 8, \quad (11.41)$$

instead of $q \ge 6$ as in (11.22) for the boundaryless case when $n = 2$.

Let us turn to positive results now. We shall first indicate how one can obtain sharp pointwise estimates for eigenfunctions and then discuss recent joint work with H. Smith [18] that shows that when $n = 2$ the optimal estimates, i.e., (11.41), are valid.

The pointwise estimate says that for compact Riemannian manifolds with boundary we have

$$\|\chi_\lambda f\|_\infty \le C(1 + \lambda)^{\frac{n-1}{2}} \|f\|_2. \tag{11.42}$$

By (11.38), this estimate is valid if and only if the kernels of the spectral projections have the following bounds when evaluated along the diagonal

$$\chi_\lambda(x, x) = O(\lambda^{n-1}). \tag{11.43}$$

Next, by the arguments from Section 11.2, this holds if

$$\tilde{\chi}_\lambda(x, x) = O(\lambda^{n-1}), \tag{11.44}$$

where

$$\tilde{\chi}_\lambda(x, x) = \int \rho(t) \cos t\lambda \, U_{\mathcal{D}}(t; x, x) \, dt,$$

with $\rho \in C_0^\infty(\mathbb{R})$ being a fixed function supported in a small neighborhood of the origin.

It turns out that the parametrix for the wave kernel at the diagonal, $U_{\mathcal{D}}(t; x, x)$ only allows one to show (11.44) when $d(x) \ge c\lambda^{-1}$ for some fixed constant $c > 0$, where $d(x)$ is the geodesic distance of x from the boundary. Thus, by using wave equation techniques that are more technical but similar to the ones described in Section 11.2, one can show the following special case of (11.42)

$$\chi_\lambda(x, x) \le C\lambda^{n-1}, \quad \text{if } d(x) \ge c\lambda^{-1}, \tag{11.45}$$

where c and C are uniform constants which are independent of λ.

To prove the bounds for the missing case, a $O(\lambda^{-1})$ neighborhood of ∂M, it turns out that one can use a maximum principle argument. This observation goes back to Grieser [12] for the case of eigenfunctions, and Grieser's argument can be modified to handle the case of functions whose spectrum lies in unit bands $[\lambda, \lambda + 1]$. In the latter case, one can use a variant of the maximum principle to see that if c is fixed then

$$\sup_{\{x:\, d(x) \le c\lambda^{-1}\}} \chi_\lambda(x, x) \le C \sup_{\{x:\, d(x) = c\lambda^{-1}\}} \chi_\lambda(x, x),$$

which, by (11.45) means that (11.43) must hold and thus completes the proof of (11.42).

Using the arguments presented earlier, the pointwise bounds (11.42) can be used to show that the operators S_λ^δ are uniformly bounded on $L^1(M)$ and $L^\infty(M)$ when $\delta > (n-1)/2$. Recently, X. Xu [29] has proven more refined pointwise estimates that include sharp pointwise bounds for the gradient of eigenfunctions on compact Riemannian manifolds with boundary. Using these estimates, he was able to show that the Hörmander multiplier theorem extends to this setting.

Let us turn to the other estimate, (11.41). In a work in progress, Smith and Sogge [18] have shown that the estimate (11.41) holds for general two-dimensional Riemannian manifolds with boundary on the range $q \geq 8$. Interpolation with the trivial boundedness of χ_λ on $L^2(M)$ then yields L^q estimates on spectral clusters which are the best possible, as shown by Grieser's observation. The proof depends on the fact that, for functions with microlocal support disjoint from a thin set in phase space consisting of geodesics tangent to the boundary, the full spectral estimates hold. This is because the wave group for transverse reflections has the same essential properties as the free wave group.

To handle the contribution of directions in phase space that are nearly tangent to the boundary, Smith and Sogge exploited ideas of Smith [17] and Tataru [27] developed to handle wave speed metrics of low regularity. The latter work involved a combination of paradifferential and frequency dependent scaling arguments to show that functions similar to the Rayleigh whispering gallery modes are the worst case. Interpolating between the tangent and transverse reflection cases yields the desired L^8 estimates for all functions.

One can obtain estimates for eigenfunctions on manifolds with boundary from appropriate estimates for Lipschitz metrics since one can reflect the eigenfunctions and metric normally across the boundary to obtain equivalent L^p estimates for the resulting Lipschitz metrics. Fortunately, the problems are tractable, at least in two-dimensions, since the metrics one obtains by doubling are piecewise smooth with special types of Lipschitz singularities contained in the image of the boundary.

To motivate this proof, let us see what happens for the analogous result in $\mathbb{R}^2$, which of course is much simpler. We recall from Section 11.2 that the euclidean analog of the dual form of the $L^2(M) \to L^8(M)$ estimates for the χ_λ operators would be the $L^{8/7}(\mathbb{R}^2) \to L^2(S^1)$ restriction theorem for the Fourier transform,

$$\|\hat{f}\|_{L^2(S^1)} \leq C\|f\|_{L^{8/7}(\mathbb{R}^2)}, \quad f \in \mathcal{S}(\mathbb{R}^2). \tag{11.46}$$

Let us see how one can give a simple proof of this estimate. If we square the left side, and use Hölder's inequality, we get

$$\|\hat{f}\|^2_{L^2(S^1)} = \int_{S^1} \hat{f}\overline{\hat{f}}d\theta = \int f(x)\,\overline{(f * \widehat{d\theta})}(x)\,dx$$

$$\leq \|f\|_{L^{8/7}(\mathbb{R}^2)}\|f * \widehat{d\theta}\|_{L^8(\mathbb{R}^2)}.$$

Thus, (11.46) would hold if

$$\|f * \widehat{d\theta}\|_{L^8(\mathbb{R}^2)} \leq C\|f\|_{L^{8/7}(\mathbb{R}^2)}.$$

But, $\widehat{d\theta} \approx \cos|x|/|x|^{1/2}$, and so this estimate would be a consequence of

$$\|f * |x|^{-1/2}\|_{L^8(\mathbb{R}^2)} \leq C\|f\|_{L^{8/7}(\mathbb{R}^2)}, \tag{11.47}$$

which follows from the classical Hardy-Littlewood-Sobolev theorem for fractional integrals.

Estimate (11.46) and the above proof is due to Stein (unpublished), and this was the first restriction theorem for the Fourier transform. Earlier Schwartz had noticed that the restriction to the circle of the Fourier transform of an $L^p(\mathbb{R}^2)$, $p < 4/3$, makes sense as a distribution, and K. DeLeeuw raised the question of whether this distribution was actually a function. Stein's result of course answered this in the affirmative when $n = 2$ for exponents $1 \leq p \leq 8/7$. Stein's $L^{8/7}(\mathbb{R}^2)$ theorem was followed by much activity, including the optimal L^2 restriction theorems of C. Fefferman, P. Tomas, and Strichartz, and the flurry of activity in the 1990's on trying to sharpen these results and prove the higher dimensional versions of the sharp two-dimensional restriction theorem, which is due to Zygmund [30] and says that, for $p < 4/3$, $f \in L^p(\mathbb{R}^2)$ has Fourier transform which restricts as a function to $L^q(S^1)$, $q = p'/3$.

The situation for compact manifolds with boundary studied by Smith and the author [18] is much more technical. However, the fact that the estimate (11.46) follows from estimate (11.47) which does not involve oscillation implicitly carries over to this setting. Indeed a key fact in the proof of (11.41) is that after microlocally breaking up the operators that arise according to the angle from tangency to the boundary one can add up the contributions of the various pieces and still get (11.41). Heuristically, this works for the same reason that in two-dimensions it is a special property of $L^{8/7}$ that the estimate (11.46), which seems to be an estimate involving oscillatory integrals, actually follows from estimate (11.47) which of course does not.

11.6. Riemannian manifolds with maximal eigenfunction growth

We shall discuss some joint work with Steve Zelditch [24]. Recall that the eigenfunction estimates

$$L^p(\lambda, g) = \sup_{\|\phi_\lambda\|_2 = 1} \|\phi_\lambda\|_p = O(\lambda^{\sigma(p)}), \tag{11.48}$$

with $p > 2$ and

$$\sigma(p) = \max\left\{n(1/2 - 1/p) - 1/2, \tfrac{n-1}{2}(1/2 - 1/p)\right\}, \tag{11.49}$$

for eigenfunctions on n-dimensional compact Riemannian manifolds cannot be improved. This is because, as mentioned before, $L^p(\lambda, g) \approx \lambda^{\sigma(p)}$ when (M, g) is the round sphere S^n. On the other hand, there are plenty of examples, such as tori with flat metrics, where the bounds (11.48) for individual eigenfunctions can be improved. The goal of [24] was to characterize those Riemannian manifolds for which, like the sphere, $L^p(\lambda, g) = \Omega(\lambda^{\sigma(p)})$, where $\Omega(\lambda^{\sigma(p)})$ means $O(\lambda^{\sigma(p)})$ but not $o(\lambda^{\sigma(p)})$.

The main result, Theorem 11.48, implies a necessary condition on compact boundaryless Riemannian manifolds (M, g) with maximal eigenfunction growth: there must exist a point $x \in M$ for which the set

$$\mathcal{L}_x = \{\xi \in S_x^* M : \exists T : \exp_x T\xi = x\} \tag{11.50}$$

of directions of geodesic loops at x has positive measure. Here $\exp$ is the exponential map for geodesic flow, and the measure $|\Omega|$ of a set Ω is the one induced by the metric g_x on the fiber $T_x^* M$ over x of the cotangent bundle, $T^* M$. For instance, the poles x_N, x_S of a surface of revolution (S^2, g) satisfy $|\mathcal{L}_x| = 2\pi$. Note also that the geodesic loops do not have to close smoothly.

Theorem 11.5. *Suppose that $|\mathcal{L}_x| = 0$. Then given $\varepsilon > 0$ there exists a neighborhood $\mathcal{N} = \mathcal{N}(\varepsilon)$ of x and a positive number $\Lambda = \Lambda(\varepsilon)$, so that*

$$\sup_{\phi \in V_\lambda} \frac{\|\phi\|_{L^\infty(\mathcal{N})}}{\|\phi\|_{L^2(M)}} < \varepsilon \lambda^{\frac{n-1}{2}}, \quad \lambda \in spec\sqrt{-\Delta} \geq \Lambda, \tag{11.51}$$

if V_λ is the space of eigenfunctions, $-\Delta\phi = \lambda^2\phi$. Furthermore, if $|\mathcal{L}_x| = 0$ for every $x \in M$ then

$$\sup_{\phi \in V_\lambda} \frac{\|\phi\|_{L^p(M)}}{\|\phi\|_{L^2(M)}} = o(\lambda^{\sigma(p)}), \quad p > \frac{2(n+1)}{n-1}. \tag{11.52}$$

The part of the theorem corresponding to $p = \infty$ can be thought of as the natural pointwise analog of the Duistermaat-Guillemin [8] theorem, and the proof relies on ideas of Ivrii [14].

The main part of Theorem 11.5 says that if $L^\infty(\lambda, g) = \Omega(\lambda^{\frac{n-1}{2}})$ then there must be an $x \in M$ so that $|\mathcal{L}_x| > 0$. However, the naive converse to this sup-norm result is simply false. This is because there are examples of Riemannian manifolds (M, g) where there is a x so that $|\mathcal{L}_x| > 0$ but $L^\infty(\lambda, g) = o(\lambda^{\frac{n-1}{2}})$. In [24] a C^∞ torus of revolution with this property was constructed.

We also remark that the bounds in (11.52) all follow from the special case where $p = \infty$. One sees this just by interpolating with the estimate that says that $L^p(\lambda, g) = O(\lambda^{\sigma(p)})$ when $p = \frac{2(n+1)}{n-1}$. An interesting open question would be to determine a condition that would imply that $L^p(\lambda, M) = o(\lambda^{\sigma(p)})$ for $p = \frac{2(n+1)}{n-1}$. By interpolation, this would lead to $o(\lambda^{\sigma(p)})$ bounds for $2 < p < \frac{2(n+1)}{n-1}$.

The author and Zelditch were also able to show that generically $L^\infty(\lambda, g) = o(\lambda^{\frac{n-1}{2}})$. This follows from Theorem 11.48 and the following result which was also proved in [24].

Theorem 11.6. *There exists a residual set $\mathcal{R}$ on the space $\mathcal{G}$ of C^∞ metrics with the Whitney C^∞ topology such that $|\mathcal{L}_x^g| = 0$ for every $x \in M$ when $g \in \mathcal{R}$.*

Corollary 11.7. $L^\infty(\lambda, g) = o(\lambda^{\frac{n-1}{2}})$ *for a generic Riemannian metric on any manifold.*

Thus, even though the sphere S^n with the round metric, or more generally surfaces of revolution have $L^\infty(\lambda, g) = \Omega(\lambda^{\frac{n-1}{2}})$, a generic metric on the sphere will have $L^\infty(\lambda, g) = o(\lambda^{\frac{n-1}{2}})$.

Even though the standard 2-torus $\mathbb{T}^2 = \mathbb{R}^2/\mathbb{Z}^2$ certainly satisfies $L^\infty(\lambda, g) = o(\lambda^{1/2})$, one could ask whether there are other metrics on $\mathbb{T}^2$ for which this quantity is $\Omega(\lambda^{1/2})$. It turns out that if one considers real analytic metric the answer is no. Indeed, we have the following result from [24] that characterizes real analytic manifolds with maximal eigenfunction growth:

Theorem 11.8. *Suppose that (M, g) is a real analytic manifold and that $L^\infty(\lambda, g) = \Omega(\lambda^{\frac{n-1}{2}})$. Then (M, g) is a Y_ℓ^m-manifold, i.e., a pointed Riemannian manifold (M, m, g) such that all geodesics issuing from the point m*

return to m at time ℓ. In particular, if dim $M = 2$, then M is topologically the two-sphere or two-dimensional projective space.

For the definition and properties of Y_ℓ^m manifolds, we refer to [4] (Chapter 7). By a theorem due to Bérard-Bergery (see [3], [4]) Y_ℓ^m manifolds M satisfy $\pi_1(M)$ is finite and $H^*(M, \mathbb{Q})$ is a truncated polynomial ring in one generator. This implies that in 2-dimensions M must be the sphere or projective space. We remark that the loops are not assumed to close up smoothly. An interesting example of this is the tri-axial ellipsoid, $E_{a_1,a_2,a_3} = \{(x_1, x_2, x_3) \in \mathbb{R}^3 : \frac{x_1^2}{a_1^2} + \frac{x_2^2}{a_2^2} + \frac{x_3^2}{a_3^2} = 1\}$ with the standard metric and $a_1 < a_2 < a_3$. For this manifold all geodesics leaving the 4 umbilical points return at the same time, but all but one do not loop back smoothly (see [1]).

References

[1] V.I. Arnold, *Mathematical methods of classical mechanics*, Graduate Texts in Mathematics, 60. Springer-Verlag, New York, 1998.

[2] V. I. Avakumovič, Über die Eigenfunktionen auf geshchlossenen Riemannschen Mannigfaltigkeiten. Math. Z. **65** (1956), 327–344.

[3] L. Bérard-Bergery, Quelques exemples de variétés riemanniennes où toutes les géodésiques issues d'un point sont fermées et de même longueur, suivis de quelques résultats sur leur topologie, Ann. Inst. Fourier (Grenoble) **27** (1977), 231–249.

[4] A. L. Besse, *Manifolds all of whose geodesics are closed. With appendices by D. B. A. Epstein, J.-P. Bourguignon, L. Bérard-Bergery, M. Berger and J. L. Kazdan.* Ergebnisse der Mathematik und ihrer Grenzgebiete [Results in Mathematics and Related Areas], 93. Springer-Verlag, Berlin-New York, 1978.

[5] J. Bourgain, Besicovitch type maximal operators and applications to Fourier analysis, Geom. Funct. Anal. **1** (1991), 147–187.

[6] J. Bourgain, L^p-estimates for oscillatory integrals in several variables, Geom. Funct. Anal. **1** (1991), 321–374.

[7] L. Carleson and P. Sjölin, Oscillatory integrals and a multiplier problem for the disc, Studia Math. **44** (1972), 287–299.

[8] J. J. Duistermaat and V. W. Guillemin, The spectrum of positive elliptic operators and periodic bicharacteristics. Invent. Math. **29** (1975), 39–79.

[9] C. Fefferman, Inequalities for strongly singular convolution operators. Acta Math. **124** (1970), 9–36.

[10] C. Fefferman, The multiplier problem for the ball. Ann. of Math. **94** (1971), 330–336.

[11] D. Grieser, L^p Bounds for Eigenfunctions and Spectral Projections of the Laplacian Near Concave Boundaries, PhD Thesis UCLA, 1992.

[12] D. Grieser, Uniform bounds for eigenfunctions of the Laplacian on manifolds with boundary, Comm. Partial Differential Equations **27** (2002),, 1283–1299.

[13] L. Hörmander, The spectral function of an elliptic operator. Acta Math. **121** (1968), 193–218.

[14] V. Ivriĭ, The second term of the spectral asymptotics for a Laplace Beltrami operator on manifolds with boundary. (Russian) Funktsional. Anal. i Prilozhen. **14** (1980), 25–34.

[15] B. M. Levitan, On the asymptoptic behavior of the spectral function of a self-adjoint differential equaiton of second order. Isv. Akad. Nauk SSSR Ser. Mat. **16** (1952), 325–352.

[16] W. Minicozzi and C. D. Sogge, Negative results for Nikodym maximal functions and related oscillatory integrals in curved space. Math. Res. Lett. **4** (1997), 221–237.

[17] H. Smith, A parametrix construction for wave equations with $C^{1,1}$ coefficients. Ann. Inst. Fourier (Grenoble) **48** (1998), no. 3, 797–835.

[18] H. Smith and C. D. Sogge, in preparation.

[19] C. D. Sogge, Oscillatory integrals and spherical harmonics. Duke Math. J. **53** (1986), 43–65.

[20] C. D. Sogge, Concerning the L^p norm of spectral clusters for second order elliptic operators on compact manifolds. J. Funct. Anal. **77** (1988), 123–134.

[21] C. D. Sogge, On the convergence of Riesz means on compact manifolds. Ann. of Math. (2) 126, no. 2 (1987), 439–447.

[22] C. D. Sogge, *Fourier integrals in classical analysis*, Cambridge Univ. Press, Cambridge, 1993.

[23] C. D. Sogge, Eigenfunction and Bochner Riesz estimates on manifolds with boundary, Math. Res. Lett. **9** (2002), 205–216.

[24] C. D. Sogge and S. Zelditch, Riemannian manifolds with maximal eigenfunction growth, Duke Math. J. **114** (2002), 387–437.

[25] E. M. Stein, Oscillatory integrals in Fourier analysis, in *Beijing lectures in harmonic analysis (Beijing, 1984)*, 307–355, Ann. of Math. Stud., 112, Princeton Univ. Press, Princeton, NJ, 1986.

[26] P. Tomas, A restriction theorem for the Fourier transform, Bull. Amer. Math. Soc. **81** (1975), 477–478.

[27] D. Tataru, Strichartz estimates for operators with nonsmooth coefficients and the nonlinear wave equation. Amer. J. Math. **122** (2000), 349–376.

[28] T. Tao, From rotating needles to stability of waves: emerging connections between combinatorics, analysis, and PDE, Notices Amer. Math. Soc. **48** (2001), 294–303.

[29] X. Xu, PhD Thesis, Johns Hopkins University (2004), to appear.

[30] A. Zygmund, On Fourier coefficients and transforms of functions of two variables, Studia Math. **50** (1974), 189–201.

Chapter 12

Five lectures on harmonic analysis

Fernando Soria

Department of Mathematics
Universidad Autónoma de Madrid
28049 Madrid (Spain)

Contents

12.1. Introduction

The aim of these lectures is to present some concrete examples which show the interplay between geometry and certain aspects of harmonic analysis. To that end, let us consider the following initial value problems defined on the half space $\mathbb{R}^{n+1}_+ = \mathbb{R}^n \times \mathbb{R}_+$:

Heat Equation

$$\partial_t u(x,t) = \Delta_x u(x,t), \qquad (x,t) \in \mathbb{R}_+^{n+1}$$
$$u(x,0) = u_0(x), \qquad\qquad x \in \mathbb{R}^n,$$

Laplace Equation

$$\partial_{tt} u(x,t) + \Delta_x u(x,t) = 0, \qquad (x,t) \in \mathbb{R}_+^{n+1}$$
$$u(x,0) = u_0(x), \qquad\qquad x \in \mathbb{R}^n,$$

Wave Equation

$$\partial_{tt} u(x,t) = \Delta_x u(x,t), \qquad (x,t) \in \mathbb{R}_+^{n+1}$$
$$u(x,0) = u_0(x), \qquad\qquad x \in \mathbb{R}^n$$
$$\partial_t u(x,0) = v_0(x), \qquad\qquad x \in \mathbb{R}^n.$$

Schrödinger Equation

$$\partial_t u(x,t) = \frac{i}{4\pi^2} \Delta_x u(x,t), \qquad (x,t) \in \mathbb{R}_+^{n+1}$$
$$u(x,0) = u_0(x), \qquad\qquad x \in \mathbb{R}^n.$$

The Fourier transform for an integrable function f is defined by

$$\widehat{f}(\xi) = \int_{\mathbb{R}^n} f(x)\, e^{-2\pi i <x,\xi>} dx.$$

The Fourier transform relates to differential operator via the identity

$$\widehat{P(D)f}(\xi) = P(2\pi i \xi)\widehat{f}(\xi),$$

where P is a polynomial in n variables and f is an appropriate smooth function. By using this transform on the x-variable in the above equations we find that they can be transformed, respectively, into the following o.d.e.'s in the variable t for each fixed ξ, with $y(t) = \widehat{u^x}(\xi,t)$;

$$y'(t) = -4\pi^2|\xi|^2 y(t), \qquad y(0) = \widehat{u_0}$$
$$y''(t) = 4\pi^2|\xi|^2 y(t), \qquad y(0) = \widehat{u_0}$$
$$y''(t) = -4\pi^2|\xi|^2 y(t), \qquad y(0) = \widehat{u_0}, \qquad y'(0) = \widehat{v_0}$$
$$y'(t) = -2\pi i|\xi|^2 y(t), \qquad y(0) = \widehat{u_0},$$

and so, the corresponding solution can be given by the inverse Fourier transform, by

$$u(x,t) = \int_{\mathbb{R}^n} e^{-4\pi^2 t|\xi|^2}\widehat{u_0}(\xi)\, e^{2\pi i x\xi}d\xi$$

$$u(x,t) = \int_{\mathbb{R}^n} e^{-2\pi t|\xi|}\widehat{u_0}(\xi)\, e^{2\pi i x\xi}d\xi$$

$$u(x,t) = \int_{\mathbb{R}^n} \cos(2\pi t|\xi|)\widehat{u_0}(\xi)\, e^{2\pi i x\xi}d\xi + \int_{\mathbb{R}^n} \frac{\sin(2\pi t|\xi|)}{2\pi|\xi|}\widehat{v_0}(\xi)\, e^{2\pi i x\xi}d\xi$$

$$u(x,t) = \int_{\mathbb{R}^n} e^{-2\pi i t|\xi|^2}\widehat{u_0}(\xi)\, e^{2\pi i x\xi}d\xi.$$

We want to study the properties of the flux map

$$\Phi_t : \{\text{space of initial data}\} \to \{\text{space of solutions}\},$$

that is, its regularity, the conservation laws that inherits, the existence of limits as $t \to 0^+$, etc. It is clear that the first two examples have "symbols" which are rapidly decreasing for large values of the variable ξ, whereas the other two simply oscillate and have no decay at all. This is determinant for the properties of Φ_t and, in particular, for the size of the space of initial data so that a proper solution exists.

12.1.1. *The role of maximal functions*

When dealing with pointwise convergence, the natural objects that arise are the so called maximal functions. Let us start with the classical example of Lebesgue differentiation theorem:

Define for $f \in L^1(\mathbb{R}^n)$, the average means

$$u(x,t) = m_t f(x) = \frac{1}{|B_t(x)|}\int_{B_t(x)} f(y)\, dy.$$

($B_t(x)$ represents the ball centered at $x \in \mathbb{R}^n$ and radious $t > 0$.) The theorem says that we can recover the function f pointwise as $t \to 0^+$, that is

$$\lim_{t \to 0^+} m_t f(x) = f(x), \qquad a.e.$$

The result is clearly true for a dense class of functions, for instance the Schwartz class, $\mathcal{S}$ (here, the class of continuous functions with compact support suffices). Moreover, the maximal operator

$$M f(x) = \sup_{t>0} |m_t f(x)| = \sup_{t>0} \frac{1}{|B_t(x)|} \left| \int_{B_t(x)} f(y) \, dy \right|$$

satisfies the remarkable inequality

$$|\{x \in \mathbb{R}^n : M f(x) > \lambda\}| \le \frac{C}{\lambda} \int_{\mathbb{R}^n} |f(y)| dy, \qquad \forall \lambda > 0.$$

To show the convergence property, suffices to prove that

$$\limsup_{t \to 0^+} m_t f(x) = \liminf_{t \to 0^+} m_t f(x), \qquad a.e.,$$

or equivalently, that the set

$$A_\lambda = \{x : \limsup_{t \to 0^+} m_t f(x) - \liminf_{t \to 0^+} m_t f(x) > \lambda\}$$

has measure 0 for every $\lambda > 0$. If g is any "nice" function in our dense class, then one has

$$A_\lambda = \{x : \limsup_{t \to 0^+} m_t(f - g)(x) - \liminf_{t \to 0^+} m_t(f - g)(x) > \lambda\},$$

and since

$$A_\lambda \subset \{x \in \mathbb{R}^n : 2M(f - g)(x) > \lambda\},$$

we conclude

$$|A_\lambda| \le \frac{2C}{\lambda} \int_{\mathbb{R}^n} |f(y) - g(y)| dy.$$

Now, the density of our class does the rest.

This argument, due to Hardy and Littlewood, is standard in our approach to the problem of the a.e convergence to the initial datum in the above examples: the boundedness of the maximal function

$$U^*(x) = \sup_{t>0} |u(x, t)| = \sup_{t>0} |\Phi_t u_0(x)|$$

on certain class of functions $\mathcal{L}$, in which $\mathcal{S}$ is dense, implies the a.e. convergence for all initial values $u_0 \in \mathcal{L}$.

The maximal function associated to the heat and Laplace equations are pointwise majorized by the Hardy-Littlewood maximal function. It fact, it is known that if the function K has an L^1 radial decreasing majorant, K^*, then $|K * f(x)| \leq \|K^*\|_{L^1} M f(x)$ ([53]). The solution to the heat and Laplace equations are given, respectively, by the convolution

$$G_t * u_0(x), \qquad P_t * u_0(x),$$

where G_1 is the Gaussian kernel and $G_t = t^{-n/2} G_1(\cdot/t^{1/2})$ and P_1 is the Poisson kernel, with $P_t = t^{-n} P_1(\cdot/t)$. Therefore,

$$\sup_{t>0} |G_t * u_0(x)|, \quad \sup_{t>0} |P_t * u_0(x)| \leq C M u_0(x).$$

Hence, a.e. convergence holds in all spaces $L^p(\mathbb{R}^n)$, $1 \leq p$.

M controls, with more generality, the convergence phenomena associated to the so called Calderón-Zygmund singular integrals (see Section 2).

The wave and the Schrödinger equations have maximal functions bounded on smaller classes of functions. For example, in the case of the Schrödinger equation, L. Carleson [15] obtained, in dimension $n = 1$, the result of the pointwise convergence of the solution to the initial datum for all functions $u_0 \in H^{1/4}(\mathbb{R})$ via the corresponding estimate of the maximal function in that space. Here, H^s denotes the space of s "derivatives" in L^2; that is

$$H^s(\mathbb{R}^d) = \{\, u_0 \in L^2(\mathbb{R}^d)\,;\, \int_{\mathbb{R}^d} |\widehat{u_0}(\xi)|^2 (1 + |\xi|)^{2s}\, d\xi < +\infty \,\}.$$

Also, the homogeneous Sobolev space $\dot{H}^s$ is defined as

$$\dot{H}^s(\mathbb{R}^d) = \{\, u_0 \in L^2(\mathbb{R}^d)\,;\, \int_{\mathbb{R}^d} |\widehat{u_0}(\xi)|^2 |\xi|^{2s}\, d\xi < +\infty \,\}.$$

The exponent $1/4$ is sharp. It is conjectured that $1/4$ of derivative in L^2 suffices in any dimension for the a.e. convergence in this problem.

Convergence and boundedness of the associated maximal function are in general equivalent, as was shown by Stein ([54], [32]).

As an example of this, the fact that the double iterated convergence in Lebesgue theorem fails in $L^1(\mathbb{R}^2)$ can be viewed as the failure of the strong maximal function of Jessen-Marcinckiewicz-Zygmund, M_s, to be bounded $L^1(\mathbb{R}^2) \to L^{1,\infty}(\mathbb{R}^2)$ ([35]).

In many cases, the singularities appear along significative directions in space. This motivates the study of maximal functions associated to a given set of unit vectors. As we will see, the boundedness properties depend largely on the geometrical structure of this set.

Another interesting feature of maximal functions is the control they have on weighted inequalities for a given singular integral T; that is, estimates of the form

$$\int |Tf|^2 w \le \int |f|^2 Pw,$$

where w is a "weight" (a positive, locally integrable function). The boundedness of $w \to Pw$ gives a complete profile of the properties of T, via the extrapolation theorem of Rubio de Francia ([31]).

For example, when T is a Calderón-Zygmund singular integral, one can take (see [31])

$$Pw = (Mw^\alpha)^{1/\alpha}, \quad \alpha > 1.$$

On the other hand, it is not known whether M_s controls the L^2-inequalities of multi-parameter C-Z singular integrals (Product Domains).

Also, $\mathcal{K}_\infty$, the Kakeya maximal function (that is, averages on rectangles in any direction) is expected to control "strongly" singular operators, like the Disc Multiplier. This is the so called Stein Conjecture.

A less standard operator is the so called spherical maximal function of Stein and Bourgain:

$$M_\sigma f(x) = \sup_{t>0} \left| S_t f(x) \right|,$$

where

$$S_t f(x) = \int_{|y|=1} f(x - ty) d\sigma(y),$$

and $d\sigma$ denotes the Lebesgue measure on the unit sphere. Observe that $\widehat{S_t f}(\xi) = \hat{f}(\xi)\widehat{d\sigma}(t\xi)$. In dimension $n = 3$ this is related to the wave equation since in that case

$$\widehat{d\sigma}(\xi) = \frac{\sin(2\pi|\xi|)}{2\pi|\xi|}.$$

In the next lectures we will study the boundedness of some of these maximal functions as well as the boundedness of the singular operators whose a.e. convergence phenomena they control.

12.2. Second lecture: spherical maximal function

The (Stein) spherical maximal function is defined as

$$M_\sigma f(x) = \sup_{t>0} \left| \int_{|y|=1} f(x - ty) d\sigma(y) \right|,$$

where $d\sigma$ denotes the Lebesgue measure on the unit sphere.

Spheres are sets of null Lebesgue measure and, for a general function $f \in L^p$, the integral could be undefined. We will see, however, that $M_\sigma f(x) < \infty$ a.e. for all $f \in L^p(\mathbb{R}^n)$, $p_n < p \leq \infty$, p_n depending on the dimension. This is shown by proving "a priori" estimates $\|M_\sigma f\|_{L^p} \leq C_{p,n}\|f\|_p$ for $f \in C_0^\infty$ and then extending the estimates to all of L^p by density.

Remarks: 1) Let us take $n = 1$. Then,

$$M_\sigma f(x) = \sup_{t>0} \frac{1}{2}\{|f(x + t)| + |f(x - t)|\}$$

So, if f is unbounded in only one point, $Mf(x) = \infty$, for all x. Actually, for $f \in C_0^\infty(\mathbb{R})$ $M_\sigma f(x) \approx \|f\|_{L^\infty}$.

2) For $n > 1$, if $f = \mathcal{X}_{B(0,1)}$ we have for $|x| \geq 1$,

$$M_\sigma f(x) \sim \frac{1}{|x|^{n-1}}.$$

Hence, if $M_\sigma f \in L^p(\mathbb{R}^n)$, then $p(n - 1) > n$. Thus, a necessary condition for M_σ to be bounded on $L^p(\mathbb{R}^n)$ is $p > \frac{n}{n-1}$. This condition turns out to be also sufficient. This was proven by Stein for $n \geq 3$ and by Bourgain in the case of $\mathbb{R}^2$.

Theorem 12.1 (E. M. Stein). *If $n \geq 3$ and $p > \frac{n}{n-1}$, then $M_\sigma : L^p(\mathbb{R}^n) \to L^p(\mathbb{R}^n)$.*

Theorem 12.2 (J. Bourgain). *If $p > 2$, then $M_\sigma : L^p(\mathbb{R}^2) \to L^p(\mathbb{R}^2)$.*

Theorem 12.1 follows from a more general result due to J. L. Rubio de Francia ([47]).

Proposition 12.1. *Let σ be a compactly supported Borel measure with $|\hat{\sigma}(\xi)| \leq C|\xi|^{-a}$. Define T_t as*

$$\widehat{T_t f}(\xi) = \hat{f}(\xi)\hat{\sigma}(t\xi)$$

and $T_* f(x) = \sup_{t>0} |T_t f(x)|$. Then, we have,

$$T^* : \; L^p(\mathbb{R}^n) \to L^p(\mathbb{R}^n) \quad \text{for } p > p_a = \frac{2a+1}{2a} \quad \text{and} \quad a > 1/2.$$

Examples. If $d\sigma$ is the Lebesgue measure on S^{n-1}, then $|\widehat{d\sigma}(\xi)| \leq \dfrac{C}{|\xi|^{\frac{n-1}{2}}}$ (see [51], chapter 4). In general, if σ is smooth, compactly supported in a hypersurface of $\mathbb{R}^n$ with k non-vanishing principal curvatures $(n \geq k > 1)$, then $|\widehat{d\sigma}(\xi)| \leq \dfrac{C}{|\xi|^{k/2}}$.

Notice that $a = \frac{n-1}{2} > 1/2$ if and only if $n > 2$. So, Proposition 12.1 does not apply to the case of Theorem 12.2.

Proof of Proposition 12.1. Set $\phi_0 \in C_0^\infty$ such that $\phi_0(0) = 1$. We define a partition of the unity:

$$\psi_j(\xi) = \phi_0(2^{-j}\xi) - \phi_0(2^{-j+1}\xi) \qquad \text{for } j \in \mathbb{Z}.$$

Then $\sum_{j=1}^{\infty} \psi_j(\xi) + \phi_0(\xi) \equiv 1$. We now define a partition of the operator T_t. Set

$$\widehat{T_t^j f}(\xi) = \psi_j(t\xi)\hat{\sigma}(t\xi)\hat{f}(\xi) \qquad \text{for } j = 1, 2, 3, \ldots, \qquad\qquad \text{and}$$
$$\widehat{T_t^0 f}(\xi) = \phi_0(t\xi)\hat{\sigma}(t\xi)\hat{f}(\xi).$$

Set also $T_*^j f(x) = \sup_{t>0} |T_t^j f(x)|$. Then we have

$$T_* f(x) = \sup_{t>0} |(\widehat{d\sigma}(t\xi)\hat{f}(\xi))^{\vee}(x)| \leq \sum_{j=0}^{\infty} T_*^j f(x).$$

$\square$

Lemma 12.1. $T_*^0 f(x) \leq CM f(x)$ *and therefore $T_*^0 f : \; L^p \to L^p$ for all $1 < p \leq \infty$.*

To prove this lemma we just have to observe that the kernel of this operator, $K_0(x) = \check{\phi}_0 * d\sigma(x)$, belongs to the Schwartz class. Therefore it has an L^1 radial decreasing majorant. By a remark in the introduction, it follows that $T_*^0 f(x) \leq CMf(x)$, where M is here the Hardy-Littlewood maximal function.

We need to obtain a deeper estimate for the operators T_*^j, $j \geq 1$.

Lemma 12.2. $T_*^j : L^2 \to L^2$ *has norm less than or equal to* $C2^{-j(a-1/2)}$.

Lemma 12.3. $T_*^j : L^1 \to L^{1,\infty}$ *with constant not bigger than* $Cj2^j$.

Using Marcinkiewick's interpolation theorem we obtain that T_*^j is a bounded operator in L^p, $1 < p \leq 2$, with norm not bigger than $C_p(j2^j)^{\frac{2-p}{p}} 2^{-j(2a-1)\frac{p-1}{p}}$. Therefore,

$$\|T_* f\|_{L^p} \leq \|T_*^0 f\|_{L^p} + \sum_{j=1}^{\infty} \|T_*^j f\|_{L^p}$$

$$\leq \Big[C_p + \sum_{j=1}^{\infty} C_p(j2^j)^{\frac{2-p}{p}} 2^{-j(2a-1)\frac{p-1}{p}} \Big] \|f\|_{L^p} \leq C_p \|f\|_{L^p},$$

provided

$$(2a-1)(p-1) > 2 - p, \qquad \text{i.e.} \qquad p > \frac{2a+1}{2a}.$$

Proof of Lemma 12.2. We use the Fundamental Theorem of Calculus. Given a differentiable function $F : \mathbb{R} \to \mathbb{R}$ such that $F(0) = 0$, we have,

$$(F(t))^2 \leq 2 \int_0^{\infty} F(s)F'(s)$$

$$\leq 2 \left(\int_0^{\infty} |F(s)|^2 \frac{ds}{s} \right)^{1/2} \left(\int_0^{\infty} |sF'(s)|^2 \frac{ds}{s} \right)^{1/2}.$$

Hence

$$|T_*^j f(x)|^2 \leq 2 \left(\int_0^{\infty} |T_t^j f(x)|^2 \frac{dt}{t} \right)^{1/2} \left(\int_0^{\infty} |t\frac{d}{dt} T_t^j f(x)|^2 \frac{dt}{t} \right)^{1/2}.$$

In general, for m a bounded multiplier, we define

$$G_m f(x) = \left(\int_0^{\infty} |(m(t\cdot) * \hat{f})\check{}(x)|^2 \frac{dt}{t} \right)^{1/2}.$$

Then, by Plancherel

$$\|G_m f\|_{L^2} = \left(\int_0^{\infty} |m(t\xi)|^2 \frac{dt}{t} \right)^{1/2} \|f\|_{L^2}.$$

(Note that the integral is independent of $\xi \neq 0$.)

With this notation,

$$|T_*^j f|^2 \leq 2(G_{m_j} f)(G_{\tilde{m}_j} f),$$

where

$$m_j(\xi) = \psi_j(\xi)\widehat{d\sigma}(\xi) \qquad \text{and} \qquad \tilde{m}_j(\xi) = <\xi, \nabla m_j(\xi)>.$$

The support of m_j is contained in $\{\xi \in \mathbb{R}^n : |\xi| \sim 2^j\}$. Moreover, $\|m_j\|_\infty \leq 2^{-ja}$; therefore, $\left(\int_0^\infty |m_j(t\xi)|^2 \frac{dt}{t}\right)^{1/2} \leq 2^{-ja}$. $\tilde{m}_j$ is also supported around $|\xi| \sim 2^j$. In this case $\|\tilde{m}_j\|_\infty \leq 2^j 2^{-ja}$, so that $\left(\int_0^\infty |\tilde{m}_j(t\xi)|^2 \frac{dt}{t}\right)^{1/2} \leq 2^{-j(a-1)}$.

Hence, using Cauchy-Schwarz's inequality,

$$\|T_*^j f\|_{L^2} \leq 2\|G_{m_j} f\|_{L^2}^{1/2}\|G_{\tilde{m}_j} f\|_{L^2}^{1/2} \leq 2^{-j(a-1/2)}\|f\|_{L^2}.$$

We have to justify a couple of details. First, $T_0^j f(x) = 0$ for all $x \in \mathbb{R}^n$ whenever $\hat{f} \in L^1$; this is an application of Lebesgue convergence theorem. The claim that $\|m_j\|_{L^\infty} \leq 2^{j-ja}$ comes from the estimate $|\nabla \hat{\sigma}(\xi)| \leq C|\xi|^{-aj}$. To prove this one, note that $\partial_i \hat{\sigma}(\xi) = \hat{\sigma} * \hat{\Phi}(\xi)$, for any $\Phi \in C_0^\infty$ so that $\Phi(x) = x_i$ for $|x| \leq 2$. $\qquad\qquad \square$

The argument above proves a more general statement ([47]):

Corollary 12.1. *Let* $m \in L^\infty(\mathbb{R}^n)$. *Set* $m_t(\xi) = m(t\xi)$. *If* m *satisfies* $|m(\xi)| \leq \frac{1}{(1+|\xi|)^a}$ *and* $|\nabla m(\xi)| \leq \frac{1}{(1+|\xi|)^b}$ *then, the maximal operator*

$$T^* f(x) = \sup_t |\tilde{m}_t * f(x)|$$

is bounded on L^2 *provided* $a + b > 1$.

Proof of Lemma 12.3. Fix $f \in L^1$ and $\lambda > 0$. Consider the **Calderón-Zygmund decomposition** of f at λ ([53]). We write $f = g + \sum b_\beta$ with

- $|g| \leq C\lambda, \int |g| \leq C \int |f|$
- $\text{supp} \, b_\beta \subset Q_\beta$, where $\{Q_\beta\}$ is a family of disjoint dyadic cubes.
- $\int b_\beta = 0$ and $\sum \|b_\beta\|_{L^1} \leq C \int |f|$.
- $|\cup Q_\beta| \leq C \int \frac{|f|}{\lambda}$.

From the L^2 estimate we deduce

$$|\{T_*^j g > \lambda\}| \le C \int \frac{|g|^2}{\lambda^2} \le \frac{C}{\lambda} \int |f|.$$

As in the theorems for the classical singular integral operators, we only need to estimate

$$\left|\left\{x \in [\cup(3Q_\beta)]^c : T_*^j\left(\sum b_\beta\right) > \lambda\right\}\right|$$
$$\le \frac{C}{\lambda}\sum\int_{(3Q_\beta)^c} |T_*^j b_\beta(x)|\, dx$$
$$= \frac{C}{\lambda}\sum_\beta\int_{(3Q_\beta)^c}\sup_t\left|\int (K_t^j(x-y) - K_t^j(x-y_\beta))b_\beta(y)\,dy\right|dx,$$

where $K^j = \check{\psi}_j * d\sigma$, $K_t^j(x) = \frac{1}{t^n}K^j(\frac{x}{t})$ and y_β is the center of Q_β. The last expression is bounded by

$$\frac{C}{\lambda}\sum_\beta\int_{Q_\beta} |b_\beta(y)|\left(\int_{(3Q_\beta)^c}\sup_{t>0} |K_t^j(x-y) - K_t^j(x-y_\beta)|dx\right)dy.$$

It suffices to prove that

$$I = \int_{|x|>2|y|}\sup_{t>0} |K_t^j(x-y) - K_t^j(x)|dx \le Cj2^j.$$

(This is a Maximal Hörmander condition.) This is a tedious but not too difficult computation. We use that

$$K_t^j(x) = \int_0^t \frac{d}{ds}(K_s^j(x))ds$$
$$= \int_0^t \left\{-ns^{-n-1}K^j\left(\frac{x}{s}\right) - s^{-n-1}\left\langle\nabla K^j\left(\frac{x}{s}\right), \frac{x}{s}\right\rangle\right\} ds.$$

Then

$$I \le C\int_0^\infty \int_{|x|>2|y|} |K_t^j(x-y) - K_t^j(x)|dx\frac{dt}{t}$$
$$+2^j C\int_0^\infty \int_{|x|>2|y|} |\tilde{K}_t^j(x-y) - \tilde{K}_t^j(x)|dx\frac{dt}{t},$$

where $\tilde{K}_t^j(x) = \frac{1}{t^n}\left\langle 2^{-j}\nabla K^j\left(\frac{x}{t}\right), x\right\rangle$. Each term is bounded by $C_n j$. We will show this for the first term (the second one is similar). For a fix y we use the mean value theorem to write

$$\int_0^\infty \int_{|x|>2|y|} |K_t^j(x-y) - K_t^j(x)| dx \frac{dt}{t}$$

$$= \int_0^\infty \int_{|x|>2|y|} \left| K^j\left(\frac{x-y}{t}\right) - K_t^j\left(\frac{x}{t}\right) \right| dx \frac{dt}{t^{n+1}}$$

$$\leq 2 \int_0^{|y|2^j} \int_{|x|>|y|} \left| K^j\left(\frac{x}{t}\right) \right| dx \frac{dt}{t^{n+1}}$$

$$+ \int_{|y|2^j}^\infty \int_{|x|>2|y|} \int_0^1 \frac{|y|}{t} \left| \nabla K^j\left(\frac{x-ry}{t}\right) \right| dr\, dx \frac{dt}{t^{n+1}}$$

$$\leq 2 \int_0^{|y|2^j} \int_{|x|>|y|/t} |K^j(x)| dx \frac{dt}{t}$$

$$+ \int_{2^j|y|}^\infty \int_{|x|>|y|/t} |y||\nabla K^j(x)|\, dx \, \frac{dt}{t^2}$$

$$\leq 2 \int_{2^{-j}}^\infty \int_{|x|>s} |K^j(x)| dx \frac{ds}{s}$$

$$+ \int_0^{2^{-j}} \int_{|x|>s} |\nabla K^j(x)|\, dx\, ds.$$

Now, recall the definition of K_j. Note that it is essentially supported in $||x| - 1| < 2^{-j}$ and that $\|K_j\|_{L^\infty} \leq 2^j$. To be more precise, $|K_j(x)| \leq C_l \frac{2^j}{(1+2^j||x|-1|)^l}$ for all $l \in \mathbb{N}$. Therefore,

$$\int_{|x|>t} |K^j(x)| dx \leq \frac{C}{1+|t|}.$$

Hence,

$$\int_{2^{-j}}^\infty \int_{|x|>t} |K^j(x)| dx \frac{dt}{t} \leq C \log 2^j = Cj.$$

A similar argument gives the bound 1 for the other term. $\qquad\qquad\square$

12.2.1. *Spherical dyadic maximal function*

We define the spherical dyadic maximal operator, M_d as

$$M_d f(x) = \sup_{j\in\mathbb{Z}} \left| \int_{|y|=1} f(x - 2^j y) d\sigma(y) \right|.$$

We will see that the behavior of this operator is much better than that of M_σ. In order to do that we need to introduce new machinery.

12.2.2. *Littlewood-Paley Theory*

Roughly speaking this theory consists in estimating the L^p behavior of

$$f \to (\sum |f_k|^2)^{1/2}$$

where f_k collects the spectrum (frequencies) of f in certain region A_k. A necessary and sufficient condition for its L^2 boundedness is

$$\sum_k \mathcal{X}_{A_k}(y) \le C < \infty.$$

Theorem 12.3 ("Dyadic" Littlewood-Paley). *Let* $\Psi \in C_0^\infty([\frac{1}{4}, 2])$. *Set*

$$\widehat{S_j f}(\xi) = \Psi(2^{-j}\xi)\hat{f}(\xi).$$

Then

$$\|(\sum_j |S_j f|^2)^{1/2}\|_{L^p} \le C_p \|f\|_{L^p}, \qquad 1 < p \le 2.$$

Moreover, if $\sum_{j \in \mathbb{Z}} |\Psi(2^{-j}t)|^2 \equiv 1$, $t \ne 0$, *then*

$$c_p \|(\sum_j |S_j f|^2)^{1/2}\|_{L^p} \le \|f\|_{L^p} \le C_p \|(\sum_j |S_j f|^2)^{1/2}\|_{L^p}, \quad 1 < p < \infty.$$

To prove this result we need a generalization of the classical Calderón-Zygmund theory.

12.2.3. *Vector valued Calderón-Zygmund theory*

In a previous result we were estimating

$$\int_{|x|>2|y|} \|K_t(x - y) - K_t(x)\|_{L^\infty(dt)} dx.$$

This generalizes to other Banach spaces. Let B be a Banach space ($B = \mathbb{C}$ represents the "classical" theory). Denote

$$L_B^p(\mathbb{R}^n) = \{f : \mathbb{R}^n \to B : \ \|f\|_{(L_B^p(\mathbb{R}^n))} := \left(\int_{\mathbb{R}^n} \|f(x)\|_B^p \, dx\right)^{1/p} < \infty\}.$$

Given a kernel $K : \mathbb{R}^n \to B$, we define the operator

$$T_K f(x) = pv \int K(x - y)f(y)dy = K * f(x) \in B.$$

Theorem 12.4. *Assume that $T_K : L^2(\mathbb{R}^n) \to L^2_B(\mathbb{R}^n)$ and that K satisfies the vector valued Hörmander condition*

$$\int_{|x|>2|y|} \|K(x-y) - K(x)\|_B dx \leq C, \qquad \text{independently of } y.$$

Then $T_K : L^p \to L^p_B(\mathbb{R}^n)$, $1 < p < \infty$ and is also of weak type 1, i.e.

$$|\{x \in \mathbb{R}^n : \|T_K f(x)\|_B > \lambda\}| \leq \frac{C}{\lambda} \|f\|_{L^1_B}, \qquad \lambda > 0.$$

The proof of this theorem is essentially the same as in the scalar case (see [31]). We are now ready to prove Theorem 12.3.

Proof of Theorem 12.3. Denote $\hat{\phi}_j(\xi) = \Psi(2^{-j}\xi)$. Define the operator

$$f \to Tf \equiv \{\phi_j * f\}_j \in \ell^2,$$

Then, $T : L^2 \to L^2_{\ell^2}$ (since $\sum_j |\Psi(2^{-j}t)|^2 \leq C$). Its kernel, $\{\phi_j\}_{j\in\mathbb{Z}}$ satisfies

$$I(y) = \int_{|x|>2|y|} \left[\sum_j \left((\phi_j(x-y) - \phi_j(x))^2 \right) \right]^{1/2} dx \leq C.$$

To see this, note that for $|x| > 2|y|$,

$$|\phi_j(x-y) - \phi_j(x)| \leq |\phi_j(x-y)| + |\phi_j(x)| \leq \frac{C_m 2^{jn}}{(2^j|x|)^m}$$

and $\qquad |\phi_j(x-y) - \phi_j(x)| \leq |\nabla(\phi_j)(\bar{x})||y| \leq \frac{C_m 2^{j(n+1)}|y|}{(2^j|x|)^m}.$

Take $m = n + \varepsilon$. Then

$$I(y) \leq C \int_{|x|>2|y|} \frac{1}{|x|^{n+\varepsilon}} dx \left(\sum_j \left[2^{-j\varepsilon} \Lambda 2^{j(1-\varepsilon)}|y| \right]^2 \right)^{1/2}$$

$$\leq C|y|^{-\varepsilon} \left(\sum_{j \geq \log_2(1/|y|)} 2^{-2j\varepsilon} \right)^{1/2}$$

$$+ C|y|^{-\varepsilon} \left(\sum_{j < \log_2(1/|y|)} (2^{j(1-\varepsilon)}|y|)^2 \right)^{1/2} \leq C.$$

If $\sum_{j\in\mathbb{Z}} |\Psi(2^{-j}t)|^2 \equiv 1$, $t \neq 0$, then, by Parseval's identity and duality

$$\|f\|_p = \sup_{\|g\|_{p'}=1} \int fg = \int \sum_j S_j f S_j g$$

$$\leq \|(\sum |S_j f|^2)^{1/2}\|_p \|(\sum_j |S_j g|^2)^{1/2}\|_{p'} \leq C_{p'} \|(\sum |S_j f|^2)^{1/2}\|_p.$$

$\square$

We now go back to the spherical dyadic operator. The measure $d\sigma$ is just one particular example for which we can prove the following result.

Theorem 12.5. *Let $d\sigma$ be a positive and finite measure with $|\widehat{d\sigma}(\xi)| \leq C|\xi|^{-a}$, for some $a > 0$. We define*

$$T_{dyad}f(x) = \sup_{j \in \mathbb{Z}} \left| \int f(x - 2^j y) \, d\sigma(y) \right|.$$

Then $T_{dyad} : L^p \to L^p$, $1 < p < \infty$.

Proof. With the notation of Proposition 12.1,

$$T_{dyad}f(x) = \sup_{k \in \mathbb{Z}} |T_{2^k} f(x)|.$$

Notice that

$$\widehat{T_{2^k}^j f} = \widehat{d\sigma}(2^k \xi)\psi(2^{-j}2^k|\xi|)\hat{f}(\xi) = \widehat{d\sigma}(2^k \xi)\psi(2^{-j}2^k|\xi|)\widehat{S_{j-k}f}(\xi),$$

where $\widehat{S_l f_l} = \Psi(2^{-l}|\xi|)\hat{f}(\xi)$, for some $\Psi = 1$ on support of ψ. There is an L^2-estimate,

$$\int |T_*^j f(x)|^2 \leq \sum_k \int |T_{2^k}^j f|^2$$

$$= \sum_k \int |\hat{d\sigma}(2^k \xi)|^2 |\psi(2^{k-j}\xi)|^2 |\widehat{S_{j-k}f}(\xi)|^2$$

$$\leq C(2^{-ja})^2 \int \sum_k |\widehat{S_{j-k}f}|^2 \leq C(2^{-ja})^2 \|f\|_2^2.$$

We want to find an L^p-estimate for these operators that allows us to sum the norms. Fix a natural number N. Since each operator T_{2^k} is bounded in L^p, $(1 < p \leq 2)$ there exists a constant $C_p(N)$, so that

$$\| \sup_{-N \leq k \leq N} |T_{2^k} f| \|_p \leq C_p(N) \|f\|_p$$

Actually, we take $C_p(N)$ as the smaller constant satisfying this inequality for all $f \in L^p$, i.e. the norm of the operator

$$f \to \sup_{-N \le k \le N} |T_{2^k}(f)|.$$

Now consider the operator

$$T: \; \{g_k\} \to \{T^j_{2^k} g_k\}.$$

$|T^j_{2^k} g(x)| \le T_{2^k}(Mg)(x)$, where M denotes the Hardy-Littlewood maximal operator. Thus

$$\| \sup_{-N \le k \le N} |T^j_{2^k} g_k| \|_p \le \| \sup_{N \le k \le N} T_{2^k}(M[\sup_{N \le k \le N} |g_k|]) \|_{L^p}$$
$$\le C_p C_p(N) \| \sup_{-N \le k \le N} |g_k| \|_{L^p}.$$

This means that T is bounded on $L^p_{\ell^\infty}$ with norm less than or equal to $C_p C_p(N)$.

We have also an $L^p_{\ell^p}$-estimate. Since $T^j_{2^k}$ is bounded in L^p with norm not bigger than $C 2^{-ja/p'}$ (this is easy to see for $p = 1$ and $p = 2$ and then by interpolation), we have

$$\|T^j_{2^k} g_k\|_{L^p_{\ell^p}} \le C 2^{-ja/p'} \|g_k\|_{L^p_{\ell^p}}.$$

By interpolation we obtain an $L^p_{\ell^2}$-estimate. Namely,

$$\|T^j_{2^k} g_k\|_{L^p_{\ell^2([-N,N])}} \le C_p C_p(N)^{1-p/2} 2^{-ajp/p'} \|g_k\|_{L^p_{\ell^2([-N,N])}}.$$

Finally, using this estimate and Theorem 5, we obtain

$$\| \sup_{-N \le k \le N} T_{2^k} f \|_p \le \sum_{j \ge 0} \| \sup_{-N \le k \le N} |T^j_{2^k} f| \|_p$$
$$\le \sum_{j \ge 0} \|(\sum_{-N}^{N} |T^j_{2^k} S_{j-k} f|^2)^{1/2}\|_{L^p}$$
$$\le C_p C_p(N)^{1-p/2} \sum_{j \ge 0} 2^{-ja(p-1)} \|(\sum_{k=-N}^{N} |S_{j-k} f|^2)^{1/2}\|_{L^p}$$
$$\le C_p C_p(N)^{1-p/2} \|f\|_{L^p}.$$

But, since $C_p(N)$ is the smallest constant satisfying such an inequality, then

$$C_p(N) \le C_p C_p(N)^{1-p/2}.$$

Hence, $C_p(N) \le \tilde{C}_p$, independent of N. $\qquad\qquad \square$

12.3. Third lecture: more on maximal functions

12.3.1. *The maximal operator along a set of directions*

We will consider variations of the Hardy-Littlewood maximal function M. The first one is the **strong maximal operator**

$$M_s f(x) = \sup_{x \in R} \frac{1}{|R|} \int_R |f(y)|\, dy,$$

where the supremum is now taken on all the rectangles with sides parallel to the coordinate axes, containing the point x.

This operator is bounded in $L^p(\mathbb{R}^n)$ for $1 < p \le \infty$ but is not of weak type (1,1). The counterexample is easy to construct: take f the characteristic function of the unit square $[0, 1] \times [0, 1]$. For any $0 < \lambda < 1$, the set $\{x \in \mathbb{R}^2 \ / \ M_s f(x) > \lambda\}$ contains all rectangles $[0, a) \times [0, 1/(\lambda a))$ for $1 < a < 1/\lambda$. Hence

$$|\{u \in \mathbb{R}^2 \ / \ M_s f(u) > \lambda\}| \ge |\{(x, y) : \ 1 < x < 1/\lambda, \ x \cdot y < \frac{1}{\lambda}\}|$$

$$\ge \int_1^{\frac{1}{\lambda}} \frac{1}{\lambda x}\, dx = \frac{1}{\lambda} \log \frac{1}{\lambda}.$$

The proof of the L^p-boundedness is just an easy observation. Define the **directional maximal operators**

$$M_1 f(x_1, x_2) = M(f(\cdot, x_2))(x_1), \qquad \text{for } f \in \mathcal{S}(\mathbb{R}^2).$$

Analogously define

$$M_2 f(x_1, x_2) = M(f(x_1, \cdot))(x_2).$$

Then we have the pointwise majorization $M_s f(x) \le M_2 M_1 f(x)$. It is a consequence of Fubini's theorem and the boundedness of M that

$$\|M_i f\|_p \le C_p \|f\|_p,$$

for $i = 1, 2$ and $p > 1$. This is enough to prove that

$$\|M_s f\|_{L^p} \le C_p \|f\|_{L^p} \qquad 1 < p < \infty.$$

In general, given v a unit vector in $\mathbb{R}^2$, we define the **directional maximal operator**

$$\mathcal{M}_v f(x) = \sup_{s > 0} \frac{1}{2s} \int_{|t| < s} |f(x - tv)|\, dt, \qquad \text{for } f \in \mathcal{S}(\mathbb{R}^2).$$

Let Σ be a set of unit vectors in $\mathbb{R}^2$. We consider the family $\mathcal{B}_\Sigma$ of all rectangles R whose longest side is parallel to a vector of Σ. Associated to this family we define the maximal operator

$$\mathfrak{M}_\Sigma f(x) = \sup_{x \in R \in \mathcal{B}_\Sigma} \frac{1}{R} \int_R |f(y)|\, dy.$$

Obviously $\|\mathfrak{M}_\Sigma f\|_\infty \leq \|f\|_\infty$. For $1 < p < \infty$, the boundedness or unboundedness and the norm of $\mathfrak{M}_\Sigma$ (in the first case) will depend on the set Σ. This is the subject of this lecture.

Let us first see that $\mathfrak{M}_\Sigma$ may sometimes be an unbounded operator. We will consider a set of **uniformly distributed directions,**

$$\Sigma = \Sigma_N = \left\{ \left(\cos \frac{2\pi j}{N}, \sin \frac{2\pi j}{N} \right) : j = 1, 2, \ldots N/8 \right\}.$$

This particular example is closely related to the Kakeya maximal operator $\mathcal{K}_{1/N}$ considered later, since one has:

$$\mathcal{K}_{1/N} f(x) \leq 10\mathfrak{M}_{\Sigma_N} f(x).$$

One can construct a **Kakeya set** P_N (this example is due to C. Fefferman [30]) that satisfies

    i) $|P_N| \leq C \log \log N$ for some universal C.

    ii) $|\{x \,/\, \mathfrak{M}_{\Sigma_N} \chi_{P_N}(x) > 1/10\}| \geq C \log N$.

From the second estimate, we deduce that $\|\mathfrak{M}_{\Sigma_N} \chi_{P_N}\|_p \geq C(\log N)^{1/p}$ and

$$\frac{\|\mathfrak{M}_{\Sigma_N} \chi_{P_N}\|_p}{\|\chi_{P_N}\|_p} \geq C \left(\frac{\log N}{\log \log N} \right)^{1/p}.$$

So, the norm of the operator depends on the cardinality of Σ. Moreover, if we take $\Sigma = S^1$ the unit sphere, then, $\mathfrak{M}_{S^1} \geq \sup \mathfrak{M}_{\Sigma_N}$ for all N and hence $\mathfrak{M}_{S^1}$ is unbounded.

Now, we should show some positive result. Without loss of generality, we will assume $\Sigma \subset \{(\cos\theta, \sin\theta) : 0 \leq \theta \leq \frac{\pi}{4}\}$. Observe first that if Σ is a finite set, $\Sigma = \{v_1, v_2, \ldots, v_N\}$, then

$$\mathfrak{M}_\Sigma f(x) \leq C \sup_{v \in \Sigma} \mathcal{M}_v M_2 f(x) \leq C \left[\sum_{v \in \Sigma} (\mathcal{M}_v M_2 f(x))^p \right]^{1/p}$$

for some universal constant C. Therefore, we have a trivial estimate:

$$\|\mathfrak{M}_\Sigma f\|_p \leq C_p N^{1/p}\|f\|_p.$$

This is not, by any means, sharp. In fact, there are positive results even for certain infinite set of directions: If Σ is a **lacunary set of directions,** *i.e.*,

$$\Sigma = \{v^j = (v_1^j, v_2^j)\}_{j=1}^\infty \quad \text{with} \quad 0 \leq \frac{v_2^{j+1}}{v_1^{j+1}} \leq \lambda \frac{v_2^j}{v_1^j}$$
$$\text{for some} \quad 0 < \lambda < 1 \quad \text{and all } j,$$

then $\mathfrak{M}_\Sigma$ is bounded on L^p, for all $1 < p \leq \infty$.

The first result about these "lacunary" directions is due to Strömberg [55]. Later, Córdoba and Fefferman [25] proved the L^2-theorem, and, finally, Nagel, Stein and Wainger proved the L^p-result above [45].

For the set of uniformly distributed directions, Córdoba [21] proved that there is a constant $C > 0$, independent of N, such that, $\|\mathfrak{M}_{\Sigma_N} f\|_2 \leq C(1 + \log N)^4\|f\|_2$ for all $f \in L^2$. Strömberg [56] obtained the sharp result,

$$\|\mathfrak{M}_{\Sigma_N} f\|_2 \leq C(1 + \log N)\|f\|_2.$$

To see that Strömberg's estimate is sharp, one just have to apply the operator to the function

$$f(x) = \frac{1}{|x|}\chi_{1 \leq |r| \leq N}.$$

Note that this function does not show that the norm of the operator depends on N for $p > 2$. This is due to the fact that this is a radial function. We will speak about this later on.

In some sense, the uniformly distributed case seems to be the worst one. There was a conjecture saying that there are positive constants C and α, such that, for every set Σ with N elements, $\|\mathfrak{M}_\Sigma f\|_2 \leq C(1 + \log N)^\alpha\|f\|_2$.

J. A. Barrionuevo [4] proved that for every set Σ with N elements,

$$\|\mathfrak{M}_\Sigma f\|_2 \leq C N^{2/\sqrt{\log N}}\|f\|_2.$$

Note that this is a great improvement of our trivial result because $C N^{2/\sqrt{\log N}} \leq C_\epsilon N^\epsilon$ for any $\epsilon > 0$.

A bit later N. Katz [39] proved finally the conjecture: There is a positive constant C such that, for every set Σ with N elements, $\|\mathfrak{M}_\Sigma f\|_2 \leq$

$C(1 + \log N)\|f\|_2$. We will present here an simple proof of this result due to Alfonseca, Soria and Vargas ([2], [3]).

Before doing that, we would like to go back to some question mentioned earlier, the case of radial functions. When these operators act on radial functions the behavior is much better. A. Carbery, E. Hernández and F. Soria [11], showed that for any radial f and any $p > 2$, we have $\|\mathfrak{M}_{S^1} f\|_p \leq C\|f\|_p$.

For a set Σ the action of the operator on radial functions is determined by its Minkowsky dimension, $d(\Sigma)$. This is defined as follows:

$$d(\Sigma) = \limsup_{\delta \to 0^+} \frac{\log \mathcal{N}(\delta)}{-\log \delta},$$

where $\mathcal{N}(\delta)$ is the minimum number of closed intervals of length δ needed to cover Σ. If Σ has positive Lebesgue measure, $d(\Sigma) = 1$. If Σ has zero measure and we write $S^1 \setminus \Sigma$ as the union of a sequence of disjoint open intervals, $\{I_j\}$, then $d(\Sigma) = \inf\{\alpha \geq 0 : \sum_j |I_j|^\alpha < \infty\}$, where $|I_j|$ denotes the length of I_j. The Minkowsky dimension is an upper bound for the Hausdorff dimension but they are different in general. Nevertheless, they coincide for self-similar sets (see [28], p. 118). In the case of the ordinary Cantor set one has $d(\Sigma) = \log 2/\log 3$.

Given any set Σ, $\mathfrak{M}_\Sigma$ is bounded on L^p_{rad} if $p > 1 + d(\Sigma)$ and unbounded if $p < 1 + d(\Sigma)$. (J. Duoandikoetxea and A. Vargas [27]).

It was believed that this result was true for general functions in the case of the the Cantor set, giving rise to yet another example of an infinite set whose maximal function would be bounded on L^2. However, it was proved by N. Katz [38] that this is not the case. The operator associated to the Cantor set is unbounded in L^2 (and therefore in L^p for $p < 2$.).

12.3.2. *A quasi-orthogonality principle*

We change slightly our previous notation. Let Ω be a subset of $[0, \pi)$. Associated to Ω we consider the basis $\mathcal{B}$ of all rectangles in $\mathbb{R}^2$ whose longest side forms an angle θ with the x-axis, for some $\theta \in \Omega$. The maximal operator associated with the set Ω is defined by

$$M_\Omega f(x) = \sup_{x \in R \in \mathcal{B}} \frac{1}{|R|} \int_R |f(y)|\, dy.$$

The study of directional maximal functions began in the 1970's, with important contributions by Strömberg [55], Córdoba and Fefferman [25], Nagel, Stein and Wainger [45], Sjögren and Sjölin [48]. More recently, the interest on these problems was renewed with the results of Barrionuevo [4] and Katz [38, 39]. Nevertheless, only the operators associated to some particular sets Ω are well understood. Namely, the cases of lacunary sets of directions ([45] and [48]) and of finite sets [39]. In [2] and [3] we proposed a new method to study this operators. We decomposed Ω into several consecutive blocks, Ω_j. We proved an almost-orthogonality principle that essentially meant that the weak L^2-norm of M_Ω is the supremum of the norms of the operators M_{Ω_j}, plus a term associated to the sequence of end-points of the blocks. Let us explain this.

Without loss of generality, we can assume that $\Omega \subset [0, \pi/4)$. Let $\Omega_0 = \{\theta_1 > \theta_2 > ... > \theta_j > ...\}$ be an ordered subset of Ω. We take $\theta_0 = \frac{\pi}{4}$ and consider, for each $j \geq 1$, sets $\Omega_j = [\theta_j, \theta_{j-1}) \cap \Omega$, such that $\theta_j \in \Omega_j$ for all j. Assume also that $\Omega = \cup \Omega_j$.

To each one of the sets Ω_j, $j = 0, 1, 2, \ldots$, we associate the corresponding basis $\mathcal{B}_j$, and define the maximal operators associated to the sets Ω_j by

$$M_{\Omega_j} f(x) = \sup_{x \in R \in \mathcal{B}_j} \frac{1}{|R|} \int_R |f(y)|\, dy, \quad j = 0, 1, 2, \ldots$$

Then we have the following result.

Theorem 12.6. *There exists a constant C independent of Ω such that*

$$\|M_\Omega\|_{L^2 \to L^2} \leq \sup_{j \geq 1} \|M_{\Omega_j}\|_{L^2 \to L^2} + C \|M_{\Omega_0}\|_{L^2 \to L^2},$$

where $\|T\|_{L^2 \to L^2}$ denotes the "strong type $(2, 2)$" norm of the operator T.

The proof relies on geometric arguments like the ones used in [2], and on a covering lemma by Carbery [10]. A version of this principle for general p, $1 < p \leq \infty$ can be found in [1].

It is worth noting that in the Theorem, the constant multiplying the supremum of the norms of the Ω_j is 1. As a corollary of Theorem 12.6, we give a simple proof of the result by Katz [39].

Corollary 12.2. *There exists a constant K such that, for any set $\Omega \subset [0, \frac{\pi}{4})$ with cardinality $N > 1$, one has*

$$\|M_\Omega\|_{L^2 \to L^2} \leq K(\log N).$$

Proof. We can assume that $N = 2^m$. We use induction on m. If m is small, then, for K big enough, the estimate follows from the boundedness of the strong maximal function. Now assume that is true for all $k < m$. If the elements of Ω are ordered, $\{\phi_1 > \phi_2 > \ldots > \phi_N\}$, we define Ω_0 to be the set consisting only on ϕ_N and the middle element $\phi_{\frac{N}{2}}$. In this way, there are only two sets Ω_1 and Ω_2. Each one of them has $N/2$ elements. So by Theorem 12.6 and induction hypothesis,

$$\|M_\Omega\|_{L^2 \to L^2} \leq K \log \frac{N}{2} + 2C = K \log N - K \log 2 + 2C.$$

If we take $K = \frac{2C}{\log 2}$, we have the result. $\qquad\qquad\qquad\square$

12.3.3. *Kakeya maximal operator*

Definition 12.1. For $0 < \delta < 1$, and $f \in L^1_{loc}(\mathbb{R}^n)$, we define the Kakeya maximal function of eccentricity δ

$$\mathcal{K}_\delta f(x)(x) = \sup \frac{1}{|R_x|} \int_{R_x} |f(y)|\, dy$$

where the 'sup' is taken over all rectangles in $\mathbb{R}^n$ homothetic to $[0,\delta] \times [0,\delta] \times \cdots \times [0,1]$ of arbitrary direction and containing x.

It has been conjectured that the the Kakeya maximal operator is bounded in $L^n(\mathbb{R}^n)$ with norm $C_n(1 + \log \frac{1}{\delta})^{a_n}$. In the two–dimensional case, $\mathcal{K}_\delta f(x) \leq C\mathfrak{M}_\Sigma f(x)$ when Σ is the uniformly distributed set of $1/\delta$ directions, and so the result is true (the original proof of this is due to A. Córdoba and it does not use this argument). In higher dimensions, if the conjecture were true, we would obtain, by interpolation, the following L^p estimates:

$$\|\mathcal{K}_\delta f\|_{L^p(\mathbb{R}^n)} \leq \begin{cases} C_n\left(1 + \log \frac{1}{\delta}\right)^{a_{n,p}} \|f\|_{L^p(\mathbb{R}^n)} & \text{for } p \geq n \\ C_{n,p}\left(\frac{1}{\delta}\right)^{\frac{n}{p}-1}\left(1 + \log \frac{1}{\delta}\right)^{a_{n,p}} \|f\|_{L^p(\mathbb{R}^n)} & \text{for } 1 < p < n. \end{cases}$$

To see it, note that the Kakeya operator is bounded on $L^\infty(\mathbb{R}^n)$ with constant 1, and is of weak–type 1 with constant $C_n\left(\frac{1}{\delta}\right)^{n-1}$. (The latter follows from the classical result for the Hardy–Littlewood maximal operator.)

Only partial results are known for $n \geq 3$. M. Christ, J. Duoandikoe-txea and J. L. Rubio de Francia proved the conjecture for $p \leq (n+1)/2$. (We will give a proof of this fact due to J. Bourgain [5].) J. Bourgain improved this for $p < (n+1)/2 + \epsilon_n$, for some positive $\epsilon_n \to 0$ as $n \to \infty$ and later, T. Wolff proved the theorem for $p \le (n+2)/2$. There are recent results due to Iosevich, Katz, Laba, Tao and others that improve even more the end point for which the result holds. All these results go beyond the scope of the present lectures.

12.3.4. *Annexes*

Proof of the result for Kakeya in $\mathbb{R}^n$ for $p \le \frac{n+1}{2}$. We will show the case $p = (n+1)/2$. The other cases follow by interpolation. Given $0 < \delta < \frac{1}{2}$, $f \in L^1_{loc}(\mathbb{R}^n)$ we define the Kakeya maximal function without dilations as

$$f^*_\delta(x) = \sup_\tau \frac{1}{|\tau|} \int_\tau f(y)\, dy$$

where the supremum is taken in all parallelepipeds τ centered at x of dimensions
$$1 \times \delta \times \delta \times \cdots \times \delta.$$

1) A sieve argument (see [20]) would give us the result once we prove the analogous estimate for f^*_δ, namely

$$\|f^*_\delta\|_p \le C_{n,p} \left(\frac{1}{\delta}\right)^{\frac{n}{p}-1} \left(\log \frac{1}{\delta}\right)^{\omega_n} \|f\|_p \qquad \text{for all} \qquad 1 < p \le \frac{n+1}{2}.$$

2) Reduction to a restricted type inequality. In order to prove this, it is enough to see that for any measurable set $A \subset B(0,1)$ and for all $0 \le \sigma \le 1$, we have:

$$|A| \ge C_n \delta^{n-p} \sigma^p |\{(\chi_A)^*_\delta > \sigma\}| \qquad \text{with} \qquad p = \frac{n+1}{2}.$$

Actually this would be equivalent to:

$$\|f^*_\delta\|_{L^{p,\infty}} \le C_d \delta^{-\left(\frac{n}{p}-1\right)} \|f\|_{L^{p,1}}.$$

Moreover,

$$\|f^*_\delta\|_{L^\infty} \le \|f\|_{L^\infty}$$

and from the properties of the Hardy–Littlewood maximal operator,

$$\|f^*_\delta\|_{L^{1,\infty}} \le C\delta^{-(n-1)} \|Mf\|_{L^{1,\infty}} \le C\delta^{-(n-1)} \|f\|_{L^1} = C\left(\frac{1}{\delta}\right)^{\frac{n}{1}-1} \|f\|_{L^1}$$

By interpolation (see [11] Proposition 5)

$$\|f_\delta^*\|_p \leq C_{n,p} \left(\frac{1}{\delta}\right)^{\frac{n}{p}-1} \left(\log\frac{1}{\delta}\right)^{\omega} \|f\|_p \qquad \text{for all} \qquad 1 < p \leq \frac{n+1}{2}.$$

3) Linearization. We first of all linearize the operator: we set a grid of cubes of side lengths δ in $\mathbb{R}^n$, $\{Q_i\}$. We associate a parallelepiped τ_i to each cube, so that $Q_i \cap \tau_i \neq \emptyset$. We define a new operator (which we denote with the same symbol)

$$f_\delta^*(x) = \left(\frac{1}{|\tau_i|} \int_{\tau_i} f(y)\, dy\right) \chi_{Q_i}(x)$$

We have to prove an estimate independent of the choice of the parallelepipeds τ_i.

4) The bushes. Denote $\mathcal{D} = \{(\chi_A)_\delta^* > \sigma\}$ and $\mathcal{E} = \{i : Q_i \subset \mathcal{D}\}$.

 By definition $\#\mathcal{E} = \frac{|\mathcal{D}|}{\delta^n}$ and for all $i \in \mathcal{E}$, $|\tau_i \cap A| > \sigma|\tau_i| = \sigma\delta^{n-1}$. Thus,

$$\int_A \left(\sum_{i\in\mathcal{E}} \tau_i\right) = \sum_{i\in\mathcal{E}} |\tau_i \cap A| \geq \sigma\delta^{n-1}\,(\#\mathcal{E}) \geq \frac{c_n\sigma|\mathcal{D}|}{\delta}.$$

Therefore, there is some $x_o \in A$ so that

$$\#\{i \in \mathcal{E} : x_o \in \tau_i\} \geq \frac{\sigma|\mathcal{D}|}{\delta|A|}.$$

We can pick $\mathcal{F}_o \subset \mathcal{E}$, such that

$$x_o \in \tau_i \qquad \text{for all} \qquad i \in \mathcal{F}_o$$

$$\text{angle}\,(\tau_i, \tau_j) \geq \frac{10\delta}{\sigma} \quad \text{if } i \neq j,\ i,j \in \mathcal{F}_o$$

and

$$\#\mathcal{F}_o \geq \frac{|\mathcal{D}|\sigma^n}{|A|}.$$

We define *the bush*

$$\mathcal{B}_o = \bigcup_{i\in\mathcal{F}_o} \tau_i.$$

We have now two estimates of the measure of $\mathcal{B}_o$. If $i \in \mathcal{F}_o$, then

$$|A \cap \tau_i| > \sigma \delta^{n-1} \qquad \text{and}$$

$$|A \cap \tau_i \cap B\left(x_o, \frac{\sigma}{3}\right)| \leq |\tau_i \cap B(x_o, \frac{\sigma}{3})| \leq \frac{2}{3} \sigma \delta^{n-1}$$

Hence,

$$|A \cap \tau_i \cap B\left(x_o, \frac{\sigma}{3}\right)^c| > \frac{\sigma \delta^{n-1}}{3} = \frac{\sigma}{3}|\tau_i|.$$

The sets $\tau_i \cap B\left(x_o, \frac{\sigma}{3}\right)^c$, $i \in \mathcal{F}_o$, are pairwise disjoints. Then, we have a lower estimate

$$|\mathcal{B}_o| \geq \sum_{\mathcal{F}_o} |\tau_i \setminus B(x_o, \frac{\sigma}{3})| \geq \sum_{\mathcal{F}_o} \frac{\delta^{n-1}}{3} = (\#\mathcal{F}_o)\frac{\delta^{n-1}}{3} \geq \frac{\delta^{n-1}\sigma^n|\mathcal{D}|}{|A|}.$$

On the other hand, we have the upper estimate

$$|\mathcal{B}_o| \leq \sum_{\mathcal{F}_o} |\tau_i| \leq \sum_{\mathcal{F}_o} \frac{3}{\sigma}|A \cap \tau_i \cap B\left(x_o, \frac{\sigma}{3}\right)^c| \leq \frac{3}{\sigma}|A \cap \mathcal{B}_o|.$$

This finishes with the estimate we wanted. $\qquad\square$

Proof of Theorem 12.6. We first linearize the operators M_Ω and M_{Ω_j}. For any $\alpha \in \mathbb{Z}^2$, Q_α will denote the unit cube centered at α. Given a set $\Lambda \subset [0, \pi/4)$, for each α we choose a rectangle $R_\alpha \in \mathcal{B}_\Lambda$, such that $R_\alpha \supset Q_\alpha$. We define the operator T_Λ as

$$T_\Lambda f(x) = \sum_\alpha{}^{\backprime} \frac{1}{|R_\alpha|}\left(\int_{R_\alpha} f\right) \chi_{Q_\alpha}(x).$$

By definition, one can easily see that

$$T_\Lambda f(x) \leq M_\Lambda f(x),$$

for any choice of rectangles $\{R_\alpha\}$. On the other hand, there is a sequence of linearized operators $\{T_\Lambda f\}$, associated to grids of smaller cubes in $\mathbb{R}^2$, which converge pointwise to $M_\Lambda f$. By scaling invariance, we need only prove it with M_Ω replaced by T_Ω.

We shall show this using the following result, proved by Carbery in [10].

Theorem 12.7. *Let T_Λ be as above. Then T_Λ is of strong type (p,p) if and only if there exist a constant $C_{p'}$, such that for any sequence $\{\lambda_\alpha\} \subset \mathbb{R}_+$, we have*

$$\int \left(\sum_\alpha \lambda_\alpha \frac{1}{|R_\alpha|}\chi_{R_\alpha}\right)^{p'} \leq C_{p'} \sum_\alpha |\lambda_\alpha|^{p'}.$$

Moreover, the infimum of the constants $(C_{p'})^{1/p'}$ satisfying the inequality is $\|T_\Lambda\|_{L^p \to L^p}$.

Let us continue with the proof of the theorem. We define T_Ω for some choice of rectangles $\{R_\alpha\}$. We only need to prove that the inequality is satisfied, with $p = 2$ and $C_2^{1/2} = \sup_{j \geq 1} \|M_{\Omega_j}\|_{L^2 \to L^2} + C \|M_{\Omega_0}\|_{L^2 \to L^2}$.

Set

$$I^2 = \int \left(\sum_\alpha \lambda_\alpha \frac{1}{|R_\alpha|} \chi_{R_\alpha} \right)^2 = \int \left(\sum_l \sum_{\alpha : R_\alpha \in \Omega_l} \lambda_\alpha \frac{1}{|R_\alpha|} \chi_{R_\alpha} \right)^2$$

$$= \int \sum_l \left(\sum_{\alpha : R_\alpha \in \Omega_l} \lambda_\alpha \frac{1}{|R_\alpha|} \chi_{R_\alpha} \right)^2$$

$$+ 2 \sum_l \sum_{j < l} \int \sum_{R_\alpha \in \Omega_l} \sum_{R_\beta \in \Omega_j} \lambda_\alpha \lambda_\beta \frac{1}{|R_\alpha||R_\beta|} \chi_{R_\alpha} \chi_{R_\beta}$$

$$= A + B.$$

For the first term we obtain

$$A \leq \sum_l \|M_{\Omega_l}\|_{L^2 \to L^2}^2 \left(\sum_{\alpha : R_\alpha \in \Omega_l} |\lambda_\alpha|^2 \right)$$

$$\leq \left(\sup_l \|M_{\Omega_l}\|_{L^2 \to L^2}^2 \right) \left(\sum_l \sum_{\alpha : R_\alpha \in \Omega_l} |\lambda_\alpha|^2 \right)$$

$$\leq \left(\sup_l \|M_{\Omega_l}\|_{L^2 \to L^2}^2 \right) \left(\sum_\alpha |\lambda_\alpha|^2 \right).$$

Now we have to study B. Using the same geometric arguments as in [2], we have that there exists a constant C such that, if $R_\alpha \in \Omega_l$ and $R_\beta \in \Omega_j$ for $j < l$,

$$\frac{|R_\alpha \cap R_\beta|}{|R_\alpha||R_\beta|} \leq C \frac{|\widetilde{R}_\alpha^- \cap R_\beta|}{|\widetilde{R}_\alpha^-||R_\beta|} + C \frac{|R_\alpha \cap \widetilde{R}_\beta^+|}{|R_\alpha||\widetilde{R}_\beta^+|},$$

where $\widetilde{R}_\alpha^-$ (respectively $\widetilde{R}_\beta^+$), are rectangles of the basis $\mathcal{B}_0$ containing R_α (respectively R_β). Then,

$$B \leq 2C \sum_l \sum_{j < l} \int \sum_{R_\alpha \in \Omega_l} \sum_{R_\beta \in \Omega_j} \lambda_\alpha \lambda_\beta \frac{1}{|\widetilde{R}_\alpha^-||R_\beta|} \chi_{\widetilde{R}_\alpha^-} \chi_{R_\beta} +$$

$$+2C\sum_{l}\sum_{j<l}\int\sum_{R_\alpha\in\Omega_l}\sum_{R_\beta\in\Omega_j}\lambda_\alpha\lambda_\beta\frac{1}{|R_\alpha||\widetilde{R}_\beta^+|}\chi_{R_\beta}\chi_{\widetilde{R}_\beta^+}=$$

$$=B^-+B^+.$$

We shall only work with the B^- (the other term is analogous). So,

$$B=2C\sum_{l}\sum_{j<l}\int\sum_{R_\alpha\in\Omega_l}\sum_{R_\beta\in\Omega_j}\lambda_\alpha\lambda_\beta\frac{1}{|\widetilde{R}_\alpha^-||R_\beta|}\chi_{\widetilde{R}_\alpha^-}\chi_{R_\beta}$$

$$\leq 2C\int\left(\sum_{l}\sum_{R_\alpha\in\Omega_l}\lambda_\alpha\frac{\chi_{\widetilde{R}_\alpha^-}}{|\widetilde{R}_\alpha^-|}\right)\left(\sum_{j}\sum_{R_\beta\in\Omega_j}\lambda_\beta\frac{\chi_{R_\beta}}{|R_\beta|}\right).$$

We use Cauchy-Schwarz's inequality to bound it by

$$2C\left(\int\left(\sum_{l}\sum_{R_\alpha\in\Omega_l}\lambda_\alpha\frac{\chi_{\widetilde{R}_\alpha^-}}{|\widetilde{R}_\alpha^-|}\right)^2\right)^{1/2}\left(\int\left(\sum_{j}\sum_{R_\beta\in\Omega_j}\lambda_\beta\frac{\chi_{R_\beta}}{|R_\beta|}\right)^2\right)^{1/2}.$$

Now, notice that $\widetilde{R}_\alpha^-\in\Omega_0$ for all α. Hence, we can majorize the first integral using again Theorem 12.7 and obtain

$$B^-\leq 2C\|M_{\Omega_0}\|_{L^2\to L^2}\left(\sum_\alpha|\lambda_\alpha|^2\right)^{1/2}I.$$

Combining all this we get

$$I^2\leq\left(\sup_{l}\|M_{\Omega_l}\|_{L^2\to L^2}^2\right)\left(\sum_\alpha|\lambda_\alpha|^2\right)+C\|M_{\Omega_0}\|_{L^2\to L^2}\left(\sum_\alpha|\lambda_\alpha|^2\right)^{1/2}I.$$

This implies

$$I\leq\left(\sup_{l}\|M_{\Omega_l}\|_{L^2\to L^2}+C\|M_{\Omega_0}\|_{L^2\to L^2}\right)\left(\sum_\alpha|\lambda_\alpha|^2\right)^{1/2}.$$

This finishes the proof of Theorem 12.6. $\qquad\square$

12.4. Fourth lecture: Bochner-Riesz and the cone multipliers

Our motivation here is the problem of inversion of the Fourier transform, *i.e.* in which sense the inversion formula

$$f(x) = \int_{\mathbb{R}^n} \hat{f}(\xi)e^{2\pi ix\xi}\,dx$$

holds. In general, the function $\hat{f}$ does not belong to L^1. To avoid this difficulty, we can use different summability methods:

$$M_{\epsilon,\phi}f(x) = \int_{\mathbb{R}^n} \phi(\epsilon\xi)\hat{f}(\xi)e^{2\pi ix\xi}\,dx,$$

for functions ϕ having some decay. This is the case of the Abel means, corresponding to $\phi(x) = e^{-|x|}$, the Gauss-Weierstrass means, with $\phi(x) = e^{-|x|^2}$, the Cesaro means, $\phi(x) = (1 - |x|)_+$, or our subject, the Bochner–Riesz means, with $\phi(x) = (1 - |x|^2)_+^\lambda$, for $\lambda > 0$. We are interested in the convergence

$$M_{\epsilon,\phi}f \xrightarrow{\ \epsilon\to 0\ } f$$

in the L^p norm and pointwise, almost everywhere.

The one–dimensional case is well–known. M. Riesz proved that

$$\int_{|\xi|\leq 1/\epsilon} \hat{f}(\xi)e^{2\pi ix\xi}\,dx \xrightarrow{\ L^p \text{ norm}\ } f.$$

A consequence of this theorem is the L^p-convergence for the summability methods above. Kolmogorov proved that there is a function $f \in L^1$ whose Fourier series diverges almost everywhere. Carleson [14] proved that for $f \in L^2$

$$\int_{|\xi|\leq 1/\epsilon} \hat{f}(\xi)e^{2\pi ix\xi}\,dx \longrightarrow f(x) \qquad \text{for almost every } x.$$

Hunt [37] extended the result to L^p, $1 < p < \infty$.

The convergence problem admits different generalizations to higher dimensions. We could consider, for instance, "rectangular convergence"

$$\tilde{S}_m f(x) = \sum_{|k_1|\leq m_1, |k_2|\leq m_2,\dots} \hat{f}(k_1, k_2, \dots)e^{2\pi i[x_1k_1 + x_2k_2 + \cdots]},$$

$m = (m_1, m_2, \ldots, m_n) \in \mathbb{Z}^n$, or "spherical convergence"

$$S_m f(x) = \sum_{|k_1|^2 + |k_2|^2 + \cdots \le m^2} \hat{f}(k_1, k_2, \ldots) e^{2\pi i [x_1 k_1 + x_2 k_2 + \cdots]}, \qquad m \in \mathbb{N}.$$

The convergence in the rectangular setting is easily deduced from the results in dimension 1. The case of the spherical convergence is completely different. S_1 is called the disk multiplier operator. It is a convolution operator,

$$S_1 f(x) = \left(\chi_{B_1(0)} \right)^{\vee} * f(x).$$

The Fourier transform of the disk multiplier $\chi_{B_1(0)}$ can be described in terms of a Bessel function (see [51] pp. 155, 158)

$$\left(\chi_{B_1(0)} \right)^{\vee} = C \frac{1}{|x|^{n/2}} \mathcal{J}_{n/2}(2\pi |x|) \sim \frac{e^{i|x|}}{1 + |x|^{(n+1)/2}}.$$

This is not a Calderón–Zygmund kernel. It does not satisfy Hörmander or size conditions. Actually, S_1 is not bounded in some L^p spaces. Note that if $S_1 : L^p \to \mathcal{S}'$ were bounded, then, by duality, $S_1 : \mathcal{S} \to L^{p'}$ would be also bounded. But, if we take $f \in \mathcal{S}$ so that $\hat{f} = 1$ in $B_1(0)$, then $S_1 f = \left(\chi_{B_1(0)} \right)^{\vee} \sim \frac{e^{i|x|}}{1 + |x|^{(n+1)/2}}$ and this is an $L^{p'}$–function only if $p' > \frac{2n}{n+1}$. Hence, if S_1 is bounded in L^p, then $p < \frac{2n}{n-1}$. Again by duality, this implies also that $p > \frac{2n}{n+1}$. In conclusion: a necessary condition for S_1 to be bounded is $\frac{2n}{n+1} < p < \frac{2n}{n-1}$.

We define the Bochner- Riesz multiplier operators

$$B^\lambda f(x) = \int_{\mathbb{R}^n} (1 - |\xi|^2)_+^\lambda \hat{f}(\xi) e^{2\pi i x \xi} \, dx.$$

We are interested in the boundedness of these operators and the associated convergence problem. These are convolution operators with kernels (see [51])

$$C_\lambda \frac{1}{|x|^{n/2+\lambda}} \mathcal{J}_{n/2+\lambda}(2\pi x).$$

When $\lambda > \frac{n-1}{2}$, this kernel belongs to L^1, and therefore B^λ is bounded on L^p, $1 \le p \le \infty$. Moreover we can define the maximal operator

$$B_*^\lambda f(x) = \sup_R |B_R^\lambda f(x)|,$$

where

$$B_R^\lambda f(x) = \int_{\mathbb{R}^n} \left(1 - \left|\frac{\xi}{R}\right|^2\right)_+^\lambda \hat{f}(\xi) e^{2\pi i x \xi} \, dx.$$

B_*^λ is also a bounded operator on L^p, $1 < p < \infty$ and is of weak type (1,1). Therefore $B_R^\lambda f \longrightarrow f$ for $f \in L^p$ in norm and pointwise (almost everywhere).

But for $\lambda \le \frac{n-1}{2}$ things are different. The exponent $\frac{n-1}{2}$ is for this reason called *the critical index*. The argument given before for the disk multiplier shows that a necessary condition for B^λ to be bounded in L^p is

$$\frac{2n}{n+1+2\lambda} < p < \frac{2n}{n-1-2\lambda}.$$

It has been conjectured that this is also a sufficient condition.

This problem became even more interesting after the result of C. Fefferman [30] proving that the disk multiplier operator is bounded in L^p only if $p = 2$. This implies also that for any $p < 2$, there are functions in L^p so that their spherical sums do not converge pointwise.

The conjecture about the Bochner–Riesz multipliers was shown to be true in $\mathbb{R}^2$ by Carleson and Sjölin [16]. There are also proofs due to Hörmander [36], C. Fefferman [29] and A. Córdoba [20].

Theorem 12.8. *For* $\lambda \le \frac{1}{2}$, B^λ *is bounded on* $L^p(\mathbb{R}^2)$ *if and only if* $\frac{4}{3+2\lambda} < p < \frac{4}{1-2\lambda}$.

Proof. The proof we present here is due to A. Córdoba. We break the multiplier,

$$m_\lambda(\xi) = \sum_{k=1}^\infty 2^{-k\lambda} m_{k,\lambda}(|\xi|) + \tilde{m}_\lambda(\xi)$$

where $\tilde{m}_\lambda \in C_o^\infty(|\xi| \le \frac{7}{8})$, and

$$\text{supp } m_{k,\lambda} \subset \{1 - 2^{-k} < s < 1 - 2^{-k-2}\}$$

and there are constants $C_{\gamma,\lambda}$, independent of k, so that:

$$|D^\gamma m_{k,\lambda}| \le C_{\gamma\lambda} 2^{k\gamma}.$$

To define $m_{k,\lambda}$ take, for instance, $\phi, \psi \in C_0^\infty([-1/2, 1/2])$ so that

$$\sum_{j=1}^\infty \psi\left(\frac{1-|t|}{2^{-j}}\right) + \phi(t) \equiv 1 \quad \text{on} \quad [-1,1].$$

Then

$$m_\lambda(\xi) = \sum_{j=1}^\infty 2^{-j\lambda} \left(\frac{1-|\xi|}{2^{-j}}\right)_+^\lambda \psi\left(\frac{1-|\xi|}{2^{-j}}\right) + \phi(|\xi|)(1-|\xi|)_+^\lambda.$$

The operator $f \longrightarrow \widehat{\tilde{m}_\lambda \hat{f}}$ has a kernel in L^1 and hence is bounded in L^p, $1 \le p \le \infty$. The main work of this proof is to obtain good estimates for the other part. We want to show

Theorem 12.9. *For $0 < \delta < 1/2$ and given a radial function $m = m^\delta \in C_0^\infty(\mathbb{R}^2)$ so that supp $m \subset \{1 - \delta < |x| < 1 - \delta/4\}$ and $|D^\gamma m| \le C_\gamma \delta^{-\gamma}$, we define de operator $\widehat{T_m f}(x) = m(x)\hat{f}(x)$. Then*

$$\|Tf\|_4 \le C\left(\log\frac{1}{\delta}\right)^\alpha \|f\|_4$$

for some fixed α, C independent of δ and f.

It is enough to show that the Bochner–Riesz multipliers are bounded in L^4. To obtain the case $4/(3 + 2\lambda) < p < 4/(1 - 2\lambda)$ one can use the interpolation theorem of analytic families of operators of Stein ([51]).

Proof.
1) A further decomposition of the operator. We define the angular sectors

$$\Gamma_l \equiv \{re^{i\theta} \in \mathbb{R}^2 / |\theta - l2\pi\delta^{1/2}| < \delta^{1/2}\}, \qquad l = 1, 2, \ldots, \delta^{-1/2}.$$

Write

$$m = \sum_{l=1}^{\delta^{-1/2}} m_l$$

where supp $m_l \subset \Gamma_l$ and

$$|D_r^\gamma D_{tang}^\beta m_l| \le C_{\gamma,\beta} \delta^{-\gamma} \delta^{-\frac{1}{2}\beta}.$$

(For instance, take a smooth $\psi_2 \in C_0^\infty([-1,1])$ so that

$$\sum_{l=1}^{\infty} \psi_2(x - 2\pi l) \equiv 1,$$

and write $m_l(\xi) = m(\xi)\psi_2\left(\frac{\theta}{\delta^{1/2}} - 2\pi l\right)$, where $\xi = |\xi|e^{i\theta}$.)

2) Quasi-orthogonality and a square function.

Write $\sum = \sum^1 + \sum^2 + \sum^3 + \sum^4$ each one corresponding to the rectangles contained in a particular quadrant. Then, for any of the four sums,

$$\left\|\sum T_l f\right\|_4^4 = \int \left|\sum T_l f\right|^4 = \int \left|\sum_{l,k}(T_l f)(T_k f)\right|^2.$$

By Plancherel

$$= \int \left|\sum_{l,k}\widehat{T_l f} * \widehat{T_k f}\right|^2.$$

The support of $\widehat{T_l f} * \widehat{T_k f}$ is contained in $S_l + S_k$. Using Cauchy-Schwarz, we obtain

$$\leq \int \left(\sum_{l,k}|\widehat{T_l f} * \widehat{T_k f}|^2\right)\left(\sum_{l,k}\chi_{S_l + S_k}\right).$$

Those sets are almost disjoint. This is the statement of

Lemma 12.4. $\sum_{l,k}\chi_{S_l + S_k} \leq C.$

Proof. Denote $\varepsilon_l = e^{2\pi i l \delta^{1/2}}$. Observe that $S_l + S_k$ is contained in a rectangle centered at $\varepsilon_l + \varepsilon_k$, of dimensions $|l - k|\delta \times \delta^{1/2}$ and with main axis orthogonal to $\varepsilon_l + \varepsilon_k$. Observe also that $|\varepsilon_l + \varepsilon_k| - 2 = [2 + 2\cos((l - k)\delta^{1/2})]^{1/2} - 2 \approx (l - k)^2\delta$. Consider two cases:

1) When $\varepsilon_l + \varepsilon_k$ is not parallel to $\varepsilon_{l'} + \varepsilon_{k'}$, or more precisely, when the angle between $\varepsilon_l + \varepsilon_k$ and $\varepsilon_{l'} + \varepsilon_{k'}$ is bigger than $10\delta^{1/2}$. Then $S_l + S_k$ is disjoint with $S_{l'} + S_{k'}$.

2) When $\varepsilon_l + \varepsilon_k$ is parallel to $\varepsilon_{l'} + \varepsilon_{k'}$. Then, the distance between the centers, $|(l - k)^2 - (l' - k')^2|\delta \geq [|l - k| + |l' - k'|]\delta$, so the corresponding regions are disjoints. $\qquad\square$

With the help of this lemma we conclude

$$\|\sum T_l f\|_{L^4}^4 \leq C \int \sum_{l,k} |\widehat{T_l f} * \widehat{T_k f}|^2 = C\|(\sum |T_l f|^2)^{1/2}\|_4^4.$$

$\square$

3) Duality and a maximal function. We have to look at T_l more carefully. Denote by ϕ_l the associated kernel, *i.e.* $\widehat{\phi_l} = m_l$. When $l = 0$ one can easily check that for $\alpha, \beta \in \mathbb{N}$,

$$|\partial_{\xi_1}^\alpha \partial_{\xi_2}^\beta m_0| \leq C_{\alpha,\beta} \delta^{-\beta/2} \delta^{-\alpha}.$$

In other words,

$$|\partial_{\xi_1}^\alpha \partial_{\xi_2}^\beta [m_0(\delta\xi_1, \delta^{1/2}\xi_2)]| \leq C_{\alpha,\beta}.$$

Taking Fourier transforms, we see that

$$\delta^{-3/2}\big|\phi^0\big(\frac{x_1}{\delta}, \frac{x_2}{\delta^{1/2}}\big)\big| \leq \frac{C_N}{|x|^N}, \qquad \text{for all } N \in \mathbb{N}.$$

In particular $\|\phi^0\|_{L^1} \leq C$ and

$$\delta^{-3/2}\big|\phi^0\big(\frac{x_1}{\delta}, \frac{x_2}{\delta^{1/2}}\big)\big| \leq C\sum_{k=0}^\infty 2^{-k}\frac{1}{2^{2k}}\chi_{[-2^k,2^k]\times[-2^k,2^k]}(x_1, x_2).$$

Hence,

$$\big|\phi^0(x_1, x_2)\big| \leq C\sum_{k=1}^\infty 2^{-k}\frac{1}{|R_k|}\chi_{R_k}(x_1, x_2),$$

where $R_k = [-2^k\delta, 2^k\delta] \times [-2^k\delta^{1/2}, 2^k\delta^{1/2}]$. By rotation, we obtain similar estimates for ϕ^l.

We now take a family of functions h_l to be determined later, so that $h_l = 1$ on $\Gamma_l \cap \operatorname{supp} m$ and define $\widehat{f_l} = h_l \hat{f}$, so that $T_l f = T_l f_l$. Note also that, since $\|\phi^l\|_1 \leq C$, we have by Jensen's inequality

$$|\phi_l * f|^2(x) \leq |\phi_l| * (|f|^2)(x).$$

By Riesz Representation Theorem

$$\|(\sum |T_l f|^2)^{1/2}\|_4^2 = \|\sum |T_l f_l|^2\|_2 \leq \|\sum |\phi^l| * (|f_l|^2)\|_2$$

$$= \sum_l \int \phi^l * (|f_l|^2) w = \sum_l \int |f_l|^2(|\phi^l| * w),$$

for some $w \in L^2$ of unit norm.

The maximal function which controls this weighted inequality is

$$\Phi^* w = \sup_l (|\phi^l| * w).$$

Recall the estimates on ϕ^l. They say that

$$\Phi^* w(x) \le C \sum_{k=1}^\infty 2^{-k} \sup \frac{1}{|R_x|} \int_{R_x} |w(y)|\, dy,$$

where the supremum is taken in all the rectangles of dimensions $2^k \delta \times 2^k \delta^{1/2}$ containing the point x.

Recall the definition of the Kakeya Maximal Function of eccentricity δ. Then, with our notation,

$$\Phi^* \omega(x) \le C \mathcal{K}_{\delta^{1/2}} w(x)$$

Using again duality and the boundedness of $\mathcal{K}_\delta$,

$$\sum_l \int |T_l f_l|^2 w \le C \int \left(\sum |f_l|^2 \right) \Phi^* w$$

$$\le C \|(\sum |f_l|^2)^{1/2}\|_4^2 \, \|\Phi^* w\|_{L^2}$$

$$\le C(\log(1/\delta)) \, \|(\sum |f_l|^2)^{1/2}\|_{L^4}^2.$$

4) A Littlewood-Paley decomposition of f. Let us determine now the family of functions $\{h_l\}_l$ by considering the following grid of squares of side $\delta^{1/2}$: Take

$$h \in C_0^\infty([-4,4] \times [-4,4]), \quad h \equiv 1 \text{ in } [-2,2] \times [-2,2]$$

and define

$$h_l(x) = h\left(\frac{x - x_l^0}{\delta^{1/2}} \right),$$

where $x_l^0/\sqrt{\delta} \in \mathbb{Z}^2$. Set as before $\hat{f_l} = h_l \hat{f}$. The following result concludes the proof of Theorem 12.9. $\qquad\qquad\square$

Theorem 12.10.

$$\|(\sum |f_l|^2)^{1/2}\|_{L^p} \le C_p \|f\|_{L^p} \qquad \textit{for all } p \ge 2.$$

Proof. The case $p = 2$ is obvious by the finite overlapping property of the squares. The case $p > 2$ follows from the remarkable pointwise inequality found by Rubio de Francia, [46],

$$\left(\sum_l |f_l|^2\right)^{1/2} \le C(M(|f|^2)(x))^{1/2},$$

where M is the Hardy-Littlewood maximal function. $\qquad\square$

12.4.1. *Stein's Conjecture*

Based on the arguments of the preceding proof, E. Stein proposed in 1978 (AMS Conference at Williamstown) to study the following:

Conjecture 12.1. *Let, as before, $T = T_\delta$ be the multiplier associated with the annulus $\{1 - \delta < |\xi| < 1\}$. Then, for every $\alpha > 1$ there exists a constant C_α, so that the following weighted inequality holds*

$$\int |T_\delta f|^2\, w \le C_\alpha \int |f|^2 (\mathcal{K}_{\delta^{1/2}} w^\alpha)^{1/\alpha}.$$

The extreme case corresponds to the disc multiplier:

Conjecture 12.2. *Let S_1 be the multiplier associated with the unit ball. Then, for every $\alpha > 1$ there exists a constant C_α, so that the following weighted inequality holds*

$$\int |S_1 f|^2\, w \le C_\alpha \int |f|^2 (\mathcal{K}_\infty w^\alpha)^{1/\alpha}.$$

The conjecture has been solved in the positive only for radial weights $w(x) = w_0(|x|)$ by A. Carbery, E. Romera and F. Soria [12]. As a consequence of it, one gets the following mixed-norm inequality for the disc multiplier (see [40], [23]):

$$\left(\int_0^\infty \left(\int_{S^{n-1}} |S_1 f(r\omega)|^2 d\sigma(\omega)\right)^{p/2} r^{n-1} dr\right)^{p/2}$$

$$\le C_p \left(\int_0^\infty \left(\int_{S^{n-1}} |f(r\omega)|^2 d\sigma(\omega)\right)^{p/2} r^{n-1} dr\right)^{p/2},$$

$$\text{for } \frac{2n}{n+1} < p < \frac{2n}{n-1}.$$

The conjecture says that the Kakeya maximal function "controls" the Bochner-Riesz transforms. In two dimensions there is a result related to this which says that Kakeya "controls" Kakeya.

Theorem 12.11 (D. Müller, F. Soria, [44]). *With the previous notation,*

$$\int_{\mathbb{R}^2} |\mathcal{K}_\delta f|^2\, w \le C \left(1 + \log \frac{1}{\delta}\right)^2 \int_{\mathbb{R}^2} |f|^2 \mathcal{K}_\delta w.$$

As a consequence, we have the square function estimate:

$$\left\| \left(\sum_l |\mathcal{K}_\delta f_l|^2 \right)^{1/2} \right\|_{L^4} \le C \left(1 + \log \frac{1}{\delta}\right)^{3/2} \left\| \left(\sum_l |f_l|^2 \right)^{1/2} \right\|_{L^4}.$$

12.4.2. *Circular maximal function of Bourgain*

Let us recall the statement of

THEOREM 12.2. $M_\sigma : L^p(\mathbb{R}^2) \to L^p(\mathbb{R}^2)$ *if* $2 < p \le \infty$.

A nice approach to the proof of this result is due to Mockenhaupt, Seeger and Sogge, [42]. It involves arguments closely related to the so called "cone" multiplier and to the notion of "local smoothing", that is, the fact that averages regularize operators. We will present here some of these arguments. The complete proof to Bourgain's theorem is too long to be included in these notes.

Let us first observe that

$$\widehat{A_t f}(\xi) = 2\pi J_0(2\pi t|\xi|)\hat{f}(\xi).$$

We define a family of operators depending on a complex parameter, α,

$$\widehat{A_t^\alpha f}(\xi) = \frac{1}{\pi^{\alpha-}} \frac{J_\alpha(2\pi t|\xi|)}{(t|\xi|)^\alpha} \hat{f}(\xi).$$

Set $A_*^\alpha f = \sup_t |A_t^\alpha f|$.

Mockenhaupt, Seeger and Sogge proved that

$A_*^\alpha : L^2(\mathbb{R}^2) \to L^2(\mathbb{R}^2)$, for $Re\,\alpha > 0$ and

$A_*^\alpha : L^4(\mathbb{R}^2) \to L^4(\mathbb{R}^2)$, for $Re\,\alpha > -1/8$.

The theorem of Interpolation of Analytic Families of Operators of Stein (see [51] pp. 205-209) allows us to conclude that A_*^α is bounded on the region

$$4 \le p < \infty \Rightarrow Re\,\alpha > -\frac{1}{2p}$$

$$2 < p \le 4 \Rightarrow Re\,\alpha > -\frac{1}{4} + \frac{1}{2p}.$$

These operators are connected with the Bochner-Riesz means: A_t^α can be written as (see [51] pg. 171)

$$A_t^\alpha f(x) = \frac{1}{\Gamma(\alpha)} m_t^{\alpha-1} * f(x),$$

where

$$m^\lambda(y) = (1 - |y|^2)_+^\lambda \qquad \text{and} \qquad m_t(y) = \frac{1}{t^n} m\left(\frac{y}{t}\right).$$

This is in some sense the dual problem in space variable to the Bochner-Riesz means of order λ (in $\mathbb{R}^2$) which are defined by $\widehat{B_\lambda f}(\xi) = m^\lambda(\xi)\widehat{f}(\xi)$. Another related operator is the Cone multiplier (in $\mathbb{R}^3$) which we now describe.

12.4.3. *The Cone Multiplier Operator*

The multiplier is defined as

$$m^\lambda(\xi,\tau) = \Phi(\tau)\left(1 - \frac{|\xi|}{\tau}\right)_+^\lambda$$

where $\Phi \in C_0^\infty([1,2]), (\xi,\tau) \in \mathbb{R}^2 \times \mathbb{R}$. The associated operator is $\widehat{C^\lambda f} = m^\lambda \widehat{f}$. It is conjectured that C^λ is bounded on $L^p(\mathbb{R}^3)$ for $\frac{4}{3+2\lambda} < p < \frac{4}{1-2\lambda}$ (This is the same range as B^λ on $\mathbb{R}^2$.) The following result is due to G. Mockenhaupt [41].

Theorem 12.12.
  If $\lambda > 1/8$, then $\| C^\lambda f \|_4 \le C \| f \|_4$.

Proof. We just give a very brief sketch of the proof. As before,

$$C^\lambda = \sum_{j=0}^{\infty} 2^{-\lambda j}\, T^j,$$

with T^j associated to $\psi\left(\frac{1-\frac{|\xi|}{\tau}}{2^{-j}}\right)\Phi(\tau)$.

Set $\delta = 2^{-j}$. It suffices to show, that for T_δ defined by

$$\widehat{T_\delta f}(\xi) = \psi\left(\frac{1 - \frac{|\xi|}{\tau}}{\delta}\right)\Phi(\tau)\hat{f}(\xi),$$

we have

$$\| T_\delta f \|_4 \leq C\delta^{-1/8}\left(\log\frac{1}{\delta}\right)^\alpha \| f \|_4.$$

We proceed as in Bochner-Riesz. We break T into $\delta^{-1/2}$ angular pieces $T = \sum_{l=1}^{\delta^{-1/2}} T_l$.

Then, we use quasi-orthogonality. In this case we have,

$$\left\| \sum T_l f \right\|_{L^4} \leq \delta^{-1/8} \left\| \left(\sum |T_l f|^2\right)^{1/2} \right\|_{L^4}.$$

There is also a maximal operator associated to this problem, $\tilde{K}_N$. It is defined as the supremum of averages over rectangles of dimensions $1 \times N \times N^2 (\equiv 1 \times \delta^{-1/2} \times \delta^{-1})$ and normal to the cone in the leading direction. It can be shown that

$$\| \tilde{K}_N f \|_{L^2} \leq C(\log N)^\alpha \| f \|_{L^2}.$$

We conclude that

$$\left\| \left(\sum |T_l f|^2\right)^{1/2} \right\|_{L^4} \leq C(\log N)^\alpha \left\| \left(\sum |f_l|^2\right)^{1/2} \right\|_4.$$

Finally, we use Littlewood-Paley theory adapted to the support, S_l, of the multiplier of T_l. (This brings another $(\log N)^\beta$.)

Everything is the same as for the Bochner-Riesz multiplier, except quasi-orthogonality, since the overlapping of supports gives

$$\sum_{l,k} \chi_{S_k + S_l} \leq C\, \delta^{-1/2} \qquad \text{(and this is sharp).} \qquad \square$$

12.4.4. *Maximal Bochner-Riesz in* $\mathbb{R}^2$

Another element of the proof for the circular maximal function is the so-called local "smoothing", i.e. the fact that average regularizes operators. The meaning of this sentence may be clarified with the next example.

We define the maximal Bochner-Riesz operator,

$$B_*^\lambda f(x) = \sup_{t>0} |(\breve{m}_\lambda)_t * f(x)|.$$

Theorem 12.13 (A. Carbery, [9]). B_*^λ *is bounded on* $L^p(\mathbb{R}^2)$ *for* $2 \le p < \frac{4}{1-2\lambda}$, $\lambda < 1/2$.

By breaking the multiplier as in the case of B^λ, we obtain

$$B_*^\lambda \le \sum_{j \ge 0} 2^{-j\lambda} \, T_*^j.$$

We see that it suffices then to prove the L^4 estimate,

Theorem 12.14. *Let* $\psi \in C_0^\infty \left(\left[\frac{1}{2}, 1 \right] \right), 0 < \delta < 1/2$, *and set*

$$\widehat{a_t^\delta f}(\xi) = \psi \left(\frac{1 - t|\xi|}{\delta} \right) \hat{f}(\xi) \qquad a_*^\delta f(x) = \sup_{t>0} |a_t^\delta \, f(x)|.$$

Then,

$$\| \, a_*^\delta f \, \|_{L^4} \le C \left(\log \frac{1}{\delta} \right)^\beta \| \, f \, \|_4 \, .$$

We give an idea of the proof. By the Fundamental Theorem of Calculus,

$$|a_s^\delta f(x)|^4 = 4 \left| \int_0^s (a_t^\delta f(x))^3 \left(t \frac{d}{dt} a_t^\delta f(x) \right) \frac{dt}{t} \right|$$

$$\le 4 \left(\int_0^\infty |a_t^\delta f(x)|^4 \frac{dt}{t} \right)^{3/4} \left(\int_0^\infty \left| t \frac{d}{dt} a_t^\delta f(x) \right|^4 \frac{dt}{t} \right)^{1/4}$$

$$=: 4(G_1(x))^3 (G_2(x)).$$

We apply Hölder's inequality and get $\| \, a_*^\delta f \, \|_4 \le 4 \| \, G_1 \, \|_4^{3/4} \| \, G_2 \, \|_4^{1/4}$.

The multiplier associated to a_t^δ is $\psi \left(\frac{1-t|\xi|}{\delta} \right)$ whereas the one associated to $t\frac{d}{dt} a_t^\delta$ is

$$\frac{t|\xi|}{\delta} \psi' \left(\frac{1 - t|\xi|}{\delta} \right) = - \left(\frac{1 - t|\xi|}{\delta} \right) \psi' \left(\frac{1 - t|\xi|}{\delta} \right) + \frac{1}{\delta} \psi' \left(\frac{1 - t|\xi|}{\delta} \right)$$

The first term is like the one of a_t^δ (with ψ replaced by $u\psi'(u)$) and so is the second too, but we have to pay an extra $\frac{1}{\delta}$. Hence, the theorem follows from

$$\| \, G_1 \, \|_4 \le C \, \delta^{1/4} \left(\log \frac{1}{\delta} \right)^{\beta/2} \| \, f \, \|_4 \, .$$

What Carbery proved is

$$\left\| \left(\int_0^\infty |a_t^\delta f|^2 \frac{dt}{t} \right)^{1/2} \right\|_{L^4} \leq C \, \delta^{1/2} \left(\log \frac{1}{\delta} \right)^\alpha \| f \|_4 \, .$$

By Littlewood-Paley theory the estimate for G_1 will be a consequence of the inequality

$$\left\| \left(\int_1^2 |a_t^\delta f|^4 \frac{dt}{t} \right)^{1/4} \right\|_{L^4} \leq C \, \delta^{1/4} \left(\log \frac{1}{\delta} \right)^{\beta/2} \| f \|_4 \, .$$

If you try to prove an estimate of this type for each t separately and then finish using Fubini, you fail miserably. You cannot get the $\delta^{1/4}$ term this way. The integration in t is what produces (local) "smoothing".

12.5. Fifth lecture: restriction theorems and applications

Let $S \subset \mathbb{R}^n$ be a surface with non-vanishing Gaussian curvature and denote by $d\sigma$ the standard surface measure. We will say that the surface satisfies a **Restriction theorem** if for some values of p and q and for any compact subset $S_0 \subset S$, the operator

$$L^p(\mathbb{R}^n) \longrightarrow L^q(S_0)$$
$$f \longrightarrow \hat{f}|_{S_0}$$

is bounded.

By duality this is equivalent to an **Extension theorem:**

$$L^{q'}(S_0) \longrightarrow L^{p'}(\mathbb{R}^n)$$
$$f \longrightarrow \widehat{f d\sigma}$$

Not all the possible restriction theorems are true. There is a necessary condition which forces some relationship between p, q and n namely

$$\frac{1}{q} \geq \frac{n+1}{n-1} \frac{1}{p'}$$

and $p < 2n/(n+1)$. To see this, consider the following

Counterexample: Take

$$f = \phi \left(\frac{x_1}{N}, \frac{x_2}{N}, \ldots \frac{x_{n-1}}{N}, \frac{x_n}{N^2} \right),$$

with $\phi \in \mathcal{S}$, $\hat{\phi} \geq 0$, $\hat{\phi}(\xi) \geq 1$ for all $|\xi| \leq 1$. Take S a sphere of radius 1 and centered at $(0, 0, \ldots, 0, 1)$. Then $\hat{f}|_S \geq N^{n+1}$ on a "square" of sidelenght $1/N$. Therefore, we have $\|f\|_p = C N^{(n+1)/p}$, while $\|\hat{f}|_S\|_q \geq C N^{n+1} N^{(1-n)/q}$. If we impose $\|\hat{f}|_S\|_q \leq C\|f\|_p$ for big N we obtain the above inequality.

Moreover, when $S = S^{n-1}$, the Fourier transform of $d\sigma$ can be explicitly computed (see [51])

$$\widehat{d\sigma}(\xi) = \frac{2\pi}{|\xi|^{(n-2)/2}} \mathcal{J}_{(n-2)/2}(2\pi|\xi|) \sim \frac{e^{i|\xi|}}{|\xi|^{(n-1)/2}}.$$

Using this in the extension problem, we see that $\widehat{d\sigma} \in L^{p'}$ implies $p' > 2n/(n-1)$ which was the second condition.

In general, one has from the stationary phase lemma([52])

Lemma 12.5. *For any smooth function ϕ, supported in S, we have* $|\widehat{\phi \, d\sigma}(\xi)| \leq C|\xi|^{-(n-1)/2}$.

The above conditions for the restriction of the Fourier transform are known to be sufficient in dimension $n = 2$. In higher dimensions, we have only partial results, although the case $q = 2$ is well settled.

Theorem 12.15 ([49], [61], [52]).

$$\|\hat{f}|_{S_0}\|_{L^2(S_0)} \leq C\|f\|_{L^p(\mathbb{R}^n)}, \qquad \textit{for all} \quad p \leq 2\frac{n+1}{n+3}.$$

The proof of Theorem 12.15 is presented at the end of the section.

Considerable effort has been done during the last decade or so to improve the range of p and $q \neq 2$ for which the restriction theorem holds. For an update of recent results we recommend [59]

12.5.1. *Application to Bochner–Riesz*

As an application of the restriction theorem, we can give a proof of the boundedness of the Bochner–Riesz multipliers with $\lambda > (n-1)/(2(n+1))$ in higher dimensions. First, we use the fact that the "singularity" of $m_\lambda(\xi) = (1 - |\xi|^2)_+^\lambda$ comes from $|\xi| = 1$. Take $\eta \in \mathcal{S}$ so that supp $\hat{\eta} \subset B_{3/4}(0)$ and

$\eta(\xi) = 1$ for $|\xi| \leq 1/2$. Define $\tilde{m}_\lambda(\xi) = \eta(\xi)m_\lambda(\xi) \in \mathcal{C}_0^\infty$ and $\tilde{K}_\lambda = (\tilde{m}_\lambda)\check{}$. Then,

$$|\tilde{K}_\lambda(x)| \leq C_{\lambda,N}(1 + |x|^N)^{-1}$$

for all $N \geq 1$. Therefore the convolution operator $f \to \tilde{B}^\lambda f = \tilde{K}_\lambda * f$ is bounded on L^p, $1 \leq p \leq \infty$. We are now reduced to study $B^\lambda - \tilde{B}^\lambda$. Take $\phi \in \mathcal{C}_0^\infty(B_2(0))$ radial and positive, so that $\phi(x) = 1$ for $|x| \leq 1$. Define $\psi_j(x) = \phi(x/2^j) - \phi(x/2^{j-1})$, for $j = 1, 2, \ldots$. We decompose the operator

$$B^\lambda f - \tilde{B}^\lambda f = T_0 + \sum_{j=1}^\infty T_j$$

where T_j is the convolution operator with kernel $K_j = (K^\lambda - \tilde{K}^\lambda)\psi_j$ and T_0 has kernel $(K^\lambda - \tilde{K}^\lambda)\phi$. ($K^\lambda$ is the kernel of T^λ.)

It is easy to see that $T_0 f(x) \leq C\chi_{B_2(0)} * f(x)$. Therefore, it will be enough to show that there is some $\eta > 0$ so that, for any cube Q of side-length 2^{j+2} and any function f with $\text{supp} f \in Q$ we have

$$\|T_j f\|_{L^p(Q)} \leq C2^{-j\eta}\|f\|_p.$$

Denote R_j the kernel of T_j. Then, by the asymptotic properties of the Bessel functions, $|R_j(x)| \leq C(1 + |x|)^{\lambda + (n+1)/2}\psi_j(x) \leq C2^{-j(\lambda + (n+1)/2)}$. Note also that

$$\int_{|x|\leq 1/8} |\widehat{R_j}(x)|^2 + \int_{|x|\geq 10} |\widehat{R_j}(x)|^2 \leq C_m 2^{-mj}$$

for all $m \in \mathbb{N}$. This is because

$$|\widehat{R_j}(\xi)| = |(m_\lambda - \tilde{m}_\lambda) * \hat{\psi}_j(\xi)| \leq 2^{jn}\int_{|\xi-\eta|\geq 1/2} (1 - |\xi - \eta|^2)_+^\lambda |\hat{\psi}(2^j\eta)|\,d\eta.$$

If $|\xi| \geq 2$ and $|\xi - \eta| \leq 1$, then $|\eta| \geq |\xi|/2$. Thus,

$$|\widehat{R_j}(\xi)| \leq 2^{jn}\int_{|\xi-\eta|\leq 1} C_m(2^j|\xi|)^{-m-n} \leq C_m 2^{-jm}|\xi|^{-m-n}.$$

If $|\xi| \leq 1/8$ and $|\xi - \eta| \geq 1/2$, then $|\eta| \geq 1/4$. Hence,

$$|\widehat{R_j}(\xi)| \leq 2^{jn}\int_{|\xi-\eta|\leq 1} C_m(2^j)^{-m-n} \leq C_m 2^{-jm}.$$

Now we can compute the norm of $T_j f$. By Hölder's inequality, for $p < 2$,

$$\|T_jf\|_{L^p(Q)}^2 \leq |Q|^{(2-p)/p}\|T_jf\|_{L^2(Q)}^2$$

$$= 2^{nj(\frac{2}{p}-1)}\int |\widehat{R_j}|^2|\hat{f}|^2$$

$$= 2^{nj(\frac{2}{p}-1)}\int_0^\infty r^{n-1}|\widehat{R_j}(r)|^2\int_{S^{n-1}}|\hat{f}(r\omega)|^2 d\sigma(\omega)\,dr$$

$$\leq C2^{nj(\frac{2}{p}-1)}\int_0^\infty r^{n-1}|\widehat{R_j}(r)|^2\|f\|_p^2 r^{-n/p'}\,dr$$

$$\leq C2^{nj(\frac{2}{p}-1)}\int_{r\sim 1} r^{n-1}|\widehat{R_j}(r)|^2\|f\|_p^2 r^{-n/p'}\,dr$$

$$+C2^{nj(\frac{2}{p}-1)}\int_{r\leq 1/8 \text{ or } r\geq 10} r^{n-1}|\widehat{R_j}(r)|^2\|f\|_p^2 r^{-n/p'}\,dr$$

$$\leq C2^{nj(\frac{2}{p}-1)}\int |R_j(x)|^2\|f\|_p^2\,dx + C_m 2^{-mj}\|f\|_p^2$$

$$\leq C2^{nj(\frac{2}{p}-1)}\|f\|_p^2\|R_j\|_2^2 + C_m 2^{-mj}\|f\|_p^2$$

$$\leq C2^{j(2n/p-n-1-2\lambda)}\|f\|_p^2.$$

Observe that we have used the restriction theorem in the second inequality. $\qquad\qquad\square$

12.5.2. *Schrödinger (and wave) equations*

Our second application will be to study the a.e. convergence to the initial datum of the solution to the Schrödinger equation.

Recall that the solution to the free Schrödinger equation,

$$\partial_t u(x,t) = \frac{i}{4\pi^2}\Delta_x u(x,t) \qquad (x,t)\in\mathbb{R}^d\times\mathbb{R}=\mathbb{R}^n$$

$$u(x,0) = u_0(x) \qquad x\in\mathbb{R}^d,$$

is given by

$$e^{it\Delta}u_0 = u(x,t) = S_t f(x) = \int_{\mathbb{R}^d} e^{2\pi i x\xi - i\pi t|\xi|^2}\widehat{u_0}(\xi)\,d\xi.$$

We want to study the pointwise convergence of the solution to the initial datum u_0. L. Carleson [15] obtained the result when $n = 1$, assuming $u_0 \in H^{1/4}(\mathbb{R})$ (see section 1). This condition was found to be also necessary by B. Dahlberg and C. Kenig [26]. Later on, P. Sjölin [50] and L. Vega [62] obtained $s > 1/2$ as a sufficient condition in any dimension. More recently J. Bourgain proved in [8] that there exists a small $\epsilon > 0$ such that if $u_0 \in H^{1/2-\epsilon}(\mathbb{R}^2)$ the convergence follows. This result has been improved by A. Moyua, A. Vargas and L. Vega [43], T. Tao and A. Vargas [60] and finally by T. Tao [58], who proved the case $u_0 \in H^s(\mathbb{R}^2)$ with $s > 0.4$. It is conjectured that the exponent $1/4$ is sufficient in any dimension.

We present here a simple proof of the $1/2 + \epsilon$ result[a]. Remember that this always follows from the corresponding boundedness of the maximal function. The argument that we give is also valid for the case of the **wave equation** with initial derivative equal to zero.

Proposition 12.2. *Let $m : \mathbb{R} \times \mathbb{R} \times \mathbb{R}^n \to \mathbb{C}$ be a measurable, bounded function and consider the Fourier multiplier operator*

$$f \to U(x,t) = \int_{\mathbb{R}^n} m(t, |\xi|, x) e^{ix \cdot \xi} \widehat{f}(\xi) \, d\xi.$$

Then we have

$$\int_{|x| \leq R} \sup_{t \in \mathbb{R}} |U(x,t)|^2 \, dx \leq CR \int_{\mathbb{R}^n} |\widehat{f}(\xi)|^2 (1 + |\xi|)^{1+\epsilon} d\xi.$$

Proof. Writing $U(x,t)$ in polar coordinates we observe that

$$|U(x,t)| = \left| \int_0^\infty m(t, s, x) \int_{\mathbf{S}^{n-1}} e^{isx \cdot \omega} \widehat{f}(s\omega) d\omega \, s^{n-1} ds \right|$$

$$\leq C \int_0^\infty \left| \int_{\mathbf{S}^{n-1}} e^{isx \cdot \omega} \widehat{f}(s\omega) d\omega \right| s^{n-1} ds.$$

Observe that the variable t is gone! Using Cauchy-Schwarz and then Fubini, we have

$$\int_{|x| \leq R} \sup_{t \in \mathbb{R}} |U(x,t)|^2 \, dx$$

$$\leq C \int_{|x| \leq R} \int_0^\infty \left| \int_{\mathbf{S}^{n-1}} e^{isx \cdot \omega} \widehat{f}(s\omega) d\omega \right|^2 s^{2n-2} (1+s)^{1+\epsilon} ds dx$$

$$= C \int_0^\infty \int_{|x| \leq sR} \left| \int_{\mathbf{S}^{n-1}} e^{ix \cdot \omega} \widehat{f}(s\omega) d\omega \right|^2 dx \, s^{n-2} (1+s)^{1+\epsilon} ds.$$

---

[a]This argument is implicit in the (unpublished) Ph.D. thesis of L. Vega. It was also discovered by A. Soljanik (personal communication to P. Sjölin).

To finish, we will use the following extension theorem:

Theorem 12.16.

$$\frac{1}{R}\int_{B(0,R)}|\widehat{g\,d\sigma}|^2 \le C\int_{\mathbf{S}^{n-1}}|g|^2.$$

Applying the result with $g(\omega) = \hat{f}(s\omega)$ we obtain

$$\int_{|x|\le R}\sup_{t\in\mathbb{R}}|U(x,t)|^2\,dx$$

$$\le C\int_0^\infty sR\int_{\mathbf{S}^{n-1}}\left|\widehat{f}(s\omega)\right|^2\,d\omega\,s^{n-2}(1+s)^{1+\epsilon}ds$$

$$= CR\int_{\mathbb{R}^n}|\widehat{f}(\xi)|^2(1+|\xi|)^{1+\epsilon}d\xi.$$

$\square$

Proof of Theorem 12.16. Write $x' = (x_1, x_2, \ldots, x_{n-1})$ and assume that g is supported in a spherical cap, $S_0 \subset S^{n-1}$ parametrized as $S_0 = \{(y', \Phi(y')) : |y' \le 1/2\}$.

$$\int_{B_R(0)}|\widehat{g\,d\sigma}|^2\,dx \le \int_{|x_n|\le R}\int_{\mathbb{R}^{n-1}} \cdot$$

$$\left|\int_{|y'|\le 1/2}g(y',\Phi(y'))e^{2\pi i(<x',y'>+x_n\Phi(y'))}J_\Phi(y')\,dy'\right|^2\,dx'\,dx_n$$

By Plancherel,

$$= \int_{|x_n|\le R}\left(\int_{|x'|\le 1/2}|g(x',\Phi(x'))J_\Phi(x')|^2dx'\right)dx_n \le CR\|g\|_{L^2}^2.$$

$\square$

$$**********************$$

Let us define $S^*f(x) = \sup_{t>0}|S_tf(x)|$, $x \in \mathbb{R}^n$. All the known cases about convergence mentioned above are obtained via a maximal inequality

$$\left(\int_{|x|\le 1}|S^*f(x)|^pdx\right)^{1/p} \le C\|f\|_{H^\alpha}, \qquad f \in \mathcal{S}(\mathbb{R}^n),$$

for different values of $p \in [1, 2]$.

Together with G. Gigante, we have investigated in [34] whether this inequality holds if we replace S^* by a spherical average operator; namely we look at the maximal square function

$$Q^* f(x) = \sup_{t>0} \left(\frac{1}{\sigma(S^{n-1})} \int_{S^{n-1}} |S_t f(|x|\omega)|^2 \, d\sigma(\omega) \right)^{1/2}.$$

Clearly, one has the inequality $\int_{|x|\leq 1} |Q^* f(x)|^2 dx \leq \int_{|x|\leq 1} |S^* f(x)|^2 dx$, and therefore the boundedness of S^* would imply a corresponding inequality for Q^*. The known counterexamples show that the smoothness condition $\alpha \geq 1/4$ is still necessary for the boundedness of this operator. The following result shows that $\alpha = 1/4$ is also sufficient for the boundedness of Q^*.

Theorem 12.17. *The operator Q^* is bounded from the (homogeneous) Sobolev space $\dot{H}^{1/4}(\mathbb{R}^n)$ into $L^2(\{|x| \leq 1\})$ in any dimension n; that is, there is a positive constant C, independent of the dimension, such that*

$$\left(\int_{|x|\leq 1} |Q^* f(x)|^2 dx \right)^{1/2} \leq C \|f\|_{\dot{H}^{1/4}}, \qquad \forall f \in \mathcal{S}(\mathbb{R}^n).$$

In particular, this gives us that $1/4$ of smoothness suffices for the *a.e.* convergence with respect to quadratic spherical means. The precise statement is contained in the following corollary.

Corollary 12.3. *If $f \in H^{1/4}(\mathbb{R}^n)$, then, for every $x_0 \in \mathbb{R}^n$ we have*

$$\lim_{t\to 0^+} \int_{S^{n-1}} |S_t f(x_0 + r\omega) - f(x_0 + r\omega)|^2 d\sigma(\omega) = 0, \quad a.e.[r]$$

Let us make a reformulation of our problem. Observe that if $\{\mathcal{Y}_k\}$ is an orthonormal basis of spherical harmonics in $L^2(S^{n-1})$, and $\hat{f}(\xi) \sim \sum_k f_k(|\xi|)\mathcal{Y}_k(\xi/|\xi|)$ denotes the corresponding expansion of $\hat{f}$ with respect to this basis, then

$$Q^* f(x) = \sup_{t>0} \left(\frac{1}{\sigma(S^{n-1})} \sum_k \frac{1}{|x|^{n-1}} \left| Q_{\nu(k)}^t \left(f_k(s) s^{(n-1)/2} \right) (|x|) \right|^2 \right)^{1/2},$$

where

$$Q_\nu^t g(r) = \int_0^\infty e^{-its^2} \widetilde{J}_\nu(2\pi r s) g(s) ds$$

and $\nu(k) = (n-2)/2 + degree(\mathcal{Y}_k)$. Here, J_ν denotes the Bessel function of order ν and $\widetilde{J}_\nu(t) = \sqrt{t} J_\nu(t)$ for $t \geq 0$. Using that the norm in $\dot{H}^{1/4}$ of f with respect to the above expansion is given by $\|f\|_{\dot{H}^{1/4}} = \sum_k \int_0^\infty |f_k(r)|^2 r^{1/2} r^{n-1} dr$, and "cancelling out the $\sum$ signs", the inequality

$$\int_{|x| \leq 1} |Q^* f(x)|^2 dx \leq C \|f\|_{\dot{H}^{1/4}}^2$$

is equivalent to the estimate

$$\int_0^1 \sup_{t > 0} |Q_\nu^t g(r)|^2 dr \leq C \int |g(r)|^2 r^{1/2} dr,$$

uniformly in the index ν too.

We can now follow Carleson's approach (see [15], [26]). First we linearize our maximal operator, by making t into a function of r, $t(r)$. Next we may assume that g is supported on a fixed interval I (as long as the final constant C is independent of I). "Moving" the smoothness to the other side (that is, redefining $g(r) r^{1/4}$ as g again), we consider instead the linear operator

$$T_\nu g(r) = \int_I \frac{e^{is^2 t(r)} \widetilde{J}_\nu(rs)}{s^{1/4}} g(s) ds.$$

Then what we have to show is

$$\int_0^1 |T_\nu g(r)|^2 dr \leq C \int_I |g(s)|^2 ds,$$

with C independent of $g \in L^2(I)$, of the interval I, of the measurable function $t(r)$ and of $\nu \in \mathbb{N}/2$. This is proved in [34].

Let us bring here a related result obtained by the authors. In [33] it was proved that the uniform estimate

$$\int_I e^{ias^2} J_\nu(s) \frac{ds}{s^\beta} = O(1),$$

independent of $\nu \in \mathbb{N}/2$, the interval I and $a \in \mathbb{R}$, holds (for $\beta < 1$) if and only if $\beta \geq 1/6$. This expression appears in a natural way as the leading term (using the product formula for Bessel functions) of the expansion of the kernel associated to $T_\nu T_\nu^*$ but replacing the "smoothness" $1/4$ by the generic smoothness α with $2\alpha - 1/2 = \beta$. This could be interpreted as an indication that the uniform estimate of the operators Q_ν by this method would only be possible on the class $\dot{H}^{1/3}$ ($\alpha = 1/3$ corresponds to the case $\beta = 1/6$). Our theorem shows however that an additional cancellation of the rest of terms in the expansion of the kernel is possible so that, as the theorem says, the result holds indeed on $\dot{H}^{1/4}$.

12.5.3. *Annex*

Proof of Theorem 12.15. Note that, since S_0 is compact, the case $p = 1$ is trivial. Hence, it suffices to prove the theorem for $p = 2\frac{n+1}{n+3}$.

We write $\mathbb{R}^n = \mathbb{R}^{n-1} \times \mathbb{R} = \{(x,t) : x \in \mathbb{R}^{n-1}, t \in \mathbb{R}\}$ and assume that $S = \{(u, \Phi(u)) : u \in \mathbb{R}^{n-1}, |u| < 1\}$ where Φ is a smooth function with non-vanishing hessian determinant. Using Plancherel's identity and Hölder's inequality,

$$\|\hat{f}\|^2_{L^2(S_0, d\sigma)} = \int |\hat{f}(\xi)|^2 \, d\sigma = \int \overline{f(x)} f * \widehat{d\sigma}(x) \, dx$$

$$\leq \|f\|_{L^p(\mathbb{R}^n)} \|f * \widehat{d\sigma}\|_{L^{p'}(\mathbb{R}^n)}.$$

It will be enough to show that

$$\|f * \widehat{d\sigma}\|_{p'} \leq C \|f\|_p.$$

We fix t and want to compute $\|f * \widehat{d\sigma}(\cdot, t)\|_{L^{p'}(\mathbb{R}^{n-1})}$. The kernel of this convolution is

$$K_t(x) = \widehat{d\sigma}(x, t) = \int e^{-2\pi i x\xi - 2\pi i t\Phi(\xi)} \, d\xi$$

and hence,

$$\widehat{K_t}^x(\xi) = e^{-2\pi i t\Phi(\xi)},$$

where $\widehat{F}^x$ denotes the Fourier transform with respect to the x variables. Therefore, if F is a function on the x variable, we have

$$\|K_t * F\|_2 = \|\hat{F}\|_2 \|\widehat{K_t}^x\|_\infty = \|\hat{F}\|_2.$$

On the other hand,

$$|K_t(x)| = |\widehat{d\sigma}(x, t)| \leq \frac{C}{|t|^{(n-1)/2}},$$

so that

$$\|K_t * F\|_\infty \leq \frac{C}{|t|^{(n-1)/2}} \|F\|_1.$$

We interpolate both inequalities and obtain, for $1 \leq p \leq 2$,

$$\|K_t * F\|_{L^{p'}(\mathbb{R}^{n-1})} \leq \frac{C}{|t|^{(n-1)(\frac{1}{p} - \frac{1}{2})}} \|F\|_{L^p(\mathbb{R}^{n-1})}.$$

We apply this to the function $f(\cdot, t)$ and use Minkowsky's inequality

$$\|f * \widehat{d\sigma}\|_{L^{p'}(\mathbb{R}^n)} = \| \, \| \int f(\cdot, \tau) *_x \widehat{d\sigma}(\cdot, t - \tau) \, d\tau\|_{L^{p'}(\mathbb{R}^{n-1})}\|_{L^{p'}(\mathbb{R})}$$

$$\leq \| \int \|f(\cdot, \tau) *_x \widehat{d\sigma}(\cdot, t - \tau)\|_{L^{p'}(\mathbb{R}^{n-1})} d\tau\|_{L^{p'}(\mathbb{R})}$$

$$\leq \left\| \int \frac{\|f(\cdot, \tau)\|_{L^p(\mathbb{R}^{n-1})}}{|t - \tau|^{(n-1)(\frac{1}{p} - \frac{1}{2})}} \right\|_{L^{p'}(\mathbb{R})}.$$

Consider now the **fractional integral operator**, defined in $\mathbb{R}^k$ by

$$I_\alpha g(x) = \int_{\mathbb{R}^k} \frac{f(y)}{|x - y|^{k-\alpha}} \, dy, \qquad \text{for } 0 < \alpha < k.$$

We will use the following result (a straightforward consequence of Hedberg's inequality, $|I_\alpha(x)| \leq C\|f\|_p^\theta (\mathcal{M}f)^{1-\theta}$):

Theorem 12.18 (Hardy–Littlewood–Sobolev).

$$\|I_\alpha g\|_{L^q(\mathbb{R}^k)} \leq C_{\alpha,p}\|g\|_{L^p(\mathbb{R}^k)} \qquad \textit{for} \quad \frac{1}{p} > \frac{\alpha}{k} \quad \textit{and} \quad \frac{1}{q} = \frac{1}{p} - \frac{\alpha}{k}.$$

In our case we have $k = 1$ and $1 - \alpha = (n - 1)(\frac{1}{p} - \frac{1}{2})$ and we want $q = p'$. Then, $p = 2\frac{n+1}{n+3}$. This gives us

$$\|f * \widehat{d\sigma}\|_{L^{p'}(\mathbb{R}^{n-1})} \leq C\| \, \|f\|_{L^p(\mathbb{R}^{n-1})}\|_{L^p(\mathbb{R})} = \|f\|_{L^p(\mathbb{R}^n)}.$$

$$\square$$

References

[1] A. Alfonseca, Strong type inequalities and an almost-orthogonality principle for families of maximal operators along directions in $\mathbb{R}^2$. *J. London Math. Soc.* (2) 67 (2003), no. 1, 208–218.

[2] A. Alfonseca, F. Soria, and A. Vargas, A remark on maximal operators along directions in $\mathbb{R}^2$. *Math. Res. Lett.* 10 (2003), no. 1, 41–49

[3] A. Alfonseca, F. Soria, and A. Vargas, An almost-orthogonality principle in L^2 for directional maximal functions. *Contemp. Math.*, 320, Amer. Math. Soc., Providence, RI, 2003, 1–7

[4] J. A. Barrionuevo, A Note on the Kakeya Maximal Operator, *Math. Res. Lett.* **3** (1995), 61–65.

[5] J. Bourgain, Besicovitch Type Maximal Functions and Applications to Fourier Analysis, *Geom. Funct. Anal.* **1** (1991).

[6] J. Bourgain, On the restriction and multiplier problems in $\mathbb{R}^3$, *GAFA Springer Lecture Notes in Math* **1469** (1991), 179–191.

[7] J. Bourgain, Some new estimates on oscillatory integrals, *Essays on Fourier Analysis in Honor of Elias M. Stein.* C. Fefferman, R. Fefferman and S. Wainger, Eds. Princeton U. Press (1995), 83–112.

[8] J. Bourgain, A remark on Schrödinger operators, *Israel J. Math.* **77** (1992), 1–16.

[9] A. Carbery, The boundedness of the maximal Bochner–Riesz operator in $L^4(\mathbb{R}^2)$, *Duke Math. J.* **50** (1983), 409–416.

[10] A. Carbery, Covering lemmas revisited. *Proc. Edinburgh Math. Soc.* (2) 31 (1988), no. 1, 145–150.

[11] A. Carbery, E. Hernández, and F. Soria, Estimates for the Kakeya Maximal Operator on Radial Functions in $\mathbb{R}^n$, in *Harmonic Analysis, ICM–90 Satellite Conference Proceedings.* S. Igari Ed. Springer Verlag, Tokio (1991), 41–50.

[12] A. Carbery, E. Romera, and F. Soria, Radial weights and mixed norm inequalities for the disc multiplier. *J. Funct. Anal.* 109 (1992), no. 1, 52–75.

[13] A. Carbery, J. L. Rubio de Francia, and L. Vega, Almost everywhere summability of Fourier integrals, *J. London. Math. Soc. (2)* **38** 3 (1988), 513–524.

[14] L. Carleson, On convergence and growth of partial Fourier series, *Acta Math.* **116** (1966), 135–157.

[15] L. Carleson, Some analytical problems related to statistical mechanics, *Euclidean Harmonic Analysis, Lecture Notes in Math.* **779** (1979), 5–45.

[16] L. Carleson, and P. Sjölin, Oscillatory Integrals and a Multiplier Problem for the Disc, *Studia Math.* **44** (1972), 287–299.

[17] M. Christ, On almost everywhere convergence of Bochner–Riesz means in higher dimensions, *Proc. Amer. Math. Soc.* **95** (1985), 16–20.

[18] M. Christ, Weak type endpoint bounds for Bochner–Riesz multipliers, *Rev. Mat. Iberoamericana* **3** (1987), 25–31.

[19] M. Christ, J. Duoandikoetxea, J. L. Rubio de Francia, Maximal Operator Related to the Radon Transform and the Calderon-Zygmund Method of Rotations, *Duke Math. J.* **53** (1986), 189–209.

[20] A. Córdoba, The Kakeya Maximal Function and the Spherical Summation Multipliers, *Amer. J. Math.* **99** (1977), 1–22.

[21] A. Córdoba, The Multiplier Problem for the Polygon, *Ann. of Math.* **105** (1977), 581–588.

[22] A. Córdoba, A note on Bochner-Riesz Operators, *Duke Math. J.* **46** (1979), 505–511.

[23] A. Córdoba, The disc multiplier. *Duke Math. J.* 58 (1989), no. 1, 21–29.

[24] A. Córdoba, and R. Fefferman, On the Equivalence Between the Boundedness of Certain Classes of Maximal and Multiplier Operators in Fourier Analysis, *Proc. Natl. Acad. Sci. USA* **74** (1977), 423–425.

[25] A. Córdoba, and R. Fefferman, On Differentiation of Integrals, *Proc. Natl. Acad. Sci. USA***74** (1977), 2211–2213.

[26] B. Dahlberg, and C. E. Kenig, A note on the almost everywhere behaviour of solutions to the Schrödinger equation, *Harmonic Analysis, Lecture Notes in Math.*, Springer Verlag **908** (1982), 205–208.

[27] J. Duoandikoetxea, and A. Vargas, Directional operators and radial func-

tions on the plane, *Ark. Mat.* (1995), 281–291.

[28] K. Falconer, *Fractal Geometry: Mathematical Foundations and Applications*, John Wiley& Sons, Chichester (1990).

[29] C. Fefferman, A Note on the Spherical Summation Multiplier, *Israel J. Math.* **15** (1973), 44–52.

[30] C. Fefferman, The Multiplier Problem for the Ball, *Ann. of Math.* **94** (1972), 330–336.

[31] J. García-Cuerva, and J. L. Rubio de Francia, *Weighted norm inequalities and related topics*. North-Holland Mathematics Studies, 116. Mathematical Notes, 104. North-Holland Publishing Co., Amsterdam, 1985

[32] A. Garsia, *Topics in almost everywhere convergence*. Lectures in Advanced Mathematics, 4 Markham Publishing Co., Chicago, Ill. 1970

[33] G. Gigante, and F. Soria, A Note on Oscillatory Integrals and Bessel Functions, *Contemp. Math.* **320**, American Mathematical Society, Providence, RI (2003), 157–172.

[34] G. Gigante, and F. Soria, On the Boundedness in $H^{1/4}$ of the Maximal Square Function Associated with the Schroedinger Equation. Math.CA/0307109

[35] M. de Guzmán, *Differentiation of integrals in R^n*. Lecture Notes in Mathematics, Vol. 481. Springer-Verlag, Berlin-New York, 1975

[36] L. Hörmander, Oscillatory Integrals and Multipliers on FL^p, *Ark. Mat.* **11** (1973), 1–11.

[37] R. Hunt, On the convergence of Fourier series, *Proc. Conf. at Southern Illinois Univ.*, Southern Illinois Univ. Press, Edwardsville **67**, 235–255.

[38] N. H. Katz, A Counterexample for Maximal Operators over a Cantor Set of directions, *Math. Res. Lett.* **3** 4 (1996), 527–536.

[39] N. H. Katz, Maximal Operators over Arbitrary Sets of Directions, *Duke Math. J.* **97** 1 (1999), 67–79.

[40] G. Mockenhaupt, On radial weights for the spherical summation operator. *J. Funct. Anal.* 91 (1990), no. 1, 174–181.

[41] G. Mockenhaupt, A note on the cone multiplier. *Proc. Amer. Math. Soc.* 117 (1993), no. 1, 145–152.

[42] G. Mockenhaupt, A. Seeger, and C. Sogge, Wave front sets, local smoothing and Bourgain's circular maximal theorem. *Ann. of Math.* (2) 136 (1992), no. 1, 207–218.

[43] A. Moyua, A. Vargas, and L. Vega, Schrödinger maximal function and restriction properties of the Fourier transform, *Internat. Math. Res. Notices* **36** (1996), 793–815.

[44] D. Müller and F. Soria, A double-weight L^2-inequality for the Kakeya maximal function *J. Fourier Anal. Appl.* 1995, Special Issue, 467–478.

[45] A. Nagel, E. M. Stein, and S. Wainger, Differentiation in Lacunary Directions, *Proc. Natl. Acad. Sci. USA* **75** (1978), 1060–1062.

[46] J. L. Rubio de Francia, Estimates for some Square Functions of Littlewood-Paley Type, *Publ. Mat.* **27** (1983), 81–108.

[47] J. L. Rubio de Francia, Maximal functions and Fourier transforms. *Duke Math. J.* 53 (1986), no. 2, 395–404.

[48] P. Sjögren, and P. Sjölin, Littlewood-Paley decompositions and Fourier multipliers with singularities on certain sets. *Ann. Inst. Fourier* (Grenoble) 31 (1981), no. 1, vii, 157–175.

[49] P. Sjölin, Unpublished manuscript.

[50] P. Sjölin, Regularity of solutions to Schrödinger equations, *Duke Math. J.* **55** (1987), 699–715.

[51] E. M. Stein, and G. Weiss, *Introduction to Fourier Analysis on Euclidean Spaces*, Princeton University Press (1971).

[52] E. M. Stein, *Oscillatory Integrals in Fourier Analysis* Beijing Lectures in Harmonic Analysis. Princeton University Press (1986).

[53] E. M. Stein, *Singular integrals and differentiability properties of functions.* Princeton Mathematical Series, No. 30 Princeton University Press, Princeton, N.J. 1970

[54] E. M. Stein, *Harmonic analysis: real-variable methods, orthogonality, and oscillatory integrals.* Princeton Mathematical Series, 43. Monographs in Harmonic Analysis, III. Princeton University Press, Princeton, NJ, 1993

[55] J. O. Strömberg, *Maximal Functions for Rectangles with Given Directions*, Thesis, Mittag–Leffler Institute, Djursholm, Sweden (1976).

[56] J. O. Strömberg, Maximal Functions Associated to Rectangles with Uniformly Distributed Directions, *Ann. of Math.* **107** (1978), 309–402.

[57] T. Tao, Weak–type endpoint bounds for Riesz means, *Proc. Amer. Math. Soc.* **124** (1996), 2797–2805.

[58] T. Tao, A sharp bilinear restrictions estimate for paraboloids. *Geom. Funct. Anal.* 13 (2003), no. 6, 1359–1384.

[59] T. Tao, Some recent progress on the Restriction conjecture. Math.CA/0303136

[60] T. Tao and A. Vargas, A bilinear approach to cone multipliers. I. Restriction estimates. *Geom. Funct. Anal.* 10 (2000), no. 1, 185–215.

[61] P. Tomas, A Restriction Theorem for the Fourier Transform, *Bull. Amer. Math. Soc.* **81** (1975), 477–478.

[62] L. Vega, Schrödinger equations: pointwise convergence to the initial data, *Proc. Amer. Math. Soc.* **102** (1988), 874–878.

[63] S. Wainger, Applications of Fourier Transform to Averages over Lower Dimensional Sets, *Pro. Sym. Pure. Math.* **35** , part 1 (1979), 85–94.

[64] T. Wolff, An improved bound for Kakeya type maximal functions, *Rev. Mat. Iberoamericana* **11** (1995), 651–674.

Chapter 13

Fractal analysis, an approach via function spaces

Hans Triebel

Mathematisches Institut
Fakultät für Mathematik und Informatik
Friedrich-Schiller-Universität Jena
*D-07737 Jena, Germany * e-mail: triebel@minet.uni-jena.de*

Contents

---

*This paper is a modified English translation of [49]. I wish to thank the B.G. Teubner GmbH Stuttgart-Leipzig-Wiesbaden 2001, holding the copyright of [49], for granting permission to publish an English translation in this volume.

13.1. Introduction

A. S. Besicovitch developed (together with only a few co-workes) in some forty years, from the end of the twenties up to the sixties of the last century *The Geometry of Sets of Points* (Title of a book planned by him which remained unfinished when he died in 1970). On this background a new branch of mathematics emerged in the last 20 or 25 years, called nowadays *fractal geometry*, as it may be found in the monographs [17], [18], [29], [19] (to mention only a few). The word *fractal* was coined by B. B. Mandelbrot in 1975. In contrast to Besicovitch, who developed a pure inner-mathematical theory of non-smooth structures (complementing differentiable analysis and geometry), Mandelbrot propagated the idea that many objects in nature have a fractal structure which cannot be described appropriately in terms of (differentiable) analysis and geometry. In particular he suggested the notion of *self-similarity*, which found its rigorous mathematical definition in [24] (1981). Together with its affine generalisation (*IFS: iterated function systems*) it is a corner stone of the recent fractal geometry and fractal analysis. There are many attempts to say under which circumstances a set (maybe a set of points in $\mathbb{R}^n$) should be called *fractal*. But there is no satisfactory definition, and *fractal* is now widely accepted as a somewhat vague synonym for *non-smooth* (either the objects themselfs or the ingredients admitted to deal with them). One has to say from case to case which specific assumptions are made.

In the last decade, and in particular in the last few years, there are many new developments in fractal geometry and now also in fractal analysis. But so far no dominating directions and techniques emerged. Asking for adequate geometries *completely distinct from Euclidean geometry* one finds in [38], Preface:

They suggest rather strong shifts in outlook, for what kind of geometries are really around, what one might look for, how one might work with them, and so on.

This has been complemented by [13], Preface:

The subject remains a wilderness, with no central zone, and many paths to try. The lack of main roadways is also one of the attractions of the subject.

One aspect culminates in the question how much geometry one really needs in order to prove well-known significant theorems in analysis. This resulted in an analysis on metric spaces (occasionally with additional assumptions), where [23] may be considered as the first comprehensive treatment.

One instrument to develop an analysis on non-smooth structures is the theory of the *Dirichlet forms (quadratic forms)* and of the operators generated by them. A recent report on diverse aspects has been given in [25], where we refer in particular to [30].

A central problem which has been considered in the last few years with great intensity can be formulated as follows:

What is a Laplacian on a fractal structure and what can be said about its spectrum?

The state of art of this question and of neighbouring problems may be found in [26], where we quote from as follows:

Why do you only study self-similar sets? The reason is that self-similar sets are perhaps the simplest and most basic structures in the theory of fractals. They should give us much information on what would happen in the general case of fractals. Although there have been many studies on analysis on fractals, we are still near the beginning in the exploration of this field. ([26], Introduction, p. 5.)

So far the outlined descriptions given above focus on operators (either directly defined or generated via Dirichlet forms) on fractal structures. But there is a second (seemingly opposite) point of view: One deals with fractal (differential-)operators on smooth structures such as $\mathbb{R}^n$, or bounded C^∞ domains Ω in $\mathbb{R}^n$. Proto-types are

$$A = -\Delta + \mu \quad \text{and} \quad B = (-\Delta)^{-1} \circ \mu, \qquad \Delta = \sum_{j=1}^{n} \frac{\partial^2}{\partial x_j^2}, \qquad (13.1)$$

where μ might be a Radon measure on $\mathbb{R}^n$ with compact support and $\mu(\mathbb{R}^n) < \infty$. *Schrödinger operators* of type A in $\mathbb{R}^n$ are of relevance in quantum mechanics. They have a long history which goes back to Fermi (1936), especially in the case of μ being a finite sum of Dirac measures. A comprehensive treatment of this subject may be found in [2]. Operators of type B emerge in connection with drums (especially in the plane $\mathbb{R}^2$) having fractal membranes, where Ω is a bounded C^∞ domain and $(-\Delta)^{-1}$ is the inverse of the Dirichlet Laplacian in Ω. Furthermore the support of μ is assumed to be a compact set in Ω. These operators are subject of the later considerations in this paper.

Function spaces are a decisive instrument of the recent analysis on quasi-metric spaces (X, ϱ, μ) where X is a set, ϱ stands for a quasi-metric, and μ is a measure. As far as Sobolev spaces $W_p^1(X)$ with $1 < p < \infty$ are concerned we refer to [20] and [23]. First comprehensive treatments of the emerging theory of the spaces of type

$$B_{pq}^s(X) \quad \text{and} \quad F_{pq}^s(X) \quad \text{with} \quad |s| \leq \theta \leq 1,\ 1 < p < \infty,\ 1 \leq q \leq \infty,$$

have been given in [21] and [22]. Fractals of diverse types can be interpreted as quasi-metric spaces. They can be treated afterwards with the help of the indicated theory of function spaces. For example, a far-reaching spectral theory of Riesz operators can be developed in this way, [51]. But this is not subject of our considerations in this paper.

The main aim of what follows is the description of some aspects of fractal analysis which are based on the recent theory of function spaces of type

$$B_{pq}^s(\mathbb{R}^n) \quad \text{and} \quad F_{pq}^s(\mathbb{R}^n), \quad 0 < p \leq \infty,\ 0 < q \leq \infty,\ s \in \mathbb{R}.$$

Special cases of these spaces are the Hölder-Zygmund spaces, the Sobolev spaces (of integer and fractional order of smoothness), the classical Besov spaces and the (inhomogeneous) Hardy spaces. Since some time one has for these spaces atomic decompositions and (at least for some of them) wavelet representations. This has been complemented in [41] and in [42] by quarkonial (or subatomic) decompositions. The elementary building blocks (quarks) are constructive, very simple and very flexible. In particular, they can easily be adapted to general structures, for example fractal sets in $\mathbb{R}^n$. Self-similarity is not needed. On this basis it is possible to study fractal operators, including those ones of type B in (13.1). It is our main aim in this paper to give a description of respective results obtained in the last few years. Some assertions are formulated here for the first time (without proofs). We restrict ourselves to the simplest cases (as far the formulations, not the proofs, are concerned). We shift a rough description of some ingredients of the proofs to Section 13.6, where we rely on [41] and, in particular, on [42].

Obviously, the above description of diverse aspects of fractal geometry and fractal analysis is dictated by the personal interests and by the (admittedly rather limited) knowledge of the author. For example, we did not mention the intense interplay between stochastics and fractal geometry. Furthermore as far as the spectral theory of the Laplace operator in bounded domains in $\mathbb{R}^n$ with fractal boundary is concerned we give at least a few references later on (but not more). It is the main aim of the above description to provide a possibility to locate what follows in the realm of what might be called nowadays *fractal analysis*.

13.2. Fractal operators

13.2.1. *The vibrating membrane*

Let Ω be a bounded C^∞ domain in the plane $\mathbb{R}^2$ and let Γ be a compact set in $\mathbb{R}^2$ with

$$\Gamma \subset \Omega \quad \text{and} \quad |\Gamma| = 0, \tag{13.2}$$

where $|\Gamma|$ is the Lebesgue measure of Γ. We interpret Ω as a drum and Γ as a highly irregular membrane which is fixed at the boundary $\partial\Omega$. Starting from the classical case it seems to be reasonable to consider Γ not only as a set of points, but also as the support of a measure μ,

$$\operatorname{supp} \mu = \Gamma \subset \Omega, \quad 0 < \mu(\Gamma) = \mu(\mathbb{R}^2) < \infty. \tag{13.3}$$

Of special interest (because well-adapted to the problem considered) are *isotropic* measures, this means measures for which there is a positive function $h(r)$, defined in the interval $(0, 1]$ with

$$\mu\left(B(\gamma, r)\right) \sim h(r), \quad \gamma \in \Gamma, \quad 0 < r \le 1, \tag{13.4}$$

where $B(\gamma, r)$ is a circle centred at γ and of radius r. We agree here and in the sequel that the use of $\sim$ for two positive functions $a(x)$ and $b(x)$ or for two sequences of positive numbers a_k and b_k (say with $k \in \mathbb{N}$, natural numbers), means that there are two positive numbers c und C with

$$c\, a(x) \le b(x) \le C\, a(x) \quad \text{or} \quad c\, a_k \le b_k \le C\, a_k \tag{13.5}$$

for all admitted variables x and k. This means with respect to (13.4) that the equivalence constants are independent of γ and r. Of special interest are so-called d-sets with

$$h(r) \sim r^d, \quad 0 < r \le 1, \quad \text{and} \quad 0 < d < 2. \tag{13.6}$$

Then any Radon measure μ with (13.3), (13.4), (13.6) is equivalent to $\mathcal{H}^d|\Gamma$, the restriction of Hausdorff measure $\mathcal{H}^d$ in $\mathbb{R}^2$ to Γ. If $d = 1$ then one might think of a line segment. Other examples are (one-dimensional and two-dimensional) Cantor sets, the snowflake curce, the Sierpinski gaskets etc. Many other examples of d-sets may be found in books dealing with fractal geometry, for example [18], [19], The following problem arises:

For which functions h do exist Radon measures μ with (13.3), (13.4)?

H. Triebel

13.2.2. *The classical theory*

The classical theory where the membrane is not a singular set Γ with (13.2), (13.3), but Ω or $\overline{\Omega}$ is well known. We formulate some assertions adapted to our later needs in the fractal setting. In the context of an L_2-theory one considers in the (complex) Hilbert space $L_2(\Omega)$ the self-adjoint positive-definite Dirichlet-Laplace operator $-\Delta$,

$$-\Delta u(x) = -\frac{\partial^2 u}{\partial x_1^2} - \frac{\partial^2 u}{\partial x_2^2}, \qquad \mathrm{dom}\,(-\Delta) = H_0^2(\Omega), \qquad (13.7)$$

where

$$H^2(\Omega) = W_2^2(\Omega) = \{f \in L_2(\Omega) \; : \; D^\alpha f \in L_2(\Omega),\; |\alpha| \le 2\} \qquad (13.8)$$

is the well-known Sobolev space and

$$H_0^2(\Omega) = W_{2,0}^2(\Omega) = \{f \in W_2^2(\Omega) \; : \; f|\partial\Omega = 0\}. \qquad (13.9)$$

Then $-\Delta$ is a self-adjoint positive-definite operator with pure point spectrum. The eigenfrequencies λ_k of the vibrating membrane and the related eigenfunctions u_k are given by

$$-\Delta u_k = \lambda_k^2 u_k, \quad u_k \in H_0^2(\Omega), \quad k \in \mathbb{N}. \qquad (13.10)$$

(Here $\mathbb{N}$ is the collection of all natural numbers). For our purposes it is reasonable to deal with the inverse operator $(-\Delta)^{-1}$ adapted by

$$B = (-\Delta)^{-1} = (-\Delta)^{-1} \circ \mu_L = (-\Delta)^{-1} \circ id^{\mu_L} \qquad (13.11)$$

to our later needs, where μ_L is the Lebesgue measure in the plane. Then (13.10) can be written as

$$Bu_k = \varrho_k u_k, \quad \varrho_1 > \varrho_2 \ge \cdots \ge \varrho_k \ge \cdots > 0, \quad \varrho_k \to 0$$

for $k \to \infty$, where $\varrho_k = \lambda_k^{-2}$ are the eigenvalues of the compact positive self-adjoint operator B in $L_2(\Omega)$. The following three assertions (adapted to our later needs) are well known:

(a) (H. Weyl, 1912, [52], [53]),

$$\varrho_k \sim k^{-1}, \qquad k \in \mathbb{N}. \qquad (13.12)$$

(b) (R. Courant, 1924, [12], p. 398/99). The largest eigenvalue ϱ_1 is simple and (ignoring a multiplicative constant)

$$u_1(x) > 0, \qquad x \in \Omega, \qquad (13.13)$$

Nullstellenfreiheit in the notation of Courant.

(c) (smoothness). We have (in obvious notation)

$$u_k \in C^\infty(\overline{\Omega}), \qquad k \in \mathbb{N}. \tag{13.14}$$

For the classical case described the assertions by Weyl had been sharper than (13.12) from the very beginning, including the volumes of Ω (main term) and of $\partial\Omega$ (remainder term). Since that time the spectral theory of (regular and singular elliptic) differential operators and pseudo-differential operators is an outstanding topic of the analysis during the whole last century. The state of art and in particular the techniques used nowadays may be found in [37]. Our aim can be described as follows:

What happens with the properties (a), (b), (c), this means (13.12), (13.13), (13.14), when one replaces in (13.11) the Lebesgue measure μ_L (hence the homogeneous membrane) by a measure μ with (13.3) and what is the definition of an appropriately modified operator B?

13.2.3. *The problem*

Let Ω be a bounded C^∞ domain in euclidean n-space $\mathbb{R}^n$. If one wishes to replace the n-dimensional Lebesgue measure μ_L in (13.11) by a singular measure μ with (13.2), (13.3) then it is quite clear that the L_2-theory as described in (13.7) - (13.11) is no longer adequate. In particular, instead of $H^2(\Omega)$ und $H_0^2(\Omega)$ in (13.8), (13.9) one needs now spaces of smaller smoothness. It comes out that

$$H^1(\Omega) = W_2^1(\Omega) = \left\{ f \in L_2(\Omega) \ : \ \frac{\partial f}{\partial x_j} \in L_2(\Omega) \text{ with } j = 1, \cdots, n \right\}$$

and

$$\overset{\circ}{H}{}^1(\Omega) = \overset{\circ}{W}{}_2^1(\Omega) = \{ f \in H^1(\Omega) \ : \ f|\partial\Omega = 0 \} \tag{13.15}$$

are especially well adapted to our purposes. The simplest way to define the Sobolev spaces $H^s(\mathbb{R}^n)$ is by lifting,

$$H^s(\mathbb{R}^n) = (id - \Delta)^{-\frac{s}{2}} L_2(\mathbb{R}^n), \qquad s \in \mathbb{R}. \tag{13.16}$$

Then $H^s(\Omega)$ is the restriction of $H^s(\mathbb{R}^n)$ to Ω. Since $C_0^\infty(\Omega) = D(\Omega)$ is dense in $\overset{\circ}{H}{}^1(\Omega)$ the dual of $\overset{\circ}{H}{}^1(\Omega)$ makes sense within the dual pairing $(D(\Omega), D'(\Omega))$ and one obtains

$$\left(\overset{\circ}{H}{}^1(\Omega) \right)' = H^{-1}(\Omega). \tag{13.17}$$

By [42], Proposition 20.3, p. 297, (13.17) remains valid for arbitrary bounded domains.

Proposition 13.1. *(a) Let Ω be an arbitrary bounded C^∞ domain in $\mathbb{R}^n$. Then*

$$(-\Delta)^{-1} : \quad H^{-1}(\Omega) \hookrightarrow \overset{\circ}{H}{}^1(\Omega) \tag{13.18}$$

is an isomorphic map.

(b) Let $\sigma \in \mathbb{R}$. Then

$$(id - \Delta)^{-\sigma} : \quad H^{-\sigma}(\mathbb{R}^n) \hookrightarrow H^\sigma(\mathbb{R}^n) \tag{13.19}$$

is an isomorphic map.

Remark 13.1. Part (b) is a well-known lift property, which is also an easy consequence of (13.16). As for part (a) we refer to [42], 19.2, p. 254/55, and the references given there.

In order to say what is meant by the operator B in (13.1), we introduce the trace operator tr_μ and the identification operator id_μ. Here μ is a Radon measure in $\mathbb{R}^n$ with compact support

$$\Gamma = \operatorname{supp} \mu \quad \text{and} \quad 0 < \mu(\mathbb{R}^n) = \mu(\Gamma) < \infty. \tag{13.20}$$

Then one has the continuous embeddings

$$tr_\mu : \quad S(\mathbb{R}^n) \hookrightarrow L_2(\Gamma, \mu), \tag{13.21}$$

and

$$id_\mu : \quad L_2(\Gamma, \mu) \hookrightarrow S'(\mathbb{R}^n), \tag{13.22}$$

where $S(\mathbb{R}^n)$, $S'(\mathbb{R}^n)$ (Schwartz space, space of tempered distributions) and $L_2(\Gamma, \mu)$ (complex Hilbert space L_2 with respect to μ) have the usual meaning. Here tr_μ is the pointwise trace (so far), hence

$$(tr_\mu \varphi)(\gamma) = \varphi(\gamma) \quad \text{for} \quad \varphi \in S(\mathbb{R}^n) \quad \text{and} \quad \gamma \in \Gamma,$$

and id_μ describes the usual identification of $f \in L_2(\Gamma, \mu)$ with the tempered distribution $id_\mu f \in S'(\mathbb{R}^n)$ according to

$$(id_\mu f)(\varphi) = \int_\Gamma f(\gamma)\,\varphi(\gamma)\,\mu(d\gamma), \quad \varphi \in S(\mathbb{R}^n).$$

This can be done without ambiguities since μ is a Radon measure.

Proposition 13.2. *Let μ be a Radon measure in $\mathbb{R}^n$ with compact support according to (13.20). The trace operator and the identification operator are given by (13.21), (13.22). Then*

$$tr'_\mu = id_\mu \tag{13.23}$$

for the dual tr'_μ of tr_μ.

Proof. Using standard notation, the assertion follows from

$$\left(tr'_\mu \psi\right)(\varphi) = \left(tr_\mu \varphi\right)(\psi) = \left(tr_\mu \varphi, \overline{\psi}\right)_{L_2(\Gamma,\mu)}$$
$$= \int_\Gamma \varphi(\gamma)\,\psi(\gamma)\,\mu(d\gamma)$$
$$= (id_\mu \psi)(\varphi),$$

for $\varphi \in S(\mathbb{R}^n)$ and all $\psi \in S(\mathbb{R}^n)$. $\qquad\square$

Problem 13.1. As will be shown, the question of how B from (13.1) can be defined, in generalisation of B from (13.11), can be reduced to the problem of whether tr_μ from (13.21) can be extended to a linear continuous map from $H^1(\mathbb{R}^n)$ in $L_2(\Gamma, \mu)$. Generalising this question to $H^s(\mathbb{R}^n)$ one arrives at the problem:

Let μ be a Radon measure according to (13.20). For which $s > 0$ does there exist a positive constant c with

$$\|tr_\mu \varphi \,|\, L_2(\Gamma, \mu)\| \le c\,\|\varphi\,|\,H^s(\mathbb{R}^n)\| \tag{13.24}$$

for all $\varphi \in S(\mathbb{R}^n)$?

We write here and in the sequel $\|a\,|\,A\|$ for the norm of $a \in A$ in the Banach space A. If one has (13.24), then tr_μ can be extended by completion to a linear and continuous operator

$$tr_\mu : \quad H^s(\mathbb{R}^n) \hookrightarrow L_2(\Gamma, \mu) \tag{13.25}$$

(sticking at the notation tr_μ). If one has (13.25), then it follows from (13.22), (13.23) that

$$id_\mu : \quad L_2(\Gamma, \mu) \hookrightarrow H^{-s}(\mathbb{R}^n) = (H^s(\mathbb{R}^n))'. \tag{13.26}$$

Then one obtains

$$id^\mu = id_\mu \circ tr_\mu : \quad H^s(\mathbb{R}^n) \hookrightarrow H^{-s}(\mathbb{R}^n). \tag{13.27}$$

As usual, $\hookrightarrow$, means that the corresponding map is linear and continuous.

Proposition 13.3. *Let μ be a Radon measure in $\mathbb{R}^n$ with compact support according to (13.20). Let $s > 0$, such that tr_μ according to (13.25) is a linear and bounded operator. Then*

$$B_s = (id - \Delta)^{-s} \circ id^\mu \; : \quad H^s(\mathbb{R}^n) \hookrightarrow H^s(\mathbb{R}^n). \tag{13.28}$$

(b) Let Ω be a bounded C^∞ domain in $\mathbb{R}^n$. Let $(-\Delta)^{-1}$ be the inverse operator of the Dirichlet-Laplace operator according to Proposition 13.1. Let μ be a Radon measure in $\mathbb{R}^n$ with

$$\Gamma = supp\ \mu \subset \Omega, \quad 0 < \mu(\Gamma) < \infty.$$

If one has (13.25) with $s = 1$, then

$$B = (-\Delta)^{-1} \circ id^\mu \; : \quad \overset{\circ}{H}{}^1(\Omega) \hookrightarrow \overset{\circ}{H}{}^1(\Omega). \tag{13.29}$$

Proof.   Part (a) is an immediate consequence of (13.27), (13.19). Correspondingly one obtains part (b) from (13.27) with $s = 1$ and (13.18). $\square$

Remark 13.2. The operator B from (13.29) is the rigorous version of B from (13.1) and the adequate generalisation of (13.11). We refer to [42], 9.2, 122-124, where one finds further explanations with respect to the duality (13.23), and (13.25), (13.26) in more general spaces.

13.3. Traces and measures

13.3.1. *Isotropic measures*

By Proposition 13.3 in 13.2.3 the problem of the boundedness of the operators B_s and B in (13.28) and (13.29) can be reduced to the question of whether the trace operator tr_μ in (13.25) is bounded. For general finite Radon measures in $\mathbb{R}^n$ (in more general function spaces) one finds in [42], Theorem 9.3, p. 125, a somewhat implicit criterion which had been complemented in [42], Corollary 9.8, p. 129/130, by a simpler sufficient condition. As will be shown one has for *isotropic* Radon measures a satisfactory simple criterion. In generalisation of (13.4) we call a Radon measure in $\mathbb{R}^n$ with

$$\Gamma = supp\ \mu \quad compact\ and \quad 0 < \mu(\Gamma) = \mu(\mathbb{R}^n) < \infty \tag{13.30}$$

isotropic, if there is a positive function $h(\cdot)$ in the interval $(0, 1]$ such that

$$\mu(B(\gamma, r)) \sim h(r), \quad \gamma \in \Gamma, \quad 0 < r \leq 1, \tag{13.31}$$

where $B(\gamma, r)$ is a ball centred at γ and of radius r. Recall our use of $\sim$ according to (13.5).

Proposition 13.4 (M. Bricchi, [5], [6], [7], [8]). *Let h be a positive continuous monotonically increasing function in the interval $(0, 1]$. Then there is an isotropic Radon measure in $\mathbb{R}^n$ with (13.30), (13.31), if, and only if, there exists a positive continuous monotonically increasing function $\widetilde{h}$ in the interval $(0, 1]$ such that*

$$\widetilde{h}(r) \sim h(r), \quad 0 < r \leq 1,$$

and

$$\frac{\widetilde{h}(2^{-j-k})}{\widetilde{h}(2^{-j})} \geq 2^{-kn} \quad \text{for } j \in \mathbb{N}_0 \text{ and } k \in \mathbb{N}_0.$$

Remark 13.3. Here $\mathbb{N}_0 = \mathbb{N} \cup \{0\}$ is the collection of all non-negative integers. *Monotonically increasing* means not decreasing, hence it might well be possible that the function is constant on some sub-intervals. Distinguished examples are the d-sets (now in $\mathbb{R}^n$), mentioned in 13.2.1, hence

$$h(r) = r^d, \quad 0 \leq d \leq n, \quad 0 < r \leq 1. \tag{13.32}$$

Further examples are perturbations of (13.32), for instance,

$$h(r) = r^d \left| \log \frac{r}{2} \right|^b, \quad 0 < d < n, \quad b \in \mathbb{R}, \quad 0 < r \leq 1, \tag{13.33}$$

as special cases of so-called (d, Ψ)-sets. The definite version of the above proposition may be found in [8], including some non-standard examples. These results had been used to develop a corresponding theory of function spaces. In case of (d, Ψ)-sets and especially of functions h with (13.33) we refer to [42], Section 22, and the literature mentioned there, in particular [31], [32].

13.3.2. *Traces*

As indicated in the introduction, we restrict ourselves to the most interesting cases (for us). These are the operator B from (13.29) in the plane, hence $n = 2$, and the operators B_s from (13.28) with $s = \frac{n}{2}$. This is in both cases the question of whether tr_μ exists according to (13.25), (13.24) with $s = \frac{n}{2}$, this means the continuity of tr_μ. We recall that $\mathbb{N}$ is the collection of all natural numbers, whereas $\mathbb{R}^n$ stands for euclidean n-space.

Theorem 13.1. *Let μ be an isotropic measure in $\mathbb{R}^n$, $n \in \mathbb{N}$, with (13.30), (13.31). Then the following 3 assertions are equivalent to each other:*

(i) $tr_\mu \;:\;\; H^{\frac{n}{2}}(\mathbb{R}^n) \hookrightarrow L_2(\Gamma, \mu)$ is continuous,

(ii) $tr_\mu \;:\;\; H^{\frac{n}{2}}(\mathbb{R}^n) \hookrightarrow L_2(\Gamma, \mu)$ is compact,

(iii) $\displaystyle\sum_{j=0}^{\infty} h(2^{-j}) < \infty.$

Remark 13.4. A proof of this surprisingly simple and definitive theorem is published in [47]. It is a special case of the following general assertion, which may also be found in [47]. Let

$$B_p^s(\mathbb{R}^n) = B_{p,p}^s(\mathbb{R}^n) \quad \text{with } s > 0 \text{ and } 1 < p < \infty,$$

be the well-known special Besov spaces with $H^s(\mathbb{R}^n) = B_2^s(\mathbb{R}^n)$. We put as usual $\frac{1}{p} + \frac{1}{p'} = 1$. Then the trace operator tr_μ,

$$tr_\mu \;:\;\; B_p^s(\mathbb{R}^n) \hookrightarrow L_p(\Gamma, \mu),$$

is continuous if, and only if, it is compact, and this happens if, and only if,

$$\sum_{j=0}^{\infty} 2^{-jp'(s-\frac{n}{p})}\, h(2^{-j})^{p'-1} < \infty.$$

This clarifies the peculiar situation if $p = p' = 2$ and $s = \frac{n}{2}$ in the above theorem.

Theorem 13.2. *Let μ be an isotropic Radon measure in $\mathbb{R}^n$, $n \in \mathbb{N}$, with (13.30), (13.31) and*

$$\sum_{j=0}^{\infty} h(2^{-j}) < \infty. \tag{13.34}$$

(a) Then the operator

$$B_{\frac{n}{2}} = (id - \Delta)^{-\frac{n}{2}} \circ id^\mu \;:\;\; H^{\frac{n}{2}}(\mathbb{R}^n) \hookrightarrow H^{\frac{n}{2}}(\mathbb{R}^n) \tag{13.35}$$

according to (13.27), (13.28) is compact.

(b) Let in addition $n = 2$ and let Ω be a bounded C^∞ domain in the plane $\mathbb{R}^2$. Furthermore, let $\Gamma \subset \Omega$, where Γ is the support of μ according to (13.30). Then the operator

$$B = (-\Delta)^{-1} \circ id^\mu \;:\;\; \overset{\circ}{H}{}^1(\Omega) \hookrightarrow \overset{\circ}{H}{}^1(\Omega) \tag{13.36}$$

according to (13.27), (13.29) is compact.

Proof. This theorem is an immediate consequence of the above Theorem 13.1 and of the proof of Proposition 13.3 in 13.2.3. $\qquad\square$

Remark 13.5. In particular the operators $B_{\frac{n}{2}}$ and B according to (13.35) und (13.36) are compact, when Γ is a d-set according to (13.32) with $0 < d \leq n$ ($n = 2$ in case of B). This applies also to the (d,Ψ)-sets in (13.33).

Remark 13.6. By Remark 13.4,

$$tr_\mu : \quad H^s(\mathbb{R}^n) \hookrightarrow L_2(\Gamma,\mu)$$

with $s > 0$ is compact if, and only if,

$$\sum_{j=0}^\infty 2^{-2j(s-\frac{n}{2})}\, h(2^{-j}) < \infty. \tag{13.37}$$

Then by Proposition 13.3 the operator

$$B_s = (id - \Delta)^{-s} \circ id^\mu : \quad H^s(\mathbb{R}^n) \hookrightarrow H^s(\mathbb{R}^n)$$

according to (13.27), (13.28) is compact. Because of the boundedness of the sequence $h(2^{-j})$ we have always (13.37) if $s > \frac{n}{2}$. Hence only for $s \leq \frac{n}{2}$ the assertion (13.37) is an additional condition. According to Theorem 13.2 we are interested here exclusively (with exception of the above remarks) in the *limiting case* $s = \frac{n}{2}$.

13.4. The fractal drum

13.4.1. *Introduction*

The word *fractal drum* can be interpreted quite differently. One may think about a fractal or a fractal set and the spectral theory of the related Laplacian. This corresponds to [26] and the theory represented there. Better known is this notation in connection with a spectral theory of the Dirichlet-Laplacian $-\Delta$ in bounded domains Ω in $\mathbb{R}^n$ (with $n = 2$ as a distinguished case) with a fractal boundary $\partial\Omega$. The first significant related paper is due to M. L. Lapidus, [27]. Further references may be found in [41], 26.2, p. 200. For more recent results we refer to [28]. More precisely these considerations deal with

drums with fractal boundary.

Somewhat in contrast we developed in [41], [42] and develop in this paper a spectral theory of

drums with fractal membranes.

It is the spectral theory of the operator B from Proposition 13.3(b) in 13.2.3 with $n = 2$, related to bounded C^∞ domains Ω in the plane.

13.4.2. *Preparations*

Let Ω be a bounded C^∞ domain in $\mathbb{R}^n$. Then we equip the space $\overset{\circ}{H}{}^1(\Omega)$ in (13.15) with the scalar product

$$(f, g)_{\overset{\circ}{H}{}^1(\Omega)} = \sum_{j=1}^{n} \int_{\Omega} \frac{\partial f}{\partial x_j} \frac{\partial \overline{g}}{\partial x_j} \, dx. \tag{13.38}$$

Furthermore we need the Hölder-Zygmund spaces $\mathcal{C}^s(\Omega)$ with $0 < s < 2$. If $0 < s < 1$, then $\mathcal{C}^s(\Omega)$ can be normed by

$$\|f \,|\mathcal{C}^s(\Omega)\|^* = \sup_{x \in \Omega} |f(x)| + \sup \frac{|f(x) - f(y)|}{|x - y|^s} \tag{13.39}$$

with $x \in \Omega$, $y \in \Omega$, $x \neq y$, in the second supremum. If $0 < s < 2$ then one can choose

$$\|f \,|\mathcal{C}^s(\Omega)\| = \sup_{x \in \Omega} |f(x)| + \sup \frac{|f(x) - 2f(\frac{x+y}{2}) + f(y)|}{|x - y|^s} \tag{13.40}$$

as a norm where $x \in \Omega$, $y \in \Omega$, $\frac{x+y}{2} \in \Omega$, $x \neq y$, in the second supremum. (If $0 < s < 1$, then both norms in (13.39) und (13.40) are equivalent). In other words, $\mathcal{C}^s(\Omega)$ is the collection of all complex-valued continuous functions in $\overline{\Omega}$, such that the norm in (13.40) is finite.

Remark 13.7. One calls $\mathcal{C}^1(\Omega)$ also the Zygmund class. Here one needs for the first time second differences. This space has been treated in [54]. But as mentioned there it was B. Riemann 1854 in his Habilitationsschrift [36], dealing with trigonometric series, who emphasised that one should use second differences when it comes to spaces of smoothness 1.

13.4.3. *d-membranes*

This subsection deals with fractal drums (or, according to 13.4.1, with drums having a fractal membrane), where the membrane Γ is a d-set. According to the above considerations, Γ is a compact set in $\mathbb{R}^2$ furnished with an isotropic Radon measure μ in $\mathbb{R}^2$ such that

$$\Gamma = \operatorname{supp} \mu, \quad \mu(B(\gamma, r)) \sim r^d, \quad \gamma \in \Gamma, \ 0 < r \le 1, \tag{13.41}$$

where $B(\gamma, r)$ is a circle with the centre γ and of radius r. As explained in connection with (13.6) one may identify μ with $\mu = \mathcal{H}^d|\Gamma$, the restriction of the Hausdorff measure $\mathcal{H}^d$ in the plane to Γ. Then it is clear that one has $0 \leq d \leq 2$. Furthermore, let $\Gamma \subset \Omega$, where Ω is a bounded C^∞ domain in $\mathbb{R}^2$. This results in the operator B from Theorem 13.2(b) in 13.3.2. The condition (13.34) requires $d > 0$. Since we are not interested in the regulare case and assume $|\Gamma| = 0$, we exclude also $d = 2$. Then we denote Γ as an d-*membrane*. By Theorem 13.2(b) in 13.3.2 the related operator B in $\overset{\circ}{H}{}^1(\Omega)$ is compact. Let $\omega = \Omega\backslash\Gamma$ and

$$\overset{\circ}{H}{}^1(\omega) = \left\{ f \in \overset{\circ}{H}{}^1(\Omega) \ : \ tr_\mu f = 0 \right\}. \tag{13.42}$$

The trace exists since $d > 0$ and (13.34) coincides with (iii) in Theorem 13.1 in 13.3.2. Although ω is an open set with an irregular boundary $\partial\Omega \cup \Gamma$ it follows that $C_0^\infty(\omega)$ is dense in $\overset{\circ}{H}{}^1(\omega)$, what justifies the notation in (13.42). We refer to [42], 19.5, p. 260, where we used some crucial assertions from [1].

Theorem 13.3. *Let Ω be a bounded C^∞ domain in the plane $\mathbb{R}^2$ and let Γ be a compact d-set with $\Gamma \subset \Omega$ and $0 < d < 2$. Furthermore let μ be a respective Radon measure according to (13.41). Then the operator*

$$B - (\ \Delta)^{-1} \cup id^\mu \ : \quad \overset{\circ}{H}{}^1(\Omega) \hookrightarrow \overset{\circ}{H}{}^1(\Omega) \tag{13.43}$$

according to (13.36) is a non-negative compact self-adjoint operator in $\overset{\circ}{H}{}^1(\Omega)$ with null-space (kernel)

$$N(B) = \overset{\circ}{H}{}^1(\omega) \tag{13.44}$$

according to (13.42). The operator B is generated by the quadratic form

$$(Bf, g)_{\overset{\circ}{H}{}^1(\Omega)} = \int_\Gamma f(\gamma) \, \overline{g(\gamma)} \, \mu(d\gamma), \quad f \in \overset{\circ}{H}{}^1(\Omega), \quad g \in \overset{\circ}{H}{}^1(\Omega), \tag{13.45}$$

with respect to the scalar product from (13.38) with $n - 2$. Let ϱ_k be the positive eigenvalues of B, counted with respect to their multiplicites and ordered by magnitude,

$$\varrho_1 \geq \varrho_2 \geq \varrho_3 \geq \cdots, \qquad \varrho_k \to 0 \quad for \quad k \to \infty, \tag{13.46}$$

and let u_k be the respective eigenfunktions, hence

$$Bu_k = \varrho_k \, u_k, \qquad k \in \mathbb{N}. \tag{13.47}$$

The largest eigenvalue ϱ_1 (> 0) is simple and the respective eigenfunctions $u_1(x)$ have no zeroes in Ω, hence

$$u_1(x) = c\,u(x) \quad \text{with} \quad c \in \mathbb{C} \text{ and } u(x) > 0 \text{ in } \Omega \tag{13.48}$$

(Courant property). It holds

$$\varrho_k \sim k^{-1} \qquad \text{with} \quad k \in \mathbb{N} \tag{13.49}$$

(Weyl property). The eigenfunctions u_k are classical harmonic functions in $\omega = \Omega\backslash\Gamma$,

$$\Delta u_k(x) = 0 \qquad \text{for} \quad x \in \omega. \tag{13.50}$$

Furthermore,

$$u_k \in C^s(\Omega) \quad \text{if, and only if } s \le d. \tag{13.51}$$

Remark 13.8. This theorem is a special case of Theorem 19.7 in [42], p. 264/265. This applies also to its proof. We described in 13.2.2, points (a), (b), (c), the classical theory. The Weylian behaviour (13.12) of the classical case extends to d-membranes according to (13.49). It is equally remarkable that also the classical assertion in 13.2.2(b) can be carried over to d-membranes: The largest eigenvalue ϱ_1 is simple and the respective eigenfunction $u(x)$ is positive in Ω (up to a multiplicative constant). The global smoothness properties of the eigenfunctions u_k from (13.14) in case of d-membranes are now given by the sharp assertion $u_k \in C^d(\Omega)$ and by (13.50).

Remark 13.9. It is remarkable that at least for the function u from (13.48) the best possible global smoothness assertion (13.51) is also the best possible local smoothness assertion. Let $B(\gamma,r)$ be a circle centred at $\gamma \in \Gamma$ and of radius $r > 0$, with $B(\gamma,r) \subset \Omega$. Then

$$u \in C^s\left(B(\gamma,r)\right) \qquad \text{if, and only if, } s \le d. \tag{13.52}$$

The best possible local smoothness of u, hence $s = d$, reflects faithfully the fractality of Γ at any point $\gamma \in \Gamma$. We overlooked this remarkable property when we wrote down 19.7 - 19.10 in [42], p. 264-272, although it is an immediate consequence of the proof of Theorem 19.7 as it stands. We return to this point in 13.4.6.

13.4.4. *Weyl measures*

In connection with Theorem 13.3 in 13.4.3 the question arises whether the Weyl property (13.49) remains valid for other measures μ than $\mathcal{H}^d|\Gamma$ with $0 < d < 2$. The definitive solution of this problem seems to be complicated and there is no idea how a (necessary and sufficient) criterion might look like. As in 13.2.1 we again assume that Ω is a bounded C^∞ domain in the plane $\mathbb{R}^2$ and that μ is a Radon measure with

$$\operatorname{supp} \mu = \Gamma \subset \Omega, \quad |\Gamma| = 0, \quad 0 < \mu(\Gamma) < \infty. \tag{13.53}$$

The right-hand side of (13.45), now for general Radon measures with (13.53), is a continuous bilinear form in $\overset{\circ}{H}{}^1(\Omega)$ if

$$tr_\mu : \quad \overset{\circ}{H}{}^1(\Omega) \hookrightarrow L_2(\Gamma, \mu) \tag{13.54}$$

is a linear and continuous trace operator according to 13.2.3 and Proposition 13.3(b). If this is the case then one obtains also for these measures that

$$B = (-\Delta)^{-1} \circ id^\mu : \quad \overset{\circ}{H}{}^1(\Omega) \hookrightarrow \overset{\circ}{H}{}^1(\Omega), \tag{13.55}$$

and that B is a non-negative self-adjoint operator. In other words we have also in this case (13.43), (13.44), (13.45).

Definition 13.1. Let μ be a finite Radon measure in $\mathbb{R}^2$ with compact support Γ. Then μ is called a Weyl measure if for any bounded C^∞ domain Ω with (13.53) the trace operator tr_μ from (13.54) exists, the operator B according to (13.55) is compact and if one has for its positive eigenvalues ϱ_k, ordered according to (13.46), including to multiplicities,

$$\varrho_k \sim k^{-1}, \qquad k \in \mathbb{N}.$$

Remark 13.10. We have by Theorem 13.3 from 13.4.3 that $\mu = \mathcal{H}^d|\Gamma$ is a Weyl measure for any compact d-set Γ with $0 < d < 2$. Let

$$\mu = \sum_{j=1}^{N} \mu_j, \qquad \mu_j = \mathcal{H}^{d_j}|\Gamma_j, \quad j = 1, ..., N,$$

where Γ_j are compact d_j-sets with $0 < d_j < 2$. Then μ is a Weyl measure. This assertion may be found in [42], 19.12, p. 274, under the assumption that the sets Γ_j are pairwise disjoint. But this additional assumption is not necessary as has been shown in [46], Corollary 2.

Remark 13.11. If μ is an isotropic measure with (13.53), then one has by Theorem 13.1 from 13.3.2 a definite criterion under which circumstances tr_μ exists. If μ is a general (not necessarily isotropic) measure with (13.53), then there is also a necessary and sufficient criterion for the existence of tr_μ. We refer to [42], Theorem 9.3, p. 125/126. But it is somewhat implicit. But one obtains as special cases the following assertions. Let Q_{jm} be the square in $\mathbb{R}^2$ centred at $2^{-j}m$ and with side-length 2^{-j+1}. Here $j \in \mathbb{N}_0$ and $m \in \mathbb{Z}^2$, where $\mathbb{Z}^2$ stands for the lattice of the points in $\mathbb{R}^2$ with integer-valued coordinates. Furthermore, let μ be a finite Radon measure in $\mathbb{R}^2$ with (13.53). Let

$$\mu_j = \sup_{m \in \mathbb{Z}^2} \mu(Q_{jm}), \qquad j \in \mathbb{N}_0.$$

If tr_μ exists according to (13.54), then

$$\sum_{j \in \mathbb{N}_0} \sum_{m \in \mathbb{Z}^2} \mu\left(Q_{jm}\right)^2 < \infty \tag{13.56}$$

(necessary condition). Conversely, if

$$\sum_{j \in \mathbb{N}_0} \mu_j < \infty, \tag{13.57}$$

then tr_μ exists (sufficient condition). We refer to [47], Proposition 13.2, which in turn is based on [42], Theorem 9.9, p. 131 and (9.47), p. 130. It follows in particular by (13.56), that a measure with (13.54) has no atoms, hence

$$\mu(\{x\}) = 0 \qquad \text{for} \quad x \in \mathbb{R}^2. \tag{13.58}$$

Following [4], Section 5.10, p. 61, and [14], 13.18, p. 215, we call a measure without atoms according to (13.58) *diffuse*.

Definition 13.2. Let μ be a finite Radon measure in $\mathbb{R}^2$ with compact support Γ. Furthermore it is assumed that μ satisfies the doubling condition, this means that there is a number $c > 0$ such that for all $\gamma \in \Gamma$ and all r with $0 < r \leq 1$,

$$\mu\left(B(\gamma, 2r)\right) \leq c\,\mu\left(B(\gamma, r)\right). \tag{13.59}$$

Then μ is called strongly diffuse, if there is a number $\varkappa$ with $0 < \varkappa < 1$ such that

$$\mu(Q_1) \leq \frac{1}{2}\mu(Q_0), \qquad (13.60)$$

for all squares Q_0 centred at $\gamma_0 \in \Gamma$ and with side-lenght r, $0 < r \leq 1$, and all squares Q_1 centred at $\gamma_1 \in \Gamma$ with side-length $\varkappa r$ and $Q_1 \subset Q_0$.

Remark 13.12. In (13.59) again $B(\gamma, r)$ is a circle centred at γ and of radius r. Examples of strongly diffuse measures are given by $\mu = \mathcal{H}^d|\Gamma$, hence

$$\mu\left(B(\gamma, r)\right) \sim r^d, \qquad \gamma \in \Gamma, \quad 0 < r \leq 1, \qquad (13.61)$$

where Γ is a compact d-set according (13.41) with $0 < d < 2$. Obviously, the assumed compactness of the support of μ is unimportant. Furthermore in (13.61) also $d = n = 2$ can be admitted with the Lebesgue measure as proto-type. Otherwise we refer to [42], 19.15, p. 276-278, where strongly diffuse measures had been introduced and discussed.

Theorem 13.4. *Every strongly diffuse measure (in $\mathbb{R}^2$) is a Weyl measure.*

Remark 13.13. An (admittedly long) proof of this theorem may be found in [42], 19.17, pp. 280-288. Recall that we a priori assumed in the above Definitions 13.1 and 13.2 that μ is a finite Radon measure with compact support. We refer for further discussions of this and other points to [42], 19.18, pp. 288-291.

Problem 13.2. By Remark 13.12, $\mu = \mathcal{H}^d|\Gamma$ with $0 < d < 2$ is strongly diffuse. Hence the Weyl property (13.49) for d-membranes is an immediate consequence of Theorem 13.4. On the other hand, (13.60) is a weak isotropicity condition. It is well known that IFS (Iterated Function Systems), based on affine contractions, creates in $\mathbb{R}^2$ beautiful ferns, grasses etc. This may be found in many books on fractal geometry. A short description including references has been given in [41], Section 4. The respective (anisotropic or non-isotropic) Radon measures do not satisfy, in general, the condition (13.60). It is totally unclear (at least for the author) whether the corresponding (anisotropic or non-isotropic) Radon measures are also Weyl measures. This results in the following problems:

What is the music of the ferns?

and, more generally,

under which conditions is a Radon measure in $\mathbb{R}^2$ a Weyl measure?

13.4.5. *h-membranes*

In 13.4.3 we dealt with d-membranes in $\mathbb{R}^2$, where d is a number with $0 < d < 2$. Now $h(\cdot)$ is a function generating a measure,

$$\mu\left(B(\gamma, r)\right) \sim h(r), \qquad \gamma \in \Gamma, \quad 0 < r \leq 1. \tag{13.62}$$

We refer to 13.3.1. By Theorem 13.1 and in particular Theorem 13.2(b) from 13.3.2,

$$B = (-\Delta)^{-1} \circ id^\mu \; : \quad \overset{\circ}{H}{}^1(\Omega) \hookrightarrow \overset{\circ}{H}{}^1(\Omega) \tag{13.63}$$

is compact if

$$\sum_{j=0}^{\infty} h(2^{-j}) < \infty. \tag{13.64}$$

This is a special case of (13.53) - (13.55). Hence B is a compact non-negative self-adjoint operator in $\overset{\circ}{H}{}^1(\Omega)$ with (13.43) - (13.48) and (13.50).

Theorem 13.5. *Let Ω be a bounded C^∞ domain in the $\mathbb{R}^2$ and let μ be an isotropic Radon measure with (13.53), (13.62). If*

$$\sum_{j=J}^{\infty} h(2^{-j}) \sim h(2^{-J}), \qquad \text{for all} \quad J \in \mathbb{N}_0, \tag{13.65}$$

then μ is a Weyl measure according to Definition 13.1 in 13.4.4.

Remark 13.14. If $J = 0$ then (13.64) is a special case of (13.65). According to our previous agreement $\sim$ in (13.65) means that one side can be estimated by the other side at the expense of positive constants which are independent of J (one direction is trivial). In the aftermath of [41] and [42] we studied Weyl measure in [46] and [47]. The above result is covered by [47], Corollary 2. On the one hand, (13.64) is necessary and sufficient that B from (13.63) is bounded. On the other hand it is not clear whether the sufficient conditions (13.65), are also necessary to ensure that μ is a Weyl measure. Further details in a more general setting may be found in [47].

13.4.6. *Courant indicators*

As indicated in Remark 13.9 at least for the distinguished eigenfunction $u(x)$ from (13.48), the global smothness (13.51), valid for all eigenfunctions u_k, reflects also the local smoothness according to (13.52). Here Γ is a d-set with $0 < d < 2$ and $\mu = \mathcal{H}^d | \Gamma$. Hence the eigenfunction $u(x)$ represents at any point $\gamma \in \Gamma$ faithfully the local fractality (at least as far the number d is concerned). It comes out that one can use this positive (in Ω) eigenfunction $u(x)$, belonging to the largest eigenvalue ϱ_1, to extract information about the fractality of Γ at any point $\gamma \in \Gamma$. It is one of the main topics of fractal geometry to get global, local and pointwise assertions about fractal sets and measures. We refer to the books mentioned in the introduction, especially [17], [18], [29], [19]. One finds fractal characteristics from the point of view of function spaces in [44], including some references. As indicated one can use the behaviour of the eigenfunction $u(x)$ to introduce further fractal characteristics. We restrict ourselves to very first results which are directly related to the above considerations. Recall that this paper is the English translation of [49]. After [49] was completed we returned in [47], [48] (which in turn is complementary survey) and [50] to the subject of this subsection. We do not change what follows in this subsection, but we refer for complementary results and arguments to the just-mentioned papers.

Again let Ω be a bounded C^∞ domain in the plane $\mathbb{R}^2$ and let μ be a Radon measure with

$$\operatorname{supp} \mu = \Gamma \subset \Omega, \quad |\Gamma| = 0, \quad 0 < \mu(\Gamma) < \infty. \tag{13.66}$$

We assume that there are numbers $c > 0$ and $\varepsilon > 0$ with

$$\mu\left(B(\gamma, r)\right) \leq c r^\varepsilon, \qquad \gamma \in \Gamma, \quad 0 < r \leq 1. \tag{13.67}$$

Then it follows by (13.57) and (13.53) - (13.55), that

$$B = (-\Delta)^{-1} \circ id^\mu : \quad \overset{\circ}{H}{}^1(\Omega) \hookrightarrow \overset{\circ}{H}{}^1(\Omega),$$

is a non-negative self-adjoint operator. It follows from the proof of the above Theorem 13.3 as given in [42], 19.7, and from (13.67) that B is compact. The positive eigenvalues ϱ_k can be ordered according to (13.46), the largest eigenvalue ϱ_1 is simple and one has (13.48) and (13.50). Furthermore one obtains in analogy to (13.51) that

$$u_k \in C^\varepsilon(\Omega), \qquad k \in \mathbb{N}. \tag{13.68}$$

This applies in particular to the positive eigenfunction u in Ω.

Definition 13.3. (Courant indicator). Under the above assumptions we put for $\gamma \in \Gamma$,

$$\mathrm{Ind}_\infty(\gamma) = \sup\left\{s \ : \ u \in C^s\left(B(\gamma, r)\right) \ \text{with} \ 0 < r \le 1\right\}. \tag{13.69}$$

Remark 13.15. The Courant indicator measures at any point $\gamma \in \Gamma$ the best possible local smoothness of the positive eigenfunction $u(x)$ belonging to the largest eigenvalue ϱ_1 of B. By (13.68) one has at least

$$\mathrm{Ind}_\infty(\gamma) \ge \varepsilon \qquad \text{for} \quad \gamma \in \Gamma.$$

Proposition 13.5 (d-membranes). *Let Ω be a bounded C^∞ domain in the plane $\mathbb{R}^2$ and let Γ be a compact d-set with $\Gamma \subset \Omega$ and $0 < d < 2$. Furthermore let μ be a related Radon measure,*

$$\Gamma = \mathrm{supp}\,\mu, \qquad \mu\left(B(\gamma, r)\right) \sim r^d, \quad \gamma \in \Gamma, \quad 0 < r \le 1.$$

Then

$$\mathit{Ind}_\infty(\gamma) = d \qquad \text{for} \quad \gamma \in \Gamma.$$

Remark 13.16. This assertion coincides with Remark 13.9 after Theorem 13.3 in 13.4.3. In particular we have the global smoothness hypothesis (13.68) with $\varepsilon = d$. In this case the supremum in (13.69) is a maximum.

Problem 13.3. For d-membranes one has the above definitive answer. If μ is an isotropic measure generating an h-membrane according to 13.4.5, then it can be expected that Ind_∞ is constant on Γ provided that (13.68) is fulfilled. If μ is a general (not necessarily isotropic) measure, satisfying the above assumptions, for example (13.53) - (13.55) and (13.57), such that B is compact, then one can expect that Ind_∞ (if defined) reflects the local fractality in any point $\gamma \in \Gamma$. At least when stepping from isotropic measures to more general measures it becomes clear that it is desirable, maybe even necessary, to generalise the indicator Ind_∞ by the indicators

$$\mathrm{Ind}_p(\gamma) = \sup\left\{s \ : \ u \in B_p^s\left(B(\gamma, r)\right) \ \text{with} \ 0 < r \le 1\right\},$$

with $1 < p \le \infty$, where $B_p^s = B_{pp}^s$ are special Besov spaces. This is suggested by the respective fractal characteristica in [44], [45] and the references given there. In case of d-membranes according to Proposition 13.5 one obtaines

$$\mathrm{Ind}_p(\gamma) = d + \frac{2-d}{p}, \qquad \gamma \in \Gamma, \quad 1 < p \le \infty.$$

This follows from [42], Theorem 19.7, p. 264-265. The problem arises, what the so-defined fractal indicators Ind_∞ and, more general, Ind_p wish to tell us and how they are related to other fractal characteristics.

Under the above assumptions for Γ, Ω, and μ, hence (13.66), (13.67) we put

$$\dim_H(\gamma) = \lim_{j\to\infty} \dim_H\left(\Gamma \cap B(\gamma, 2^{-j})\right), \qquad \gamma \in \Gamma, \tag{13.70}$$

where $\dim_H(M)$ is the Hausdorff dimension of a set M. Furthermore let

$$\mu\left(B(\gamma', r)\right) \le c_j\, r^{d_j(\gamma)} \quad \text{for} \quad \gamma \in \Gamma, \quad |\gamma' - \gamma| \le 2^{-j} \quad \text{and} \quad 0 < r \le 2^{-j},$$

and

$$d(\gamma) = \lim_{j\to\infty} d_j(\gamma), \qquad \gamma \in \Gamma. \tag{13.71}$$

The sequence on the right-hand side (13.70) is monotonically decreasing (i.e. not increasing), the sequence on the right-hand side of (13.71) is monotonically increasing (i.e. not decreasing) and it holds

$$d(\gamma) \le \dim_H(\gamma), \qquad \gamma \in \Gamma. \tag{13.72}$$

In case of a d-set the two numbers in (13.72) are d.

Theorem 13.6. *Let Ω be a bounded C^∞ domain in $\mathbb{R}^2$ and let μ be a Radon measure with (13.66), (13.67). Then*

$$d(\gamma) \le Ind_\infty(\gamma) \le \dim_H(\gamma), \qquad \gamma \in \Gamma. \tag{13.73}$$

Remark 13.17. Proposition 13.5 is a special case of (13.73).

Problem 13.4. In connection with the introduction of Ind_∞ in Definition 13.3, the indicated generalisation Ind_p, and Theorem 13.6 one gets the following problem. So far we assumed tacitly that μ (hence also Γ) and Ω are given according to (13.66). The question arises to which extent Ind_p depends for given μ on Ω. It seems to be reasonable to indicate this possible dependence and to write

$$\mathrm{Ind}_p(\gamma) \Longrightarrow \mathrm{Ind}_p^\Omega(\gamma) \Longrightarrow \mathrm{Ind}_p^{\mu,\Omega}(\gamma), \qquad \gamma \in \Gamma = \mathrm{supp}\,\mu.$$

Afterwards one may ask of whether $\mathrm{Ind}_p^{\mu,\Omega}$ is independent of Ω. By Proposition 13.5 and Problem 13.3 this is the case when Γ is a d-set. This might be also possible for general isotropic Radon measures, if the conditions formulated in Problem 13.3 are satisfied. If μ is not isotrop, then the independence of Ω is at least questionable.

> *Is* $\mathrm{Ind}_p^{\mu,\Omega}$ *for given* μ *and given* p *independent of* Ω *if, and only if,* μ *is isotrop?*

And what tells us for given (non-isotropic) μ the collection of the indicators $\mathrm{Ind}_p^{\mu,\Omega}$ about μ and $\Gamma = \mathrm{supp}\ \mu$ and in which language? In this context we again refer to [48], [50].

13.5. n-dimensional Weyl measures

13.5.1. *Introduction and definitions*

It is almost obvious that the theory for the operator B in the plane $\mathbb{R}^2$ as outlined in Section 13.4 has an n-dimensional counterpart. One finds in [42], Theorem 19.7, p. 264/265, the n-dimensional generalisation of the above Theorem 13.3 from 13.4.3 for the operator B. Now the respective conditions and results depend on n. The couterpart of (13.49) reads now as follows,

$$\varrho_k \sim k^{-1+\frac{n-2}{d}}\ , \qquad k \in \mathbb{N}, \quad n-2 < d < n, \qquad (13.74)$$

(with $0 < d < 1$ for $n = 1$).

We are now interested in the operator B_s considered in Proposition 13.3(a). But it is not our aim to formulate most general assertions. According to Remark 13.6 we restrict ourselves to the distinguished limiting cases

$$B_{\frac{n}{2}} = (id - \Delta)^{-\frac{n}{2}} \circ id^{\mu}\ : \quad H^{\frac{n}{2}}(\mathbb{R}^n) \hookrightarrow H^{\frac{n}{2}}(\mathbb{R}^n). \qquad (13.75)$$

This means in case of $n = 2$, that the operator B given by (13.55) is replaced by the operator B_1 according to (13.75). Then the domain Ω does not appear any longer and we have no physical interpretation as a fractal drum (or a drum with fractal membrane). On the other hand, after the necessary technical modifications many assertions for the operator B from Section 13.4 can be carried over to the operator B_1 (in $\mathbb{R}^2$).

In the n-dimensional case we did the necessary preparations in the Sections 13.2 and 13.3 for the operator $B_{\frac{n}{2}}$ and more general for

$$B_s = (id - \Delta)^{-s} \circ id^{\mu}\ : \quad H^s(\mathbb{R}^n) \hookrightarrow H^s(\mathbb{R}^n),$$

$s > 0$, parallel to the operator B from Section 13.4. We refer to the Proposition 13.3(a) from 13.2.3 and to the Theorems 13.1 and 13.2(a) from

13.3.2. In analogy to (13.38) we introduce in $H^{\frac{n}{2}}(\mathbb{R}^n)$ the specific scalar product

$$(f,g)_{H^{\frac{n}{2}}(\mathbb{R}^n)} = \int_{\mathbb{R}^n} (id - \Delta)^{\frac{n}{4}} f(x) \cdot (id - \Delta)^{\frac{n}{4}} \overline{g}(x)\, dx. \tag{13.76}$$

As for the operator B from Section 13.4, the operator $B_{\frac{n}{2}}$ according to (13.75), is a non-negative compact self-adjoint operator in $H^{\frac{n}{2}}(\mathbb{R}^n)$. Its positive eigenvalues ϱ_k, counted with respect to their multiplicities, can be ordered by magnitude,

$$\varrho_1 \geq \varrho_2 \geq \varrho_3 \geq \cdots, \quad \varrho_k \to 0 \quad \text{for} \quad k \to \infty. \tag{13.77}$$

Recall that (13.75) according to (13.25) and Proposition 13.3(a) from 13.2.3 is equivalent to the assertion that the trace operator

$$tr_\mu : \quad H^{\frac{n}{2}}(\mathbb{R}^n) \hookrightarrow L_2(\Gamma, \mu) \tag{13.78}$$

exists. In analogy to Definition 13.1 from 13.4.4 we can now introduce Weyl measures in $\mathbb{R}^n$ as follows.

Definition 13.4. Let μ be a finite Radon measure in $\mathbb{R}^n$ with compact support Γ. Then μ is called a Weyl measure, if the trace operator tr_μ according to (13.78) exists, if the operator $B_{\frac{n}{2}}$ according to (13.75) is compact and if for its positive eigenvalues ϱ_k, ordered according to (13.77), including multiplicities,

$$\varrho_k \sim k^{-1}, \quad k \in \mathbb{N}.$$

Remark 13.18. It can be seen easily that μ is a Weyl measure in the plane $\mathbb{R}^2$ according to Definition 13.1 if, and only if, it is a Weyl measure according to Definition 13.4.

13.5.2. *Isotropic Weyl measures*

It comes out that Theorem 13.5 from 13.4.5 can be carried over without changes. For sake of completeness we give an explicit formulation. Let μ be a finite isotropic Radon measure in $\mathbb{R}^n$ according to (13.30), (13.31) with

$$\sum_{j=0}^{\infty} h(2^{-j}) < \infty.$$

Then $B_{\frac{n}{2}}$ from (13.75) is a compact, self-adjoint non-negative operator in $H^{\frac{n}{2}}(\mathbb{R}^n)$, equipped with the scalar product from (13.76). This follows from the considerations in 13.5.1 and from the Theorems 13.1 and 13.2(a) from 13.3.2.

Theorem 13.7. *Let μ be an isotropic Radon measure with (13.30), (13.31). If*

$$\sum_{j=J}^{\infty} h(2^{-j}) \sim h(2^{-J}), \qquad \text{for all} \quad J \in \mathbb{N}_0,$$

then μ is a Weyl measure according to Definition 13.4 from 13.5.1.

Remark 13.19. This is the direct counterpart of Theorem 13.5 from 13.4.5, of Remark 13.14 in 13.4.5 and of the explanations and references given there. In contrast to Theorem 13.5 we did not assume $|\Gamma| = 0$ where $\Gamma = \text{supp } \mu$. But this is unimportant. Theorem 13.5 remains valid without this assumption. In Section 13.4 our interest was restricted from the very beginning to singular measures with $|\Gamma| = 0$.

13.6. Methods

13.6.1. *Introduction*

The techniques which lead finally to the results as described above started in [16]. The central point in [16] are quantitative assertions on compact embedding operators between function spaces of type

$$id: \quad B^{s_1}_{p_1}(\Omega) \hookrightarrow B^{s_2}_{p_2}(\Omega), \quad s_1 - \frac{n}{p_1} > s_2 - \frac{n}{p_2}, \qquad (13.79)$$

$0 < p_1 \leq p_2 \leq \infty$, expressed in terms of entropy numbers and approximation numbers. Here Ω is a bounded domain in $\mathbb{R}^n$. The spaces $H^s = B^s_2$ and $\mathcal{C}^s = B^s_\infty$ are special cases. The precise knowledge of entropy numbers and approximation numbers had been used in [16] to develop a spectral theory of (degenerate) elliptic (pseudo-)differential operators. Assertions on embedding operators id of type (13.79) had been obtained in [16] by using the Fourier-analytic definition of the spaces $B^s_p(\mathbb{R}^n)$ and their restrictions on domains Ω in $\mathbb{R}^n$. The step from domains Ω and (degenerate) elliptic operators to fractals and related fractal elliptic operators had been done in [41]. It is based on characterisations of function spaces in terms of quarkonial decompositons. The high flexibility of this technique admits not only to discretise spaces of type $B_p^s(\Omega)$, but also to introduce corresponding spaces on fractal sets. If, for example, Γ is a compact d-set in $\mathbb{R}^n$ with $0 < d \le n$, then one can introduce spaces $B_p^s(\Gamma)$ via subatomic means and one gets quantitative assertions about entropy numbers and approximation numbers for compact embeddings of the type

$$id : \quad B_{p_1}^{s_1}(\Gamma) \hookrightarrow B_{p_2}^{s_2}(\Gamma), \quad s_1 - \frac{d}{p_1} > s_2 - \frac{d}{p_2},$$

$0 < p_1 \le p_2 \le \infty$. This results in a spectral theory of fractal elliptic operators which is the main point of the above considerations. We continued these investigations in [42] and afterwards in the papers mentioned in the references. Here we cannot give a precise description of these techniques and how to use them. We restrict ourselves in 13.6.2 to a description of the interplay between entropy numbers, approximation numbers and spectral theory, in 13.6.3 to an outline of some quarkonial decompositions and, finally, in 13.6.4 to indicate roughly how these ingredients come together in the context of the results described in this paper.

13.6.2. *Entropy numbers and approximation numbers*

Let T be a compact operator in the complex Banach space A. Furthermore, let TU_A be the pre-compact image of the unit ball U_A in A. For $k \in \mathbb{N}$ the entropy number $e_k(T)$ is defined as the infimum over all numbers $\varepsilon > 0$, such that TU_A can be covered by 2^{k-1} balls in A of radius ε. Since T is compact, the spectrum of T consists, outside of the origin on A, of at most countably many eigenvalues ϱ_k of finite algebraic multiplicity, which can be ordered, including their algebraic multiplicities, by

$$|\varrho_1| \ge |\varrho_2| \ge |\varrho_3| \ge \cdots, \quad \varrho_k \to 0 \quad \text{for} \quad k \to \infty, \tag{13.80}$$

(where the last assertion must be dropped, if there are only finitely many or no eigenvalues different from zero).

Proposition 13.6 (Carl's inequality). *Under the above assumptions one has the uncertainty relation*

$$\sqrt{2}\,|\varrho_k|^{-1}\, e_k(T) \ge 1, \qquad k \in \mathbb{N}.$$

Remark 13.20. In other words, if one knows the geometric quantities $e_k(T)$ then the absolute values of the eigenvalues can be estimated from above. This inquality goes back to [11], [9]. Proofs and further explanations may be found in [10] and [16].

In addition to the entropy numbers one needs the approximation numbers. Again let T be a compact operator in the complex Banach space A. With $k \in \mathbb{N}$,

$$a_k(T) = \inf \{ \|T - L\| \ : \ L \in L(A), \ \operatorname{rank} L < k \} ,$$

is the kth approximation number, where $L(A)$ is the collection of all linear and continuous operators in A, and $\operatorname{rank} L$ is the dimension of the image of L.

Proposition 13.7. *Let T be a compact non-negative self-adjoint operator in the Hilbert space H and let ϱ_k be its positive eigenvalues ordered according to (13.80). Then*

$$a_k(T) = \varrho_k , \qquad k \in \mathbb{N}.$$

Remark 13.21. This is a special case of a well-known classical assertion. We refer to [15], S. 91.

13.6.3. *Quarkonial decompositions*

The decisive instrument in the context of the results decribed in this paper are quarkonial decompositions in function spaces. They had been invented just for this purpose and had been presented in [41] for the first time. It came out afterwards that these techniques can also be used for other purposes. They resulted finally in [42] in a new *constructive* or *Weierstrassian approach* to the theory of function spaces, including *Taylor expansions of distributions*. The basic idea, for example, for function spaces in $\mathbb{R}^n$ consists of a combination of the well-known wavelet philosophy with the philosophy of Taylor expansions, hence,

$$\psi(x) \Longrightarrow \psi\left(2^j x - m\right) \quad \text{with} \quad \psi(x) \Longrightarrow x^\beta \psi(x), \quad x \in \mathbb{R}^n, \qquad (13.81)$$

where $j \in \mathbb{N}_0$, $m \in \mathbb{Z}^n$, $\beta \in \mathbb{N}_0^n$ and $x^\beta = x_1^{\beta_1} \cdots x_n^{\beta_n}$. Again $\mathbb{N}_0$ is the collection of all non-negative integers, $\mathbb{Z}^n$ is the lattice in $\mathbb{R}^n$ consisting of all points $x = (x_1, \ldots, x_n) \in \mathbb{R}^n$ with integer-valued components, and $\mathbb{N}_0^n$ is the collection of all multi-indices $\beta = (\beta_1, \ldots, \beta_n) \in \mathbb{R}^n$ with $\beta_j \in \mathbb{N}_0$. One asks for elementary building blocks ψ, such that one obtains with the help of (13.81) frames in the respective function spaces, where the frame-coefficients (complex numbers) allow to decide in a definite way to which function spaces the given and constructively expanded distribution belongs. These techniques are highly flexible. They allow constructive approaches not only to all function spaces of type B_{pq}^s and F_{pq}^s in $\mathbb{R}^n$ for all admitted parameters s, p, q, but also for respective spaces on domains Ω in $\mathbb{R}^n$, on manifolds and in particular on fractals. To provide a better understanding we describe a comparatively simple, but typical case.

Let

$$B_p^s(\mathbb{R}^n) = B_{pp}^s(\mathbb{R}^n) \quad \text{with} \quad s > 0 \quad \text{and} \quad 1 \le p \le \infty, \tag{13.82}$$

are the well-known special Besov spaces. We do not give an explicit definition but we recall that

$$H^s(\mathbb{R}^n) = B_2^s(\mathbb{R}^n) \quad \text{and} \quad \mathcal{C}^s(\mathbb{R}^n) = B_\infty^s(\mathbb{R}^n), \qquad s > 0,$$

are the (fractional) Sobolev spaces and Hölder-Zygmund spaces used in this paper. As for the background of the respective theory of function spaces we refer to [39], [40]. Let ψ be a non-negative C^∞ function in $\mathbb{R}^n$ with

$$\operatorname{supp} \psi \subset \{y : |y| < 2^J\}$$

and

$$\sum_{m \in \mathbb{Z}^n} \psi(x - m) = 1, \qquad x \in \mathbb{R}^n,$$

for a suitable number $J \in \mathbb{N}$ (resolution of unity in $\mathbb{R}^n$). Let

$$\psi^\beta(x) = (2^{-J}x)^\beta \psi(x), \qquad x \in \mathbb{R}^n, \quad \beta \in \mathbb{N}_0^n.$$

For given s and p according to (13.82) we introduce the elementary building blocks, β-quarks,

$$(\beta\text{-qu})_{jm}(x) = 2^{-j(s-\frac{n}{p})} \psi^\beta \left(2^j x - m\right), \quad j \in \mathbb{N}_0, \ m \in \mathbb{Z}^n, \ \beta \in \mathbb{N}_0^n. \tag{13.83}$$

The normalising factors ensure that

$$\left\| (\beta\text{-qu})_{jm} \,|\, B_p^s(\mathbb{R}^n) \right\| \sim 1, \tag{13.84}$$

where the equivalence constants in (13.84) may depend on $\beta \in \mathbb{N}_0^n$, but they are independent of $j \in \mathbb{N}_0$, $m \in \mathbb{Z}^n$. Then a function $f \in L_p(\mathbb{R}^n)$ belongs to $B_p^s(\mathbb{R}^n)$ if, and only if, it can be represented by

$$f(x) = \sum_{\beta \in \mathbb{N}_0^n} \sum_{j \in \mathbb{N}_0} \sum_{m \in \mathbb{Z}^n} \lambda_{jm}^\beta (\beta\text{-qu})_{jm}(x), \tag{13.85}$$

with $(\beta\text{-qu})_{jm}$ given by (13.83) and

$$\|\lambda\,|\ell_p\| = \left(\sum_{\beta,j,m} |\lambda_{jm}^{\beta}|^p\right)^{\frac{1}{p}} < \infty \tag{13.86}$$

(with the usual modification for $p = \infty$). Here (13.85) with (13.86) converges absolutely (and hence unconditionally) in $L_p(\mathbb{R}^n)$. Furthermore,

$$\|f\,|B_p^s(\mathbb{R}^n)\| \sim \inf \|\lambda\,|\ell_p\|, \tag{13.87}$$

where the infimum is taken over all representations (13.85) with (13.86)
(equivalent norms). There are functions $\Phi^{\beta} \in S(\mathbb{R}^n)$ (the usual Schwartz
space) and a number $\varkappa \in \mathbb{R}$, such that (13.85) with

$$\lambda_{jm}^{\beta} = \lambda_{jm}^{\beta}(f) = \left(f,\, 2^{j\varkappa}\Phi^{\beta}(2^j \cdot -m)\right)$$

is an optimal frame representation in $B_p^s(\mathbb{R}^n)$, and that (in obvious notation)

$$\|\lambda(f)\,|\ell_p\| \sim \|f\,|B_p^s(\mathbb{R}^n)\|.$$

This theory started in [41] and had been developed afterwards systematically [42]. The above version is a modification and specialisation of a
corresponding presentation in [45], where, in particular, we calculated the
functions Φ^{β} explicitly. We refer also to [43], where we surveyed diverse
types of wavelet frames. Representations of the above type exist not only
for all spaces $B_{pq}^s(\mathbb{R}^n)$ and $F_{pq}^s(\mathbb{R}^n)$, but also for respective spaces on domains and on fractals. For this extension it is crucial that the regular lattice
$\mathbb{Z}^n$ in (13.85) can be replaced by irregular lattices which can be adapted
optimally to fractal sets.

13.6.4. *Elements of proofs*

The central point of the investigations in [41], [42] and also of the above
considerations is the spectral theory of fractal operators, hence assertions of
type (13.49), the indicated generalisation (13.74), and the question under
which conditions μ is a Weyl measure. We explain the interplay of the
results described in 13.6.2 and 13.6.3 in a simplified way.

As in Theorem 13.3 from 13.4.3 we assume that Ω is a bounded C^∞ domain
in the plane $\mathbb{R}^2$ and that Γ is a compact d-Menge with $0 < d < 2$ and $\Gamma \subset \Omega$.
Let $\mu = \mathcal{H}^d|\Gamma$ be the respective measure and

$$B = (-\Delta)^{-1} \circ id_\mu \circ tr_\mu : \quad \overset{\circ}{H}{}^1(\Omega) \hookrightarrow \overset{\circ}{H}{}^1(\Omega)$$

be the operator given by (13.43). Let

$$tr_\mu \, \overset{\circ}{H}{}^1(\Omega) = H^{\frac{d}{2}}(\Gamma, \mu),$$

where we use, without further explanations, the terminology from the literature mentioned above, especially [41], [42]. Here $H^{\frac{d}{2}}(\Gamma, \mu)$ is a Sobolev space on Γ for which one has a quarkonial decomposition on Γ in analogy to (13.85) - (13.87). The embedding

$$id : \quad H^{\frac{d}{2}}(\Gamma, \mu) \hookrightarrow L_2(\Gamma, \mu) \tag{13.88}$$

is compact. In obvious generalisation of the beginning in 13.6.2 one can introduce entropy numbers for compact operators acting between Banach spaces. Then one obtains for id from (13.88) that

$$e_k(id) \sim k^{-\frac{1}{2}}, \qquad k \in \mathbb{N}. \tag{13.89}$$

The proof of (13.89) is based on quarkonial decompositions in $H^{\frac{d}{2}}(\Gamma, \mu)$ and the resulting possibility to reduce problems of type (13.89) to the calculation of entropy numbers for embeddings between weighted ℓ_2 spaces. They can be calculated. On the other hand it comes out that the positive eigenvalues of $\sqrt{B}$, hence $\varrho_k^{\frac{1}{2}}$, can be estimated from above by $e_k(id)$, using Proposition 13.6 from 13.6.2, hence

$$\varrho_k^{\frac{1}{2}} \le c \, e_k(id) \sim k^{-\frac{1}{2}}.$$

Furthermore, the corresponding approximation numbers $a_k(id)$ can be estimated from below in a similar way. Using Proposition 13.7 from 13.6.2 one gets finally

$$\varrho_k \sim k^{-1}, \qquad k \in \mathbb{N},$$

and hence (13.49). Details may be found in the quoted literature. It is the aim of the above outline to provide a first glimpse about the interplay of the results sketched in 13.6.2 and 13.6.3.

13.7. Epilogue

According to the quotations in the Introduction fractal geometry and fractal analysis search for footpaths through the fractal wilderness. We hope, supported by the above discussions, that the recent theory of function spaces, opens up a new track: somewhat away from the fashionable resorts of self-similarity and IFS (iterated function systems), preferably hiking in the isotropic area, glancing bewildered at the surrounding non-isotropic jungle.

The key equipment consists of quarkonial decompositions of type (13.85) in $\mathbb{R}^n$ and on fractal sets Γ in $\mathbb{R}^n$, which result finally in spaces of type

$$B_p^s(\Gamma), \qquad s \in \mathbb{R}, \quad 1 < p \leq \infty,$$

and in a C^∞-theory or, better, $(D(\Gamma), D'(\Gamma))$-theory on Γ. As indicated in [51], 3.3 (but not treated in detail so far) such a theory can be extended by means of *euclidean charts* (in analogy to atlasses of local charts in Riemannian geometry) and by means of *snowflaked transforms* to some quasi-metric spaces. This may serve as a basis for a respective analysis on such quasi-metric spaces. Such a way is characterised by the symbiotic relationship between

(i) *the smooth and the non-smooth,*
(ii) *the dimensionality and the fractality,*
(iii) *the fractal geometry and the fractal analysis.*

Here the left-hand sides of (i) - (iii) profit from the right-hand sides and vice versa.

One of the most spectacular proofs of one of the most famous problems in mathematics is connected with twisted (*spiralling*, [34]) or, how one could call it nowadays *fractal* embeddings of abstract structures in euclidean spaces. J. Nash proved in [34], [35], that every n-dimensional Riemannian C^∞ manifold can be isometrically embedded in an euclidean spaces $\mathbb{R}^N$ with $2N = n(n+1)(3n+11)$. We refer for details to [3], p. 123. This goal was reached in [35] based on the preceding paper [34]. The C^1-isometries treated there are related to fractal embeddings. We quote from [33], p. 158:

He (Nash) *showed that you could fold the manifold like a silk handkerchief, without distorting it. Nobody would have expected Nash's theorem to be true. In fact, everyone would have expected it to be false.* M. Gromov (Interview, 1997): *Psychologically the barrier he broke is absolutely fantastic. ... There has been some tendency in recent decades to move from harmonic to chaos. Nash says chaos is just around the corner.* J. Conway (Interview, 1994) called Nash's result *one of the most important pieces of mathematical analysis of this century.*

References

[1] D. R. Adams, L. I. Hedberg. Function spaces and potential theory. Springer, Berlin, 1996

[2] S. Albeverio, P. Kurasov. Singular perturbations of differential operators. Cambridge Univ. Press, 2000

[3] T. Aubin. Some nonlinear problems in Riemannian geometry. Springer, Berlin, 1998

[4] N. Bourbaki. Élements de Mathématiques. XXI, Livre VI, Ch. 5, Intégration des mesures. Hermann, Paris, 1956

[5] M. Bricchi. Tailored function spaces and related h-sets. PhD Thesis, Jena, 2001

[6] M. Bricchi. Existence and properties of h-sets. Georgian Math. Journ. **9** (2002), 13-32

[7] M. Bricchi. Tailored Besov spaces and h-sets. Math. Nachr. **263-264** (2004), 36-52.

[8] M. Bricchi. Complements and results on h-sets. In: Proc. Conf. *Function Spaces, Differential Operators and Nonlinear Analysis*, Teistungen 2001. Birkhäuser, Basel, 2003, 219-229

[9] B. Carl. Entropy numbers, s-numbers and eigenvalue problems. Journ. Funct. Analysis **41** (1981), 290-306

[10] B. Carl, I. Stephani. Entropy, compactness and the approximation of operators. Cambridge Univ. Press, 1990

[11] B. Carl, H. Triebel. Inequalities between eigenvalues, entropy numbers, and related quantities of compact operators in Banach spaces. Math. Ann. **251** (1980), 129-133

[12] R. Courant, D. Hilbert. Methoden der mathematischen Physik. Springer, Berlin, 1993 (4. Auflage), (1. Auflage, 1924)

[13] G. David, S. Semmes. Fractured fractals and broken dreams. Clarendon Press, Oxford, 1997

[14] J. Dieudonné. Grundzüge der modernen Analysis 2. VEB Deutscher Verl. Wissenschaften, Berlin, 1975

[15] D. E. Edmunds, W. D. Evans. Spectral theory and differential operators. Clarendon Press, Oxford, 1987

[16] D. E. Edmunds, H. Triebel. Function spaces, entropy numbers, differential operators. Cambridge Univ. Press, 1996

[17] K. J. Falconer. The geometry of fractal sets. Cambridge Univ. Press, 1985

[18] K. J. Falconer. Fractal geometry. Wiley, Chichester, 1990

[19] K. J. Falconer. Techniques in fractal geometry. Wiley, Chichester, 1997

[20] P. Hajłasz, P. Koskela. Sobolev met Poincaré. Memoirs Amer. Math. Soc. **145**, 688. Amer. Math. Soc., Providence, 2000

[21] Y. S. Han, E. T. Sawyer. Littlewood-Paley theory on spaces of homogeneous type and the classical function spaces. Memoirs Amer. Math. Soc. **110**, 530. Amer. Math. Soc., Providence, 1994

[22] Y. Han, D. Yang. New characterizations and applications of inhomogeneous Besov and Triebel-Lizorkin spaces on homogeneous type spaces and fractals. Dissertationes Math. **403** (2002)

[23] J. Heinonen. Lectures on analysis on metric spaces. Springer, New York, 2001

[24] J. E. Hutchinson. Fractals and self similarity. Indiana Univ. Math. Journ. **30** (1981), 713-747

[25] J. Jost, W. Kendall, U. Mosco, M. Röckner, K.-T. Sturm. New directions in Dirichlet forms. Amer. Math. Soc., Providence, 1998

[26] J. Kigami. Analysis on fractals. Cambridge Univ. Press, 2001

[27] M. L. Lapidus. Fractal drums, inverse spectral problems for elliptc operators and a partial resolution of the Weyl-Berry conjecture. Trans. AMS **325** (1991), 465-529

[28] M. L. Lapidus, M. van Frankenhuysen. Fractal geometry and number theory. Birkhäuser, Boston, 2000

[29] P. Mattila. Geometry of sets and measures in euclidean spaces. Cambridge Univ. Press, 1995

[30] U. Mosco. Dirichlet forms and self-similarity. In: [25], 117-155

[31] S. Moura. Function spaces of generalised smoothness, entropy numbers, applications. PhD Thesis, Coimbra, 2001

[32] S. Moura. Function spaces of generalised smoothness. Dissertationes Math. **398** (2001)

[33] S. Nasar. A beautiful mind. Faber and Faber, London, 1998

[34] J. Nash. C^1 isometric imbeddings. Ann. Math. **60** (1954), 383-396

[35] J. Nash. The imbedding problem for Riemannian manifolds. Ann. Math. **63** (1956), 20-63

[36] B. Riemann. Ueber die Darstellbarkeit einer Function durch eine trigonometrische Reihe. Habilitationsschrift, Univ. Göttingen, 1854. (In: Bernhard Riemann, Gesammelte math. Werke, wissenschaftl. Nachlass und Nachträge. Leipzig, Teubner-Verlag, 1990, 227-264)

[37] Yu. Safarov, D. Vassiliev. The asymptotic distribution of eigenvalues of partial differential operators. Amer. Math. Soc., Providence, 1997

[38] S. Semmes. Some novel types of fractal geometry. Clarendon Press, Oxford, 2001

[39] H. Triebel. Theory of function spaces. Birkhäuser, Basel, 1983

[40] H. Triebel. Theory of function spaces II. Birkhäuser, Basel, 1992

[41] H. Triebel. Fractals and spectra. Birkhäuser, Basel, 1997

[42] H. Triebel. The structure of functions. Birkhäuser, Basel, 2001

[43] H. Triebel. Towards a Gausslet analysis: Gaussian representations of functions. In: *Function Spaces, Interpolation Theory and Related Topics*, Proc. Conf. Lund 2000. W. de Gruyter, Berlin, 2002, 425-449

[44] H. Triebel. Fractal characteristics of measures, an approach via function spaces. Journ. Fourier Analysis Applications **9** (2003), 411-430

[45] H. Triebel. Wavelet frames for distributions; local and pointwise regularity. Studia Math. **154** (2003), 59-88

[46] H. Triebel. The distribution of eigenvalues of some fractal elliptic operators and Weyl measures. In: Erhard Meister Memorial Volume, Birkhäuser, Basel, 2004, 457-473

[47] H. Triebel. Approximation numbers in function spaces and the distribution of eigenvalues of some fractal elliptic operators. Journ. Approximation Theory **129** (2004), 1-27

[48] H. Triebel. The fractal Laplacian and multifractal quantities. In: *Fractal Geometry and Stochastics III*, Proc. Conf. Friedrichroda, 2003. Progress in Probability **57**, Birkhäuser, Basel, 2004, 173-192

[49] H. Triebel. Fraktale Analysis aus der Sicht der Funktionenräume. Jahresbericht DMV **104** (2002), 171-199

[50] H. Triebel. Characterisation of function spaces via mollification; fractal quantities for distributions. Journ. Function Spaces Applications **1** (2003), 75-89

[51] H. Triebel, D. Yang. Spectral theory of Riesz potentials on quasi-metric spaces. Math. Nachr. **238** (2002), 160-184

[52] H. Weyl. Über die Abhängigkeit der Eigenschwingungen einer Membran von deren Begrenzung. J. Reine Angew. Mathematik **141** (1912), 1-11

[53] H. Weyl. Das asymptotische Verteilungsgesetz der Eigenwerte linearer partieller Differentialgleichungen. Math. Ann. **71** (1912), 441-479

[54] A. Zygmund. Smooth functions. Duke Math. Journ. **12** (1945), 47-76

Author Index